# HUMAN CELL CULTURE

Volume III: Cancer Cell Lines Part 3

# Human Cell Culture

Volume 3

*The titles published in this series are listed at the end of this volume.*

# Human Cell Culture

## Volume III

## Cancer Cell Lines

## Part 3: Leukemias and Lymphomas

edited by

**John R.W. Masters**
*University College London, UK*

and

**Bernhard O. Palsson**
*Department of Bioengineering, University of California San Diego, USA*

**KLUWER ACADEMIC PUBLISHERS**
DORDRECHT / BOSTON / LONDON

A C.I.P. Catalogue record for this book is available from the Library of Congress

ISBN-0-7923-6225-X

Published by Kluwer Academic Publishers,
P.O. Box 17, 3300 AA Dordrecht, The Netherlands.

Sold and distributed in North, Central and South America
by Kluwer Academic Publishers,
101 Philip Drive, Norwell, MA 02061, U.S.A.

In all other countries, sold and distributed
by Kluwer Academic Publishers,
P.O. Box 322, 3300 AH Dordrecht, The Netherlands

*Printed on acid-free paper*

Printed in the Netherlands

# Contents

# Chapter 1

# Human Leukemia-Lymphoma Cell Lines: Historical Perspective, State of the Art and Future Prospects

Hans G. Drexler[1] and Jun Minowada[2]
[1]*DSMZ-German Collection of Microorganisms and Cell Cultures, Department of Human and Animal Cell Cultures, Braunschweig, Germany;* [2]*Fujisaki Cell Center, Hayashibara Biochemical Laboratories, Okayama, Japan. Tel: +49-531-2616-160; Fax: +49-531-2616-150; E-mail: hdr@dsmz.de*

## 1. BURKITT CELL LINES WERE THE FIRST HUMAN HEMATOPOIETIC CELL LINES

In 1951 at Johns Hopkins University (Baltimore, Maryland, USA), Gey et al. established the first continuously growing human cell line (HeLa) from uterine cervix carcinoma [1]. The HeLa cell line and most other human cell lines subsequently established from various solid tumors adhere to the culture vessel growing in monolayers. In 1963 at the University of Ibadan, Nigeria, Pulvertaft established the first continuous human hematopoietic cell lines, a series of cell lines derived from Nigerian patients with Burkitt lymphoma in a suspension-type cell culture: RAJI is the best known cell line of this panel [2] (Table 1). In suspension cultures, these cells are free-floating, singly or in clusters, in the nutrient medium.

Electron microscopic analysis of these and subsequent Burkitt lymphoma-derived cell lines led to the identification of herpes-type virus particles which were later designated Epstein-Barr virus (EBV) [4,5]. The first leukemia-derived cell line was thought to be RPMI 6410, established from an American patient with acute myeloid leukemia (AML) containing similar herpes-type virus (EBV) particles in the cells [3]; however, it was shown later that this cell line was derived from normal bystander B-cells immortalized spontaneously by EBV infection and not from the leukemia cells.

The etiological significance of EBV for lymphomagenesis was questioned by several findings. Several hundred lymphoblastoid cell lines (LCLs) were established from the peripheral blood of patients with leukemias, lymphomas, other malignant tumors, and even from many healthy individuals [7–9,60]. However, EBV was detected in every cell line irrespective of the blood

1

*Table 1.*  Historical milestones in the establishment of hematopoietic cell lines (chronological by the year of publication)

| Year of publication | Year est. | Cell line | Investigator | Remarks: Type of cell line, derivation from malignancy, unique features | Ref. |
|---|---|---|---|---|---|
| 1964 | 1963 | RAJI, JIJOYE, OGUN, KUDI | Pulvertaft | B-cell lines (EBV+) from African Burkitt lymphoma | 2 |
| 1964 | 1964 | RPMI 6410 | Iwakata & Grace | 'AML-derived' cell line (EBV+) → isolation of herpes-type virus | 3 |
| 1965 | 1964 | EB-1, -2, -3 | Epstein et al. | B-cell lines (EBV+) from African Burkitt lymphoma → EBV isolation | 4, 5 |
| 1965 | 1964 | CCRF-CEM | Foley et al. | T-cell line (EBV−) from T-ALL, E-rosette-negative T-cells | 6 |
| 1967 | 1966 | RPMI 7206 + 18 B-LCLs | Moore et al. | EBV+ B-LCL from normal blood donors → EBV immortalizes B-LCL | 7, 8 |
| 1967 | 1966 | RPMI 8226 | Moore et al. | Plasma cell line (EBV−) from multiple myeloma | 9, 10 |
| 1972 | 1971 | MOLT-1, -2, -3, -4 | Minowada et al. | T-cell lines (EBV−) from T-ALL → E-rosetting of T-cells | 11 |
| 1973 | 1970 | K-562 | Lozzio & Lozzio | Myeloid-erythroid cell line from Ph+ CML-blast crisis → t(9;22) *BCR-ABL* (M-bcr) fusion gene | 12–14 |
| 1974 | 1972 | U-698-M | Nilsson & Sundström | B-cell line (EBV−) from B-cell lymphoma | 15 |
| 1974 | 1973 | SU-DHL-1 | Epstein & Kaplan | ALCL cell line from malignant histiocytosis → t(2;5) *NPM-ALK* fusion gene | 16 |
| 1975 | 1973 | BJA-B | Menezes et al. | B-cell line (EBV−) from African Burkitt lymphoma | 17 |
| 1976 | 1973 | SU-AmB-1, -2 | Epstein et al. | B-cell lines (EBV−) from American Burkitt lymphoma → t(8;14) *MYC-IGH* fusion gene | 18 |
| 1976 | 1974 | U-937 | Sundström & Nilsson | Monocytic cell line from histiocytic lymphoma | 19 |
| 1977 | 1974 | REH | Rosenfeld et al. | Precursor B-cell line from cALL → t(12;21) *ETV6/TEL-AML1* fusion gene | 20 |
| 1977 | 1975 | JBL | Miyoshi et al. | B-cell line (EBV−) from Japanese Burkitt lymphoma | 21 |
| 1977 | 1975 | NALM-1 | Minowada et al. | Precursor B-cell line from Ph+ CML-blast crisis | 22–24 |
| 1977 | 1976 | BALM-1, -2 | Minowada et al. | B-cell lines (EBV+) from B-ALL | 25 |
| 1977 | 1976 | JURKAT/JM | Schneider et al. | T-cell line from T-cell NHL → IL-2 producer | 26,27 |
| 1977 | 1976 | HL-60 | Collins et al. | Promyelocytic cell line from AML M2 | 28 |

*Continued on next page*

*Table 1.* (continued)

| Year of publication | Year est. | Cell line | Investigator | Remarks: Type of cell line, derivation from malignancy, unique features | Ref. |
|---|---|---|---|---|---|
| 1978 | 1977 | KG-1 | Koeffler & Golde | Myelocytic cell line from AML | 29 |
| 1979 | 1976 | NALM-6 to -15 | Minowada et al. | Precursor B-cell lines from pre B-ALL | 30 |
| 1979 | 1977 | SKW-3 | Hirano et al. | T-cell line from T-CLL | 31 |
| 1979 | 1978 | L-428 | Schaadt et al. | (Non-T non-B type) Hodgkin cell line from nodular sclerosis Hodgkin's disease | 32,33 |
| 1980 | (?) | H9/HUT 78 | Gazdar et al. | T-cell line from Sézary syndrome → HIV isolation | 34,35 |
| 1980 | 1978 | CTCL-2 | Poiesz et al. | T-cell line (HTLV-1+) from Sézary syndrome → HTLV-1 isolation | 36,37 |
| 1980 | 1978 | MT-1 | Miyoshi et al. | T-cell line (HTLV-1+) from ATL → immortalizing T-cells by co-culture | 38 |
| 1980 | 1978 | THP-1 | Tsuchiya et al. | Monocytic cell line from AML M5 | 39 |
| 1981 | 1979 | JOK-1 | Andersson et al. | B-cell line from HCL | 40 |
| 1982 | 1979 | 697 | Findley et al. | Precursor B-cell line from ALL → t(1;19) *E2A-PBX1* fusion gene | 41 |
| 1982 | 1980 | HEL | Martin & Papayannopoulou | Erythroid cell line from AML M6 | 42 |
| 1985 | (?) | RS4;11 | Stong et al. | Precursor B-cell line from ALL → t(4;11) *MLL-AF4* fusion gene | 43 |
| 1985 | 1981 | KU-812 | Kishi et al. | Basophilic cell line from CML-blast crisis | 44 |
| 1985 | 1983 | MEG-01 | Ogura et al. | Megakaryocytic cell line from CML-blast crisis | 45 |
| 1985 | 1983 | YT | Yodoi et al. | T-/NK cell line from thymoma → NK activity | 46 |
| 1985 | 1984 | EoL-1 | Saito et al. | Eosinophilic cell line from eosinophilic leukemia | 47 |
| 1986 | 1982 | HDLM-1, -2, -3 | Lok et al. | (T-type) Hodgkin cell line from nodular sclerosis Hodgkin's disease | 48 |
| 1987 | 1983 | TOM-1 | Okabe et al. | Precursor B-cell line from Ph+ ALL → t(9;22) *BCR-ABL* (m-bcr) fusion gene | 49 |

*Continued on next page*

Table 1. (continued)

| Year publication | Year of est. | Cell line | Investigator | Remarks: Type of cell line, derivation from malignancy, unique features | Ref. |
|---|---|---|---|---|---|
| 1988 | (?) | HMC-1 | Butterfield et al. | Mast cell line from mast cell leukemia | 50 |
| 1988 | 1987 | M-07e | Avanzi et al. | Megakaryocytic cell line from AML M7 → constitutively growth factor-dependent | 51 |
| 1989 | 1987 | TF-1 | Kitamura et al. | Erythroid cell line from AML M6 → constitutively growth factor-dependent | 52 |
| 1991 | 1988 | ME-1 | Yanagisawa et al. | Monocytic cell line from AML M4eo → inv(16) CBFB-MYH11 fusion gene | 53 |
| 1991 | 1989 | Kasumi-1 | Asou et al. | Myelocytic cell line from AML M2 → t(8;21) AML1-ETO fusion gene | 54 |
| 1991 | 1989 | NB4 | Lanotte et al. | Promyelocytic cell line from AML M3 → t(15;17) PML-RARA fusion gene | 55 |
| 1991 | 1990 | DoHH2 | Kluin-Nelemans et al. | B-cell line from B-cell NHL → t(14;18) IGH-BCL2 fusion gene | 56 |
| 1994 | 1991 | MDS92 | Tohyama et al. | Myelocytic cell line from MDS (RARS/RAEB) | 57 |
| 1995 | 1992 | BC-1 | Cesarman et al. | B-cell line from AIDS-PEL → HHV-8 isolation | 58 |

Modified from reference [59]. This selective and arbitrary list summarizes historical milestones in the establishment of human leukemia-lymphoma cell lines and is not intended to be comprehensive. Since the first publication in 1964 more than 1000 individual cell lines have been described. It is intended to list the first cell lines for the various types and subtypes of diseases and for the different categories of cell lineages, and the first cell lines with unique or specific features (e.g. chromosomal translocations leading to fusion genes).

AIDS – acquired immunodeficiency syndrome; ALCL – anaplastic large cell lymphoma; ALL – acute lymphoblastic leukemia; AML – acute myeloid leukemia (M2, myeloblastic; M3, promyelocytic; M4eo, myelomonocytic with eosinophils; M5, monocytic; M6, erythroid; M7, megakaryocytic; AML1-ETO – fusion gene in t(8;21)(q22;q22); ATL – adult T-cell leukemia; B-cell – B-lymphocyte; BCR-ABL – fusion gene in t(9;22)(q34;q11); cALL – common ALL; CBFB-MYH11 – fusion gene in inv(16)(p13q22); CLL – chronic lymphocytic leukemia; CML – chronic myelocytic leukemia; E2A-PBX1 – fusion gene in t(1;19)(q23;p13); EBV – Epstein-Barr virus; E-rosetting – formation of rosettes with sheep erythrocytes (T-cell marker); ETV6/TEL-AML1 – fusion gene in t(12;21)(p13;q22); HCL – hairy cell leukemia; HHV-8 – human herpesvirus type-8; HIV – human immunodeficiency virus; HTLV-1 – human T-leukemia virus-1; IGH-BCL2 = fusion gene in t(14;18)(q32;q21); IL-2 – interleukin-2; LCL – lymphoblastoid (normal B-) cell line; M-bcr/m-bcr – major/minor breakpoint cluster region in t(9;22)(q34;q11); MDS – myelodysplastic syndromes; MLL-AF4 = fusion gene in t(4;11)(q21;q23); MYC-IGH – fusion gene in t(8;14)(q24;q32); NHL – non-Hodgkin's lymphoma; NK – natural killer; NPM-ALK – fusion gene in t(2;5)(p23;q35); PEL – primary effusion lymphoma; Ph – Philadelphia chromosome; PML-RARA – fusion gene in t(15;17)(q22;q11); pre B – precursor B-lymphocyte; RARS/RAEB – refractory anemia with ring sideroblasts/excess of blasts; T-cell – T-lymphocyte.

donor's health status. Furthermore, Moore and his colleagues established the first proven human myeloma cell line, RPMI 8226, which was ironically free from EBV infection [10,61,62]. Most of the Burkitt lymphoma cell lines derived from North American [18,63] and Japanese patients [21] lacked EBV genomes. Moreover, the African Burkitt lymphoma cell line BJA-B was found to be free from EBV infection [17]. Thus the mere presence of EBV does not prove the neoplastic nature of the infected cells. Further studies established that EBV infection is capable of immortalizing certain B-cell subsets in normal human leukocyte cultures, hence the designation 'EBV+ B-lymphoblastoid cell lines (B-LCL)'; this term was adopted to define this type of non-malignant lymphoid cells [8].

Cytogenetic analysis of Burkitt cell lines led to the demonstration of specific chromosome translocations, t(8;14) or t(2;8) involving either heavy- or light-chain immunoglobulin genes and the c-myc oncogene [64–67]. Thus, the availability of these early hematopoietic cell lines stimulated the initial and subsequent research in diverse areas of biomedical sciences.

## 2. THE FIRST T LYMPHOCYTE CELL LINES

In 1971, the hematopoietic cell lines MOLT-1, -2, -3, and -4 were established from the peripheral blood of a patient in relapse from acute lymphoblastic leukemia (ALL) by Minowada et al. [11]. These cells lacked surface and cytoplasmic immunoglobulins (as signs of B-cell lineage commitment) and EBV infection. The most distinctive characteristic of the MOLT-1, -2, -3 and -4 cells was their rosette-forming ability with sheep, goat, horse and pig erythrocytes. This new immunological test referred to as the 'E-rosette test' was then found to be a specific normal human T-cell membrane marker, now termed CD2 [11].

Prior to the establishment of the MOLT 1–4 cell lines, Foley et al. had established a cell line, CCRF-CEM, from the peripheral blood of a patient with ALL, but at that time they were not able to prove its commitment to the T-cell lineage which was later shown by immunophenotyping [6].

## 3. ESTABLISHMENT OF CELL LINES FROM ALL HEMATOPOIETIC CELL LINEAGES

In the 1970s several still widely used and hence extremely important leukemia-lymphoma cell lines were established; just to name a few: K-562 as the first myeloid cell line and as a paradigm for a pluripotential cell line [12]; HL-60, KG-1, U-937 and THP-1 as myelocytic and monocytic model cell lines with differentiation potential [19,28,29,39]; REH, NALM-1 and NALM-6 as precursor B-cell lines derived from patients with common ALL, pre B-ALL and lymphoid blast crisis of CML [20,25,30].

Table 1 shows the chronological steps in the establishment of the first cell lines representing each of the respective subtypes of human leukemias and lymphomas and several other cell lines that were instrumental in the detection of significant new scientific information, e.g. for the isolation of viruses, cloning of chromosomal translocation breakpoints and their new fusion genes, etc. In recent years, a number of excellent reviews have summarized and presented in great detail specific groups or types/subtypes of human leukemia-lymphoma cell lines [68–100].

The development of several public cell line banks with the ensuing availability of large panels of human leukemia-lymphoma cell lines has tremendously enhanced research in this area. A primary function of cell line banks is to provide authenticated, clean and well-characterized cell material [101,102]. The most extensive public, non-profit collection of leukemia-lymphoma cell lines has been established at the DSMZ-German Collection of Microorganisms and Cell Cultures (Braunschweig, Germany; website: <www.dsmz.de>). Other cell banks such as ATCC-American Type Culture Collection (Manassas, Virginia, USA; website: <www.atcc.org>), JCRB-Japanese Collection of Research Bioresources (Tokyo, Japan; website: <http://cellbank.nihs.go.jp/defaulte.htm>) and RIKEN (Tsukuba, Ibaraki, Japan; website: <www.rtc.riken.go.jp>) also hold a limited number of the most often used and best-known leukemia-lymphoma cell lines.

## 4. COMMON CHARACTERISTICS OF LEUKEMIA-LYMPHOMA CELL LINES

Leukemia-lymphoma cell lines demonstrate the following three common characteristics:
- monoclonal origin
- differentiation arrest at a discrete stage during maturation in each lineage
- sustained proliferation of the cultured cells.

Until the availability of first unpurified and later recombinant growth factors in the 1980s, cell line proliferation was operationally 'growth factor-independent', i.e. no external cytokines were added and cells grew autonomously; however, it must be assumed that fetal bovine serum and human serum which are commonly employed as medium supplements contain certain growth-enhancing molecules. Furthermore, the autocrine stimulation of some cell lines by elaboration of known (and possibly so far unknown) growth factors has been shown [103,104]. Constitutively growth factor-dependent cell lines represent a new category of leukemia cell lines which were first developed deliberately in the late 1980s [99,105].

The analysis of the wide range of diverse leukemia-lymphoma cell lines has allowed for the recognition of a number of facts about human hematopoietic malignancies:

- There is no specific common marker for leukemias and lymphomas in a strict sense, although operationally there are markers and/or marker profiles specific for certain subtypes of leukemias and lymphomas, such as the expression of the leukemia subtype-specific hybrid or fusion gene products.
- Multiple marker profiles of cultured leukemia-lymphoma cell lines are very similar if not identical to the marker profiles of corresponding fresh leukemia and lymphoma cells.
- There is considerable heterogeneity or near individuality in the marker profiles of cultured leukemia-lymphoma cell lines; this appears to reflect differentiation arrest of the malignant cells at various stages of normal hematopoietic cell differentiation in a general sense.
- The availability and utilization of these leukemia-lymphoma cell lines have greatly facilitated the steady progress that has been made in molecular biological and molecular genetic studies of human hematopoietic malignancies. This has been particularly evident with regard to the possible association of unique cytogenetic markers with immunoglobulin and T-cell receptor genes and a vast array of cellular oncogenes [75,106]; these findings have been extended readily to fresh leukemias and lymphomas.
- Cell lines are extremely useful for a nearly unlimited variety of practical purposes; including screening of monoclonal antibodies, pharmaceutical drugs and hormones in order to evaluate their antigenic specificity and differential effects in a variety of subtypes of leukemias and lymphomas [107]; selection of subclones based on specific features such as drug-resistance or additional chromosomal or molecular aberrations [108]; cloning of translocation breakpoints and analysis of the incidence of fusion genes, point mutations and deletions [75,109]; use of cytokine-dependent cell lines for the establishment of bioassay systems [110]; examination of expression of cytokine receptors and proliferative response to cytokines [111,112]; studies on virus susceptibility and propagation [77].

## 5. CATEGORIZATION AND CLASSIFICATION OF LEUKEMIA-LYMPHOMA CELL LINES

There are various possibilities for categorizing human leukemia-lymphoma cell lines, for instance according to the diagnosis of the patient or according to immunophenotypes, specific cytogenetic alterations, functional characteristics or other features of the cultured cells. The most often used categorization of these cell lines is based on the physiological spectrum of the normal hematopoietic cell lineages: firstly, lymphoid versus myeloid and secondly T-cell, B-cell, NK cell versus myelocytic, monocytic, megakaryocytic and erythrocytic (in addition to specific subtypes such as myeloblastic, promyelocytic, eosinophilic, basophilic for the myelocytic cell lineage; precursor B-cell, mature B-cell, plasma cell for the B-cell lineage; and immature and mature T-cells) (Table 2). It is still a matter of debate whether natural killer (NK) cells represent a third lymphoid lineage [119] or whether they are a branch of the T-cell lineage.

The most useful technique for assigning a given cell line to one of the major cell lineages is undoubtedly immunophenotyping. The more extensive and complete the immunoprofile, the more precise is the categorization and classification of the cell lineage-derivation and status of arrested differentiation along this cell axis. Other techniques may add highly valuable information in cases of uncertainty of cell lineage assignment (see below).

Commonly this assignment of any given leukemia-lymphoma cell line to a cell lineage and stage of arrested differentiation, based on its immunological and other phenotypes, does not present any problems. Exceptions to this rule are the Hodgkin's disease (HD) and anaplastic large cell lymphoma (ALCL)-derived cell lines. Although the lymphoid nature of Hodgkin-Reed-Sternberg cells (the presumed neoplastic cells in HD) appears to be established [83,118] and thus HD-derived cell lines may be assigned to lymphoid T- or B-cell categories, their uniqueness and the fact that such cell lines display very unusual and often asynchronous marker profiles which are not found in normal cells, justifies a separate category for HD cell lines and the equally unusual ALCL lines. Continuous human dendritic cell lines have not as yet been described; such presumably unparalleled cell lines with specific profiles will require their own category.

## 6. CHARACTERIZATION OF LEUKEMIA-LYMPHOMA CELL LINES

Cell lines originating from different cell lineages are, for the most part, impossible to distinguish by morphology alone. Since leukemia cell lines

*Table 2.* Categorization of leukemia-lymphoma cell lines

| Main type | Physiological hematopoietic cell lineage | Subtype of neoplastic cell line | Prototype of cell line[a] |
|---|---|---|---|
| Lymphoid | B-cell | Precursor B-cell line | REH |
| | | B-cell line | U-698-M |
| | | Plasma cell line | RPMI-8226 |
| | T-cell | Immature T-cell line | CCRF-CEM |
| | | Mature T-cell line | SKW-3 |
| | NK cell | NK cell line | YT |
| Myeloid | Myelocytic | Myelocytic cell line | HL-60 |
| | | Promyelocytic cell line | NB4 |
| | | Eosinophilic cell line | EoL-1 |
| | | Basophilic cell line | KU-812 |
| | Monocytic | Monocytic cell line | U-937 |
| | Erythrocytic | Erythrocytic cell line[b] | K-562 |
| | Megakaryocytic | Megakaryocytic cell line[b] | MEG-01 |
| Dendritic | Lymphoid? myelomonocytic? | Dendritic cell line[c] | –[c] |
| Hodgkin/ALCL | Lymphoid? other? | Hodgkin cell line[d] | L-428 |
| | T-cell/null cell | ALCL cell line | SU-DHL-1 |

Modified from references [113,114].

[a] The best known cell lines for each of these categories are often the 'oldest' cell lines (nearly all available from cell line banks DSMZ or ATCC); see Table 1 for the historical perspective.

[b] It is often difficult to assign cell lines to either the erythrocytic or megakaryocytic cell lineage as most of these cell lines express features of both lineages, e.g. (hemo)globin, specific transcription factors, surface antigens, differentiation potential, etc.; thus, it may be preferable to use the term 'erythrocytic-megakaryocytic cell line'.

[c] At present, no continuous human dendritic cell line has been published; see references [115–117] for reviews on this controversial cell type.

[d] See references [69,118] for reviews on this controversial cell type.

commonly grow as single or clustered cells in suspension or only loosely adherent to the flask, single cell populations can be easily prepared and the cells can thus be characterized and classified. Table 3 lists a variety of parameters useful for the description of the cells and a panel of possible tests applicable for the phenotypic and functional characterization of most cell lines. This necessary multiparameter examination of the cellular phenotype provides important information on the likely cell of origin, the variable stringency of maturation arrest, and the predominantly normal pattern of gene expression. The list is not intended to cover comprehensively all possible informative parameters, as with new techniques becoming available and research areas extending to new avenues, other or entirely new features might be of interest to scientists. Thus, only some of the features of the phenotypic profiles of cell lines which are most often studied are highlighted. Immunophenotypic analysis and cytogenetic karyotyping appear to be the most important and informative examinations (Table 3). It is also important to indicate when in the life of a cell line particular data were generated and also whether alterations in the phenotypic features of the cells might occur during prolonged culture.

While the extent of the analytical characterization of leukemia cell lines is variable, a minimum data set is obligatory and essential for the identification, description and culture of a cell line; these data include the clinical and cell culture description of the cell line (the example of the human leukemia cell line HL-60 is given in Table 4). Clearly, the origin of an established cell line must be documented sufficiently.

## 7. ESTABLISHMENT AND DESCRIPTION OF NEW LEUKEMIA-LYMPHOMA CELL LINES

We discerned six cardinal requirements for the description and publication of new leukemia-lymphoma cell lines (Table 5).

### 7.1. Immortality

First, a cell line should be grown in continuous culture for at least 6 months, even better for more than a year. Upon addition of growth factors, primary neoplastic cells or normal cells can be kept in culture for several months before proliferation ceases; these cultures cannot be regarded as 'continuous cell lines'. Continuous cell lines have been defined as cultures that apparently are capable of an unlimited number of population doublings (immortalization); it should be recognized that an immortalized cell is not necessarily one that is neoplastically or malignantly transformed [121].

### 7.2. Verification of Neoplastic Origin

A cell line established from a patient with leukemia is not necessarily a 'leukemia cell line'. For instance, it appears to be 10- to 100-fold easier to establish an EBV-transformed B-lymphoblastoid cell line (EBV+ B-LCL) from a patient with leukemia than a neoplastic leukemia cell line [122]. Thus, the neoplastic nature of the cell line should be demonstrated by functional assays or by the detection of clonal cytogenetic abnormalities. With regard to the latter point, it is of note that among 596 well- or partly characterized leukemia-lymphoma cell lines (excluding sister cell lines, subclones, EBV+

*Table 3.*    Analytical characterization of leukemia-lymphoma cell lines

| Parameter | Details and examples |
|---|---|
| **Most important data** | |
| Clinical data | • Patient's data (see Table 4) |
| In vitro culture | • Growth kinetics, proliferative characteristics (see Table 4) |
| Immunophenotyping | • Surface marker antigens (fluorescence microscopy, flow cytometry) |
| | • Intracytoplasmic and nuclear antigens (immunoenzymatic staining) |
| Cytogenetics: | • Structural and numerical abnormalities |
| | • Specific chromosomal markers |
| **Further characterization** | |
| Morphology | • In-situ (flask, plate) under inverted microscope |
| | • Light microscopy (May-Grünwald-Giemsa staining) |
| | • Electron microscopy (transmission, scanning) |
| Cytochemistry | • Acid phosphatase, ($\alpha$-naphthyl acetate esterase, others) |
| Genotyping | • Southern blot analysis of T-cell receptor (TCR) and immunoglobulin (Ig) heavy and light chain gene rearrangements |
| | • Northern blot analysis of expression of TCR and Ig transcripts |
| Cytokines | • Production of cytokines |
| | • Expression of cytokine receptors |
| | • Response to cytokines, dependency on cytokines |
| Functional aspects/ specific features | • Phagocytosis |
| | • Antigen presentation |
| | • Immunoglobulin production and secretion |
| | • (Hemo)globin synthesis |
| | • Capacity for (spontaneous or induced) differentiation |
| | • Positivity for EBV or HTLV-1 or other viruses |
| | • Heterotransplantability into mice or other animals |
| | • Colony formation in agar/methylcellulose – clonogenicity |
| | • Production/secretion of specific proteins |
| | • Natural killer cell activity |
| | • Oncogene expression |
| | • Transcription factor expression |
| | • Unique point mutations |
| Date of analysis | • Age of cell line at time of analysis |
| | • Possible changes in the specific marker profile over prolonged culture |

Adapted from references [113,114,120].

*Table 4.*   Sets of clinical and cell culture data required for leukemia-lymphoma cell lines

| Parameter | Example |
|---|---|
| **Identification** | |
| Name of cell line | HL-60 |
| Cell phenotype | Myelocytic cell |
| **Clinical data** | |
| Original disease of patient | Initially AML M3, later corrected to AML M2 |
| Disease status | At diagnosis |
| Patient data (age, race, sex) | 35-year-old Caucasian woman |
| Source of material | Peripheral blood |
| Year of establishment | 1976 |
| **Cell culture data** | |
| Culture medium | 90% RPMI 1640 + 10% FBS |
| Subcultivation routine | Maintain at $0.1–0.5 \times 10^6$cells/ml, split ratio 1:5 to 1:10 every 2–3 days |
| Minimum cell density | $0.5–1.0 \times 10^5$ cells/ml |
| Maximum cell density | $1.0–1.5 \times 10^6$ cells/ml |
| Doubling time | 24–36 hours |
| Cell storage conditions | 70% RPMI 1640 + 20% FBS + 10% DMSO |
| In situ morphology | Round, single cells in suspension |
| Mycoplasma contamination | None – checked with PCR |
| EBV status | Negative – checked by PCR |

Modified from references [113,114,120].

B-LCLs, Burkitt and ATL cell lines) for which karyotypes have been published in the literature, only six cell lines (1%) showed a normal karyotype without any structural or numerical aberrations (five precursor B-/mature B-cell lines and one immature T-cell line). Colony formation in methylcellulose or agar or heterotransplantability into immunosuppressed mice are often regarded as signs suggestive of neoplasticity.

## 7.3.  Authentication

The origin of a new cell line must be proven by authentication, i.e. it must be shown that the cultured cells are indeed derived from the presumed patient's tumor. It has been estimated that 10–20% of human leukemia-lymphoma cell lines are misidentified or cross-contaminated, thus 'false cell lines'. The method of choice for identity control is forensic-type DNA fingerprinting [123]. Microsatellite analysis does not appear to be sufficient as the loci seem

*Table 5.*   Cardinal requirements for new leukemia-lymphoma cell lines

- Immortality
- Verification of neoplastic origin
- Authentication
- Scientific significance
- Characterization
- Availability

Modified from references [113,114].

to be unstable; immunophenotyping will not suffice either as cell lines of the same category will often have similar if not identical immunoprofiles. Unique cytogenetic marker chromosomes or molecular biological analyses (e.g. identical clonal gene rearrangement patterns on Southern blots) might also provide unequivocal evidence for the derivation of the cell line from the patient.

## 7.4.  Scientific Significance and Characterization

For the sake of the necessities important for scientific publications, namely novelty and scientific significance, the new cell line should have features not yet detected in previously established cell lines. A thorough characterization of the cells (see above and Table 3) will often detect unique characteristics of cell lines proving their scientific importance.

## 7.5.  Availability

The availability of cell lines to other qualified investigators upon request is of utmost importance. Some journals have adopted the policy that any readily renewable resources, including cell lines published in the journal, shall be made available to all qualified investigators in the field, if not already obtainable from commercial sources. The policy stems from the long-standing scientific principle that authenticity requires reproducibility. While cell lines are proprietary and unique, suitable material transfer agreements can be drawn up between the provider and requester [124]. By providing authenticated and unique biological material, cell line banks play a major role in leukemia-lymphoma research [101,102].

## 8. SUCCESS RATE OF ESTABLISHING LEUKEMIA-LYMPHOMA CELL LINES

The deliberate establishment of new leukemia-lymphoma cell lines remains by and large an unpredictable random process (except for EBV+ Burkitt lymphoma and HTLV-1+ ATL cell lines). Few systematic attempts to develop continuous new cell lines have been reported, thus a reasonable estimate of the success is in the range of 1–10% for myeloid cell lines and 10–20% for lymphoid cell lines [120]. It appears somewhat easier to establish precursor B-cell lines where success rates of up to 66% have been reported, however only when applying special techniques and culture conditions (reviewed in ref. [95]). It should be noted that some of these latter cell lines have extremely long doubling times (10–14 days) which clearly limits their usefulness [125]. Furthermore, a method for establishing T-cell lines from pediatric T-ALL cases with a very high success rate was described [126]. The reproducibility of this method in other hands and the long-term growth (immortalization) of these cultures is not known and remains to be established.

Difficulties in establishing continuous human leukemia-lymphoma cell lines may also originate from the inappropriate selection of nutrients and growth factors for these cells. Thus, a suitable microenvironment for hematopoietic cells, whether they are malignant or normal, cannot yet be created; it should be remembered that normal cells cannot be cultured long-term without EBV or HTLV-1 infection. In a strict sense, they are also no longer considered normal cells. A much higher percentage of leukemia samples can be grown in vivo in immunodeficient mice (athymic, SCID, NOD/SCID, etc.) after xenotransplantation than in vitro [127]; these murine microenvironments are, however, not likely to be entirely representative of microenvironments in human organs [120]. Many types of neoplastic cells may not be capable of indefinite proliferation. The success rate for establishing a hematopoietic cell line appears to be higher for more immature than for more mature cells, and for lymphoid (T- and B-cell precursor) than for myeloid (including monocytic, erythrocytic and megakaryocytic) cell lines. Specimens from patients at relapse and specimens obtained from ascites, pleural effusion and the leukemic phase of lymphoma may represent more suitable material for cell culture attempts.

## 9. FUTURE PROSPECTS

We estimate that more than 1000 leukemia-lymphoma cell lines have been established. However, only a relatively small percentage of these cell lines has been sufficiently well described and characterized; in many cases it is not known whether these cell lines are truly continuously proliferating and whether these cell lines do still exist. Efforts should be undertaken so that publications of new cell lines fulfil the requirements listed in Table 5. Furthermore, the awareness of scientists of the benefits of institutionalized cell culture collections should be heightened; otherwise, substantial numbers of unique and potentially important cell lines might be lost. Despite more than 35 years of hematopoietic cell culture, it is not understood why certain cells start to multiply indefinitely in culture and others do not. This enigma shows that our present in vitro culture conditions by no means accurately reflect the physiological in vivo microenvironment. Much further work is required to achieve significant improvements in the efficiency of immortalization. The use of certain cytokines might permit the long-term culture of many leukemia cells giving these cells the possibility to adapt to in vitro conditions or to give the few cells that a priori are amenable to in vitro growth the necessary time to multiply to sufficient cell numbers; subsequently, in some cases, continuous cell lines might evolve. Future technical innovations, e.g. transfection with oncogenes such as the anti-apoptotic gene bcl-2 or with a mutant p53 gene, might also improve the success rate enhancing the frequency of immortalization (although such cell lines are unique, they may not reflect the pathophysiology of the real leukemia in patients).

The availability of large numbers of continuous leukemia- lymphoma cell lines has facilitated clinical and immunobiological studies of hematopoietic malignancies. As with all in vitro studies, one should be cautious before extrapolating the data gathered from such studies to the in vivo situation. Thus, it is of paramount importance that the findings generated by cell line studies should be substantiated by studies with fresh leukemia cells. Despite this limitation, it is obvious that continuous leukemia-lymphoma cell lines have played major roles in the advancement of leukemia research. Undoubtedly, the acquisition of new information about human hematopoietic malignancies will be greatly furthered by continued research utilizing leukemia-lymphoma cell lines, optimally combined with studies on primary material.

# References

1. Gey GO, Coffman WD, Kubicek MT. *Cancer Res* 12: 264–265, 1952.
2. Pulvertaft RJV. *Lancet* i: 238–240, 1964.
3. Iwakata S, Grace JT Jr. *NY State J Med* 64: 2279–2282, 1964.
4. Epstein MA, Barr YM. *J Natl Cancer Inst* 34: 231–240, 1965.
5. Epstein MA, Achong BG, Barr YM. *Lancet* i: 702–703, 1964.
6. Foley GE, Lazarus H, Farber S et al. *Cancer* 18: 522–529, 1965.
7. Moore GE, Gerner RE, Franklin HA. *J Am Med Assoc* 199: 519–524, 1967.
8. Moore GE, Minowada J. *In Vitro* 4: 100–114, 1969.

9. Belpomme D, Minowada J, Moore GE. *Cancer* 30: 282–287, 1972.
10. Matsuoka Y, Moore GE, Yagi Y et al. *Proc Soc Exp Biol Med* 125: 1246-1250, 1967.
11. Minowada J, Ohnuma T, Moore GE. *J Natl Cancer Inst* 49: 891–895, 1972.
12. Lozzio CB, Lozzio BB. *Blood* 45: 321–334, 1975.
13. Lozzio B, Lozzio CB. *Leukemia Res* 3: 363–370, 1979.
14. Andersson L, Nilsson K, Gahmberg CG. *Int J Cancer* 23: 143–147, 1979.
15. Nilsson K, Sundström C. *Int J Cancer* 13: 808–823, 1974.
16. Epstein AL, Kaplan HS. *Cancer* 34: 1851–1872, 1974.
17. Menezes J, Leibold W, Klein G et al. *Biomedicine* 22: 276–284, 1975.
18. Epstein AL, Henle W, Henle G et al. *Proc Natl Acad Sci USA* 73: 228–232, 1976.
19. Sundström C, Nilsson K. *Int J Cancer* 17: 565–577, 1976.
20. Rosenfeld C, Goutner A, Choquet C et al. *Nature* 267: 841–843, 1977.
21. Miyoshi I, Hiraki S, Kubonishi I et al. *Cancer* 40: 2999–3003, 1977.
22. Minowada J, Tsubota T, Greaves MF et al. *J Natl Cancer Inst* 59: 83–87, 1977.
23. Minowada J, Koshiba H, Janossy G et al. *Leukemia Res* 3: 261–266, 1979.
24. LeBien TW, Hozier J, Minowada J et al. *New Engl J Med* 301: 144–147, 1979.
25. Minowada J, Oshimura M, Tsubota T et al. *Cancer Res* 37: 3096–3099, 1977.
26. Schneider U, Schwenk HU, Bornkamm G. *Int J Cancer* 19: 621–626, 1977.
27. Gillis S, Watson J. *J Exp Med* 152: 1709–1719, 1980.
28. Collins SJ, Gallo RC, Gallagher RE. *Nature* 270: 347–349, 1977.
29. Koeffler HP, Golde DW. Science 200: 1153–1154, 1978.
30. Hurwitz R, Hozier J, LeBien T et al. *Int J Cancer* 23: 174–180, 1979.
31. Hirano T, Kishimoto T, Kuritani T et al. *J Immunol* 123: 1133–1140, 1979.
32. Schaadt M, Fonatsch C, Kirchner H et al. *Blut* 38: 185–190, 1979.
33. Schaadt M, Diehl V, Stein H et al. *Int J Cancer* 26: 723–731, 1980.
34. Gazdar AF, Carney DN, Bunn PA et al. *Blood* 55: 409–417, 1980.
35. Popovic M, Sarngadharan MG, Read E et al. *Science* 224: 497–500, 1984.
36. Poiesz BJ, Ruscetti FW, Mier JW et al. *Proc Natl Acad Sci USA* 77: 6815–6819, 1980.
37. Poiesz BJ, Ruscetti FW, Gazdar AF et al. *Proc Natl Acad Sci USA* 77: 7415–7419, 1980.
38. Miyoshi I, Kubonishi I, Sumida M et al. *Gann* 71: 155–156, 1980.
39. Tsuchiya S, Yamabe M, Yamaguchi Y et al. *Int J Cancer* 26: 171–176, 1980.
40. Andersson L, Gahmberg CG, Jansson SE et al 1981 In: Knapp W (ed.): *Leukemia Markers*. Academic Press, London, 297–300.
41. Findley Jr HW, Cooper MD, Kim TH et al. *Blood* 60: 1305–1309, 1982.
42. Martin P, Papayannopoulou T. *Science* 216: 1233–1235, 1982.
43. Stong RC, Korsmeyer SL, Parkin JL et al. *Blood* 65: 21–31, 1985.
44. Kishi K. *Leukemia Res* 9: 381–390, 1985.
45. Ogura M, Morishima Y, Ohno R et al. *Blood* 66: 1384–1392, 1985.
46. Yodoi J, Teshigawara K, Nikaido T et al. *J Immunol* 134: 1623–1630, 1985.
47. Saito H, Bourinbaiar A, Ginsburg M et al. *Blood* 66: 1233–1240, 1985.
48. Drexler HG, Gaedicke G, Lok MS et al. *Leukemia Res* 10: 487–500, 1986.
49. Okabe M, Matsushima S, Morioka M et al. *Blood* 69: 990–998, 1987.
50. Butterfield JH, Weiler D, Dewald G et al. *Leukemia Res* 12: 345–355, 1988.
51. Avanzi GC, Lista P, Giovinazzo B et al. *Brit J Haematol* 69: 359–366, 1988.
52. Kitamura T, Tange T, Terasawa T et al. *J Cell Physiol* 140: 323–334, 1989.
53. Yanagisawa K, Horiuchi T, Fujita S. *Blood* 78: 451–457, 1991.
54. Asou H, Tashiro S, Hamamoto K et al. *Blood* 77: 2031–2036, 1991.
55. Lanotte M, Martin-Thouvenin V, Najman S et al. *Blood* 77: 1080–1086, 1991.

56. Kluin-Nelemans HC, Limpens J, Meerabux J et al. *Leukemia* 5: 221–224, 1991.
57. Tohyama K, Tsutani H, Ueda T et al. *Brit J Haematol* 87: 235–242, 1994.
58. Cesarman E, Moore PS, Rao PH et al. *Blood* 86: 2708–2714, 1995.
59. Drexler HG, Minowada J. *Leukemia Lymphoma* 31: 305–316, 1998.
60. Prempree T, Marz T. *Nature* 216: 810–811, 1966.
61. Moore GE, Kitamura H. *NY State J Med* 68: 2054–2060, 1968.
62. Minowada J, Nonoyama M, Moore GE et al. *Cancer Res* 34: 1898–1903, 1974.
63. Pagano J, Huang CH, Levine P. *New Engl J Med* 289: 1395–1399, 1973.
64. Manolov G, Manolova Y. *Nature* 237: 33–34, 1972.
65. Zech L, Haglund U, Nilsson K et al. *Int J Cancer* 17: 47–56, 1976.
66. Kaiser-McCaw B, Epstein AL, Kaplan HS et al. *Int J Cancer* 19: 482–486, 1977.
67. Dalla-Favera R, Bregni M, Erikson J et al. *Proc Natl Acad Sci USA* 79: 7824–7827, 1982.
68. Collins SJ. *Blood* 70: 1233–1244, 1987.
69. Diehl V, von Kalle C, Fonatsch C et al. *Semin Oncol* 17: 660–672, 1990.
70. Drexler HG, Amlot PL, Minowada J. *Leukemia* 1: 629–637, 1987.
71. Drexler HG, Jones DB, Diehl V et al. *Hematol Oncol* 7: 95–113, 1989.
72. Drexler HG, Minowada J. *Human Cell* 5: 42–53, 1992.
73. Drexler HG. *Leukemia Lymphoma* 9: 1–26, 1993.
74. Drexler HG. *Leukemia Res* 18: 919–927, 1994.
75. Drexler HG, MacLeod RAF, Borkhardt A et al. *Leukemia* 9: 480–500, 1995.
76. Drexler HG, Quentmeier H, MacLeod RAF et al. *Leukemia Res* 19: 681–691, 1995.
77. Drexler HG, Uphoff CC, Gaidano G et al. *Leukemia* 12: 1507–1517, 1998.
78. Drexler HG, MacLeod RAF, Uphoff CC. *Leukemia Res* 23: 207–215, 1999.
79. Ferrero D, Rovera G. *Clin Haematol* 13: 461–487, 1984.
80. Haluska FG, Brufsky AM, Canellos GP. *Blood* 84: 1005–1019, 1994.
81. Hassan HT, Freund M. *Leukemia Res* 19: 589–594, 1995.
82. Hassan HT, Drexler HG. *Leukemia Lymphoma* 20: 1–15, 1995.
83. Herbst H, Stein H, Niedobitek G. *Crit Reviews Oncogenesis* 4: 191–239, 1993.
84. Hoffman R. *Blood* 74: 1196–1212, 1989.
85. Hozumi M. *CRC Crit Reviews Oncol/Hematol* 3: 235–277, 1985.
86. von Kalle C, Diehl V. *Int Review Exp Pathol* 33: 185–203, 1992.
87. Keating A. *Baillière's Clin Haematol* 1: 1021–1029, 1987.
88. Koeffler HP, Golde DW. *Blood* 56: 344–350, 1980.
89. Koeffler HP. *Blood* 62: 709–721, 1983.
90. Koeffler HP. *Semin Hematol* 23: 223–236, 1986.
91. Lübbert M, Koeffler HP. *Blood* Reviews 2: 121–133, 1988.
92. Lübbert M, Koeffler HP. *Cancer Reviews* 10: 33–62, 1988.
93. Lübbert M, Herrmann F, Koeffler HP. *Blood* 77: 909–924, 1991.
94. Matsuo Y, Minowada J. *Human Cell* 1: 263–274, 1988.
95. Matsuo Y, Drexler HG. *Leukemia Res* 22: 567–579, 1998.
96. Minowada J. *Cancer Reviews* 10: 1–18, 1988.
97. Nilsson K. *Human Cell* 5: 25–41, 1992.
98. Nilsson K. 1994 In: Dammacco F, Barlogie B (eds.): *Challenges in Modern Medicine*, Vol 4 – *Multiple Myeloma and Related Disorders*. Ares-Serono Symposia Publications, 83–92.
99. Oval J, Taetle R. *Blood Reviews* 4: 270–279, 1990.
100. Schaadt M, von Kalle C, Tesch H et al. *Cancer Reviews* 10: 108–122, 1988.
101. Hay RJ, Reid YA, McClintock PR et al. *J Cell Biochem* 24: 107–130, 1996.

102. Drexler HG, Dirks W, MacLeod RAF, Quentmeier H, Steube K, Uphoff CC (eds.): *DSMZ Catalogue of Human and Animal Cell Cultures* (7th Edition), Braunschweig, 1999.
103. Freedman MH, Grunberger T, Correa P et al. *Blood* 81: 3068–3075, 1993.
104. Pietsch T. *Nouv Rev Fr Hematol* 35: 285–286, 1993.
105. Drexler HG, Zaborski M, Quentmeier H. *Leukemia* 11: 701–708, 1997.
106. Drexler HG, Borkhardt A, Janssen JWG. *Leukemia Lymphoma* 19: 359–380, 1995.
107. Chabner BA. *J Natl Cancer Inst* 82: 1083–1085, 1990.
108. Nielsen D, Skovsgaard T. *Biochim Biophys Acta* 1139: 169–183, 1992.
109. Drexler HG. *Leukemia* 12: 845–859, 1998.
110. Mire-Sluis AR, Page L, Thorpe R. *J Immunol Methods* 187: 191–199, 1995.
111. Drexler HG. *Leukemia* 10: 588–599, 1996.
112. Drexler HG, Quentmeier H. *Leukemia* 10: 1405–1421, 1996.
113. Drexler HG, Matsuo Y, Minowada J. *Human Cell* 11: 51–60, 1998.
114. Drexler HG, Matsuo Y. *Leukemia* 13: 835–842, 1999.
115. Hart DNJ. *Blood* 90: 3245–3287, 1997.
116. Banchereau J, Steinman RM. *Nature* 392: 243–252, 1998.
117. Robinson SP, Saraya K, Reid CDL. *Leukemia Lymphoma* 29: 477–490, 1998.
118. Drexler HG. *Leukemia Lymphoma* 8: 283–313, 1992.
119. Spits H, Lanier LL, Phillips JH. *Blood* 85: 2654–2670, 1995.
120. Drexler HG, Gignac SM, Minowada J. 1994 Hematopoietic cell lines. In: Hay RJ, Park JG, Gazdar A (eds.): *Atlas of Human Tumor Cell Lines*. Academic Press, Orlando, 213–250.
121. Schaeffer WI. *In Vitro Cell Dev Biol* 26: 97–101, 1990.
122. Nilsson K, Pontén J. *Int J Cancer* 15: 321–341, 1975.
123. Häne B, Tümmler M, Jäger K et al. *Leukemia* 6: 1129–1133, 1992.
124. Kaushansky K. *Blood* 91: 1–2, 1998.
125. Zhang LQ, Downie PA, Goodell WR et al. *Leukemia* 7: 1865–1874, 1993.
126. Smith SD, McFall P, Morgan R et al. *Blood* 73: 2182–2187, 1989.
127. Uckun FM. *Blood* 88: 1135–1146, 1996.

Chapter 2

# B-Cell Precursor Cell Lines

Stefan Faderl and Zeev Estrov
*Department of Bioimmunotherapy, Box 302, M. D. Anderson Cancer Center, 1515 Holcombe
Blvd., Houston, TX 77030, USA. Tel: +1-713-794-1675; Fax; +1-713-745-2374; E-mail:
zestrov@notes.mdacc.tmc.edu*

## 1. INTRODUCTION

Lymphopoiesis is a complex process. It originates from a pool of uncom-
mitted stem cells capable of self-renewal and, at the same time, giving rise
to early progenitors which develop along an orderly path of differentiation
through precursor cell stages of increasing maturation committed to produce lymphocytes. The concept of a lymphoid stem cell has long been a
matter of debate [1]. However, experiments in cell cultures and with leth-
ally irradiated mice where fetal bone marrow cells expressing high levels of
CD34 were able to reconstitute human B and T cells, point toward a population of CD34- and TdT (terminal deoxynucleotide transferase)-expressing
pluripotent hematopoietic cells that can serve as lymphoid progenitors [2].

B-cell ontogeny occurs in two phases. Antigen-dependent expansion and
differentiation occurs in spleen and lymph nodes and is mediated by the
interaction of surface immunoglobulin (sIg)-positive B cells with access-
ory signals from T cells, macrophages, natural killer (NK) cells, and other
antigen-presenting cells such as dendritic cells. Antigen-independent maturation occurs in the bone marrow and precedes these events. This phase of B
cell development is characterized by rearrangement of the Ig heavy and light
chain genes and sequential expression of their transcription products in the
cytoplasm and on the surface of B cells. It eventually results in the production of sIg-expressing B lymphocytes. In this sense, B-cell precursors (BCP)
represent stages in B cell maturation prior to expression of sIg, and various
stages within BCP are distinguished by the status of their Ig rearrangements
and antigen expression (CD markers).

The Ig heavy and light chain loci are organized into multiple genes that
result in the transcription of the variable (V) and constant (C) regions of
the Ig proteins. Additional genes code for the joining (J) and diversity (D)

19

 *J.R.W. Masters and B.O. Palsson (eds.), Human Cell Culture Vol. III, 19–59.*
© 2000 *Kluwer Academic Publishers. Printed in the Netherlands.*

regions, the latter in the heavy chain locus only, that encode the hypervariable region of the V segments [3]. The genes are separated by noncoding introns of varying lengths. For the B lymphocyte to produce and excrete functional Ig proteins, an orderly progression of gene rearrangements takes place that starts at the $\mu$ heavy chain locus and from there progresses to the $\kappa$ and $\lambda$ light chain loci. Initial steps involve joining of $D_H$ and $J_H$ regions followed by $V_H \rightarrow DJ_H$ joining by splicing out intervening intron sequences. Crucial enzymes catalyzing these steps are TdT which adds nontemplate-directed nucleotides in between the joining regions [4], and proteins encoded by the genes *RAG* (recombination activating gene)-*1* and *RAG-2* [5]. A B-cell at this stage is called a pro-B cell, i.e., it is capable of completing a functional rearrangement of the Ig gene locus. Once cytoplasmic Ig$\mu$ is expressed, the BCP is termed a pre-B cell and rearrangement of the light chain loci begins [1]. Until expression of a functional $\mu/\kappa$ or, in the case of nonfunctional $\kappa$ rearrangements, $\mu/\lambda$ sIg, a surrogate light chain (SLC) molecule composed of the covalently linked $\lambda 5$ and the noncovalently linked $V_{pre-B}$ protein is attached to Ig$\mu$.

At every stage along the developmental path of lymphopoiesis, expansion of a monoclonal population of cells with a phenotype of immature cells manifests most frequently as acute lymphoblastic leukemia (ALL). Leukemic lymphoblasts share many of the phenotypic, immunological, and genetic properties of their normal B-cell counterparts. ALL is the most common form of childhood neoplasia. Only 20% of ALL occur in patients older than 17 years of age [6]. ALL can be classified according to morphology, histocytochemistry, immunophenotype and, increasingly, molecular abnormalities and cytogenetics [7]. The French–American–British (FAB) Cooperative Group distinguishes three morphological groups according to cell size, nuclear/cytoplasmic (N/C) ratio, presence or absence of nucleoli, and cytoplasmic vacuolation [8]. L1 morphology, most frequent in childhood ALL, is thus characterized by a population of small, homogenous cells with a relatively high N/C ratio and indistinct nucleoli. L2 cells are more heterogenous in size, have a lower N/C ratio and express several, often large nucleoli. L3 morphology is almost exclusively a feature of mature B-cell leukemia (Burkitts leukemia) associated with characteristic chromosomal translocations.

A more widespread system used for the classification of ALL is based on expression of distinct surface and cytoplasmic markers detected by immunophenotyping. The European Group for the Immunological Classification of Leukemias (EGIL) has established guidelines for the immunophenotypic diagnosis of ALL [9]. B lineage ALL is defined by the expression of at least two of the three early B cell markers CD19, CD22, and CD79a. According to the degree of B lymphoid differentiation of the leukemic cells, BCP ALL is further defined as pro-B-ALL (B-I) without further expression of other B-cell antigens, common ALL (B-II) if CD10-positive, and pre-B-ALL if cIg$\mu$ can be detected (B-III). The expression of sIg or cytoplasmic/surface light chains qualifies the leukemia as a mature B-cell ALL and is not the subject of this chapter.

The first BCP ALL cell line, REH, was established in 1974 from the leukemic cells of a girl with ALL in relapse [10]. Since then, more than 150 BCP ALL cell lines have been described, although only a handful are well characterized [11]. The establishment of lymphoid leukemic cell lines is still a challenge and is successful in at most 10% of cases [11]. BCP ALL cell lines have been established mainly from children with ALL, most frequently in relapse or with resistant disease. Some are derived from patients with chronic myelogenous leukemia (CML) in lymphoid blast crisis and some have been classified as acute undifferentiated leukemias (AUL). Many cell lines harbor typical leukemic cytogenetic abnormalities such as translocations t(1;19) and t(9;22) with their respective fusion gene products. These karyotype aberrations are associated with poor clinical outcome and therefore cell lines with these changes enable the study of these neoplasms. It should also be noted, however, that many cell lines have additional cytogenetic aberrations, some of which occur in vitro. Many cell lines are found to respond to cytokines by either proliferation or inhibition of proliferation, and identification of cytokine receptors on the surface of some of the cell lines has further facilitated the study of cytokine interactions and signal transduction pathways in BCP ALL. The growing repertoire of BCP cell lines creates more opportunities to analyze the unique biological properties of the vast heterogeneity of human BCP ALL.

## 2. CLINICAL CHARACTERIZATION

More than 80 BCP cell lines are summarized in Table 1, according to the degree of B-lymphoid differentiation (B-I, B-II, and B-III). In nine of the cell lines (G2, IARC-318, JM, Km-3, LILA-1, LK-63, Tree92, Z-119, and Z-181), expression of cytoplasmic Ig is unknown or not clearly identified to warrant categorization at either B-II or B-III stage of differentiation. Tree92 is unusual as it is derived from a patient with mature B-cell ALL (ALL L3) and shows faint expression of surface Igs, otherwise displaying markers of a BCP line. The majority of cell lines were derived from patients with ALL (78/85, 92%). Two cell lines were derived from patients with lymphoid blast crisis of CML (CML-BC). Three patients were classified as acute undifferentiated leukemia (AUL) and one as lymphoblastic lymphoma (LL). No information was given as to the diagnosis of the patient in one case. The sex distribution is 42 from male patient and 40 from females. No information was provided in three cases. Not surprisingly, and in keeping with the distribution of incidence rates according to age in ALL, the majority of BCP cell lines were derived from children of 15 years of age or less (51/78, 65%). Twenty cell lines (26%) were established from samples of adults with ALL. In seven cases, the age of the patient is unknown. Among the 51 children, 10 (20%) were infants less than one year of age.

FAB morphology has largely lost its role in classification of ALL and its prognostic significance, except for ALL L3. Nevertheless, in cases where information as to the FAB type was given, the majority of patients with ALL displayed ALL FAB type L1 (14/21, 67%). FAB type L3 is usually associated with mature B-cell ALL and not BCP ALL. However, cell line Tree92, although consistent with BCP characteristics, was derived from a patient with L3 morphology. Most cell lines were developed from patients at relapse or with resistant disease (36/85, 31%). Twenty-three samples (27%) were established from diagnostic specimens. Two patients were in lymphoid blast crisis of CML. Information is lacking in 25 cases (29%). Nearly equal numbers of cell lines were derived from peripheral blood (30/85, 35%) or bone marrow (36/85, 42%). One line was developed from both peripheral blood and bone marrrow sources, and one line from peritoneal fluid (PF). No source was provided in 17 (20%) cases. The majority of cell lines were established in the 1980s (21/47, 45%). Equal numbers, where a reliable date could be ascertained, were developed in the 1970s and 1990s (13/47, 28%, each). The preferred choice of medium is RPMI-1640 supplemented with either FCS or FBS.

Sister cell lines were established from several patients. In the case of NALM-19, NALM-20, NALM-24/-25, and NALM-29, EBV-positive non-leukemic B lymphoid cell lines were developed. In other cases, longitudinal lines were established from the same patient at first, second, or further relapse (NALM-6 to NALM-13, NALM-21/-22/-23, NALM-30/-31/-32, SUP-B19, SUP-B28, SUP-B31, UoC-B5, UoC-B6, KH-3A and KH-3B, PC-53A).

## 3.  IMMUNOPHENOTYPICAL CHARACTERIZATION

Precursor B-cell development (Table 2) is characterized by sequential expression of cytoplasmic and surface marker antigens that allow classification into three broad categories. Guidelines for the classification of acute lymphoblastic leukemias were proposed by the European Group for the Immunological Classification of Leukemias [9]. Accordingly, BCP cell lines can be divided into pro-B lines (B-I), common-B lines (B-II), and pre-B lines (B-III). Ten of the 87 (11%) cell lines described here belong to the pro-B cell

*Table 1.* Precursor B-cell lines: clinical characteristics

| Cell line | Patient age/sex | Diagnosis | Treatment status | Specimen site | Year est. | Culture medium | Ref. |
|-----------|-----------------|-----------|------------------|---------------|-----------|----------------|------|
| **Pro-B cell lines (B-I)** | | | | | | | |
| A-1 | ?/M | ALL | Relapse | PB | | $\alpha$-MEM + 10% FBS | 12,13 |
| B1 | 14/M | ALL | Relapse | BM | 1990 | $\alpha$-MEM + 10% FBS | 14 |
| HBL-3 | 9/F | ALL | Diagnosis | BM | 1985 | RPMI-1640 + 20% FBS | 15 |
| JKB-1 | 16/F | ALL | Relapse | BM | 1992 | RPMI-1640 + 10% FCS | 16 |
| KH-4 | 11/M | ALL | Resistant | BM | 1982 | RPMI-1640 + 20% FBS | 17 |
| KOCL-51 | <1/M | ALL (L1) | | | | | 18 |
| NALM-19[a] | 26/M | AUL | Diagnosis | PB | 1988 | | 19 |
| RS4;11 | 32/F | ALL (L2) | Relapse | BM | 1983 | RPMI-1640 + 5% FCS | 20 |
| SEM | 5/F | ALL | Relapse | PB | | IMDM + 10% FCS | 21 |
| TOM-1 | 54/F | ALL | Resistant | BM | 1983 | RPMI-1640 + 10% FBS | 22 |

*Continued on next page*

*Table 1.* (continued)

| Cell line | Patient age/sex | Diagnosis | Treatment status | Specimen site | Year est. | Culture medium | Ref. |
|---|---|---|---|---|---|---|---|
| Common-B cell lines (B-II) | | | | | | | |
| BV173 | 45/M | CML-BC | Blast crisis | PB | 1980 | RPMI-1640 + 10% FCS | 23 |
| EU-1 | 16/M | ALL | Relapse | BM | NA | RPMI-1640 + 10% FCS | 24,25 |
| HAL-01 | 17/F | ALL (L2) | | PB | 1990 | RPMI-1640 + 25% FCS | 26 |
| HOON | 9/M | ALL | | | 1982 | RPMI-1640 + 4% FCS | 27 |
| HYON | 11/M | ALL | | | 1982 | RPMI-1640 + 4% FCS | 27 |
| MHH-CALL-2 | 15/F | ALL | Diagnosis | PB | | b | 28 |
| MHH-CALL-3 | 11/F | ALL | Diagnosis | BM | | b | 28 |
| KH-3c | 14/F | ALL | Relapse | BM | 1983 | RPMI-1640 + 20% FBS | 17 |
| Kid-92 | 28/M | ALL | Resistant | PB | 1992 | | 29 |
| KOPN-1 | <1/F | ALL (L1) | | | | | 30 |
| KOPN-K | 5/M | ALL | | | 1985 | | 11 |
| LC4-1d | 13/F | ALL (L1) | | PB/BM | | IMD medium + 15% FCS | 31 |
| MIELIKI | 1/F | ALL (L1) | Diagnosis | BM | | RPMI-1640 + 10% FCS | 32 |
| MR-87 | 4/M | ALL | Diagnosis | BM | | RPMI-1640 + 10% FCS | 33 |
| NALL-1 | 14/M | ALL | Relapse | PB | 1976 | RPMI-1640 + 20% HCS | 34 |
| NALM-16 | 12/F | ALL | Relapse | PB | 1977 | RPMI-1640 + 10% FCS | 35,36 |
| NALM-20e,f | 62/M | AUL | Diagnosis | PB | 1989 | RPMI-1640 + 10% FCS | 37 |
| NALM-24/-25g | 42/F | ALL | Diagnosis | PB | 1990 | RPMI-1640 + 10% FCS | 38 |
| NALM-27/-28 | 38/M | ALL | Diagnosis | PB | 1995 | RPMI-1640 + 10% FCS | 11 |

*Continued on next page*

*Table 1.* (continued)

| Cell line | Patient age/sex | Diagnosis | Treatment status | Specimen site | Year est. | Culture medium | Ref. |
|---|---|---|---|---|---|---|---|
| Common-B cell lines (B-II) (continued) | | | | | | | |
| NALM-29[h,i] | 46/M | ALL | Diagnosis | PB | 1995 | | 11 |
| NALM-33/-34 | 72/M | ALL | Diagnosis | PB | 1987 | | 11 |
| OM9;22 | 19/F | ALL[j] | Relapse | BM | 1987 | RPMI-1640 + 20% FCS | 39 |
| PC-53 | 33/F | ALL (L2)[k] | Relapse | BM | 1985 | IMDM + 10% FCS | 40 |
| RCH-ACV | 8/F | ALL | Relapse | BM | | RPMI-1640 + 10% FCS | 41 |
| REH | 15/F | ALL | Relapse | PB | 1975 | RPMI-1640 | 10 |
| SUP-B2 | 5/M | ALL | Relapse | | | Modified McCoy 5A medium+15% FCS | 42 |
| SUP-B7 | 2/F | ALL (L1)[l] | Diagnosis | BM | 1983 | NA | 43 |
| SUP-B26[m] | 5/M | ALL | Relapse | BM | | Modified McCoy 5A medium + 15% FCS | 42 |
| TC78 | 7/M | ALL (L2) | Relapse | BM | | RPMI-1640 + 10% FCS | 44 |
| UoC-B1 | 15/F | ALL | Relapse | BM | | Modified McCoy 5A medium+15% FCS | 42 |
| UoC-B4[n] | 3/F | ALL | Relapse | PF | | Modified McCoy 5A medium + 15% FCS | 42 |
| UoC-B7 | 5/M | ALL | Relapse | BM | | Modified McCoy 5A medium + 15% FCS | 42 |
| UoC-B8 | 3/M | ALL | Diagnosis | BM | | Modified McCoy 5A medium + 15% FCS | 42 |
| UoC-B9 | 9/F | ALL | Diagnosis | BM | | Modified McCoy 5A medium + 15% FCS | 42 |
| UoC-B10 | 26/M | ALL | Diagnosis | BM | | Modified McCoy 5A medium+15% FCS | 42 |

*Continued on next page*

*Table 1.* (continued)

| Cell line | Patient age/sex | Diagnosis | Treatment status | Specimen site | Year est. | Culture medium | Ref. |
|---|---|---|---|---|---|---|---|
| **Pre-B cell lines (B-III)** | | | | | | | |
| 207 | 10/M | ALL | Relapse | BM | 1980 | Modified McCoy 5A medium + 20% FCS | 45 |
| 697 | 12/M | ALL | Relapse | BM | 1979 | Modified McCoy 5A medium+20% FCS | 45 |
| BLIN-1 | 11/M | ALL | Diagnosis | BM | 1988 | RPMI-1640+25% FBS+10%v/v L-BCGF | 46 |
| HPB-NULL | 47/M | ALL | | PB | 1978 | | 11 |
| INC | 68/F | ALL | | | 1985 | | 29 |
| KLM-2 | M | ALL | | PB | | | 47 |
| KM3 | | | | | | | 48 |
| KMO-90 | 12/F | ALL | Diagnosis | BM | 1990 | RPMI-1640 + 10% FCS | 49 |
| KOCL-33 | <1/F | ALL (L1) | | | | | 50 |
| KOCL-44 | <1/F | ALL (L1) | | | | | 50 |
| KOCL-45 | <1/M | ALL (L1) | | | | | 50 |
| KOCL-50 | <1/F | ALL (L1) | | | | | 50 |
| KOCL-58 | <1/M | ALL (L1) | | | | | 50 |
| KOPB-26 | 1/F | ALL (L1) | | | | | 50 |
| KOPN-8 | <1/F | ALL | | PB | 1977 | | 11 |
| LAZ-221 | 24/F | ALL | Diagnosis | PB | 1977 | Medium 199 + 5% FBS | 51 |
| NALM-1[o] | 3/F | CML-BC | Blast crisis | PB | 1975 | RPMI-1640 + 10% FCS | 52 |
| NALM-6[p] | 19/M | ALL | Relapse | PB | 1976 | RPMI-1640 + 10% FCS | 53 |

*Continued on next page*

*Table 1.* (continued)

| Cell line | Patient age/sex | Diagnosis | Treatment status | Specimen site | Year est. | Culture medium | Ref. |
|---|---|---|---|---|---|---|---|
| Pre-B cell lines (B-III) (continued) | | | | | | | |
| NALM-17/-18 | 9/M | ALL | | PB | 1978 | | 11 |
| NALM-26 | 24/M | ALL | Diagnosis | PB | 1992 | | 54 |
| P30/OHKUBO | 11/F | ALL (L2) | Relapse | BM | 1980 | | 55 |
| PER-278 | 10/M | ALL (L1) | Diagnosis | BM | 1987 | | 56 |
| PRE-ALP | 6/F | ALL (L1) | Diagnosis | BM | | RPMI-1640 + 10% FCS | 57 |
| SMS-SB | 16/F | LL | Relapse | PB | 1977 | RPMI-1640 + 15% FBS | 58 |
| SUP-B15 | 9/M | ALL | Relapse | BM | | | 59 |
| SUP-B16[q] | 10/F | ALL | Relapse | BM | | Modified McCoy 5A medium + 15% FCS | 42 |
| SUP-B24[r] | 3/M | ALL | Diagnosis | BM | | Modified McCoy 5A medium + 15% FCS | 42 |
| SUP-B27 | 15/M | ALL | Relapse | BM | | Modified McCoy 5A medium + 15% FCS | 42 |
| Tahr-87 | 27/M | AUL | | BM | 1980 | | 60 |
| TS-2 | 3/F | ALL (L1) | Relapse | PB | 1994 | RPMI-1640 + 10% FBS | 61 |
| UoC-B3[s] | 14/F | ALL | Diagnosis | BM | | Modified McCoy 5A medium + 15% FCS | 42 |

*Continued on next page*

*Table 1.*   (continued)

| Cell line | Patient age/sex | Diagnosis | Treatment status | Specimen site | Year est. | Culture medium | Ref. |
|---|---|---|---|---|---|---|---|
| **CyIgM status unknown (B-II/III)** | | | | | | | |
| G2 | ?/F | ALL | Relapse | PB | | α-MEM + 10% FBS | 13 |
| IARC-318 | | ALL | | | | | 28 |
| JM | 14/M | ALL | Relapse | PB | 1977 (?) | RPMI-1640 + 10% FCS | 62 |
| Km-3 | 12/M | ALL | Relapse | PB | 1977 (?) | RPMI-1640 + 10% FCS | 62 |
| LILA-1 | | ALL | | PB | | RPMI-1640 + 10% FCS | 63 |
| LK-63 | F | ALL | | PB | | RPMI-1640 + 10% FCS | 63 |
| Tree92 | 9/M | ALL (L3) | | | 1992 | | 29 |
| Z-119 | 25/F | ALL (L2) | Relapse | BM | 1990 | RPMI-1640 + 10% FCS | 64 |
| Z-181 | 32/M | ALL | Relapse | BM | 1991 | RPMI-1640 + 10% FCS | 64 |

BM – bone marrow, PB – peripheral blood, PF – peritoneal fluid.
[a] B239 and B240 (EBV-positive B-LCL) were established from the same leukemic sample. [b] Cell lines were grown in basal growth media (RPMI 1640, Dulbecco's MEM, Iscove's MEM, α-MEM, McCoy's 5A, Gibco BRL) supplemented with 5–20% FBS. [c] Two cell lines (KH-3A and KH-3B) were established from the same patient. KH-3A at first, and KH-3B at second relapse. Two more clones (KH-3A-2 and KH-3A-3) were established from cell line KH-3A [17]. [d] subclone of LC4 cells with most favorable growth characteristics. [e] B250 (EBV-positive, non-leukemic B-LCL) was established from the same leukemic sample. [f] NALM-21/-22/-23 were established from the same patient as NALM-20, however at relapse. [g] B262 (EBV-positive, non-leukemic B-LCL) was established from the same leukemic sample. [h] B391 (EBV-positive B-LCL) was established from the same leukemic sample. [i] NALM-30/-31/-32 were established from the same patient as NALM-27/-28, however at relapse. [j] cell lines post-BMT. [k] PC-53A was established from the same patient later during her terminal relapse. [l] Cell line was predominantly FAB L2 morphology. [m] SUP-B28 was established from the same patient at second relapse. [n] UoC-B6 was established from the same patient at second relapse. [o] initial in vitro testing of NALM-1 cells after establishment of the cell line did not reveal cytoplasmic Ig. However, in further experiments in 1978, the authors detected cytoplasmic fluorescence with affinity purified anti-μ [66]. [p] NALM-6 is one of 8 leukemia cell lines (NALM-6 to NALM-13) which were derived from the same patient at relapse. The data here refer to NALM-6 which was further characterized in Minnesota as NALM-6-M1. [q] SUP-B19 was established from the same patient at second relapse. [r] SUP-B31 was established from the same patient at first relapse. [s] UoC-B5 was established from the same patient at first relapse.

category. However, except for two lines (NALM-19, SEM), expression of CD19 in association with absence of expression of CD10, cytoplasmic Ig, and surface membrane Ig, have been the basis for classification into category B-I. Expression of CD79a has been very sporadically tested. Common B-lines are distinguished by positivity for CD10 (common acute lymphoblastic leukemia antigen, CALLA) in the absence of cytoplasmic or surface immunoglobulin markers. Thirty-eight of the 87 cell lines (44%) meet these criteria. Finally, pre-B lines, i.e., cell lines with expression of cytoplasmic Ig with or without coexpression of CD10, account for 32 of 87 lines (37%). In a number of cases (9/87, 10%), the status of expression of cytoplasmic Ig is unclear. Coexpression of myeloid markers on otherwise ALL cells has been described in up to 20% of cases of ALL. Coexpression of at least one myeloid marker (CD13 or CD33) has been described in 18 of 50 cell lines tested (36%), with expression of two myeloid markers in another 6 cell lines (12%). Coexpression of T cell markers (CD5 and/or CD7) was found in only a minority of cell lines tested (7/70, 10%).

## 4. CYTOKINE-RELATED CHARACTERIZATION

Cytokine receptor expression, production of cytokines, and response to cytokines in terms of proliferation and differentiation are summarized in Table 3. Rarely are cell lines systematically screened with regard to response to cytokines. In most cases, cytokines are chosen according to personal interest, creating a patchy picture of influence of cytokines and biological response modifiers in the literature. Frequently tested cytokines are the interleukins (IL) IL-1, IL-3, IL-4, IL-6, and IL-7, the interferons, TNF-$\alpha$, and hematopoietic growth factors such as G-CSF and GM-CSF. IL-1$\alpha$ and 1-$\beta$ were found to stimulate proliferation of B1 cells [14]. IL-4 was mainly inhibitory for the proliferation of SEM, KM3, MIELIKI, and REH cells [21,32,67]. IL-7 caused variable responses. Growth inhibition in B1, 697, NALM-6, and MIELIKI cells [11,32,42,68–71] was partly due to upregulation of apoptotic pathways mediated by Fas-ligand/APO-1 signaling cascades. Promotion of cell proliferation has been observed in JKB-1, PRE-ALP, NALM-20, NALM-21/-22/-23, NALM-24/-25, and OM9;22 cells [16,21,39,57]. Receptors for cytokines and production of cytokines have been described for a few cell lines (Table 3). Growth-factor dependent cell lines are infrequent among BCP lines.

*Table 2.*  Precursor B-cell lines: immunophenotypical characterization

| Cell line | HLA-DR | CD 19 | CD 22 | CD 79a | CD 10 | CyIgM | SmIg | CD 20 | CD 5 | CD 7 | CD 13 | CD 33 | CD 34 | CD 85 | CD 19 | CD 127 | CD 135 | Ref. |
|---|---|---|---|---|---|---|---|---|---|---|---|---|---|---|---|---|---|---|
| Pro-B cell lines (B-I) | | | | | | | | | | | | | | | | | | |
| A-1 | + | + | | | – | – | – | – | – | – | – | | | | | | | 12,13 |
| B1 | + | + | | | – | – | – | – | – | – | – | + | | | | | | 14 |
| HBL-3 | + | + | | | – | – | – | – | – | – | – | – | | | | | | 15 |
| JKB-1 | + | + | – | | – | – | – | – | – | – | – | – | – | | | | | 16 |
| KH-4 | + | + | | | – | – | – | – | – | | | | | | | | | 17 |
| KOCL-51 | + | + | | | – | – | – | – | – | – | + | | | | | | | 50 |
| NALM-19 | + | + | + | + | – | – | – | – | + | – | + | – | – | + | + | + | + | 11 |
| RS4;11 | + | + | + | | – | | – | – | – | – | | | | | | | | 20 |
| SEM | + | + | + | | – | – | – | | – | – | + | + | – | | | | | 21 |
| TOM-1 | + | + | | | –ᵃ | – | – | | – | | | | | | | | | 22 |

*Continued on next page*

*Table 2.* (continued)

| Cell line | HLA-DR | CD 19 | CD 22 | CD 79a | CD 10 | CyIgM | SmIg | CD 20 | CD 5 | CD 7 | CD 13 | CD 33 | CD 34 | CD 85 | CD 19 | CD 127 | CD 135 | Ref. |
|---|---|---|---|---|---|---|---|---|---|---|---|---|---|---|---|---|---|---|
| **Common-B cell lines (B-II)** | | | | | | | | | | | | | | | | | | |
| BV173 | + | + | + | + | + | – | – | – | – | – | +[b] | + | + | + | + | – | + | 11 |
| EU-1 | + | + | + | + | + | – | – | | – | – | +[b] | +[b] | – | | + | – | | 24,25 |
| HAL-01 | + | + | + | + | + | – | – | – | – | – | + | – | – | + | + | – | + | 26 |
| HOON | + | + | | | + | – | – | – | | | | | | | | | | 27 |
| HYON | + | + | | | + | – | – | – | | | | | | | | | | 27 |
| KH-3 | + | + | | | + | – | – | – | – | | | | | | | | | 17 |
| Kid-92 | + | + | + | + | + | – | – | – | – | – | + | – | – | + | + | + | + | 11 |
| KOPN-1 | + | + | | | + | – | – | | – | – | – | + | | | | | | 50 |
| KOPN-K | + | + | + | + | + | – | – | – | – | – | – | – | – | – | + | – | + | 11 |
| LC4-1 | + | + | | | + | – | – | + | – | – | – | – | | | | | | 31 |
| MHH-CALL-2 | + | + | | + | + | – | – | | | – | | | | | | | | 28 |
| MHH-CALL-3 | + | + | + | + | + | – | – | | | – | | | | | | | | 28 |
| MIELIKI | + | + | | + | + | – | – | + | – | – | –[c] | –[c] | | | | | | 32 |
| MR-87 | + | + | + | + | + | – | – | – | – | – | + | + | – | | | | | 33 |
| NALL-1 | + | + | + | + | + | – | – | + | + | – | – | + | – | – | + | + | + | 11 |
| NALM-16 | + | + | + | + | + | – | – | + | + | – | – | – | – | – | + | + | + | 11 |
| NALM-20 | + | + | + | + | + | – | – | – | – | – | + | – | + | + | + | + | + | 11,37 |
| NALM-21/-22/-23 | + | + | + | + | + | – | – | – | – | – | + | – | + | – | + | + | + | 11,37 |

*Continued on next page*

*Table 2.* (continued)

| Cell line | HLA-DR | CD 19 | CD 22 | CD 79a | CD 10 | CyIgM | SmIg | CD 20 | CD 5 | CD 7 | CD 13 | CD 33 | CD 34 | CD 85 | CD 19 | CD 127 | CD 135 | Ref. |
|---|---|---|---|---|---|---|---|---|---|---|---|---|---|---|---|---|---|---|
| **Common-B cell lines (B-II) (continued)** | | | | | | | | | | | | | | | | | | |
| NALM-24/-25 | + | + | + | + | + | – | – | – | – | – | + | – | + | + | + | + | + | 11,38 |
| NALM-27/-28 | + | + | + | + | + | – | – | – | – | – | + | – | + | + | – | – | + | 11 |
| NALM-29 | + | + | + | + | + | – | – | – | – | – | + | – | – | + | + | + | + | 11 |
| NALM-30/-31/-32 | + | + | + | + | + | – | – | – | – | – | + | – | – | – | + | + | + | 11 |
| NALM-33/-34 | + | + | + | + | + | – | – | – | – | – | + | – | – | + | + | – | + | 11 |
| OM9;22 | + | + | + | + | + | – | – | – | – | – | + | + | + | – | – | – | + | 11,39 |
| PC-53 | + | + | | | + | – | – | – | – | – | | – | – | | | | | 40 |
| RCH-ACV | + | + | + | | + | – | – | – | – | – | | – | – | – | | | | 41 |
| REH | + | + | | + | + | – | – | – | – | – | – | – | – | – | – | – | + | 11 |
| SUP-B2 | + | + | | | + | – | – | – | – | – | | | – | | | | | 42 |
| SUP-B7 | + | + | | | + | – | – | – | – | – | | | – | | | | | 43 |
| SUP-B26 | + | + | | | + | – | – | – | – | – | | | | – | | | | 42 |
| TC78 | + | | | | + | – | – | – | – | – | – | – | | | | | | 44 |
| UoC-B1[d] | + | + | | | + | – | – | – | – | – | | | | | | | | 42 |
| UoC-B4[d] | + | + | | | + | – | – | –[e] | – | – | | | – | | | | | 42 |
| UoC-B7[d] | + | + | | | + | – | – | – | – | – | | | – | | | | | 42 |
| UoC-B8[d] | + | + | | | + | – | – | – | – | – | | | | | | | | 42 |
| UoC-B9[c] | + | + | | | + | – | – | + | – | – | | | | | | | | 42 |
| UoC-B10 | + | + | | | + | – | – | + | – | – | | | | | | | | 42 |
| UoC-B11 | + | + | | | + | – | – | – | – | – | | | – | – | | | | 65 |

*Continued on next page*

Table 2. (continued)

| Cell line | HLA-DR | CD 19 | CD 22 | CD 79a | CD 10 | CyIgM | SmIg | CD 20 | CD 5 | CD 7 | CD 13 | CD 33 | CD 34 | CD 85 | CD 19 | CD 127 | CD 135 | Ref. |
|---|---|---|---|---|---|---|---|---|---|---|---|---|---|---|---|---|---|---|
| **Pre-B cell lines (B-III)** | | | | | | | | | | | | | | | | | | |
| 207 | + | + | | | + | + | +[f] | | – | | | | | | | | | 45 |
| 697 | + | + | | | + | + | +[f] | – | | | | | | | | | | 45 |
| BLIN-1 | + | + | | | + | + | – | +/– | + | | | | | | | | | 46 |
| HPB-NULL | + | + | + | + | – | + | – | | + | – | – | – | – | + | + | + | + | 11 |
| INC | + | + | + | + | + | + | – | + | – | – | + | – | – | + | + | + | + | 11 |
| KLM-2 | + | + | + | + | + | + | – | + | – | – | – | – | – | + | + | – | – | 11 |
| KMO-90 | + | + | | + | + | + | – | – | – | – | – | – | | | | | | 49 |
| KM3 | + | + | | | + | + | – | – | | | | | | | | | | 67 |
| KOCL-33 | + | + | + | + | – | + | – | | | – | | – | | | | | | 50 |
| KOCL-44 | + | + | + | + | + | + | – | | | – | | + | | | | | | 50 |
| KOCL-45 | + | + | + | + | – | + | – | | | – | | – | | | | | | 50 |
| KOCL-50 | + | + | + | + | – | + | – | | | + | | + | | | | | | 50 |
| KOCL-58 | + | + | + | + | – | + | – | | | – | | – | | | | | | 50 |
| KOPB-26 | + | + | + | + | – | + | – | | | – | | – | | | | | | 50 |
| KOPN-8 | + | + | + | | | | | | – | – | – | – | – | + | + | + | + | 11 |
| LAZ-221 | + | + | + | | | | | + | + | – | – | – | – | + | + | + | + | 11 |
| NALM-1 | + | + | + | | | | | + | – | – | – | – | + | + | + | + | + | 11 |
| NALM-6 | + | + | – | | | | | – | – | – | – | – | – | + | + | + | – | 11 |
| NALM-17/-18 | + | + | + | | | | | – | – | – | – | – | + | + | + | + | + | 11 |
| NALM-26 | + | + | – | | | | | – | – | – | – | – | – | – | + | + | + | 11 |

Continued on next page

*Table 2.* (continued)

| Cell line | HLA-DR | CD 19 | CD 22 | CD 79a | CD 10 | CyIgM | SmIg | CD 20 | CD 5 | CD 7 | CD 13 | CD 33 | CD 34 | CD 85 | CD 19 | CD 127 | CD 135 | Ref. |
|---|---|---|---|---|---|---|---|---|---|---|---|---|---|---|---|---|---|---|
| Pre-B cell lines (B-III) (continued) | | | | | | | | | | | | | | | | | | |
| P30/OHKUBO | + | + | + | + | + | + | − | − | − | − | − | − | − | + | + | + | + | 11 |
| PER-278 | + | + | + | | + | + | − | − | − | − | − | − | − | + | + | + | + | 56 |
| PRE-ALP | + | + | | | + | + | − | + | − | − | | | − | | | | | 57 |
| SMS-SB | + | | | | − | + | − | | | − | | | | | | | | 58 |
| SUP-B13 | + | + | | | + | + | − | | − | − | | | +/− | | | | | 65 |
| SUP-B15 | + | + | | | + | + | − | | | | | | | | | | | 59 |
| SUP-B16 | + | + | | | − | + | − | + | − | − | | | − | | | | | 42 |
| SUP-B24 | + | + | | | + | + | − | − | − | − | | | − | | · | | | 42 |
| SUP-B27 | + | − | | | + | + | − | − | − | − | | | − | | | | | 42 |
| Tahr-87 | + | + | + | + | + | + | − | − | − | − | + | − | + | + | + | + | | 11 |
| TS-2 | + | + | + | | − | + | − | − | − | | | − | | | | + | + | 61 |
| UoC-B3 | + | + | | | + | + | − | | − | − | | | | | | | | 42 |

*Continued on next page*

*Table 2.* (continued)

| Cell line | HLA-DR | CD 19 | CD 22 | CD 79a | CD 10 | CyIgM | SmIg | CD 20 | CD 5 | CD 7 | CD 13 | CD 33 | CD 34 | CD 85 | CD 19 | CD 127 | CD 135 | Ref. |
|---|---|---|---|---|---|---|---|---|---|---|---|---|---|---|---|---|---|---|
| CyIgM status unknown (B-II/III) | | | | | | | | | | | | | | | | | | |
| G2 | + | + | | | − | ? | − | − | | | | | | | | | | 13 |
| IARC-318 | + | + | | | + | ? | − | | | | | | | | | | | 62 |
| JM | | | | | | ? | − | | | | | | | | | | | 62 |
| KM-3 | | | | | | ? | − | | | | | | | | | | | 62 |
| LILA-1 | + | + | | | + | ? | − | − | | | | | | | | | | 63 |
| LK-63 | − | + | + | | + | ? | −g | − | | | | | | | | | | 63 |
| Tree 92 | + | + | | | − | ? | −h | − | − | − | − | | | | | | | 29 |
| Z-119 | + | + | + | | + | ? | NA | + | − | − | + | + | + | | | | | 64 |
| Z-181 | + | + | + | | + | ? | NA | + | − | − | + | − | + | | | | | 64 |

a Up to three months after the beginning of culture cells expressed CD10, but lost positivity for CALLA thereafter. b Expression of myeloid markers CD13, CD14, CD33 only after treatment with DMSO for 3 days. c Cell line expressed CD15. d All cell lines are positive for expression of CD15. e CD20 expression is positive in cell line UoC-B6 from same patient at second relapse. f SIg staining faint by immunofluorescence. g Lk-63 has been described to express membrane IgM by the authors. The origin of the cell line of a typical CD10-expressing childhood ALL with a t(1;19) prompted classification as a pre-B-cell line. h Limited expression of sIg (μ, 19%; λ, 6%).

*Table 3.*　Precursor B-cell Lines: cytokine-related characterization

| Cell lines | Cytokine receptor expression | Cytokine production | Proliferation response to cytokines | Differentiation response to cytokines | Dependency on cytokines | Ref. |
|---|---|---|---|---|---|---|
| Pro-B cell lines (B-I) | | | | | | |
| B1 | IL-1, IL-6, IL-7, $\gamma$ IFN, TNF$\alpha$ | IL-1$\alpha$ and IL-1$\beta$ | IL-1$\alpha$ and IL-1$\beta$ stimulate cell growth. IL-6, IL-7, $\gamma$ IFN, and TNF are potent inhibitors of B1 cell growth | | Autocrine stimulation of growth by IL-1 | 14 |
| JKB-1 | | | Increased proliferation with SCF and IL-3 or IL-7. SCF and IL-3/IL-7 act synergistically. Decreased proliferation by IL-6 | | | 16 |
| RS4;11 | Constitutive expression of CD95/APO-1 | | | | | 70 |
| SEM | Expresses mRNA for IL-7 receptor | | IL-7 enhances proliferation of SEM cells; IL-4, TNF-$\alpha$, IFN-$\alpha$, IFN-$\gamma$ inhibit cell growth | | | 21 |

*Continued on next page*

*Table 3.* (continued)

| Cell lines | Cytokine receptor expression | Cytokine production | Proliferation response to cytokines | Differentiation response to cytokines | Dependency on cytokines | Ref. |
|---|---|---|---|---|---|---|
| Common-B cell lines (B-II) | | | | | | |
| EU-1 | | TNF-α (mRNA, protein) | Resistant to inhibition by TNF-α | | | 24 |
| HAL-01 | | | Proliferation of growth is suppressed by IL-3 in a dose-dependent fashion | | | 26 |
| HOON | Express endoglin as part of TGF-β receptor complex | | | | | 72 |
| KOPN-1 | | | Growth inhibition by rhIFNα | | | 68 |
| MIELIKI | CDw124 and CDw127 are associated with high affinity binding for IL-4 and IL-7 | | IL-4 and IL-7 mediate growth suppression of cell line | | | 32 |
| MR-87 | Expresses receptors for IL-7 and IL-4 | | No proliferative response to IL-7 | | | 39 |
| NALM-16 | Constitutive expression of CD95/APO-1 on cell surface | | | | | 70 |

*Continued on next page*

*Table 3.* (continued)

| Cell lines | Cytokine receptor expression | Cytokine production | Proliferation response to cytokines | Differentiation response to cytokines | Dependency on cytokines | Ref. |
|---|---|---|---|---|---|---|
| Common-B cell lines (B-II) | | | | | | |
| NALM-20 | | | Enhanced colony growth with IL-7 | | | 39 |
| NALM-21/-22/-23 | | | Enhanced colony growth with IL-7 | | | 39 |
| NALM-24/-25 | | | Enhanced colony growth with IL-7 | | | 39 |
| OM9;22 | Receptors for IL-7 and IL-4 | | Enhanced colony growth by IL-7 and decreased growth by IL-4 | Treatment with IL-7 induces expression of CD20 | | 39 |
| REH | | | Decreased proliferation of cells with exposure to rIL-4; rIL-4 downregulates IL-3-induced proliferation; decreased cell proliferation by IFN-$\gamma$ | Induction of marker maturation of leukemic cells by incubation with rIL-4; IFN-$\gamma$ induces expression of class I MHC antigens, whereas TPA induces expression of class II MHC antigens | | 48,67,73 |

*Continued on next page*

*Table 3.* (continued)

| Cell lines | Cytokine receptor expression | Cytokine production | Proliferation response to cytokines | Differentiation response to cytokines | Dependency on cytokines | Ref. |
|---|---|---|---|---|---|---|
| Pre-B cell lines (B-III) | | | | | | |
| 697 | Constitutive expression on of CD95/APO-1 cell surface | | IL-7 enhances susceptibility of cells to CD95/APO-1 mediated apoptosis | | | 70 |
| BLIN-1 | High affinity IL-3 receptors | | Increased proliferation with L-BCGF, IL-3 | | Cells are dependent on L-BCGF for optimal growth | 46 |
| HPB-NULL | Expression of IL-1 receptors | IL-2 | No proliferative response to rIL-1 | | | 69,71 |
| INC | | IL-2 | | | | 71 |
| uKM3 | | | rIL-4 decreases proliferation of cells in culture; decreased cell proliferation by IFN-$\gamma$ | Induction of surface marker maturation (increased expression of CD20, CD23) | | 48,67 |
| LAZ-221 | | IL-2 | | | | 71 |
| NALM-1 | | IL-2 | | | | 71 |

*Continued on next page*

*Table 3.* (continued)

| Cell lines | Cytokine receptor expression | Cytokine production | Proliferation response to cytokines | Differentiation response to cytokines | Dependency on cytokines | Ref. |
|---|---|---|---|---|---|---|
| Pre-B cell lines (B-III) (continued) | | | | | | |
| NALM-6 | Express endoglin as part of TGF-$\beta$ receptor complex; expression of IL-1 receptors; constitutive expression of CD95/APO-1 on cell surface | IL-2 | Growth inhibition by rhIFNa; no proliferative response to stimulation with rIL-1; IL-7 enhances susceptibility of cells to CD95/APO-1 induced apoptosis | | | 68–72 |
| PRE-ALP | | | Incubation with IL-7 causes cell proliferation in dose-dependent manner | No induction of TPA, differentiation by IL-7, and other cytokines | | 57 |
| CyIgM status unknown (B-II/III) | | | | | | |
| Z-119 | No binding site for GM-CSF | IL-1$\beta$, G-CSF, GM-CSF | | | | 64 |
| Z-181 | High-affinity receptors for GM-CSF; RT-PCR shows transcripts present for GM-CSF receptor | IL-1$\beta$, G-CSF, GM-CSF | Anti-GM-CSF antibodies significantly suppress colony formation in a dose-dependent fashion | | | 64 |

## 5. CYTOGENETIC AND MOLECULAR CHARACTERIZATION

As in vivo, cell lines in vitro are characterized by complex chromosomal aberrations, both structural and numerical [74]. Karyotypes of BCP cell lines that have been characterized and described are summarized in Table 4. Besides a wide array of modal numbers, three structural chromosomal aberrations are prominent, including translocations t(9;22), the Philadelphia chromosome, t(4;11), and t(1;19). Not surprisingly, and reflective of in vivo incidence rates, the Philadelphia translocation is the most common. Translocation t(9;22)(q34;q11) occurs in 12 of the cell lines included. In most cases, the *BCR-ABL* fusion product has been characterized. Next most frequent was translocation t(1;19) resulting in the characteristic *E2A-PBX1* fusion product. Interestingly, BCP cell line TS-2 is carrying a t(1;19) abnormality, but no evidence of the *E2A-PBX1* gene rearrangement was found by RT-PCR. Cell line KMO-90 is also harboring a point mutation at codon 177 of the p53 gene as identified by SSCP analysis [49]. Translocation t(4;11) was found in four cell lines.

Deletions of the *CDKN2* gene ($p16^{INK4a}/p14^{ARF}$) have been identified in some BCP cell lines [80]. In some cases, cell lines have been used for the cloning of novel fusion genes and the localization of genes at or near chromosomal breakpoints. The sequence of a putative tumor suppressor gene, i.e. *p16/p14* and *p15*, on chromosome 9p21 has been facilitated by analysis of various tumor cell lines [81,82]. The localization of these genes, encoding cyclin-dependent kinase inhibitor proteins, has also been established in cell lines [83,84]. The sequence of the *MLL* gene spanning the breakpoint at chromosome 11q23 was determined by analysis of the BCP cell line RS4;11 [85]. The fusion of the *TEL* gene on chromosome 12 to the *AML1* gene on chromosome 21, as it occurs in up to 30% of childhood B-lineage ALL, has also been demonstrated in tumor cell lines [86]. Identification of the same cytogenetic and/or molecular profile between patient specimen and cell line provides a good means to establish authenticity of the cell line created.

## 6. FUNCTIONAL CHARACTERIZATION

Doubling times of BCP cell lines vary widely, ranging from less than 24 hours to one week. Most cell lines have been tested for EBV infection and found to be negative (Table 5). Not surprisingly, BCP cell lines are MPO-negative, although in some cases MPO expression at the mRNA level was demonstrated. The significance of this finding in vivo is unclear [99]. The

*Faderl and Estrov*

*Table 4.* Precursor B-cell lines: cytogenetic characterization

| Cell lines | Karyotype | Characteristic translocation | Genetic rearrangements | Ref. |
|---|---|---|---|---|
| Pro-B cell lines (B-I) | | | | |
| A-1 | 45, XY with multiple structural and numerical abnormalities | | | 13 |
| B1 | 45, XY, der(1)t(1;8)(p36;q13), −4, +6, −9, der(10)t(1;10)(q15;p15), der(11)t(4;11)(q21;q23) | t(4;11)(q21;q23) | | 14 |
| HBL-3 | 46, XX, −3, −9, −9, +der(3) t(3;?), +der(9) t(9;?), +der(9) t(1;9) | | | 15 |
| JKB-1 | 46, XX, t(9;14)(p21;q32) | | | 16 |
| KH-4 | 46, XY; 45, XY, −E | | | 17 |
| KOCL-51 | 46, XY, del(11)(q23) | | 11q32 | 75 |
| NALM-19 | 46, XY, −11, +der(11)t(11;?)(11pter-11q23?)/46, XY | | | 19,37,38 |
| RS4;11 | 46, XX, i (7q), t(4;11)(q21;q23) | t(4;11)(q21;q23) | | 20 |
| SEM | 46, XX, t(4;11)(q21;q23), del(7)(p15), del(13)(q12) | t(4;11)(q21;q23) | | 21 |

*Continued on next page*

*Table 4.* (continued)

| Cell lines | Karyotype | Characteristic translocation | Genetic rearrangements | Ref. |
|---|---|---|---|---|
| Common-B cell lines (B-II) | | | | |
| BV173 | 47(46–48)<2n>X/XY, +mar, der(22)t(9;22)(q34;q11) | t(9;22)(q34;q11) | BCR-ABL (b2/a2) | 76 |
| EU-1 | 45, X, t(8;21), del(3p), del(4q), del(5q), der(12), der(16), der(16) | | | 24 |
| HAL-01 | 46, XX, t(1;17)(p34;q21), t(17;19)(q21;p13) | t(17;19)(q21;p13) | E2A-HLF | 26 |
| KH-3 | 46, XX; 46, XX, +2q; 45, XX, –E; 47, XX, +13 | | | 17 |
| Kid-92 | t(4;11)(q21;q23) | t(4;11)(q21;q23) | MLL-AF4 | 29 |
| KOPN-1 | 46, XX, t(11;19)(q23;p13) | | | 50,75 |
| LC4-1 | 46, XX | | | 31 |
| MIELIKI | 45, XX, t(7;9) | | | 32 |
| MR-87 | 46, XY, 9p–, 17p?, t(9p?q+, 22q–) | t(9;22) | BCR-ABL (e1a2) | 33 |
| NALL-1 | 43, –X, –Y | | | 11 |
| NALM-16 | 27, X, +10, +14, +18, +21, 7p+ | | | 29 |
| NALM-20 | 46, XY, +2, –8, t(9;22)(q34;q11) | t(9;22)(q34;q11) | BCR-ABL | 37 |
| NALM-21 | 49, XY, –3, +4, +5, +18, –19, +mar, t(9;22)(q34;q11) (2) | t(9;22)(q34;q11) | BCR-ABL | 37 |
| NALM-24 | 45, XX, –15, –20, +mar, t(9;22)(q34;q11), del(9)(p21) | t(9;22)(q34;q11) | BCR-ABL | 38 |
| NALM-27 | 46, XY, t(9;22;10)(q34;q11;q22) | t(9;22)(q34;q11) | BCR-ABL (b3/a2) | 11 |
| NALM-29 | 46, XY, del(6)(q15q21), t(9;22)(q34;q11) | t(9;22)(q34;q11) | BCR-ABL | 11 |
| NALM-30 | 46-47, XY, –14, –15, –20, +4mar, add(3)(q11), del(6)(q15q21), del(6)(q15q21), t(9;22;10)(q34;q11;q22) | t(9;22)(q34;q11) | BCR-ABL | 11 |
| NALM-33 | 48, XY, +1, +21, t(8;14)(?q24;?q32), del(15)t(15;20)(p12;p/q11) | | | 11 |

*Continued on next page*

*Table 4.* (continued)

| Cell lines | Karyotype | Characteristic translocation | Genetic rearrangements | Ref. |
|---|---|---|---|---|
| Common-B cell lines (B-II) (continued) | | | | |
| OM9;22 | 45, X, –X, –4, –8, –15, –16, –17, +4mar, +der(1)t(1:?)(q25:?), del(3)(q26), t(9;22)(q34:q11) | t(9:22)(q34;q11) | *BCR-ABL* (e1/a2) | 39 |
| PC-53 | 45, XX, –1, –1, –3, –14, –17, der(1)(1pter-q32 or 44::1p31-pter), +M1, +M2, +M3 | | | 40 |
| RCH-ACV | 47, XX, +8, t(1:19)(q23;p13.3) | t(1;19)(q23;p13.3) | | 41 |
| REH | 46, X, –X, +16, del(3)(p21.3p24), der(4)(4pter-q32::?16q24.3-qter), t(5;12)(q31.2:p13), der(12)(12qter-q23::12p13-q23::4q32-qter), der(16)t(16;21)(q24.3:q22) (2)/der(21)t(12;21)(p13:q22) | t(12;21)(p13;q22) | *TEL/ETV6-AML1* | 28 |
| SUP-B2 | 44, XY, t(2;4)(q13;q25), t(3:7)(q25;p15), inv(5)(q13q33), del(6)(q23q27), t(11;17)(p11.2;q11.2), –14, del(17)(p11p13), dic(21;21)(q21;q21)/88 | | | 43 |
| SUP-B7 | 46, XX, del(3)(q26q28) | | | |
| SUP-B26 | 48, XY, der(5)(5:?9)(q33;q12), del(8)(q22), der(12)t(?5;12)(q33;p13), +21, +21 | | | 42 |
| TC78 | 48, XY, 2q+, 5q+, 6q–, 7q–, 8q+, 9p–, 12p–, +mar(2) | | | 44 |
| UoC-B1 | 46, X, –X, –1, –6, t(7;11)(q22;q13), add(12)(q24), der(12)t(1;12)(q21;q13), –17, der(19) t(17;19)(q22;p13), +4 mar | (17;19)(q21/22;p13) | *E2A-HLF* | 42 |
| UoC-B4 | 47, XX, t(2;14)(p11;q32), del(6)(q15q22), t(7;15)(q32;q15 or q21), add(16)(p13), +21 (4)/47, XX, del(8)(p11p21 or p21p23) | | | 42 |
| UoC-B7 | 46, XY, del(4)(q21q31), del(7)(q32q36), del(9)(p21 or p21p24), del(17)(p11p13) | | | 42 |
| UoC-B8 | 46, XY(13)/46, XY, dup(1)(q21q42), add(21)(p11) (6) | | | 42 |
| UoC-B9 | 46, XX | | | 42 |
| UoC-B10 | 46, XY, dic(8;22)(q24;p11), del(11)(p12), del(11)(q13), +i (12p), t(13;13)(q10;q10), der(20)t(17;20)(q11;q13.3) | | | 42 |

*Continued on next page*

*Table 4.* (continued)

| Cell lines | Karyotype | Characteristic translocation | Genetic rearrangements | Ref. |
|---|---|---|---|---|
| Pre-B cell lines (B-III) | | | | |
| 207 | 46, XY | | | 45 |
| 697 | 46, XY, t(7;19)(q11;q13) | | | 45 |
| BLIN-1 | 46, XY, −9, +der(9)t(8;9)(q?21.2:p?2?2) | | | 46 |
| KLM-2 | 47, XY, −4, −6, +7, −8, −9, −9, −14, +16, −17, −18, +der(4)t(4;?)(pter-q3?1::?), +der(6)t(6;18)(6qter-6p21::18q1?q-18qter), +der(9)t(9;17)(9qter-9p2?1::17q21-17qter), +der(9)t(9;11?)(9pter-9q34::11?q1?3-11?qter), +der(14or8)t(14;8)(14qter-14p1?1::8q1?1-8q24), +der(14)t(14;8)(14pter-14q32::8q24-8qter), mar+ | | | 11 |
| KMO-90* | 48, XX, +8, +19, t(1;19)(q23;p13) | t(1;19)(q23;p13) | *E2A-PBX1* | 49 |
| KOCL-33 | 46, XX, t(11;19)(q23;p13) | | | 50,75 |
| KOCL-44 | 46, XX, t(11;19)(q23;p13) | | | 50,75 |
| KOCL-45 | 46, XY, t(4;11) | | | 50 |
| KOCL-50 | 46, XXX, t(7;11) | | | 50 |
| KOCL-58 | 46, X?, t(4;11) | | | 50 |
| KOPB-26 | 46, XX, t(11;?) | | | 50 |
| KOPN-8 | 45, XX, −1, −13, −14, t(11;19)(q23;p13), t(8;13)(q24;q22), +der(1)t(1;?)(?;?), +der(13)t(13;14)(p11;q11) | t(11;19)(q23;p13) | *ENL-MLL* | 11 |

*Continued on next page*

*Table 4.* (continued)

| Cell lines | Karyotype | Characteristic translocation | Genetic rearrangements | Ref. |
|---|---|---|---|---|
| Pre-B cell lines (B-III) (continued) | | | | |
| LAZ-221 | 45, XX, −9, −12, +t(9q12q) | | | 51 |
| NALM-1 | 45(42-47)<2>X, −X, der(9)t(9;22)(q24;q12), dup(13)(q21), der(22)t(9;22)(q34;q12) | t(9;22)(q34;q12) | *BCR-ABL* | 76,77 |
| NALM-6 | 46(43-47)<2n>XY, t(5;12)(q33.2;q13.2) | | | 53,76,78 |
| NALM-26 | 45, X, −9, −19, +mar, dic(1;11)(p13;q25), +del(1)(p13), +del(1)(q21), i(9)(p10), add(14)(q24) | | | 79 |
| P30/OHKUBO | 46(45), X(X), del(2)(p23), del(9)(p12-q31), t(11;12)(q25;q13), inv(12)(q13q24) | | | 11 |
| PER-278 | 46, XY, −9, −19, +der(9) t(1;9)(q23;p13), +der(19) t(1;19)(q23;p13) | t(1;19)(q23;p13) | | 56 |
| PRE-ALP | 45, XX, −2, t(1;19)(q23;p13.3) | t(1;19)(q23;p13.3) | *E2A-PBX1* | 57 |
| SMS-SB | 47, XX, +mar | | | 58 |
| SUP-B16 | 46, XX, t(4;9)(p12;q12), del(9)(q12q22), del(13)(q12q14) | | | 42 |
| SUP-B24 | 47, XY, −2, der(3), −5, inv(7)(p13p22), −9, del(12)(p11.2p13), −15, der(16)(q12q24), −17, +mar | | | 42 |
| SUP-B27 | 47, XY, t(1;19)(q23;p13.3), +8 | t(1;19)((q23p13.3) | | 42 |
| TS-2 | 47, X, −X, t(1;19)(q23:13), add(3)(q22), del(5)(q13), +6, del(6)(q12q23), del(9)(q13q22), del(11)(p11), add(15)(q22), +mar1 | t(1;19)(q23;p13) | No evidence of *E2A-PBX1* gene rearrangement by RT-PCR | 61 |
| UoC-B3 | 46, XX, der(9)t(9;?9)(p1?2;q22), der(19)t(1;19)(q23;p13.3) | t(1;19)(q23;p13.3) | | |

*Continued on next page*

*Table 4.* (continued)

| Cell lines | Karyotype | Characteristic translocation | Genetic rearrangements | Ref. |
|---|---|---|---|---|
| CyIgM status unknown (B-II/III) | | | | |
| G2 | 46, XX, der(5) t(5;?)(q15;?), der(9) t(9;?)(p13;?), −11, der(14) t(14;?)(q22;?), +M1 | | | 13 |
| LILA-1 | 46, XX, −9, t(1;19)(q23;p13), +der(9)(1;9)t(q11;p11) | t(1;19)(q23;p13) | | 42,63 |
| LK-63 | 45, X, −X (or Y), +8, −9, −17, t(1;19)(q23;p13), +der(9)t(9;17)(p11;p11) | t(1;19)(q23;p13) | | 63 |
| Tree92 | (8;14)((q24;q32) | | | 29 |
| Z-119 | 45X, −?X/Y, der(1) t(1;8)(q24;q13), t(7;12)(q11.2;q24.3), t(9;22)(q34;q11.2) | t(9;22)(q34;q11) | *BCR-ABL* (e1a2) | 64 |
| Z-181 | 45, X−?X/Y, t(6;9;22)(q25;q34;q11.2), del(7)(p11.2;p22), +8, der(9) t(6;9;22) | t(9;22)(q34;q11) | *BCR-ABL* (e1a2) | 64 |

* Point mutation at codon 177 of the p53 gene identified by SSCP analysis.

*Table 5.* Precursor B-cell Lines: functional characterization

| Cell lines | Doubling time (hrs) | EBV status | Cytochemistry | Inducibility of differentiation | Engraftment into mice | Other features | Ref. |
|---|---|---|---|---|---|---|---|
| Pro-B cell lines (B-I) | | | | | | | |
| A-1 | 48 | Negative | | | Can engraft into SCID mice (BM and peripheral blood) | | 12, 13 |
| B1 | 40–50 | Negative | PAS+, AP+, NSE+, SBB− | | | | 14 |
| HBL-3 | 48–72 | Negative | TdT+, MPO−, AP−, esterase- | | Failed to inoculate into nude mice | | 15 |
| JKB-1 | 24 | Negative | PAS−, MPO−, NBE−, CAE− | Induction of differentiation after incubation with irradiated bone marrow stromal cells[a] | | | 16 |
| KH-4 | 72 | Negative | TdT+, AP+, MPO−, SBB−, NSE− | Differentiation into B-cell lineage by treatment with TPA | | | 17 |
| KOCL-51 | | | MPO− | | | | 50 |
| NALM-19 | | | TdT+, MPO− | | | | 11 |
| RS4;11 | 60–70 | Negative | TdT+, MPO−, SBB−, AP+, NSE+/− | TPA induces monocyte-like phenotype | | Induction of apoptosis and decrease in cell viability by cyclosporin A | 20, 59 |
| SEM | 48 | Negative | TdT+, MPO−, SBB−, NSE−, PAS− | | | | 21 |

*Continued on next page*

*Table 5.* (continued)

| Cell lines | Doubling time (hrs) | EBV status | Cytochemistry | Inducibility of differentiation | Engraftment into mice | Other features | Ref. |
|---|---|---|---|---|---|---|---|
| Common-B cell lines (B-II) | | | | | | | |
| BV173 | 30–35 | Negative | TdT+, MPO− (protein), +(mRNA), PAS− esterase− | No induction of differentiation by TPA, retinoic acid, or butyric acid | | | 23, 87 |
| EU-1 | | Negative | TdT+, MPO−, SBB−, esterase− | Can be induced to differentiate along myeloid pathway by DMSO | | | 24, 25 |
| HAL-01 | 32 | Negative | TdT+, MPO− (protein), +(mRNA), SBB−, PAS−, esterase− | | Cells are transplantable into nude mice (ascites tumor, multiorgan infiltration) | | 26 |
| HOON | 24 | | MPO−, SBB−, PAS−, NSE−, AP+ | | | Good stimulatory capacity in MLR | 27 |
| HYON | 36 | | MPO−, SBB−, PAS−, NSE−, AP+ | | | Does not stimulate lymphocytes in MLR | 27 |
| KH-3 | 48 | Negative | TdT−, AP+, MPO−, SBB−, NSE−, PAS− | Can be induced to differentiate into T or B cells after treatment with TPA | | | 17 |

*Continued on next page*

*Table 5.* (continued)

| Cell lines | Doubling time (hrs) | EBV status | Cytochemistry | Inducibility of differentiation | Engraftment into mice | Other features | Ref. |
|---|---|---|---|---|---|---|---|
| Common-B cell lines (B-II) (continued) | | | | | | | |
| Kid-92 | 36.5 | Negative | TdT+, MPO− (protein), + (mRNA) | | | Growth inhibition by dexamethasone | 29 |
| KOPN-1 | | | MPO− | | | | 50,68 |
| KOPN-K | | | TdT+, MPO− | | | | 11 |
| LC4-1 | | Negative | MPO− | Treatment with PMA induces loss of CD10 expression | | | 31 |
| MIELIKI | 120–144 | | TdT+ | | | | 32 |
| MR-87 | 120–144 | Negative | TdT+, MPO+, SBB+, PAS-, NSE-[b] | | | | 27 |
| NALL-1 | 60–90 | Negative | TdT+, MPO− (protein)[c], +(mRNA) | Differentiative response to TPA as reduction in expression of CD10 and decrease in proliferation | | Stimulate allogeneic lymphocytes in one-way MLR reactions | 34, 88 |
| NALM-16 | 36.7 | Negative | TdT+, MPO− | Differentiative response to TPA as reduction in expression of CD10 and decrease in proliferation | | No growth inhibition by dexamethasone; expresses 140 kD isoform of NCAM protein on cell surface | 29, 36, 88 |

*Continued on next page*

*Table 5.* (continued)

| Cell lines | Doubling time (hrs) | EBV status | Cytochemistry | Inducibility of differentiation | Engraftment into mice | Other features | Ref. |
|---|---|---|---|---|---|---|---|
| Common-B cell lines (B-II) (continued) | | | | | | | |
| NALM-20 | 72 | | TdT+, MPO− (protein), +(mRNA) | | | | 37 |
| NALM-21/-22/-23 | 72 | | TdT+, MPO− (protein), +(mRNA) | | | | 37 |
| NALM-24/-25 | 72 | | TdT+, MPO− (protein), +(mRNA) | | | | 38 |
| NALM-27/-28 | | | TdT+, MPO− (protein)[c], +(mRNA) | | | | 11 |
| NALM-29 | | | TdT+, MPO− (protein), +(mRNA) | | | | 11 |
| NALM-30/-31/-32 | | | TdT+, MPO− (protein) [c], +(mRNA) | | | | 11 |
| NALM-33/-34 | | | TdT+, MPO− | | | | 11 |
| OM9;22 | 80 | Negative | TdT+, MPO− (protein), +(mRNA), PAS−, esterase− | | | | 39 |
| PC-53 | 70–80 | Negative | TdT+, MPO−, PAS+, SBB−, esterases− | | | Secretion of autostimulatory factors suggested | 40 |
| RCH-ACV | 36 | Negative | | | | | 41 |

*Continued on next page*

*Table 5.* (continued)

| Cell lines | Doubling time (hrs) | EBV status | Cytochemistry | Inducibility of differentiation | Engraftment into mice | Other features | Ref. |
|---|---|---|---|---|---|---|---|
| Common-B cell lines (B-II) (continued) | | | | | | | |
| REH | 34.7 | Negative | TdT+, MPO− (protein), +(mRNA) | Differentiative response to TPA as reduction in expression of CD10 and decrease in proliferation; maturation by TPA to intermediate stages of differentiation[d] | | Resistant to growth inhibition by dexamethasone, induction of apoptosis and decrease in cell viability by cyclosporin A; LIF activity in culture supernatant[e] | 29, 53, 86, 87, 89 |
| SUP-B2 | | Negative | TdT+ | | | | 72 |
| SUP-B7 | 168 | Negative | TdT+, PAS+, AP+, NSE−, CE−, SBB− | | | | 43 |
| SUP-B26 | | Negative | TdT+ | | | | 42 |
| TC78 | | Negative | TdT+, PAS+, MPO−, esterase− | | Tumorigenicity was established in NIH Swiss *nu/nu* mice | | 90 |
| UoC-B1 | | Negative | TdT+ | | | | 42 |
| UoC-B4 | | Negative | TdT+ | | | | 42 |
| UoC-B7 | | Negative | TdT+ | | | | 42 |
| UoC-B8 | | Negative | TdT+ | | | Growth inhibited by exogenous $PGE_2$ | 65 |
| UoC-B9 | | Negative | TdT+ | | | | 42 |
| UoC-B10 | | Negative | TdT− | | | | 42 |

*Continued on next page*

*Table 5.* (continued)

| Cell lines | Doubling time (hrs) | EBV status | Cytochemistry | Inducibility of differentiation | Engraftment into mice | Other features | Ref. |
|---|---|---|---|---|---|---|---|
| Pre-B cell lines (B-III) | | | | | | | |
| 207 | 60–70 | Negative | TdT+, MPO–, esterase–, AP+, PAS+ | | | | 45 |
| 697 | 60–70 | Positive[f] | TdT+, MPO–, esterase-, AP+, PAS+ | | | | 45 |
| BLIN-1 | 43[g] | Negative | TdT–, NSE–, MPO–, SBB–, PAS+ | Differentiation into sIgκ-positive B cells under serum-free conditions | | | 46 |
| HPB-NULL | | | TdT–, MPO– | | | | 11 |
| INC | 44.4 | Negative | TdT+, MPO– (protein), +(mRNA | | | Sensitive to growth inhibition by dexamethsone | 29 |
| KLM-2 | | | TdT–, MPO– | Macrophage differentiation in the presence of TPA and CSF | | | 91 |
| KM-3 | | | | Differentiation to intermediate stage of B-cell differentiation by TPA[f] | | | 62, 87 |
| KMO-90 | 72 | Negative | AP+, MPO–, esterase–, PAS– | | | | 49 |

*Continued on next page*

*Table 5.* (continued)

| Cell lines | Doubling time (hrs) | EBV status | Cytochemistry | Inducibility of differentiation | Engraftment into mice | Other features | Ref. |
|---|---|---|---|---|---|---|---|
| Pre-B cell lines (B-III) (continued) | | | | | | | |
| KOCL-33 | | | MPO– | | | | 50 |
| KOCL-44 | | | MPO– | | | | 50 |
| KOCL-45 | | | MPO– | | | | 50 |
| KOCL-50 | | | MPO– | | | | 50 |
| KOCL-58 | | | MPO– | | | | 50 |
| KOPB-26 | | | MPO– | | | | 50 |
| LAZ-221 | 96 | Negative | TdT+, MPO– | | | | 11 |
| NALM-1 | 72–120 | Negative | TdT+, MPO– (protein), +(mRNA), SBB–, PAS–, NSE | | | Cells continue to grow up to 3 weeks with good viability without additional medium supplement | 52 |
| NALM-6 | 40.2 | Negative | TdT+, MPO– | Differentiation to intermediate stage of B-cell differentiation by TPA[f] | Engrafts into SCID mice | Sensitive to growth inhibition by dexamethasone; expresses CD95 and undergoes apoptosis after exposure to anti-CD95; induction of apoptosis and decrease in cell viability by cyclosporin A | 59, 87, 92–95 |

*Continued on next page*

*Table 5.* (continued)

| Cell lines | Doubling time (hrs) | EBV status | Cytochemistry | Inducibility of differentiation | Engraftment into mice | Other features | Ref. |
|---|---|---|---|---|---|---|---|
| Pre-B cell lines (B-III) (continued) | | | | | | | |
| NALM-17/-18 | | | TdT+, MPO− (protein), +(mRNA) | | | | 11 |
| NALM-26 | | | TdT+, MPO− | | | | 11 |
| P30/OHKUBO | | | TdT+, MPO− (protein), +(mRNA) | | | | 11 |
| PER-278 | 62–68 | EBNA+ | TdT+, PAS+, AP−, MPO−, NSE− | | | | 56 |
| PRE-ALP | 24 | Negative | MPO− | | | | 57 |
| SMS-SB | 48 | Negative | TdT− | | | Overexpression of c-fos protooncogene, addition of soluble CD23 rescues cells from apoptosis | 96, 97 |
| SUP-B15 | | Negative | | | | Induction of apoptosis and decrease in cell viability by cyclosporin A | 59 |
| SUP-B16 | | Negative | TdT+ | | | | 42 |
| SUP-B24 | | Negative | TdT+ | | | | 42 |
| SUP-B27 | | Negative | TdT+ | | | | 42 |
| Tahr-87 | | | TdT+, MPO+ (protein) and mRNA | | | | 60 |
| TS-2 | | | MPO−, PAS−, α-NBE−, naphthol ASD chloro-acetate esterase− | | | | 61 |
| UoC-B3 | | Negative | TdT+ | | | | 42 |

*Continued on next page*

Table 5. (continued)

| Cell lines | Doubling time (hrs) | EBV status | Cytochemistry | Inducibility of differentiation | Engraftment into mice | Other features | Ref. |
|---|---|---|---|---|---|---|---|
| CyIgM status unknown (B-II/III) | | | | | | | |
| G2 | 72 | Negative | | | | | 13 |
| JM | | Negative | AP+, PAS-, NSE-, MPO- | | | | 62 |
| KM-3 | | Negative | PAS+, AP-, NSE-, MPO- | Differentiative response to TPA as reduction in expression of CD10 and decrease in proliferation. Downregulation of PKC with TPA | | | 62 88 98 |
| LILA-1 | | Negative | | TPA induces appearance of sCD20 antigen on all cells | | | 63 |
| LK-63 | 13 | Negative | | TPA induces appearance of sCD20 antigen on all cells | | | 63 |
| Tree92 | 35.4 | Negative | TdT+ | | | | 29 |
| Z-119 | 20–30 | Negative | TdT-, PAS+ | | | | 64 |
| Z-181 | 30–40 | Negative | TdT+, MPO-, NASD-, NSE-, PAS- | | | | 64 |

a As evidenced by expression of common-B-cell antigens (CD10, CD20), cyIgM, light chain gene rearrangement, and disappearance of TdT. b 5% of cells were reactive to MPO cytochemically. More than 90% were positive for MPO when examined by EM. c MPO positive with monoclonal antibody staining. d assessed by isoenzyme expression profile [87]. e after 48-hour incubation in serum free medium (RPMI 1640); effect potentiated by pulsing with PHA [89]. f EBNA and capsid antigens. g doubling time under optimal growth conditions, up to 96 hours without L-BCGF.

most consistent cytochemical profile in BCP ALL, however, is MPO-, SBB-, esterase-, AP+/−, PAS+/−, TdT+.

Differentiation has been described for a variety of BCP cell lines (Table 5). Exposure, in most cases to TPA or PMA, results in maturation along the B-lineage differentiation pathway, as is evident by acquisition of surface or cytoplasmic differentiation markers as well as intracellular isoenzyme expression patterns (KH-4, RS4;11, REH, BLIN-1, KLM-2, KM-3, NALM-6, KH-3, LC4-1, NALL-1, NALM-16, KM-3, LILA-1, LK-63). Cell line EU-1 can be induced to differentiate along the myeloid pathway by DMSO [25,80]. Engraftment into mice and establishment of the leukemic potential of BCP lines in vivo has been successfully done with A-1, HAL-01, NALM-6 and TC-78 cells.

Differences have been observed between sublines grown in different laboratories. For example, in a comparison of HL-60 promyelocytic leukemia cells from different institutions, it was observed that the morphology, cell surface markers and DNA histograms were indistinguishable. However, when induced to differentiate with TPA, only one subline of HL-60 cells differentiated into mature granulocytes.


## 7. CONCLUSIONS

Many BCP cell lines have been described, but most are used only for a specific purpose, and few have been adequately characterized. Cell lines provide a useful laboratory tool, but scientists should be aware of their limitations and care should be taken in extrapolating data to in vivo situations.


## References

1. LeBien TW, in Henderson ES, Lister AT, Greaves MF (eds.) 1996 *Leukemia*, W.B. Saunders, Philadelphia.
2. Gore SD, Kastan MB, Civin CL. *Blood* 77: 1681, 1991.
3. Abbas AK, Lichtman AH, Pober JS (eds.) 1997 *Cellular and Molecular Immunology*, W.B. Saunders, Philadelphia.
4. Komori T, Okada A, Stewart V, et al. *Science* 261: 1171, 1993.
5. Schatz DG, Oettinger MA, Schlissel MS. *Annu Rev Immunol* 10: 359, 1992.
6. NCI 1989
7. Faderl S, Kantarjian HM, Talpaz M, Estrov Z. *Blood* 91: 3995–4019, 1998.
8. Bennett JM, Catovsky D, Daniel MT, et al. *Br J Haematol* 47: 553–561, 1981.
9. EGIL. *Leukemia* 9: 1783–1786, 1995.
10. Rosenfeld C, Goutner A, Choquet C, et al. *Nature* 267: 841–843, 1977.
11. Matsuo Y, Drexler HG. *Leukemia Res* 22: 567–579, 1998.
12. Kamel-Reid S, Letarte M, Sirard C, et al. *Science* 246: 1597–1600, 1989.

13.  Kamel-Reid S, Dick JE, Greaves A, et al. *Leukemia* 6: 8–17, 1991.
14.  Cohen A, Grunberger T, Vanek W, et al. *Blood* 78: 94–102, 1991.
15.  Abe 1999.
16.  Urashima M, Hasegawa N, Kamijo M, et al. *Am J Hematol* 46: 112–119, 1994.
17.  Nagasaka-Yabe M, Maeda S, Mabuchi O, et al. *Jpn J Cancer Res* 79: 59–68, 1988.
18.  Yamamoto 1993.
19.  Matsuo Y, Ariyasu T, Ohmoto E, et al. *Cell* 4: 257–260, 1991a.
20.  Stong RC, Korsmeyer SJ, Parkin JL, et al. *Blood* 65: 21–31, 1985.
21.  Greil J, Gramatzki M, Burger R, et al. *Br J Haematol* 86: 275–283, 1994.
22.  Okabe M, Matsushima S, Morioka M, et al. *Blood* 69: 990–998, 1987.
23.  Pegoraro L, Matera L, Ritz J, et al. *J Natl Cancer Inst* 70: 447–453, 1983.
24.  Zhou M, Findley HW, Ma L, et al. *Blood* 77: 2002–2007, 1991.
25.  Zhou M, Zaki SR, Ragab AH, et al. *Leukemia* 8: 659–663, 1994.
26.  Ohyashiki K, Fujieda H, Miyauchi J, et al. *Leukemia* 5: 322–331, 1991.
27.  Okamura J, Okano H, Ikuno Y, et al. *Leukemia Res* 8: 97–104, 1984.
28.  Uphoff CC, MacLeod RAF, Denkmann SA, et al. *Leukemia* 11: 441–447, 1997.
29.  Tani A, Tatsumi E, Nakamura F, et al. *Leukemia* 10: 1592–1603, 1996.
30.  Nakazawa S, Kaneko K, Sasaki Y, et al. *Jpn J Clin Hematol* 20 (Suppl 1): 189, 1978.
31.  Yoshimura T, Mayumi M, Yorifuji T, et al. *Am J Hematol* 26: 47–54, 1987.
32.  Renard N, Duvert V, Matthews DJ, et al. *Leukemia* 9: 1219–1226, 1995.
33.  Okamura J, Yamada S, Ishii E, et al. *Blood* 72: 1261–1268, 1988.
34.  Miyoshi I, Hiraki S, Tsubota T, et al. *Nature* 267: 843–844, 1977.
35.  Kohno S, Minowada J, Sandberg AA. *J Natl Cancer Inst* 64: 485–493, 1980.
36.  Patel K, Frost G, Rossell R, et al. *Leukemia Res* 16: 307–315, 1992.
37.  Matsuo Y, Ariyasu T, Ohmoto E, et al. *Human Cell* 4: 335–338, 1991b.
38.  Matsuo Y, Ariyasu T, Adachi T, et al. *Human Cell* 4: 339–342, 1991c .
39.  Ohyashiki K, Miyauchi J, Ohyashiki JH, et al. *Leukemia* 7: 1034–1040, 1993.
40.  Tarella C, Pregno P, Gallo E, et al. *Leukemia Res* 14: 177–184, 1990.
41.  Jack I, Seshadri R, Garson P, et al. *Cancer Genet Cytogenet* 19: 261–269, 1986.
42.  Zhang LQ, Downie PA, Goodell WR, et al. *Leukemia* 7: 1865–1874, 1993.
43.  Smith SD, Morgan R, Galili N, et al. *Cancer Res* 47: 1652–1656, 1987.
44.  Jenski LJ, Lampkin BC, Goh TS, et al. *Leukemia Res* 12: 1497–1506, 1985.
45.  Findley HW, Cooper MD, Kim TH, et al. *Blood* 60: 1305–1309, 1982.
46.  Wörmann B, Anderson JM, Liberty JA, et al. *J Immunol* 142: 110–117, 1989.
47.  Fu SM, Hurley JN. *Proc Natl Acad Sci USA* 76: 6637–6640, 1979.
48.  Blomhoff HK, Davies C, Ruud E, et al. *Scan J Immunol* 22: 611–617, 1985.
49.  Sotomatsu M, Hayashi Y, Kawamura M, et al. *Leukemia* 10: 1615–1620, 1993.
50.  Iida S, Saito M, Okazaki T, et al. *Leukemia Res* 16: 1155–1163, 1992.
51.  Lazarus H, Barell EF, Krishan A, et al. *Cancer Res* 38: 1362–1367, 1978.
52.  Minowada J, Tsubota T, Greaves MF, Walters TR. *J Natl Cancer Inst* 59: 83–87, 1977.
53.  Hurwitz R, Hozier J, LeBien T, et al. *Int J Cancer* 23: 174–180, 1979.
54.  Matsuo Y, Adachi T, Tsubota T. *Human Cell* 7: 221–226, 1994.
55.  Hirose M, Minato K, Tobinai K, et al. *Jpn J Cancer Res* 73: 600–605, 1982.
56.  Kees UR, Lukeis R, Ford J, et al. *Cancer Genet Cytogenet* 46: 201–208, 1990.
57.  Pandrau D, Frances V, Martinez-Valdez H, et al. *Leukemia* 7: 635–642, 1993.
58.  Smith RG, Dev VG, Shannon WA. *J Immunol* 126: 596–602, 1981.
59.  Ito C, Ribeiro RC, Behm FG, et al. *Blood* 91: 1001–1007, 1998.
60.  Yamaguchi N, Tatsumi E, Yoneda N, et al. *Blood* 76: 338a, 1990.
61.  Yoshinari M, Imaizumi M, Eguchi M, et al. *Cancer Genet Cytogenet* 10: 95–102, 1998.

62. Schneider U, Schwenk H-U, Bornkam G, et al. *Int J Cancer* 19: 621–626, 1977.
63. Salvaris E, Novotny JR, Welch K, et al. *Leukemia Res* 16: 655–663, 1992.
64. Estrov Z, Talpaz M, Zipf TF, et al. *J Cell Physiol* 166: 618–630, 1996.
65. Giordano L, Moldwin RL, Downie PA, et al. *Leukemia Res* 21: 925–932, 1997.
66. Minowada J, Koshiba H, Janossy G, et al. *Leukemia Res* 3: 261–266, 1979.
67. Ouaaz F, Mentz F, Mossalayi MD, et al. *Eur J Haematol* 51: 276–281, 1993.
68. Shibata T, Shimada Y, Shimoyama M. *Jpn J Clin Oncol* 15: 67–75, 1985.
69. Uckun FM, Myers DE, Fauci AS, et al. *Blood* 74: 761–776, 1989.
70. Levy Y, Benlagha K, Buzyn A, et al. *Clin Exp Immunol* 110: 329–335, 1997.
71. Holan V, Minowada J. *Neoplasma* 40: 3–8, 1993.
72. Zhang H, Shaw ARE, Mak A, Letarte M. *J Immunol* 156: 565–573, 1996.
73. Wang CY, Al-Katib A, Lane CL, et al. *J Exp Med* 158: 1757, 1983.
74. Drexler HG, MacLeod RAF, Janssen JWG. *Leukemia* 9: 480–500, 1995.
75. Iida S, Seto M, Yamamoto K, et al. *Jpn J Cancer Res* 84: 532–537, 1993.
76. Drexler HG, Dirks W, MacLeod RAF, et al. (eds.) 1997 *DSMZ Catalogue of Human and Animal Cell Lines*, Braunschweig, Germany: DSMZ.
77. Sonta S-I, Minowada J, Tsubota T, Sandberg AA. *J Natl Cancer Inst* 59: 833–835, 1977.
78. Wlodarska I, Aventin A, Ingléd-Esteve J, et al. *Blood* 89: 1716–1722, 1997.
79. Matsuo Y, Nakamura S, Ariyasu T, et al. *Leukemia* 10: 700–706, 1996.
80. Zhou M, Gu L, James CD, et al. *Leukemia* 9: 1159–1161, 1995.
81. Olopade OI, Jenkins RB, Ransom DT, et al. *Cancer Res* 52: 2523, 1992.
82. Liu Q, Neuhausen S, McClure M, et al. *Oncogene* 10: 1061, 1995.
83. Kamb A, Gruis NA, Weaver-Feldhaus J, et al. *Science* 264: 436, 1994.
84. Nobori T, Miura K, Wu DJ, et al. *Nature* 368: 753, 1994.
85. Ziemin-van der Poel S, McCabe NR, Gill HJ, et al. *Proc Natl Acad Sci USA* 88: 10735, 1991.
86. Romana SP, Mauchauffé M, Le Coniat M, et al. *Blood* 85: 3662, 1995.
87. Drexler HG, Gaedicke G, Minowada J. *Exp Hematol* 14: 732–737, 1986.
88. Sakagami H, Ozer H, Minowada J, et al. *Leukemia Res* 8: 187–195, 1984.
89. Rocklin RE, Blidy A, Butler J, Minowada J. *J Immunol* 127: 534–539, 1981.
90. Jerkins 1985.
91. Hiraoka A, Kubota K, Preisler H, Minowada J. *Blood* 59: 997–1000, 1982.
92. Dirks W, Schöne S, Uphoff C, et al. *Br J Haematol* 96: 584–593, 1997.
93. Arguello F, Sterry JA, Zhao YZ, et al. *Blood* 87: 4325–4332, 1996.
94. Onur EK, Reaman GH, Crankshaw DL, et al. *Leukemia Lymphoma* 28: 509–514, 1998.
95. Filshie R, Gottlieb D, Bradstock K. *Br J Haematol* 102: 1292–1300, 1998.
96. Tsai L-H, Nanu L, Smith RG, Ozanne B. *Oncogene* 6: 81–88, 1991.
97. White LJ, Ozanne BW, Graber P, et al. *Blood* 90: 234–243, 1997.
98. Ase K, Berry N, Kikkawa U, et al. *FEBS* 236: 396–400, 1988.
99. Serrano J, Román J, Jiménez A, et al. *Leukemia* 13: 175–180, 1999.

# Chapter 3

# B-Lymphoid Cell Lines

Ramzi M. Mohammad, Nathan Wall and Ayad Al-Katib
*Department of Hematology and Oncology, Wayne State University and Karmanos Cancer
Institute, Lande Research Building Room 317, 550E Canfield, Detroit, MI 48201, USA. Tel:
+1-313-577-7919; Fax: +1-313-577-7925; E-mail: mohammad@karmanos.org*

## 1. INTRODUCTION

Human B-cell tumors include a group of heterogeneous diseases with vary-
ing natural histories and responsiveness to therapy. Classic examples of
B-cell tumors are Burkitt's lymphoma (BL), chronic lymphocytic leukemia
(CLL) and multiple myeloma (MM). These tumors express the conventional B-cell markers, that is, surface and/or cytoplasmic immunoglobulins.
However, malignant transformation can affect precursors of the mature B-
lymphocytes as exemplified by the non-T cell acute lymphoblastic leukemia
(ALL). Such cases demonstrate immunoglobulin gene rearrangements and
react with monoclonal antibodies to B-cell antigens. B-cell tumors, therefore,
represent a spectrum of disorders extending from the immature stem cell to
the most mature plasma cell of the B-lineage.

Unlike the granulocytic series where each stage of differentiation has
characteristic light microscopic features, morphology is not a reliable indicator of the B-cell differentiation stage. Monoclonal antibodies to B-cell
differentiation antigens have been developed, some of which are not only
lineage-specific but also stage-restricted. The gain or loss of such markers in
response to exogenous agents can provide an objective measure of a change
in the differentiation state of the B-cells. Based on this assumption, a number
of hypothetical models for B-cell differentiation have been proposed by different groups [1,2]. A hypothetical scheme of B-cell differentiation (Figure
1) is used in our institute. The range of reactivity of each antibody is based
on antigen expression on fresh cells taken from patients with B-cell tumors
and B-cell lines [3–7].

With the exception of Burkitt lymphoma, attempts to culture lymphomas
have been mostly unsuccessful. The most common problem is the overgrowth
of Epstein-Barr virus (EBV) positive lymphoblastoid cell lines from B-

*J.R.W. Masters and B.O. Palsson (eds.), Human Cell Culture Vol. III, 61–79.*
© 2000 *Kluwer Academic Publishers. Printed in the Netherlands.*

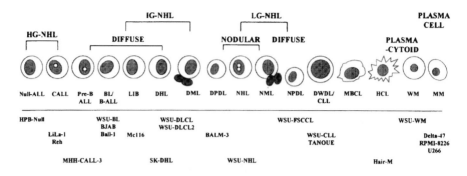

**CELL LINE(s)**

*Figure 1.* A hypothetical scheme of B-cell differentiation. Stages are arranged on top, from the most immature (stem cell) on the far left to the most mature (plasma cells) on the far right. ALL – Acute lymphoblastic leukemia; CALL – Common ALL; BL – Burkitt's lymphoma; LIB – Large cell immunoblastic lymphoma; DHL – Diffuse histiocytic lymphoma; DML – Diffuse mixed lymphoma; DPDL – Diffuse poorly differentiated lymphocytic lymphoma; NHL – Nodular histiocytic lymphoma; NML – Nodular mixed lymphoma; NPDL – Nodular poorly differentiated lymphocytic lymphoma; DWDL/CLL – Diffuse well differentiated lymphocytic lymphoma/Chronic lymphocytic leukemia; MBCL – Monocytoid B cell lymphoma; HCL – hairy cell leukemia; WM – Waldenstrom's macroglobulinemia; MM – Multiple myeloma; HG-WHL – High grade NHL; IG-NHL – Intermediate grade NHL; LG-NHL – Low grade NHL.

lymphocyte precursors contaminating the original cultured tumor cells [8,9]. Many B-cell lines reported in the literature have been established through EBV infection [10]. Although such B-cell lines may represent the original malignant phenotype, it remains unclear whether the incorporation of the EBV in the genome altered the genetic and biological characteristics. B-cell lines which are EBV-positive show features typical of lymphoblastoid cell lines [10–13]. Such concerns make EBV-transformed cell lines unsuitable for preclinical investigation [14].

Since 1986, we have established more than ten cell lines from B-cell tumors. All of these cell lines were established without the aid of exogenous mitogens, growth factors or viral transformation and all are EBV-negative. The success rate of establishing a B-cell line is approximately 10% [15]. There are no clear predictive factors for successful establishment of such a cell line. However, tumor cells derived from serous effusions appear to have a better chance of continuous growth in vitro. In our experience, 60% of the B-cell lines were established from either a pleural effusion or ascites fluid.

When a fresh specimen is received, mononuclear cells are isolated by Lymphoprep (Nycomed Pharma AS, Oslo, Norway) density centrifugation (density of 1.077 g/ml and osmolality of 280 mOsm). Cells are washed twice

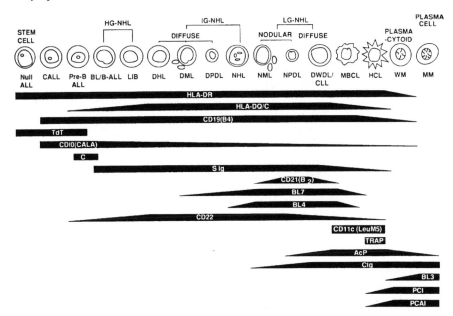

*Figure 2.* Representative cell line(s) for B-cell tumors (bottom). Stages of B-cell tumors (on top) are the same as in Figure 1.

with either phosphate buffered saline (PBS) or Hanks balanced salt solution (HBSS) and then plated at densities ranging from half a million to ten million cells per ml in RPMI-1640 medium supplemented with 5–30% fetal bovine serum (FBS) for each cell density. Cell density and FBS concentration play a major role in the establishment of a successful B-cell line.

Figure 2 shows examples of various types of B-cell tumors together with the associated stages of the B-cell differentiation pathway. We have also listed other representative line(s) for the corresponding tumor type. The scheme presented in Figure 2 is hypothetical. The nodular (follicular) B-lymphomas are believed to be more mature than the diffuse lymphomas. However, the difference between the small cell and large cell lymphomas (or the poorly differentiated lymphocytic and histiocytic, according to the Rappaport classification) may be related to transformation rather than differentiation [16].

## 2. EARLY STAGES OF B-CELLS

[Examples: Early pre-B-ALL (HPB-Null); cALL (Lila-1, MHH-CALL, REH); pre-B-ALL(Ball, Km-3, Laz 221); B-ALL (MN-60)].

The classification of B-lineage ALL can be assessed by determining the immunophenotype of cells using monoclonal antibodies to leukocyte anti-

*Table 1.*   Immunologic classification of acute lymphoblastic leukemia

| Subtype | Profile of antigen expression |
| --- | --- |
| Early pre-B | CD19+, CD22+, CD79a+, CD10+/−, CD7−, CD3−, cIgμ−, sIgκ−, sIgλ− |
| Pre-B | CD19+, CD22+, CD79a+, CD10+/−, CD7−, CD3−, cIgμ+, sIgμ−, sIκ−, sIgλ− |
| Transitional pre-B | CD19+, CD22+, CD79a+, CD10+/−, CD7−, CD3−, cIgμ+, sIgμ+, sIgκ−, sIgλ− |
| B-cell | CD19+, CD22+, CD79a+, CD10+/−, CD7−, CD3−, cIgμ+, sIgμ+, sIgκ+ or sIgλ+ |

cIg – cytoplasmic immunoglobulin; sIg – surface immunoglobulin; μ (mu) – heavy-chain protein; κ (kappa) – light-chain protein; λ (lambda) – light-chain protein.

gens known to be present during the stages of early pre-B, pre-B, transitional pre-B and B-cell lineages (Table 1).

CD7 and CD3 antibodies are used to rule out the possibility of T-ALL. In cellular differentiation along the B-lymphocyte pathway, the initial commitment of the putative pluripotent stem cell to the B-cell lineage may normally occur early in development in fetal liver [17]. Cells present in the mouse fetal liver at 12 days of gestation possess cytoplasmic immunoglobulin M (cIgM) but lack detectable surface immunoglobulins (sIg) [18].

Additional evidence indicates that these cells, which have been called pre-B cells, are direct precursors of sIg-positive cells [19]. This pattern of B-lymphocyte differentiation from pre-B cells (cyIg+, sIg−) to B cells (sIg+) is now known to occur in man [20]. The synthesis of μ-chains precedes the appearance of immunoglobulin light chains in these cells [20].

The phosphoprotein B1 is a human lymphocyte cell surface marker detected by the CD20 cluster of monoclonal antibodies [22]. During ontogeny, this molecule is first expressed by a subset of pre-B cells. Approximately 50% of pre-B-cell acute lymphoblastic leukemias (ALL) express the CD20 molecule [23]. This indicates that CD20 appears late in pre-B-cell development [22] and persists on mature B cells throughout the resting and activated stages of B-cell development. The CD20 antigen is not expressed on either normal or neoplastic plasma cells [24].

The molecular basis of ALL is likely to be more complex than previously thought. Genetic aberrations resulting in increased cellular proliferation, diminished cell differentiation, and abrogation of the normal process of programmed cell death (apoptosis) may all play a role in the development of the leukemic phenotype. Chromosome translocations have been well studied in ALL and have provided a major contribution towards understanding the biology of this malignancy. Germline mutation in known tumor suppressor

*Table 2.* Recurring chromosomal translocations in B-ALL

| Translocation | Gene | Function | Frequency |
|---|---|---|---|
| t(9;22)(q34;q11) | BCR | unknown | adults: 20% |
| | ABL | tyrosine kinase | children: 4% |
| t(1;19)(q23:p13) | E2A | bHLH transcription factor | adults: 2% |
| | PBX1 | hemeotic | children: 5% |
| t(11;v)(q23;v) | MLL | trithorax-like | adults: 3% |
| | | variable | 75% in infants |
| t(12;21)(p13;q22) | TEL | Ets-like transcription factor | adults: 3% |
| | AML1 | runt-like transcription factor | children: 25% |
| t(17;19)(q22;p13) | E2A | bHLH transcription factor | <1% |
| | HLF | bZip transcription factor | |
| t(5;14)(q31;q32) | IL-3 | cytokine | <1% |
| | IgH | Ig enhancer | |
| t(8;14)(q24;q32) | MYC | bHLH transcription factor | 2–5% |
| | IgH | Ig enhancer | |
| t(8;22)(q24;q11) | MYC | bHLH transcription factor | <1% |

genes such as RB and p53 are rare in ALL. Gene overexpression through chromosomal translocation is very common in ALL, likely due to the highly recombinogenic immunoglobulin. The best example of this is the *BCR*/ABL chimeric protein which results from the fusion of the *BCR* gene on chromosome 22 to the ABL tyrosine kinase on chromosome 9 as a consequence of the Philadelphia chromosome (t(9;22)). Other chromosomal translocations in ALL are listed below in Table 4 [25–42].

Clinically, ALL is a disease of childhood. Patients usually present with symptoms and signs of bone marrow failure. Common manifestations include fever and infection secondary to neutropenia, bleeding secondary to thrombocytopenia and weakness, tachycardia, dyspnea on exertion due to anemia. The disease has a tendency to involve the central nervous system (unlike acute myelogenous leukemia). Based on light microscopic features of lymphoblasts, ALL is classified into three categories according to the French-American- British (FAB) classification, L1, L2 and L3. ALL-L1 is the most common type accounting for 60–70% of all cases and has the best prognosis. L1 blasts are small in size with indistinct nucleoli. L2 is the next most common subtype consisting of a mixture of small (like L1) and large blasts. L3 consist of large blasts with distinct nucleoli and deep blue cytoplasm that has characteristic vacuoles. ALL-L3 cells are identical to Burkitt's lymphoma cells; it is the least common subtype accounting for 1% of all ALL cases. It

*Table 3.* EBV-negative human B cell lines: clinical characterization

| Cell line[a] | Donor age[b]/ sex | Diagnosis[c] | Specimen site | Year est. | Culture medium[d] | Source |
|---|---|---|---|---|---|---|
| Null acute lymphoblastic leukemia (null-ALL) | | | | | | |
| HPB-Null | 47 yr, M | ALL | ND | 1981 | 90% RPMI 1640 with 10% FBS | 43 |
| SUP-B15 | 10 yr, M | acute precursor ALL | bone marrow | 1988 | 80% McCoy's 5A + 20% FBS | 44 |
| Common acute lymphoblastic leukemia (cALL) | | | | | | |
| LiLa-1 | ND | cALL | peripheral blood | 1992 | 90% RPMI 1640 with 10% FBS | 22 |
| LK63 | ND | cALL | peripheral blood | 1992 | 90% RPMI 1640 with 10% FBS | 22 |
| MHH-CALL-2 | 15 yr, Caucasian | cALL | peripheral blood | 1995 | 80% RPMI 1640 with 20% FBS | 45 |
| 697 | 12 yr, M | cALL | | 1979 | 90% RPMI 1640 with 10% FBS | 46 |
| MHH-CALL-4 | 10 yr, Caucasian | cALL | peripheral blood | 1993 | 80% RPMI 1640 with 20% FBS | 45 |
| REH | 15 yr, F | cALL | peripheral blood | 1977 | 90% RPMI 1640 with 10% FBS | 47 |
| Pre-B-acute lymphoblastic leukemia (pre-B-ALL) | | | | | | |
| MHH-CALL-3 | 11 yr, F | pre-ALL | bone marrow | 1993 | 80% RPMI 1640 with 20% FBS | 45 |

*Continued on next page*

*Table 3.* (continued)

| Cell line[a] | Donor age[b]/sex | Diagnosis[c] | Specimen site | Year est. | Culture medium[d] | Source |
|---|---|---|---|---|---|---|
| B-acute lymphoblastic leukemia (B-ALL) | | | | | | |
| 380 | 15 yr, M | ALL (FAB L3) | peripheral blood | 1983 | RPMI 1640 with 10-20% FBS | 48,49 |
| Ball1 | 75 yr, M | ALL | peripheral blood | 1977 | 90% RPMI 1640 with 10% FBS | 50 |
| KM-3 | 12 yr, M | ALL | peripheral blood | 1977 | 90% RPMI 1640 with 10% FBS | 51 |
| Laz 221 | 24 yr, F | ALL | peripheral blood | 1978 | 80% RPMI 1640 with 20% FBS | 52 |
| NALL-1 | 14 yr, M | ALL | peripheral blood | 1977 | 80% RPMI 1640 with 15% FBS | 50 |
| NALM-6 | 24 yr, M | ALL | peripheral blood | 1979 | 90% RPMI 1640 with 10% FBS | 53 |
| NALM-6-B | 24 yr, M | ALL | peripheral blood | 1979 | 90% RPMI 1640 with 10% FBS | 53 |
| NALM-6-MI | 24 yr, M | ALL | peripheral blood | 1979 | 90% RPMI 1640 with 10% FBS | 53 |
| NALM-12 | ND | ALL | peripheral blood | 1979 | 90% RPMI 1640 with 10% FBS | 5,54 |
| NALM-16 | 12 yr, F | ALL | peripheral blood | 1976 | 90% RPMI 1640 with 10% FBS | 55 |
| NALM-26 | 24 yr, M | ALL | peripheral blood | 1994 | 90% RPMI 1640 with 10% FBS | 56 |
| RPMI-8382 | 16 yr, F | ALL | peripheral blood | 1972 | 90% RPMI 1640 with 10% FBS | 57,58 |
| RPMI-8392 | 16 yr, F | ALL | peripheral blood | 1972 | 90% RPMI 1640 with 10% FBS | 57,58 |
| RPMI-8422 | 16 yr, F | ALL | peripheral blood | 1972 | 90% RPMI 1640 with 10% FBS | 57,58 |
| RPMI-8432 | 16 yr, F | ALL | peripheral blood | 1972 | 90% RPMI 1640 with 10% FBS | 57,58 |
| RPMI-8442 | 16 yr, F | ALL | peripheral blood | 1972 | 90% RPMI 1640 with 10% FBS | 57,58 |
| RT | 16 yr, F | ALL | bone marrow | 1989 | 80% RPMI 1640 with 20% FBS | 59 |

*Continued on next page*

*Table 3.* (continued)

| Cell line[a] | Donor age[b]/ sex | Diagnosis[c] | Specimen site | Year est. | Culture medium[d] | Source |
|---|---|---|---|---|---|---|
| B-non-Hodgkin's lymphoma (B-NHL) refer to the next chapter (non-Hodgkin's B-lymphoma) | | | | | | |
| Hairy Cell Leukemia (HCL) | | | | | | |
| Hair-M | 86 yr,M | HCL | peripheral blood | 1983 | RPMI 1640 with 10% FBS | 56 |
| HCL-Z1 | ND | HCL | spleen | 1985 | RPMI 1640 with 10% FBS | 11 |
| Waldenstrom macroglobulinemia (WM) | | | | | | |
| WSU-WM | 51 yr, M | WM | pleural fluid | 1993 | RPMI 1640 with 20% FBS | 6,14,15,60 |
| Others | | | | | | |
| MN-60 | 20 yr, M | ALL-Burkitt's type | peripheral blood | 1982 | F-10 medium with 20% FCS | 61 |
| NC-37 | ND | lymphoblasts | peripheral blood | 1973 | McCoy's medium with 20% FCS | 62 |
| RS4;11 | 32 yr, F | lymphoid | bone marrow | 1985 | α-modified MEM with 10% FBS | 63 |
| SMS-SB | 16 yr, F, black | lymphoblasts | peripheral blood | 1977 | RPMI 1640 with 15% FBS | 64 |
| TOM-1 | 54 yr, F | leukemia | bone marrow | 1987 | RPMI 1640 with 20% FBS | 65 |
| CJ | 54 yr, F | nodular small cleaved lymphoma | lymph node | 1989 | RPMI 1640 with 20% FBS | 59 |

[a] Cell lines names are given as listed in the original literature.
[b] Age of donor at time of cell line establishment.
[c] Diagnosis is indicated as given in the original reference.
[d] Cell line might also grow in other culture medium.
BM – bone marrow, PB – peripheral blood, ND – not determined.

is associated with mature B-cell phenotype (sIgM+), t(8;14) and has the least favorable prognosis.

In general, ALL in adults has less favorable prognosis than children for reasons that remain unknown. Although earlier studies have indicated a worse outcome in T-cell ALL, others have shown equal results using intensive chemotherapy protocols. Chemotherapy is the mainstay of treatment in ALL with three distinct phases including induction, consolidation and maintenance. The clinical and immunophenotypic characterization of the ALL cell lines, listed in Tables 3–5, represent very well the early stage of B-cells. However, the genetic characterization for these ALL cell lines (table 4) is complicated and therefore it is difficult to reach a reasonable conclusion regarding each stage (early pre-B, pre-B, transitional pre-B and B-cell lineages, Table 5).

## 3. INTERMEDIATE STAGES OF B-CELLS

Certain surface antigens are expressed in intermediate stages of the B-cell differentiation pathway. These are CD21 (B2) [66], BL4 [2] and BL7 [6]. Such markers are very useful in determining the position of the cells along the B-cell differentiation pathway. The CD11c and CD22 are two markers that were first reported in 1985 [67]. While CD11c is expressed on monocytes and CD22 on B-lymphocytes, the co-expression of the two antibodies was initially thought to be specific for hairy cell leukemia (HCL). Since then, however, a new subset of non-Hodgkin's lymphoma (NHL), the monocytoid B-cell lymphoma (MBCL), has been described that also co-expresses CD11c and CD22 [68,69]. Both HCL and MBCL also express acid phosphatase (AcP). However, such expression can be inhibited by tartrate (tartrate sensitive) in MBCL but not in HCL [70]. The most common forms of NHL are follicular small cleaved cell lymphoma (about 40% of cases) and follicular mixed small cleaved and large cell lymphomas (about 20–40% of cases) [71]. These follicular lymphomas express the B-cell antigens CD19, CD20 and CD22 and are CD5-negative. The expression of antigens varies among other forms of NHL. More than 80% of splenic lymphoma with villous lymphoma (SLVL) cases are CD24+ and FMC7+ and express membrane CD22 [72]. Mantle cell NHL is almost always CD5+ and CD43+ [73] and demonstrates overexpression of cyclin D1 (unlike other B-cell NHLs except for some cases of SLVL [74]).

*Table 4.* EBV-negative human B cell lines: genetic characterization

| Cell line | Primary reference | Doubling time | EBV status | Karyotype |
|---|---|---|---|---|
| **Null acute lymphoblastic leukemia (null-ALL)** | | | | |
| HPB-Null | *Cancer* 47: 1812, 1981 | 24–30 hrs | Negative | ND |
| SUP-B15 | *Cancer Res* 48: 2876, 1988 | 60 hrs | Negative | human pseudodiploid karyotype; 46<2n>XY, der(1)t(1;1)(p11;q31), der(4)t(1;4) |
| **Common acute lymphoblastic leukemia (cALL)** | | | | |
| LiLa-1 | *Leukemia Res* 16: 655, 1992 | ND | Negative | 46XX, −9. t(1;19)(q23;p13), +der(9)t(1:9)(q11;p11) |
| LK63 | *Leukemia Res* 16: 655 1992 | 13 hrs | Negative | 45, X, −X (or Y), +8, −9, −17, t(1;19)(q23;p13), +der(9)t(9;17)(p11;p11) |
| MHH-CALL-2 | *Br J Haematol* 89: 771, 1995 | 100 hrs | Negative | human hyperdiploid karyotype with 13% polyploidy; 52(50–52)<2n>X;+21 |
| 697 | *Blood* 60: 1305, 1982 | 30–40 hrs | Negative | human near diploid karyotype; 46(45–48)<2n>XY, t(1;19)(q23;p13) |
| MHH-CALL-4 | *Br J Haematol* 89: 771, 1995 | 80–100 hrs | Negative | human hypodiploid karyotype with 4% polyploidy; 45(42–46)<2n>X |
| REH | *Nature* 267: 841, 1977 | 70 hrs | Negative | human pseudodiploid karyotype; 46(44–47)<2n>X. −X, +16, del(3), t(4;12;21;16) |

*Continued on next page*

*Table 4.* (continued)

| Cell line | Primary reference | Doubling time | EBV status | Karyotype |
|---|---|---|---|---|
| **Pre-B-acute lymphoblastic leukemia (pre-B-ALL)** | | | | |
| MHH-CALL-3 | *Br J Haematol* 89: 771, 1995 | 50 hrs | Negative | human pseudodiploid karyotype; 46(45–46)<2n>XX, del(6)(q15), der(19)t(1;19) |
| **B-acute lymphoblastic leukemia (B-ALL)** | | | | |
| 380 | *PNAS* 81: 7166, 1984 | 24–30 hrs | Negative | t(8;14;18)(q24:q32;q21), +der(14) t(8;14;18)(q24:q32:q21); corres. karyotype |
| Ball1 | *J Natl Cancer Inst* 59: 93, 1977 | 24–28 hrs | Negative | |
| KM-3 | *Int J Cancer* 19(5): 521, 1977 | 100 hrs | Negative | |
| Laz 221 | *Cancer Res* 38(5): 1362, 1978 | 80 hrs | Negative | 45, XX, −9, −12, +(9q12q) |
| NALL-1 | *Cancer* 40(5): 2131, 1977 | 72–88 hrs | Negative | chromosome # of 43 with 50 metaphases, neither X or Y chromosome |
| NALM-6 | *Int J Cancer* 23: 174, 1979 | 36 hrs | Negative | human near-diploid karyotype; 46(43–47)<2n>XY, t(5;12)(q33.2:p13) |
| NALM-6-B | *Int J Cancer* 23: 174, 1979 | 36 hrs | Negative | pseudodiploid karyotype; 46 XY, t(5q–/12p+);mar Y |
| NALM-6-MI | *Int J Cancer* 23: 174, 1979 | 36 hrs | Negative | pseudodiploid karyotype; 46 XY, del(5q);mar Y |
| NALM-12 | *J Exp Med* 158(5): 1757, 1983 | 30–36 hrs | Negative | |
| NALM-16 | *Immunology* 35(2): 333, 1978 | 30–36 hrs | Negative | |
| NALM-26 | *Hum Cell* 7: 221, 1994 | ND | Negative | |
| RPMI-8382 | *J Natl Cancer Inst* 53: 655, 1974 | ND | Negative | |
| RPMI-8392 | *J Natl Cancer Inst* 53: 655, 1975 | ND | Negative | |
| RPMI-8422 | *J Natl Cancer Inst* 53: 655, 1977 | ND | Negative | |
| RPMI-8432 | *J Natl Cancer Inst* 53: 655, 1978 | ND | Negative | |
| RPMI-8442 | *J Natl Cancer Inst* 53: 655, 1979 | ND | Negative | |
| RT | *Blood* 75: 1311;1990 | ND | Negative | 52, XY, +7, +12, +13, t(1;17)(q22;p12), t(2;12)(q37;q13), +del(3)(p22> |

*Continued on next page*

*Table 4.* (continued)

| Cell line | Primary reference | Doubling time | EBV status | Karyotype |
|---|---|---|---|---|
| B-non-Hodgkin's lymphoma (B-NHL) refer to the next chapter (non-Hodgkin's B-lymphoma) | | | | |
| Hairy cell leukemia (HCL) | | | | |
| Hair-M | *Nippon Ketsueki Gakkai Zasshi* 46: 1222, 1983 (in Japanese) | ND | Negative | 46XY, −11, −14, +2M |
| HCL-Z1 | *Exp Cell Biol* 53: 61, 1985 | ND | Negative | ND |
| Waldenstrom macroglobulinemia (WM) | | | | |
| WSU-WM | *Blood* 81: 3034, 1993 | 16 hr | Negative | 46–54, XY, t(8;14)(q24;q32), t(12;17)(q24:q21), 2P−, extra copies of 1, 3, 4, 7 |
| Others | | | | |
| MN-60 | *Leukemia Res* 6: 685, 1982 | 25 hrs | Negative | t(8q−;14q+)(1q+, 6q−) |
| NC-37 | *Lancet* i, 93, 1973 | ND | Negative | |
| RS4;11 | *Blood* 65: 21, 1985 | 60–70 hrs | Negative | 46XX, i(7q), t(4;11)(q21;q23) |
| SMS-SB | *J Immunology* 126: 596, 1981 | 48 hrs | Negative | 47XX with 1 or 2 submetacentric marker chromosomes. Ph1(−) |
| TOM-1 | *Blood* 69: 990, 1987 | 50–60 hrs | Negative | human hyperdiploid karyotype with 7% polyploidy;47(47–48)<2n>X; |
| CJ | *Blood* 75: 1311, 1990 | ND | Negative | 51, XX, +X, −4, +12, +del(2)(p21), +del(2)(q32), +del(3)(q21), +del(3)(q21) |

*Table 5.* EBV-negative human B cell lines: immunophenotypic characterization

| Cell | Immunophenotype | Availability |
|------|-----------------|--------------|
| **Null acute lymphoblastic leukemia (null-ALL)** | | |
| HPB-Null | CD3−, CD10+, CD19+, HLA-DR+ | |
| SUP-B15 | CD3−, CD10+, CD13+, CD19+, CD20−, CD34+, CD37+, HLA-DR+, sm/cylambda−, smkappa−, cykappa+ | |
| **Common acute lymphoblastic leukemia (cALL)** | | |
| LiLa-1 | CD10+, CD19+, CD20−, CD21−, CD11b−, CD54−, HLA-DR+, HLA-DQ−, smIgM− | |
| LK63 | CD10+, CD19+, CD20−, CD21−, CD11b−, CD54−, HLA-DR−, HLA-DQ−, smIgM+ | |
| MHH-CALL-2 | CD3−, CD10+, CD13+, CD19+, CD20+, CD37−, HLA−DR+, sm/cyIgG+, sm/cylambda− | DSMZ |
| 697 | CD3−, CD10+, CD13−, CD19+, CD20+, CD37+, HLA-DR+, sm/cyIgM+, sm/cylambda− | |
| MHH-CALL-4 | CD3−, CD10+, CD13+, CD19+, CD20+, CD37+, HLA-DR+, sm/cyIgG−, sm/cylambda− | DSMZ |
| REH | CD3−, CD10+, CD13−, CD19+, CD20+, CD37−, CD71−, CD138+, HLA-DR+, sm/cyIgG−, sm/cyIgM−, sm/cylambda−, sm/cykappa− | ATTC, DSMZ |
| **Pre-B-acute lymphoblastic leukemia (pre-B-ALL)** | | |
| MHH-CALL-3 | CD3−, CD10+, CD13+, CD19+, CD37−, HLA-DR+, sm/cyIgG−, sm/cyIgM− | DSMZ |

*Continued on next page*

*Table 5.* (continued)

| Cell | Immunophenotype | Availability |
| --- | --- | --- |
| B-acute lymphoblastic leukemia (B-ALL) | | |
| 380 | CD3−, CD10+, CD13−, CD19+, CD37+, HLA-DR+, sm/cyIgM−, sm/cyIgG− | ATTC, DSMZ |
| Ball1 | CD10+, CD19−, CD20+, CD21−, CD11b−, CD54+ | |
| KM-3 | ND | |
| Laz 221 | smIg− | |
| NALL-1 | ND | |
| NALM-6 | CD3−, CD10+, CD13−, CD19+, CD34−, CD37−, CD138+, HLA-DR+, sm/cyIgG−, smIgM−, cyIgM+, sm/cylambda−, sm/cykappa− | DSMZ |
| NALM-6-B | CD3−, CD10+, CD13−, CD19+, CD34−, CD37−, CD138+, HLA-DR+, sm/cyIgG−, sm/cyIgM+, sm/cylambda−, sm/cykappa− | |
| NALM-6-MI | CD3−, CD10+, CD13−, CD19+, CD34−, CD37−, CD138+, HLA-DR+, sm/cyIgG−, smIgM−, cyIgM+, sm/cylambda−, sm/cykappa− | |
| NALM-12 | CD3−, CD10+, CD13−, CD19+, CD34−, CD37−, CD138+, HLA-DR+ | |
| NALM-16 | CD3−, CD10+, CD13−, CD19+, CD34−, CD37−, CD138+, HLA-DR+ | |
| NALM-26 | T cell associated CD5 and Myeloid cell associated CD13 antigens, IL-7 receptor + (CDw127) | |
| RPMI-8382 | CD3−, CD10+, CD13−, CD19+, CD34−, HLA-DR+ | |
| RPMI-8392 | CD3−, CD10+, CD13−, CD19+, CD34−, HLA-DR+ | |
| RPMI-8422 | CD3−, CD10+, CD13−, CD19+, CD34−, HLA-DR+ | |
| RPMI-8432 | CD3−, CD10+, CD13−, CD19+, CD34−, CD37−, CD138+, HLA-DR+ | |
| RPMI-8442 | CD3−, CD10+, CD13−, CD19+, CD34−, HLA-DR+ | |
| RT | CD19+, CD20+, CD10+, cIg+, CD2−, CD3−, smIg− | |

*Continued on next page*

*Table 5.* (continued)

| Cell | Immunophenotype | Availability |
|------|-----------------|--------------|
| B-non-Hodgkin's lymphoma (B-NHL) refer to the next chapter (non-Hodgkin's B-lymphoma) | | |
| Hairy cell leukemia (HCL) | | |
| Hair-M | cy and smgIG-kappa, E−, IgGFcR−, IgMFcR−, C3R−, OKT-9+, OKT10+, OKT1+, MCS-1+, TdT−, CD10−, B1+, B7+, CD3− | |
| Waldenstrom macroglobulinemia (WM) | | |
| WSU-WM | CD3−, CD19+, CD20+, CD22+, HLA-DR+, CD10+, smIgM+, smIg-lamda+ | Author, DSMZ |
| Others | | |
| MN-60 | CD10+, HLA-DR+, HLA-A-C+, sm/cyIgG−, smIgM+, cyIgM+, smlambda+, cylambda−, smlambda+, cylambda−, sm/cykappa− | |
| NC-37 | ND | |
| RS4;11 | CD4−, CD8−, CD10−, CD19+, CD20−, CD71+, HLA-DR+, sm/cyIgG− | |
| SMS-SB | CD10−, HLA-DR+, sm/cylambda−, sm/cykappa−, smIgM−, cyIgM−, cyIgM+, sm/cyIgG− | |
| TOM-1 | CD3−, CD10+, CD13+, CD19+, CD20+, CD34+, CD37−, HLA-DR+, sm/cylambda−, sm/cykappa− | |
| CJ | CD19+, CD20+, CD10+, sIg+, HLA-DR+, CD2−, CD3−, Tac(IL-2R)− | |

(+), strong, definite protein expression (more than 50% cells are positive); (+/−), moderately (20−50% cells are positive); (−/+), weak expression (less than 20% cells are positive; (−), no protein expression.

ATTC − American Tissue Type Collection; DSMZ − Deutsche Sammlung Mikroorganismen und Zellkuturen GmbH.

## 4. MATURE B-CELLS

[Examples: Hair-M for Hairy cell leukemia; WSU-WM for Waldenstrom's Macroglobulinemia].

The conventional marker of the most mature B-cells, the plasma cells, is the intense cyIg expression and secretion. Surface antigens that are associated or restricted to the Ig-secreting stage are BL3, PC1, and PCA1 [1,75,76].

### 4.1. Hairy Cell Leukemia (HCL)

[Example, HCL-Z1 cell line].

HCL (Leukemic reticuloendotheliosis) is an indolent lymphoid malignancy first described by Bouroncle in 1958. Over the last two decades, HCL cell lines have increased dramatically the understanding of the biology of this disease. They have also led to the discovery of a new retrovirus and the isolation and cloning of granulocyte-macrophage colony-stimulating factor [77,78]. HCL is a rare disorder, it develops more often in men than in women, and its incidence peaks during the fifth decade of life. The neoplastic HCL cells are derived from B-cells [1,76]. Typical HCL display several surface antigens, including immunoglobulins of both light- and heavy-chain types, characteristic of B-cells [79]. The most specific marker is the expression of B-ly7 with CD19, which can be used to differentiate HCL from CLL. HCL also expresses CD11c with CD19 and, usually, CD25 with CD19. CD5+ cells are highly atypical in HCL [80]. Some cases of HCL show overexpression of cyclin D1 [74]. While the hair-like surface projections place them closer to the monocytic lineage [81], these unusual surface features of hairy cells are not seen in normal or neoplastic B-lymphocytes. Several cell lines from HCL patients had been established; however, all of them are EBV-positive, and therefore show features typical of lymphoblastoid cell lines. In vivo, hairy cells are usually EBV-negative. The cardinal features of HCL are splenomegaly which is usually associated with pancytopenia, and occasional characteristic lymphocytes bearing hair-like cytoplasmic extensions. Unfortunately, the clinical, genetic and immunophenotypic characterization of the two cell lines, Hairy-M and HCL-1, listed under hairy cell leukemia, do not provide enough information to consider them as a perfect model representing hairy cell leukemia.

### 4.2. Waldenstrom's Macroglobulinemia (WM)

[Example, WSU-WM cell line].

The presence of CD20 and CD22 antigens distinguishes WM from multiple myeloma (MM). As in CLL, CD19 and CD21 are expressed [58,70].

The differential diagnostic criteria are the overproduction of monoclonal IgM protein and diffuse infiltration of bone marrow and occasionally of spleen, liver or lymph nodes [82]. The histology is that of malignant B-cells with the appearance of plasmacytoid lymphocytes; however, many patients have only well-differentiated lymphocytes, as in CLL [83,84]. In 1944, Jan Waldenstrom described two patients with epistaxis, anemia, retinal hemorrhage, hyperglobulinemia, and infiltration of the bone marrow by lymphocytes [85]. The disease that he described and now bears his name results from malignant proliferation of plasmacytoid lymphocytes (or lymphocytoid plasma cells) producing a monoclonal IgM protein, now known as the macroglobulin. WM is closely related to other mature lymphocytic or plasmacytic neoplasms such as CLL, low-grade NHL, HCL, and MM [86]. Clinically, although CLL, NHL, HCL and MM have different pathologic manifestations (contributing to their different nomenclature), they are almost always of B-cell origin, affect elderly people, and have similar natural history characterized by chronic course. The median survival of patients with WM is reported to be 5 years [87]; it is almost never seen in people younger than 40 [88]. The WSU-WM cell line is EBV-negative and carries the 8;14 chromosomal translocation with a unique breakpoint of chromosome 8 downstream of exon 3 of the c-MYC proto-oncogene, which is rearranged. The cells grow in liquid culture, soft agar, athymic nude mice, and severe combined immune deficient (SCID) mice [14]. The WSU-WM cell line is the only cell line in the literature that clinically and immunophenotypically partly represents Waldenstrom's macroglobulinemia.

## ACKNOWLEDGEMENTS

This work was supported by Grant CA79837 from the US National Cancer Institute and from the Leukemia Society of America grant No. 6323-99. The authors would like to thank Dr. Ishtiaq Ahmad for his technical help.

## References

1. Anderson KC, Bates MP, Slaughenhoupt B, et al. *J Immunol* 132: 3172, 1984.
2. Hashimi L, Wang CY, Al-Katib A, et al. *Cancer Res* 46: 5431, 1986.
3. Hokland P, Ritz J, Schlossman SF, et al. *J Immunol* 135: 1746, 1985.
4. Loken M, Shah VO, Dattilio KL, et al. *Blood* 70: 1316, 1987.
5. Wang CY, Azzo W, Al-Katib A, et al. *J Immunol* 133: 684, 1984.
6. Al-Katib A, Wang CY, Bardales R, et al. *Hematol Oncol* 3: 271, 1985.
7. Small TN, Keever CA, Weiner-Fedus, et al. *Blood* 76: 1647, 1990.
8. Nilsson K, Ponten JJ. *Cancer* 15: 321, 1975.

9.  Nilsson K, Klareskog L, Ralph P, et al. *Hematol Oncol* 1: 277, 1983.
10. Findley HW Jr, Cooper MD, Kim TH, et al. *Blood* 60: 1305, 1982.
11. Lang AB, Odermatt BF, Gut DR, et al. *Exp Cell Biol* 53: 61, 1985.
12. Miyoshi I, et al. *Cancer* 47: 60, 1981.
13. Saxon A et al. *J Immunol* 120: 777, 1978.
14. Al-Katib A, Mohammad R, Hamdan M, et al. *Blood* 81: 3034; 1993.
15. Al-Katib A, Mohammad RM. *Basic and Clinical Applications of Flow Cytometry*, Valeriote FA, Nakeff A, Valdivieso M (eds.), Kluwer Academic Publishers, Boston, USA, 1992.
16. Rosenberg SA, Bernard CW, Brown BW Jr, et al. *Cancer* 49: 2112, 1982.
17. Hozier J, Lindquist L, Hurwitz R, et al. *Int J Cancer* 26: 281, 1980.
18. Raff MC, Megson M, Owen JJT, Cooper MD. *Nature* (Lond.) 259: 224, 1976.
19. Owen JJ, Wright DE, Habu S, Raff MC, Cooper MD. Studies on the generation of B lymphocytes in fetal liver and bone marrow. *J Immunol* 118(6): 2067–73, 1977.
20. Perl ER, Volger LB, Okos AJ, Crist WM, Lawton AR 3rd, Cooper MD. B lymphocyte precursors in human bone marrow: An analysis of normal individuals and patients with antibody-deficiency states. *J Immunol* 120(4): 1169–75, 1978.
21. Levitt D, Cooper MD. *Cell* 19: 617, 1980.
22. Salvaris E, Novotny JR, Welch K, et al. *Leukemia Res* 16: 655, 1992.
23. Nadler LM, Korsmeyer SJ, Anderson KC, et al. *J Clin Invest* 74: 332, 1984.
24. Boyd AW. *Pathology* 19: 329, 1987.
25. Quesnel B, Preudhomme C, Philippe et al. *Blood* 85: 657, 1985.
26. Stranks G, Height SE, Mitchell P, et al. *Blood* 85: 893, 1995.
27. Hebert J, Cayuela J, Berkely J, et al. *Blood* 84: 4038, 1994.
28. Okuda T, Shurtleff SA, Valentine MB, et al. *Blood* 85: 2321, 1995.
29. Croce C, Nowell P. *Blood* 65: 1, 1985.
30. Aplan P, Lombardi D, Ginsberg A, et al. *Science* 250: 1426, 1990.
31. Brown L, Cheng J-T, Siciliano M, et al. *EMBO J* 9: 3343, 1990.
32. Nourse J, Mellentin J, Galili N, et al. *Cell* 60: 535, 1990.
33. Raimondi SC. *Blood* 81: 2237, 1993.
34. Dedera D, Waller E, LeBrun D, et al. *Cell* 74: 833, 1993.
35. Inaba T, Inukai T, Yoshihara T, et al. *Nature* 382: 541, 1996.
36. Thirman MJ, Gill HJ, Burnet RC, et al. *N Engl J Med* 329: 909, 1993.
37. Golub T, Goga A, Barker G, et al. *Mol Cell Biol* 16: 4107, 1996.
38. Suto Y, Sato Y, Smith SD, et al. *Cancer* 18: 254, 1997.
39. Papadopoulos P, Ridge SA, Boucher CA, et al. *Cancer Res* 55: 34, 1995.
40. Buijs A, Sherr S, van Baal S, et al. *Oncogene* 10: 1511, 1995.
41. Romana SP, Poirel H, Leconiat M, et al. *Blood* 86: 4263, 1995.
42. Rubnitz JE, Downing JR, Pui CH, et al. *J Clin Oncol* 15: 1150, 1997.
43. Nakao Y, Tsuboi S, Fujita T, et al. *Cancer* 47(7): 1812, 1981.
44. Naumovski L, Morgan R, Hecht F, et al. *Cancer Res* 48: 2876, 1988.
45. Tomeczkowski J, Yakisan E, Wieland B, et al. *Br J Haematol* 89: 771, 1995.
46. Findley HW Jr, Cooper MD, Kim TH, et al. *Blood* 60: 1305, 1982.
47. Minowada J. Markers of human leukemia-lymphoma cell lines. Reflect haematopoietic cell differentiation. In: Serrou B and Rosenfeld C (eds.). *Human Lymphocyte Differentiation*. Elsevier/North-Holland Biomedical Press, 1978.
48. Pegoraro L, Palumbo A, Erikson J, et al. *PNAS* 81: 7166, 1984.
49. Pegoraro L, Malavasi F, Bellone G, et al. *Blood* 73: 1020; 1989.
50. Hiraki S, Miyoshi I, Masuji H, et al. *J Natl Cancer Inst* 59: 93, 1977.

51. Schneider U, Schwenk HU, Bornkamm G. *Int J Cancer* 19(5): 521, 1977.
52. Lazarus H, Barell EF, Krishan A, et al. *Cancer Res* 38(5): 1362, 1978.
53. Hurwitz R, Hozier J, LeBien T, et al. *Int J Cancer* 23: 174, 1979.
54. Wang CY, Al-Katib A, Lane CL, et al. *J Exp Med* 158(5): 1757, 1983.
55. Han T, Minowada J. *Immunology* 35(2): 333, 1978.
56. Matsuo Y, Adachi T, Tsubota T. *Hum Cell* 7: 221, 1994.
57. Huang CC, Hou Y, Woods LK, et al. *J Natl Cancer Inst* 53: 655, 1974.
58. Huang N, Kawano MM, Mahmoud MS, et al. *Blood* 85: 3704, 1995.
59. Ford RJ, Goodacre A, Ramirez I, et al. *Blood* 75: 1311; 1990.
60. Al-Katib A, Wang CY, Koziner B. *Cancer Res* 45: 3058, 1985.
61. Roos G, Adams A, Giovanella B, et al. *Leukemia Res* 6: 685, 1982.
62. Pattengale PK, Smith RW, Gerber P. *Lancet* 2: 93, 1973.
63. Stong RC, Korsmeyer SJ, Parkin JL, et al. *Blood* 65: 21, 1985.
64. Smith RG, Dev VG, Shannon WA Jr. *J Immunology* 126: 596, 1981.
65. Okabe M, Matsushima S, Morioka M et al. *Blood* 69: 990, 1987.
66. Nadler LM, Stashenko P, Hardy R, et al. *J Immunol* 126: 1941, 1981.
67. Schwarting R, Stein H, Wang CY. *Blood* 65: 974, 1985.
68. Sheibani K, Burke JS, Swartz WG, et al. *Cancer* 62: 1531, 1988.
69. Traweek ST, Sheibani K, Winberg CD, et al. *Blood* 73: 573, 1989.
70. Burke JS, Sheibani K. *Leukemia* 1: 298, 1987.
71. Eyre HJ, Farver ML. Holleb AI, Fink DJ, Murphy GP (eds.) *American Cancer Society Textbook of Clinical Oncology*. Atlanta, Ga: ACS Co., p377, 1991.
72. Matutes E, Morilla R, Owusu-Ankomah K, et al. *Blood* 83: 1558, 1994.
73. Weisenburger DD, Armitage JO, *Blood* 87: 4483, 1996.
74. De Boer CJ, Schuuring E, Dreef E, et al. *Blood* 86: 2715, 1995.
75. Anderson KC, Park EK, Bates MP, et al. *J Immunol* 130: 1132, 1983.
76. Anderson KC, Bates MP, Slaughehoupt BL, et al. *Blood* 63: 1424, 1984.
77. Glaspy JA, Golde DW. *Oncology* 4: 25, 1990.
78. Antman KS et al. *N Engl J Med* 319: 593, 1988.
79. Golomb HM, Braylan R, Polliak A. et al. *Br J Haematol* 29: 455, 1975.
80. Palackdharry CS. Skeel RT & Lachant NA (eds.) *Handbook of Cancer Chemotherapy*, 4th edn. Boston, Mass: Little, Brown & Co., p425, 1995.
81. Golomb HM, Braylan R, Polliak A, et al. *Br J Haematol* 29: 455, 1975.
82. *Dorlands Illustrated Medical Dictionary*, 27th edn. Philadelphia, Pa: WB Saunders Co., p965, 1988.
83. Dimopoulos MA, Alexanian R. *Blood* 83: 1452, 1994.
84. Oken MM. Skeel RT, Lachant NA (eds.) *Handbook of Cancer Chemotherapy*, 4th edn. Boston, Mass: Little, Brown & Co., 1995.
85. Waldenstrom J. *Acta Med Scand* 127: 216, 1944.
86. Bergsagel DE: Macroglobulineamia, in Williams WJ, Beutler E, Erslev AJ, Lichtman (eds.), *Hematology*, 3rd edn. New York, NY, McGraw-Hill, p1104, 1983.
87. Pappenberger R, Sommerfeld W, Hoffmann-Fezer G. *Scand J Haematol* 31: 359, 1983.
88. Kyle RA, Garton JP. *Mayo Clin Proc* 62: 719, 1987.

# Chapter 4

# Multiple Myeloma Cell Lines

Helena Jernberg-Wiklund and Kenneth Nilsson
*Section of Pathology, Rudbeck Laboratory, Uppsala University, 751 85 Uppsala, Sweden. Tel:
+46-18-611-0252; Fax: +46-18-558-931; E-mail: Helena.Jernberg_Wiklund@genpat.uu.se*

## 1. INTRODUCTION

Multiple myeloma (MM) is an incurable B cell malignancy characterized
by the clonal expansion of malignant plasmablasts/plasma cells in the bone
marrow. Among B cell tumors, MM exhibits some unique characteristics.
The immunoglobulin (Ig) isotype of MM cells is generally IgG or IgA, suggesting that MM is derived from a post- switched B cell. This and the fact
that the Ig of MM cells is somatically hypermutated suggests that the critical
transformation steps occur in the germinal centre of the lymph node, perhaps
driven by antigenic stimulation.

MM is essentially localized to the bone marrow and only rarely, and after
progression, will MM disseminate, systemically in the form of a PCL but also
to pleural or peritoneal cavities, forming effusions at these sites. MM cells
interact via cell-cell contacts or soluble factors with normal cells in the bone
marrow environment, including stromal cells, osteoblasts, osteoclasts and
cells of normal hematopoiesis. Adhesion molecules expressed on MM cells
are involved in these interactions and in the MM cell-extracellular matrix
(ECM) interactions. The area of cell-cell interaction via adhesion molecules
has recently attracted attention and a number of reports have demonstrated
the importance of cell-cell contact for the production of cytokines mediating
growth and survival in the bone marrow [1]. Since a number of MM cell lines
have been established which are highly dependent on growth factors for their
maintenance in vitro, paracrine growth/survival factor influence is a likely
mechanism crucial for the establishment of MM cell lines.

Until 1980, only two MM cell lines had been established, RPMI 8226
[2] and U-266 [3]. To date at least 81 established true MM cell lines
from 73 patients have been reported (see Table 2). The vast majority of
these cell lines have been established in non-stirred suspension cultures
from patients with advanced, often terminal, disease by explanting MM

81

*J.R.W. Masters and B.O. Palsson (eds.), Human Cell Culture Vol. III, 81–155.*
© 2000 *Kluwer Academic Publishers. Printed in the Netherlands.*

cells from bone marrow, pleural effusion, peripheral blood or ascites. From BM biopsies obtained from non-progressive MM there is usually a requirement for either feeder cell layers of autologous BM, adherent monolayers of murine plasmacytoma cell lines [4–6], conditioned medium from PHA-stimulated leukocytes (PHA-LCM) or rIL-6 [7] alone or in combination with GM-CSF [8,9].

MM bone marrow stromal cells, ECM proteins, cytokines and other soluble factors provide the necessary anchorage and cell-cell contacts for MM establishment as continuous cell lines. The major explanations for the difficulty in establishing MM cell lines are the slow proliferative activity of MM cells and the requirement for anchorage dependence on feeder layer stromal cells. Additionally, there is at least a 30% risk of the outgrowth of Epstein Barr positive lymphoblastoid cell lines (LCLs) with less stringent growth requirements, which will overgrow the MM cells [10,11]. The understanding of the nature of feeder cells and fibroblast conditioned medium, and the discovery of IL-6 as a major growth factor of MM cells in vitro, have greatly improved the previously low success rate (2–5%) in the attempts to establish MM cell lines [12–14].

In discriminating true MM cell lines from EBV positive LCLs derived from the patient's non-neoplastic B cells, several markers are available (Table 1). It is essential to examine a putative MM cell line within 3 months of primary culture, as several of the hallmarks of lymphoblastoid cell lines, including polyclonality, diploidy, and lack of tumorigenicity can alter during prolonged culture [15]. It is not sufficient that data on new cell lines are obtained only 1–2 months after setting up the culture. The cells must be checked at least at 6 and 12 months. Hallmarks for authentic MM cell lines are EBV negativity, the presence of chromosomal abnormalities identical to those of the primary tumor cells, Ig rearrangements identical to the MM clone in vivo, Ig production identical to the primary tumor, and expression of characteristic surface markers (CD19, CD20, CD138, CD11a, CD49e). A complete analysis can distinguish the authentic MM cell lines from EBV positive LCLs and define useful models for studies of the biology of MM in vitro.

Authentic MM cell lines and highly purified primary MM cells have allowed the investigation of the biology of MM. However, in several studies two types of established cell lines, authentic MM lines and EBV+ nonmalignant lymphoblastoid cell lines, have been used, making several of the conclusions in these studies irrelevant to the biology of MM [10,11,16–18].

How can we then discriminate the authentic MM cells from the EBV+ lymphoblastoid cell lines using immunophenotypical markers? Eight discriminant surface markers were selected by the European Myeloma Research Network to discriminate between MM and LCL; CD19, CD20, sIg, CD11a,

*Table 1.* Comparison of the characteristics of newly established LCLs and MM cell lines and normal plasma cells

| Markers | EBV+ LCLs | MM | Normal plasma cells |
|---|---|---|---|
| EBV status | + | − | − |
| Aneuploidy | − | + | − |
| Polyclonality | + | − | + |
| Tumorigenicity in nude mice | − | +/−[b] | − |
| Colony formation in agar | − | +/−[b] | − |
| sIg[a] | + | − | − |
| Immunophenotype: Surface markers | | | |
| CD19[a] | + | − | + |
| CD20[a] | + | − | |
| CD11a[a] | + | − | + |
| CD49e[a] | + | − | − |
| CD38[a] | −/+ | + | + |
| CD28[a] | − | +[b] | − |
| CD138[a] | − | + | + |
| CD40 | | +[b] | + |
| CD56[a] | − | +[c] | − |

[a] Selected discriminating immunophenotypical markers of EBV+ LCLs.
[b] Expression/feature related to advanced disease.
[c] Expression lost with aggressive disseminating disease.

CD49e, CD38, CD28 and CD138 [18] (see Table 1). Strong expression of CD19 and CD20 was found on the EBV+ cell lines HS- Sultan and ARH-77 and other MM-derived cell lines with a lymphoblastoid phenotype [17,19]. The LCLs, however, failed to express CD138 (syndecan-1/B-B4). This was confirmed by other studies demonstrating weak or absent staining of CD138 in LCLs (ARH-77, HS- Sultan, MC/CAR, and IM9). The LCLs also failed to express CD28, but were positive for sIg and the adhesion markers CD11a and CD49e. In addition, CD38 expression was observed in HS-Sultan and MC/CAR, while ARH-77 stained positive for PCA-1 (Tables 1 and 2b).

In contrast, the MM cell lines RPMI 8226, JJN-3, U-266, NCI-H929 as well as XG-1, XG-2, XG-6, LP-1, and OPM-2 express CD28 and CD138 but lack the expression of CD19, CD20, sIg and the adhesion molecules CD11a and CD49e. In RPMI 8226, weak expression of CD49e was detected. With the exception of U-266 and XG-6, CD38 was expressed in all the MM cell lines mentioned above (Table 3).

*Table 2a.*    Multiple myeloma cell lines: clinical characterization

| Cell line[a] | Patient age/sex | Diagnosis[b] | Treatment status[c] | Specimen site[d] | Authentication[e] | Year est. | Culture medium[f] | Ig production of cell line | Other requirements for establishment[m] | Primary ref./ availability[n] |
|---|---|---|---|---|---|---|---|---|---|---|
| ACB-885 | 39 M | IgG/κ MM | T | PE | yes | | DMEM+5%FCS | | 20% hu PI, 10% PHA-LCM later 10% huPI, 7.5% PHA-LCM | 20 (probably unavailable) |
| ACB-1085 | 39 M | IgG/κ MM, sister cell line/ same origin as of ACB-885 | T | PE | yes | | | | | 20 (probably unavailable) |
| AD3 | 69 M | IgA/κ plasmacytoma/ hyperamylasemia/subclone of FR4 | D | ascites | yes | 1986 | RPMI1640+10%FCS | IgA/κ 0.87 µg/72h | no additional feeder cells | 21 |
| AMO1 | 64 F | IgA/κ plasmacytoma | | ascites | yes | 1985 | RPMI1640+7%FCS | IgA/κ | Macrophage feeder layers later no feeder layers | 22 |
| ANBL-6 | 58 F | MM/PCL BJP | R | PB | yes* | 1992 | RPMI1640+10%FCS | | 1 ng/ml IL6 | 23 |
| ANBM-6 | 58 F | MM/PCL BJP same origin as ANBL-6/phenotypically indistinguishable from ANBL-6/ sister cell line | R | BM | yes | 1992 | RPMI1640+10%FCS | | 1 ng/ml IL6 | 23 |
| delta-47 | 72 M | IgD/λ extraosseus MM | T | ascites | yes | 1986 | RPMI1640+20%FCS | IgD/λ 2 µg/72 h/ milj cells | no feeder/growth factors | 24 |
| DOBIL-6 | 65 F | nonsecr. MM/hypercalcaemia | R | BM | yes* | 1995 | RPMI1640+10%FBS | IgA/κ nonsecretory | 1-5 ng/ml IL-6 | 25 |
| DP-6 | 78 F | IgA/λ MM | T | PB | yes | 1994 | RPMI1640+10%FCS | IgA/λ | | 26 |
| EJM | 58 F | IgG/λ extramedullary MM/ PCL | T | PE | yes | 1988 | IMDM/Ham+20%FCS | IgG/λ 5–10 µg/24 h/ milj cells | initially adherent cells or 1–2% EwSarGF (IL-1+IL-6) | 27 |
| FLAM-76 | 77 M | nonsecr. κ type MM/PCL no BJP | T | PB | yes | 1990 | RPMI1640+10%FCS | nonsecretory | 10% human cord serum, later 4 ng/ml rIL-6 | 28 |
| FR4 | 69 M | IgA/κ plasmacytoma/ hyperamylasemia | D | ascites | yes | 1986 | RPMI1640+10%FCS | IgA/κ 0.555 µg/72 h | | 21 |
| GM2132 (=RPMI 8226) | | IgG/λ MM | | PB | yes | | RPMI1640+10%FCS | | | 29 |
| HL407g | | IgA MM | D | BM | yes | 1987 | RPMI1640+10%FBS | IgA/λ | BM stromal feeder layer | 6 |
| ILKM2 | | IgG/κ MM | | BM | yes | 1987 | RPMI1640+8%FCS | IgG/κ | BM macrophage/fibroblasts 3 mo later allog. MQ/MDF/rIL-6 (2 ng/ml) | 5 |

*Continued on next page*

*Table 2a.* (continued)

| Cell line[a] | Patient age/sex | Diagnosis[b] | Treatment status[c] | Specimen site[d] | Authentication[e] | Year est. | Culture medium[f] | Ig production of cell line | Other requirements for establishment[m] | Primary ref./availability[n] |
|---|---|---|---|---|---|---|---|---|---|---|
| ILKM3 | | IgG/κ MM | | BM | yes | 1987 | RPMI1640+8%FCS | κ | BM macroph/fibrobl, 3 mo later allog. MQ/MDF/rIL-6 (2 ng/ml) | 5 |
| JIM-1[h] | 60 M | IgA/λ MM | | PE | yes | | Dexter medium | | autol. PE fluid, autol. stromal feeder layer, later IMDM (Ham's F12+20%FCS) | 30 |
| JIM-2 | | (subclone of JIM-1) | | | yes | | | | | 30 |
| JIM-3 | | (subclone of JIM-1) | | | yes | | | | | 30 |
| JIM-4 | | (subclone of JIM-1) | | | yes | | | | | 30 |
| JJN-1[i] | 57 F | IgA/κ PCL | D | BM | yes | 1987 | Isc/Ham+20%FCS RPMI1640+15%FCS | IgA/κ | coculture with XG-Ag8.653 plasmacytoma | 4 |
| JJN-2 | 57 F | IgA/κ PCL (subclone of JJN) | D | BM | yes | 1987 | Isc/Ham+20%FCS RPMI1640+15%FCS | κ | derived from JJN-1+ ESG (IL-6), 0.5% ESG (IL-6) | 4 |
| JJN-3 | | IgA/κ PCL (subclone of JJN) | | BM | yes | | | | | 27 |
| Karpas 620 | 77 F | IgG/κ PCL | D | PB | yes | 1987 | RPMI1640+20%FCS | | no feeder cells | 31 |
| Karpas 707 | 53 M | IgG/λ MM | R | BM/PB | yes | 1981 | RPMI1640+10%FBS | | | 32 |
| KAS-6/1 | 54(56) F | IgG/κ MM | T | PB | yes | 1994 | RPMI1640+10%FCS | λ ~10–50 μg/ml | 1–5 ng/ml IL-6 | 26 |
| KHM-1A | 53 M | IgA/λ MM/hyperamylasemia | T | PE | yes | 1986 | RPMI1640+10%FCS | IgG/κ | | 33 |
| KHM-1B | 53 M | IgA/λ MM/hyperamylasemia sister cell line/same origin as KHM-1A | T | PE | yes | 1986 | RPMI1640+10%FCS | IgA/λ | | 33 |
| KHM-11 | 52 M | IgA/κ BJP MM (AMHL) | T | PE | yes* | 1992 | RPMI1640+10%FCS | | | 34 |
| KMM-1 | 62 M | BJPλ plasmacytoma | T | plasma-cytoma | yes | 1980 | RPMI1640+20%FCS | λ | | 35,36,RCB |
| KMS-5 | 80 M | IgD/λ MM | ND | PE | yes | 1982 | RPMI1640+10%FCS | nonsecretory | | 36 |
| KMS-11 | 67 F | IgG/κ MM | ND | PE | yes | 1987 | RPMI1640+10%FCS | κ | | 36 |
| KMS-12-PE (same patient as KMS-12-BM) | 64 F | nonsecr. MM | ND | PE | yes | 1987 | RPMI1640+10%FCS | nonsecretory | | 36,JCRB |

*Continued on next page*

*Table 2a.* (continued)

| Cell line[a] | Patient age/sex | Diagnosis[b] | Treatment status[c] | Specimen site[d] | Authentication[e] | Year est. | Culture medium[f] | Ig production of cell line | Other requirements for establishment[m] | Primary ref./availability[n] |
|---|---|---|---|---|---|---|---|---|---|---|
| KMS-12-BM (same patient as KMS-12-PE) | 64 F | nonsecr. MM | ND | BM | yes | 1988 | RPMI1640+10%FCS | nonsecretory | | 36 JCRB |
| KMS-18 | 58 M | IgA/λ BJP MM-BJP/λ PCL | R | PB | yes | 1995 | RPMI1640+15%FCS | λ | later RPMI1640+10%FCS | 37 |
| KP-6 | 53(54) M | IgG/κ PCL | T | PE | yes | 1994 | RPMI1640+10%FCS | IgG/κ | 1–5 ng/ml IL-6 | 26 |
| KPMM2 | 76 F | IgG/λ MM plasmacytoma | R | ascites | yes | 1991 | RPMI1640+20%FCS | IgG/λ 10 μg/72h/ milj cells | 4 ng/ml rIL6 | 38 |
| L363 | 36 F | IgG PCL | D | PB | yes | 1977 | RPMI1640+20%FCS | nonsecretory | | 39,DSMZ |
| LB-831 | 36 F | IgG/κ MM stageIIIA | T | PE | yes | 1983 | M3 | IgG/κ 69 μg/72 h/ milj cells | L15, insulin, transferrin, 5%FCS, autologous adherent feeder layer | 40 |
| LB-832 | 36 F | IgG/κ MM stageIIIA sister cell line/same origin as LB-831 | T | PE | yes | 1983 | M3 | IgG/κ 69 μg/72 h/ milj cells | L15, insulin, transferrin, 5%FCS, autologous, adherent feeder layer | 40 |
| LB 84-1 | 45 M | IgA/κ MM BJP/κ | R | BM | yes* | 1983 | RPMI1640+15%FCS M3 | IgA/λ (+), κ-, nonsecr. | L15, insulin, transferrin, 5%FCS | 79 |
| LOPRA-1[j] | 62 F | IgA/κ MM Stage IIA extramedullary plasmacytoma | T | ascites | yes | 1982 | Iscove's+20%FCS | IgA/κ | autologous ascites fluid or human AB serum or CM from mouse peritoneal cells/cloned by limiting dilution | 41 |
| LP-1 | 56 F | IgG/λ MM-PCL | R | PB | yes | 1986 | IMDM+20%FCS | IgG/κ 50 μg/24 h/ milj cells | 5 μg/ml transferrin+ 5 μg/ml porcine insulin | 43,DSMZ |
| MEF-1 | 67 F | IgG/κ BJP MM | D | BM | yes* | 1986 | IMDM+20%FCS | κ | no feeder or growth factors | 44 |
| MM.1 | 42 F | IgA/λ MM | T | PB | yes | 1986 | RPMI1640+20%FCS | λ-light chain 4–9 μg/24 h/ milj cells | | 45 |
| MM5.1[k] | 55 M | IgA/κ MM stage IIIA | R | BM | yes | 1993 | RPMI1640 | κ | 12.5% horse serum, 12.5%FCS normal irr. BM/stromal cell feeder layer | 46 |
| MM5.2 | 55 M | MM (subclone of MM5.1) | R | BM | yes | 1988 | RPMI1640+10%FCS | κ | 1 year after the initiation of MM5.1 | 46 |
| MM-A1 | 39 M | IgG/λ stage IIIA MM | D | PE | yes | 1988 | RPMI1640+20% human plasma, 5%PHA-LCM or 10 μg/ml rIL-6 | IgG/λ | | 7 |
| MM-C1 | 50 M | IgG/λ stage IIIA MM | D | BM | yes | 1988 | RPMI1640 | IgG/λ | 20% huPl, 5% PHA-LCM or 10 μg/ml rIL6 | 7 |

*Continued on next page*

Table 2a. (continued)

| Cell line[a] | Patient age/sex | Diagnosis[b] | Treatment status[c] | Specimen site[d] | Authentication[e] | Year est. | Culture medium[f] | Ig production of cell line | Other requirements for establishment[m] | Primary ref./availability[n] |
|---|---|---|---|---|---|---|---|---|---|---|
| MM-M1 | | IgG/κ MM | | | | | | nonsecretory | | 47 |
| MM-S1 | 56 F | BJP/λ stage IIIA MM | T | PE | yes | 1983 | RPMI1640 | λ | 20% huPl, 5% PHA-LCM or 10 μg/ml rIL–6 later 10% FBS+ rIL–6 (10 U/ml) | 7,48 |
| MM-Y1 | 61 F | IgG/κ stageIIIA MM | T | BM | yes | 1988 | RPMI1640 | IgG/λ | 20% huPl, 5% PHA-LCM or 10 μg/ml rIL–6 | 7 |
| MT3 | 51 F | IgA/κ PCL | D | PB | yes | 1985 | RPMI1640+20%FCS | IgA/κ 40–50 μg/ 24 h/milj cells | | 49 |
| NCI-H929 | 62 F | IgA/κ MM | R | PE | yes | 1984/85 | RPMI1640+10%FCS | IgA/κ >80μg/24 h/ milj cells | or 10–20% pleural fluid or ACL-3 (RPMI1640 suppl with insulin, transferrin, EGF, BSA) | 50, DSMZ |
| NOP-2 | 47 M | BJP/λ MM extramedullary plasmacytoma | T | plasma-cytoma | yes | 1986 | RPMI1640+20%FCS | λ | later 10%FCS | 51 |
| OCI-My1 | | advanced MM | | | yes | | IDMEM | | 20% hu Plasma, PHA-LCM | 52 |
| OCI-My2 | | advanced MM | | | yes | | IDMEM | | 20% hu Plasma, PHA-LCM | 52 |
| OCI-My3 | | advanced MM | | | yes | | IDMEM | | 20% hu Plasma, PHA-LCM | 52 |
| OCI-My4 | | advanced MM | | | yes | | IDMEM | | 20% hu Plasma, PHA-LCM | 52 |
| OCI-My5 | | IgA/λ advanced MM | T | PB | yes | | IDMEM | λ | 20% hu Plasma, PHA-LCM primary cells cloned in 0.9% methylcellulose | 52–54 |
| OCI-My6 | | advanced MM | | | yes | | IDMEM | | 20% hu Plasma, PHA-LCM | 52 |
| OCI-My7 | | advanced MM | | | yes | | IDMEM | | 20% hu Plasma, PHA-LCM | 52 |
| OH-2 | 52 F | IgG/MM terminal stage | R | PE | yes | 1984 | RPMI1640+10%FCS | IgG/κ 60 μg/72h/0.5 milj cells | TNF+IL-6 (5 ng/ml), later 10% huA-serum+ IL6 (2 ng/ml) | 55 |
| OPM-1 | 56 F | IgG/λ MM – terminal PCL G/λ–λ | R | PB | yes | 1982 | RPMI1640+10%FCS | λ | | 56 |
| OPM-2 | 56 F | IgG/λ MM – terminal PCL | R | PB | yes | 1982 | RPMI1640+10%FCS | λ | | 56, DSMZ |
| PCM6 | 60 F | IgG/λ BJP MM | R | PB | yes | 1992 | IMDM+20%FCS | IgG/λ/ BJP | 30U/ml rIL-6 | 57 |
| RPMI 8226 | 61 M | IgG/λ MM BJP | | PB | yes | 1966 | RPMI1640+10%FCS | λ 15 μg/24 h/ milj cells | | 2,58, DSMZ, ATCC, JCRB |

*Continued on next page*

*Table 2a.* (continued)

| Cell line[a] | Patient age/sex | Diagnosis[b] | Treatment status[c] | Specimen site[d] | Authentication[e] | Year est. | Culture medium[f] | Ig production of cell line | Other requirements for establishment[m] | Primary ref./availability[n] |
|---|---|---|---|---|---|---|---|---|---|---|
| SK-MM-1 | 55 M | PCL | T | BM | yes | | RPMI1640+10%FCS | κ 0.9 μg/48 h/ 0.3 milj cells | no feeder | 59 |
| SK-MM-2 | 54 M | PCL | T | PB | yes | | RPMI1640+10%FCS | κ 1.1 μg/48 h/ milj cells | no feeder | 59, DSMZ |
| TH | 32 M | κ chain MM (plasmoblastic) | D | BM | yes | | IMDM+10%FCS | IgG/κ | | 60 |
| U-1957 | 60 M | IgG/κ PCL | T | PE | yes* | 1983 | RPMI1640+20%FCS | IgG/κ | feeder cells | 61 |
| U-1958 | 60 M | IgG/κ PCL sister cell line/same origin as U-1957 | T | PE | yes | 1983 | RPMI1640+20%FCS | IgG/κ 1.5 μg/24 h/ milj cells | feeder cells | 61 |
| U-1996 | 70 F | κ extramedullary MM with leukemic course (PCL) | D | ascites | yes | 1983 | RPMI1640+20%FCS | κ | feeder cells | 61 |
| U-2030 | | IgA/κ MM | | PE | yes* | 1983 | RPMI1640+10%FCS | nonsecretory | (1) 30%CM of hematopoietic cell lines, fibroblasts, (2) grid organ, (3) feeder cells | 62 |
| U266Bl (U-266)[p] | 53 M | IgE/λ MM/PCL | T | PB | yes | 1968 | F10+10%NCS | IgE/λ 4-9 μg/24 h/ milj cells | feeder cells | 3, ATCC, DSMZ |
| UCD-HL461 | | IgG/κ Stage IIIB MM | | BM | yes | 1988 | RPMI1640+10%FBS | IgG/κ 15 ng/24 h/ milj cells | stromal cell adh. later independent of adh. and dependent of CM | 63 |
| UMJF-2[o] | 74 M | IgG/λ BJP MM/diffuse plasmacytosis | | BM | yes | 1988 | IMDM+20%FCS | IgG/λ | RPMI1640+15%FBS | 64 |
| UTMC-2 | 72 F | IgA/κ MM | T | PE | yes | 1993 | RPMI1640+10%FBS | IgA/κ 4 μg/24 h/milj | | 65 |
| XG-1 | M | IgA/κ PCL (2ndary) | | PB | yes* | | RPMI1640+10%FCS | | rIL-6(1ng/ml)+GM-CSF(10ng/ml) | 9 |
| XG-2 | F | IgG/λ PCL (stageIII) | | PE | yes* | | RPMI1640+10%FCS | | rIL-6+GM-CSF | 9 |
| XG-3P | F | λ PCL (primary) | D | PB | yes* | | RPMI1640+10%FCS | | rIL-6+GM-CSF | 9 |
| XG-3E | F | λ PCL stage III | R | PE | yes* | | RPMI1640+10%FCS | | rIL-6+GM-CSF | 9 |
| XG-4 | M | IgG/κ PCL (2ndary) | | PB | yes* | | RPMI1640+10%FCS | | rIL-6+GM-CSF | 9 |
| XG-5 | F | λ PCL (2ndary) | | PB | yes* | | RPMI1640+10%FCS | | rIL-6+GM-CSF | 9 |
| XG-6 | F | IgG/λ PCL (2ndary) | | PB | yes* | | RPMI1640+10%FCS | | rIL-6+GM-CSF | 9 |
| XG-7 | F | IgA/κ PCL (2ndary) | | PB | yes* | | RPMI1640+10%FCS | | rIL-6+GM-CSF | 9 |

*Continued on next page*

*Table 2a.* (continued)

| Cell line[a] | Patient age/sex | Diagnosis[b] | Treatment status[c] | Specimen site[d] | Authentication[e] | Year est. | Culture medium[f] | Ig production of cell line | Other requirements for establishment[m] | Primary ref./ availability[n] |
|---|---|---|---|---|---|---|---|---|---|---|
| XG-8 | F | IgAλ PCL (2ndary) | | PB | yes* | | RPMI1640+10%FCS | | rIL-6+GM-CSF | 9 |
| XG-9 | M | IgAκ PCL (2ndary) | | BM | yes* | | RPMI1640+10%FCS | | rIL-6+GM-CSF | 9 |

[a] Cell lines are named as indicated in original literature (later given names in parenthesis; sister cell lines derived independently from the same patient of origin, and subclones originating from the in vitro growth of the parental cell line are indiated. [b] Diagnosis as indicated in the original literature. MM – multiple myeloma, PCL – plasma cell leukemia. [c] Treatment status as indicated in original literature. D – at diagnosis, T – during therapy, R – at relapse, ND – not determined. [d] Specimen site; BM – bone marrow, PB – peripheral blood, PE – pleural effusion. [e] Based on the criteria mentioned in Table 1, the authentication was determined by EBV negativity, and * Ig rearrangements, structural abnormalities or immunophenotype consistent with the fresh tumor cells. [f] Culture medium as indicated in original literature. [g] Subclone of HL407 (HL407E (early) was obtained during in vitro growth and progression from stroma cell dependent, feeder cell CM/IL-6 dependent growth to IL-6 independent autonomous growth in the HL407L (late) cell line. [h] Several subclones of JIM-1 were established (JIM-2, JIM-3, JIM-4 only distinguishable by the expression of bcl-2). [i] Subclones were obtained in the JJN cell line by in vitro cultivation in Ewing Sarcoma Growth Factor (EwSarGF) (i.e. IL-1+ IL-6). The JJN-3 line has retained IL-6 sensitivity. [j] LOPRA-1 was cloned by in vitro cultivation in Ewing Sarcoma Growth Factor (EwSarGF) (LOPRA-1/4 and LOPRA-1/5) differing only in the expression of cIg. LOPRA1/4 stained negative for cIg and subsequent Ig production. [k] The subclone MM5.2 was obtained at 1 year after initiation of the parental cell line MM5.1. [l] The OCI-My1-5 cell lines were obtained from BM or PB samples from advanced MM. Six of these were obtained by colony formation of fresh tumor cells in methyl cellulose, and one directly initiated in liquid suspension. No detailed data is available concerning the origin or karyotype of each individual cell line [54]. [m] huPl – human plasma; PHA-LCM – conditioned medium from PHA stimulated leukocytes; EwSarGF/ESG – Ewing sarcoma growth factor (IL1 and IL6), MQ/MDF – macrophage/MQ derived factors. [n] DSMZ – Deutsche Sammlung von Mikroorganismen und Zellkulturen. RCB – RIKEN Cell Bank. JCRB – Japanese Cell Resource Bank. ATCC – American Type Culture Collection. [o] EBV-positive (Drexler, personal communication). [p] Unfortunately U-266 has two other names. Both SKO 007 (Brodin et al. *J Immunol Methods* 60: 1, 1983) and AF10 [66] are sublines of and should be regarded as identical to U-266. Also note that GM2132 = RPMI 8226.

*Table 2b.*   EBV+ LCLs phenotypic characteristics*

| Cell line | EBV status | Doubling time/Ig production | Cytochemistry | Heterotransplantability in nude mice | B-cell/plasma cell associated marker | Activation marker/other | Adhesion marker | Cytogenetic karyotype | Ref. |
|---|---|---|---|---|---|---|---|---|---|
| ARH-77 | EBNA+ | 110 h 3.8–4.7 d IgG/λ 12.1 μg/milj cells | Acid mucopolysaccaride−, lipids−, fats−, esterase−, SBB−, ORO−, NASDCAE− | | CD10− CD19+ CD20+ CD21± CD38− PCA-1+ sIg+ | HLA-DR+ CD28(+) | CD11a+ CD11b− CD11c−CD18+ CD29+ CD40+ CD49b− CD49d+ CD49e+ CD49c+ CD49f+ CD54+ CD56− | 45–46 no marker chr | 50,51, 67–69 |
| Fr/FRAVEL | EBNA−, later EBNA+ | 62 h IgG/κ | | | CD10+ CD19+ CD20+ CD21+ CD23− CD24− CD37+ PCA-1− sIg− cIg− | HLA-DR+ | CD11a+ CD11b+ CD11c− CD18+ CD29+ CD49b− CD49d+ CD49c+ CD49f+ CD54+ CD56− | | 69–71 |
| GM1312 | EBV+ | | | | CD20+ CD38− | OKIa+ | | | 29 |
| GM1500 | EBV+ | | | | CD20+ CD38− | OKIa+ | | | 29 |
| HS-Sultan | EBV+ | | | | CD20+ CD21+ CD19+ CD10+ PCA-1+ CD38+ B-B4+ sIg+ | HLA-DR+ CD28− CD40+ | CD11a+ CD49e+ | | 17,18, 50,72 |
| IM-9 | EBV+ | | | in nude mice 2/2, clonable in agarose | CD19+ CD20+ CD38− B-B4+ | CD28− CD40(+) | CD11a(+) CD49e+ | | 17,18,72 |
| MC/CAR | EBV+ IgG/κ nonsecretory | 19.4 h | MGP+, SBB−, PAS(+), ANAE+, BGLU+, AcP+, TdT− | | CD10− CD19+ CD20+ CD23− CD24− CD38− B-B4(+) sIg+ | OKIa1+ HLA-DR+ CD25− CD28− CD45+ CD30− | CD11a+ CD49e(+) CD56− | 46, XY | 17,18, 71–73 |

* Phenotype of some frequently used EBV positive LCLs.

*Table 3.* Multiple myeloma cell lines: immunophenotypic characterization[a]

| Cell line | T-/NK-/myelomonocytic cell marker | B-cell/plasma cell associated marker | Activation marker | E/Fc/complement receptor/other | Adhesion marker | Ref. |
|---|---|---|---|---|---|---|
| ACB-885 | CD2– CD3– CD4– CD8a– | CD10– CD20– CD21– CD38+ | HLA-DR– | | | 20 |
| AD3 | | CD10(+) CD19(+) CD20+ CD38+ PCA-1+ sIg– cIg+ | HLA-DR+ | Fcγ– | | 21 |
| AMO1 | CD2– CD3– CD4+ CD5– CD8– CD13– CD33– | CD10– CD19– CD20– CD21– CD38+ PCA-1+ cIg+ | | CD34– | CD11b– CD11c– | 22 |
| ANBL-6 (ANBM-6) | CD2– CD3– CD5– | CD10+ CD19– CD20– CD38+ cIg+ CD21– sIg– CD73– CD23– | CD40+ CD25+ | CD45– | CD11a– CD11b– CD18– CD29+ CD44+ CD49a+ CD49e– CD54+ CD56– | 23 76 |
| delta-47 | CD4(+) CD5– CD8– CD13– | CD10– CD19– CD38+ PCA-1– sIg– cIg+ | HLA-DR– CD30+ | E– EA– EAC– | | 24 |
| DOBIL-6 | CD2– CD3– CD4– CD5– CD7– CD8– CD13– CD33– | CD10– CD19+ CD20+ CD21– CD38+ PCA-1+ sIg– cIg+ | HLA-DR+ CD40+ | CD45RA– CD45RO+ | CD11a/18– CD29+ CD44+ CD49d+ CD49e– CD54+ CD56– | 25 |
| DP-6 | CD2– CD4– CD5– CD14– CD16– CD25– CD33+ | CD10– CD19– CD20– CD38+ CD72– CD73+ CD80– CD81– PCA-1+ UV-3+ | CD28+ CD40+ CD40L– | CD95+ CD126+ CD130+ CD45+ | CD11a+ CD44+ CD54+ CD56+ | 26 |

*Continued on next page*

Table 3. (continued)

| Cell line | T-/NK-/myelomonocytic cell marker | B-cell/plasma cell associated marker | Activation marker | E/Fc/complement receptor/other | Adhesion marker | Ref. |
|---|---|---|---|---|---|---|
| EJM | | CD9+ CD10− CD19− CD20(+) CD21(+) CD38+ CD75+ CD78+ cIg+ | HLA-DR+ | CD32(FcRII)+ CD71+ | CD11a− CD11b− CD11c− CD29(+) | 27 |
| | | | | | CD44+ CD49b− CD49d+ CD49e− | 69 |
| | | | | | CD49c± CD49f± CD54+ CD56− | 77 |
| FLAM-76 | CD2− CD3− CD4− CD5− CD8− CD13− CD15− CD33− | CD10− CD19− CD20− CD38+ PCA-1− sIg− cIg+ | Ia− Ki67+ CD30− | CD32− | CD11a− CD18− CD44+ CD54+ | 28 |
| FR4 | | CD10+ CD19(+) CD20(+) CD38+ PCA-1+ sIg− cIg+ | HLA-DR+ | | | 21 |
| GM2132 (=RPMI 8226) | | CD20− CD38+ OKIa− | | | | 29 |
| HL407 | CD2− CD3− CD4− CD5− CD7− CD8− CD13− CD14− CD15− CD33− CD16− Leu7− | CD9− CD10− CD19− CD20− CD21− CD22− CD23− CD24− PCA-1+ MM4+ CD38+ cIg+ | HLA-DR− | CD25− FcR− | CD56− | 6 |
| ILKM2 | | CD10− CD19− CD20− CD21− OKIa-1− PCA-1+ CD38+ cIg+ | | | | 5 |
| ILKM3 | | CD10− CD19− CD20− CD21− OKIa-1− PCA-1+ CD38+ cIg+ | | | | 5 |

Continued on next page

*Table 3.* (continued)

| Cell line | T-/NK-/myelomonocytic cell marker | B-cell/plasma cell associated marker | Activation marker | E/Fc/complement receptor/other | Adhesion marker | Ref. |
|---|---|---|---|---|---|---|
| JIM-1(-2-3) | | CD38+ PCA-1+ | | | CD29+ CD44+ CD49b− CD49d+ CD49e− CD49c± CD49f− CD54+ CD56+ CD58+ | 30 |
| JJN-1 | | CD9− CD10− CD19− CD20− CD21− CD22− CD23− CD24+ CD37− CD38+ cIg/sIg+ | MHCclassII+ CD25− | CD39++ CDw40− CD45R− | | 4 |
| JJN-2 | | CD9− CD10− CD19− CD20− CD21− CD22− CD23− CD24+ CD25− CD37− CD38+ cIg/sIg+ | MHCclassII− | CD39++ CDw40− CD45R− | | 4 |
| JJN-3 | | CD10+ CD19− CD20(+) CD21− PCA-1− sIg− cIg− | HLA-DR+ | | CD11a− CD11b− CD11c− CD18− CD29+ CD49b− CD49d+ CD49e− CD49c± CD49f± CD54+ CD56− | 69 77 |
| Karpas 620 | CD1− CD3− CD4− CD5− CD7− CD8− CD14− CD15− CD33− | CD10− CD20− CD37+ CD38− | HLA-DR+ | CD45− | | 31 |
| Karpas 707 | | CD10− CD19− CD20− CD21− PCA-1+ | HLA-DR(+) | | CD11a+ CD11b− CD11c− CD18+ CD29+ CD49b− CD49d+ CD49e+ CD49f+ CD54+ CD56+ | 69 |

*Continued on next page*

*Table 3.* (continued)

| Cell line | T-/NK-/myelomonocytic cell marker | B-cell/plasma cell associated marker | Activation marker | E/Fc/complement receptor/other | Adhesion marker | Ref. |
|---|---|---|---|---|---|---|
| KAS-6/1 | CD2– CD4– CD5– CD14– CD16– CD33– | CD10– CD19– CD20– CD38+ CD72– CD73– CD80– CD81– PCA-1+ UV-3+ | CD25– CD28+ CD40+ CD40L– | CD95+ CD126+ CD130+ CD45– | CD11a– CD44+ CD54+ CD56– | 26 |
| KHM-1A | CD2– | CD10– CD20– PCA-1+ | HLA-DR+ | | | 33 |
| KHM-1B | CD2– | CD10– CD20– PCA-1+ | HLA-DR+ | | | 33 |
| KHM-11 | CD3– CD4– CD5– CD7+ CD8– CD13– | CD10– CD19– CD20–CD25– CD38+ PCA-1+ cIg+ | HLA-DR– | CD45RA+ CD45RO+ | CD49d+ CD49e– | 34 |
| KMM-1 | CD4– CD5– CD8a– | CD20+ CD21+ PCA-1+ CD38+ sIg– cIg+ | HLA-DR(+) | CD71+ E– EA-EAC– | | 35 36 |
| KMS-5 | ND | CD20(+) CD21(+) PCA-1(+) CD38(+) sIg–cIg+ CD10+ | HLA-DR(+) | CD71+ | CD11a+ CD54+ | 78 36 |
| KMS-11 | ND | CD20(+) CD21 (+) PCA-1+ CD38(+) sIg+cIg+ | HLA-DR(+) | CD71+ | | 36 |
| KMS-12-PE | ND | CD20(+) CD21(+) PCA-1+ CD38+ | HLA-DR(+) | CD71+ | | 36 |
| KMS-12-BM | ND | CD20+ CD21+ PCA-1+ CD38+ | HLA-DR(+) | CD71+ | | 36 |
| KMS-18 | CD1(+) CD2(+) CD3(+) CD4(+) CD5(+) CD7(+) CD8(+) CD13(+) CD14(+) CD33(+) CD36(+) | CD10(+) CD19(+) CD20(+) CD23+ CD38(+) | HLA-DR+ CD30(+) | CD34(+) | | 37 |

*Continued on next page*

*Table 3.* (continued)

| Cell line | T-/NK-/myelomonocytic cell marker | B-cell/plasma cell associated marker | Activation marker | E/Fc/complement receptor/other | Adhesion marker | Ref. |
|---|---|---|---|---|---|---|
| KP-6 | CD2– CD4– CD5– CD14– CD16– CD25– CD33– | CD10– CD19– CD20– CD38+ CD72+ CD73+ CD80+ CD81– PCA-1+ UV-3+ | CD40+ CD40L– CD28+ | CD95+ CD126+ CD130+ CD45+ | CD11a– CD44+ CD54+ CD56+ | 26 |
| KPMM2 | CD3– CD4– CD5– CD7– CD8– CD13– CD14– CD33– | CD10– CD19– CD20– CD22– CD38+ PCA-1+ BL3+ sIg+ | HLA-DR(+) CD25– | CD34– CD42b– CD45+ CD45RA– CD45RO– CD62– CD63+ CD69– CD71+ glycophorinA– CD61– CD62L– | CD11a– CD11b– CD11c– CD18– CD29+ CD44+ CD54+ CD56+ CD58+ CD49d+ CD49e– | 38 |
| L363 | CD2– CD7– CD13– CD14– CD15– CD16– CD17+ CD37+ | CD9– CD10– CD19– CD20± CD21– CD23– CD24– CD38+ B-B4+ PCA-1± sIg– | HLA-DR+ CD28+ CD40– CD25– CD30– | E– EA– EAC– EBNA–, EA–, VCA– CD45+ CD30– | CD11a– CD11b– CD11c– CD18– CD29+ CD49b– CD49d+ CD49e– CD49c+ CD49f+ CD54+ CD56+ | 69 17,18, 72 71 |
| LB-831 | T cell antigens– | CD10+ (later CD10–) CD20+ CD21– cIg+ | HLA-DR+ | | | 40 |
| LB-832 | T cell antigens– | CD10+ CD20+ CD21– cIg+ | HLA-DR+ | | | 40 |
| LB 84-1 | CD1– CD2– CD3– CD4+ CD5– CD7– CD8– CD13± | CD9– CD10– CD19+ CD20+ CD21– CD22– CD23– CD24– | HLA-DR+ Ki-67+ CD25– | CD45+ | CD11a– CD11b– CD11c± CD18– CD29+ CD49b– CD49d+ | 79 71 |

*Continued on next page*

*Table 3.* (continued)

| Cell line | T-/NK-/myelomonocytic cell marker | B-cell/plasma cell associated marker | Activation marker | E/Fc/complement receptor/other | Adhesion marker | Ref. |
|---|---|---|---|---|---|---|
| | CD14- CD15± CD16- CD17+ Leu M3+ Leu7- Leu8- Leu9- | CD38+ CD37+ PCA-1+ PC-1+ | | | CD49c+ CD49f+ CD54+ CD56± | 69 |
| LOPRA-1 | CD1- CD2- CD3- CD4- CD5- CD6- CD7- CD8- CD13- | PCA-1+ CD24(+) cIg+ CD10- CD19- CD20- CD21- CD22- CD23- CD37- CD38+ CD39- sIg- | HLA-ABC+ CD30- HLA-DR- CD25- CD28+ | CD45- CD71+ | | 41 |
| LP-1 | CD2- CD3- CD16- | CD10- CD19- CD20- CD21- CD38+ PCA-1- PC1- sIg+ B-B4+ | HLA-DR+ CD25- CD28+ CD40+ CD40L- | CD35(+) | CD11a- CD11b- CD11c- CD18- CD29+ CD49b- CD49d+ CD49e- CD49c+ CD49f+ CD54+ CD56+ | 43 69 17,18, 72 |
| MEF-1 | CD2- CD5- CD7- CD13- CD14- | CD19- CD20- CD21- CD22(+) CD23- CD24 (+) CD38- PCA-1- sIg- | HLA-DR- HLA-ABC+ CD28+ | CD34- CD41a- CD45(+) | CD44+ CD56- CD71+ | 44 |
| MM.1 | | CD10- CD20- CD19- PCA-1+ CD38+ sIg- T11- LN-1+ | HLA-DR+ | CD71+ | | 45 |
| MM5.1 | CD3- CD4- CD8- CD14- CD16- | CD10- CD19- CD20- CD21- CD38+ B-B4+ cIg+ | HLA-DR- B7- | CD40- CD45- | CD11a- CD11b- CD11c- CD18- CD29+ CD49a- CD49b- CD49c- CD49d+ CD49e- CD49f- CD44+ CD54+ CD56- CD58- | 46 |

*Continued on next page*

*Table 3.* (continued)

| Cell line | T-/NK-/myelomonocytic cell marker | B-cell/plasma cell associated marker | Activation marker | E/Fc/complement receptor/other | Adhesion marker | Ref. |
|---|---|---|---|---|---|---|
| MM-A1 | CD3- CD13- | CD10- CD19- CD20- CD38+ PCA-1+ cIg+ | HLA-DR- | | | 7 |
| MM-M1 | CD3- CD13- | CD10- CD19- CD20- CD38- PCA-1 (+) cIg- | HLA-DR- | | | 47 |
| MM-C1 | CD3- CD13- | CD10- CD19- CD20- CD38+ PCA-1+ cIg+ | HLA-DR- | | | 7 |
| MM-S1 | CD3- CD13- | CD10- CD19- CD20- CD38- PCA-1+ cIg+ | HLA-DR- | CD41a- glycophorinA- | | 48 7 |
| MM-Y1 | CD3- CD13- | CD10- CD19- CD20- CD38+ PCA-1+ cIg+ | HLA-DR- | | | 7 |
| MT3 | CD1a- CD3- CD7- CD13- CD14- CD33- | CD10(+) CD19- CD20- CD21- CD24- PCA-1+ CD38+ cIg/sIg+ | HLA-DR(+) | CD71+ | | 49 |
| NCI-H929 | CD5- L1- L2- L3- L4- L7- L8- L11- L12- LeuM1- LeuM3- | CD10- CD19- CD20- CD21- CD38+ B-B4+ PCA-1+ cIg+ sIg+ | HLA-DR- I2- TdT- CD25- CD28+ | CD71+ | CD11a- CD49e- | 50 17,18,72 |
| NOP-2 | CD2- CD3- CD7- CDw14- CDw15- CD24- | CD10- CD19- CD20- CD21- CD38+ PCA-1+ sIg- cIg+ | HLA-B(classI)+ HLA-DR- CD25- | | CD11- | 51 |

*Continued on next page*

*Table 3.*   (continued)

| Cell line | T-/NK-/myelomonocytic cell marker | B-cell/plasma cell associated marker | Activation marker | E/Fc/complement receptor/other | Adhesion marker | Ref. |
|---|---|---|---|---|---|---|
| OH-2 | CD2– CD4– CD5– CD14– | CD10– CD19– CD20– CD38+ PCA-1+ B-B4+ | | | CD56+ | 55 |
| OPM-1 | CD1a– CD3– CD4– CD8– | CD9– CD10– CD19– CD20– CD21– CD24– CD38+ PCA-1– sIg– | HLA-DR– | E– EA– EAC– | CD11a– CD11b– CD11c– CD18– CD29+ CD49b– CD49d+ CD49c+ CD49f+ CD54– CD56+ | 51 69 |
| OPM-2 | CD1a– CD3– CD4– CD8– | CD9– CD10– CD19– CD20– CD24– CD38+ B-B4+ sIg– | CD28+ | E– EA– EAC– | CD11a– CD49e– | 51 |
| PCM6 | CD13– | CD19– CD10– PCA-1– CD38+ | Ia– | | | 17,18,72 57 |
| RPMI8226 | CD1a– CD2– CD3– CD4– CD7– CD8– | CD9+ CD10– CD19– CD20– CD21– CD22– CD24– CD38+ OKIa-1(+) PCA-1+ B-B4+ sIg– | HLA-DR– CD40+ CD28+ | CD45– | CD11a(+) CD18± CD29+ CD49b– CD49d+ CD49e+ CD54+ CD56± | 51 71 17,18, 69,72 |
| SK-MM-1 | CD2– CD25– CD33– M195– | CD10– CD19– CD20+CD21– BL3+ CD38– PCA-1– sIg– | HLA-DR– | EAC– | | 59 |
| SK-MM-2 | CD2– CD13– CD33– | CD10– CD19– CD20+ CD21– BL3+ CD38+ PCA-1+ sIg– | HLA-DR– | EAC– | | 59 |
| TH | CD5– | CD9+ CD10(+) CD19– CD20(+) CD22– CD23– CD24– CD38+ PCA-1+ cIg+ | HLA-DR+ CD40+ CD25(+) CD28+ | CD34– CD45– CD6(+) | | 60 |

*Continued on next page*

*Table 3.* (continued)

| Cell line | T-/NK-/myelomonocytic cell marker | B-cell/plasma cell associated marker | Activation marker | E/Fc/complement receptor/other | Adhesion marker | Ref. |
|---|---|---|---|---|---|---|
| U-1957 | | CD19– CD20– CD21– CD38+ PCA-1+ PC-1– cIg+ | CD25– | E– Fcγ+ C2– | | 61 |
| U-1958 | | CD19– CD20– CD21– CD38+ PCA-1+ PC-1– cIg+ | CD25– | E– C2– | | 61 |
| U-1996 | | CD19– CD20– CD21– CD38+ PCA-1+ PC-1– cIg+ | CD25– | E– Fcγ– C2– | | 61 |
| U-2030 | | CD10– CD19– CD20– CD21– PCA-1+ | CD25+ | | | 62 |
| U-266 | CD2– CD7- CD13– CD14– CD15– CD16– CD17– | PC-1– CD38+ cIg– sIg– | HLA-DR– | CD71+ CD45– | CD11a– CD11b– CD11c– CD18– | 51 |
| | | CD10– CD19– CD20– CD21– CD23– CD24– CD37– CD38– | HLA-DR+ CD40– CD28+ | | CD29± CD49b– CD49d+ CD49e– | 50 |
| | | B-B4+ PCA-1(+) sIg– | CD25– | | CD49c± CD49f± CD54+ CD56– | 17,18,69 71,72,77 |
| UCD-HL461 | CD1– CD2– CD3– CD4– CD5+ CD7– CD8– CD14– CD13– CD16– CD33– Leu7– | CD9(+) CD10+ CD19+ CD20+ CD21(+) CD22+ CD24+ cIg– sIg+ CD38+ PCA-1+ MM4+ | HLA-DR+ I2– CD25+ | CD35(+) | CD11– CD56– | 63 |
| UMJF-2[b] | CD2– CD3– CD4+ CD5– CD7– CD8– CD13– CD16(+) CD33(+) | CD10– CD19+ CD20+ CD38+ PCA-1+ | | CD57(+) CD34– | CD11b– CD11c– | 64 |

*Continued on next page*

*Table 3.* (continued)

| Cell line | T-/NK-/myelomonocytic cell marker | B-cell/plasma cell associated marker | Activation marker | E/Fc/complement receptor/other | Adhesion marker | Ref. |
|---|---|---|---|---|---|---|
| UTMC-2 | CD3– CD4– CD5– CD8– CD13– | CD10– CD19– CD20– CD38+ sIg– cIg+ | HLA-DR– | CD45RA– | CD29– CD29+ CD54+ CD56– CD11a/CD18– | 65 |
| XG-1 | CD2– CD3– CD5– CD13– CD14– CD15– CD16– CD33– | CD10+ CD19– CD20– CD21– CD23+ CD24– CD37+ CD72– B-B4+ CD38+ cIg+ sIg– | CD25– CD28+ B7+ CTLA-4– CD30+ CD40– CD71+ HLA-I+ HLA-II– | CD32– CD57– CD65– CD77+ | CD11a– CD11b– CD11c– CD18– CD44+ CD49b– CD49d+ CD49e– CD49f+ CD54+ CD56+ CD58+ PNA+ | 9,17 18,72 |
| XG-2 | CD2– CD3– CD5– CD13– CD14– CD15– CD16– CD33+ | CD10– CD19– CD20– CD21+ CD23+ CD24– CD37– CD72– B-B4+ CD38+ cIg+ sIg– | CD25– CD28+ B7– CTLA-4– CD30– CD40+ CD71+ HLA-I+ HLA-II+ | CD57– CD65– CD77+ | CD11a+ CD11b– CD11c– CD18+ CD44+ CD49b– CD49d+ CD49e± CD49f+ CD54+ CD56– CD58+ PNA+ | 9,17 18,72 |
| XG-3P | CD2– CD3– CD5– CD13– CD14– CD15– CD16– CD33+ | CD10+ CD19– CD20– CD21+ CD23– CD24– CD37+ CD72– B-B4+ CD38+ cIg+ sIg– | CD25– CD28+ B7– CTLA-4– CD30– CD40+ CD71+ HLA-I+ HLA-II– | CD57+ CD65– CD77+ | CD11a– CD11b– CD11c– CD44+ CD49b– CD49d+ CD49e+ CD49f+ CD54+ CD56+ CD58+ PNA+ | 9,17 18,72 |
| XG-3E | CD2– CD3– CD5– CD13– CD14– CD15– CD16– CD33+ | CD10– CD19– CD20– CD21– CD23– CD24– CD37– CD72– B-B4+ CD38+ cIg+ sIg– | CD25– CD28+ B7+ CTLA-4– CD30– CD40+ CD71+ HLA-I+ HLA-II– | CD57+ CD65– CD77+ | CD11a– CD11b– CD11c– CD44+ CD49b– CD49d+ CD49e+ CD49f+ CD54+ CD56+ CD58+ PNA+ | 9,17 18,72 |

*Continued on next page*

*Table 3.* (continued)

| Cell line | T-/NK-/myelomonocytic cell marker | B-cell/plasma cell associated marker | Activation marker | E/Fc/complement receptor/other | Adhesion marker | Ref. |
|---|---|---|---|---|---|---|
| XG-4 | CD2– CD3– CD5– CD13– CD14– CD15– CD16– CD33– | CD10– CD19– CD20– CD21– CD23– CD24– CD37+ CD72– B-B4+ CD38+ cIg+ sIg– | CD25– CD28+ B7– CTLA-4– CD30– CD40+ CD71+ HLA-I+ HLA-II– | CD32– CD57+ CD65– CD77+ | CD11a– CD11b– CD18+ CD44+ CD49b– CD49d+ CD49e– CD49f– CD54+ CD56– CD58+ PNA+ | 9,17 18,72 |
| XG-5 | CD2– CD3– CD5– CD13– CD14– CD15– CD16– CD33+ | CD10+ CD19– CD20– CD21+ CD23– CD24– CD37+ CD72– B-B4+ CD38+ cIg+ sIg– | CD25– CD28+ B7+ CTLA-4– CD30+ CD40– CD71+ HLA-I+ HLA-II– | CD32– CD57– CD65– CD77+ | CD11a– CD11b– CD11c– CD18– CD44+ CD49b– CD49d+ CD49e+ CD49f– CD54+ CD56– CD58+ PNA+ | 9,17 18,72 |
| XG-6 | CD2– CD3– CD5– CD13– CD14– CD15– CD16– CD33– | CD10– CD19– CD20– CD21+ CD23– CD24– CD37+ CD72– B-B4+ CD38± cIg+ sIg– | CD25– CD28+ B7– CTLA-4– CD40+ CD71+ HLA-I+ HLA-II– | CD32– CD57– CD65– CD77+ | CD11a– CD11b– CD11c– CD18– CD44+ CD49b– CD49d+ CD49e± CD49f– CD54+ CD56– CD58+ PNA+ | 9,17 18,72 |
| XG-7 | CD2– CD3– CD5– CD13– CD14– CD15– CD16– | CD10– CD19– CD20– CD23– CD24– CD37– CD72– B-B4+ CD38+ cIg+ sIg– | CD25– CD28+ B7– CTLA-4– CD40– CD71+ HLA-I+ HLA-II– | CD32– CD57– CD65– CD77+ | CD11a– CD11b– CD11c– CD18– CD44+ CD49b– CD49d+ CD49e+ CD49f+ CD54+ CD56+ CD58+ PNA+ | 9,17 18,72 |

*Continued on next page*

Table 3. (continued)

| Cell line | T-/NK-/myelomonocytic cell marker | B-cell/plasma cell associated marker | Activation marker | E/Fc/complement receptor/other | Adhesion marker | Ref. |
|---|---|---|---|---|---|---|
| XG-8 | CD2– CD3– CD5– CD13– CD14– CD15– | CD10– CD19– CD20– CD21– CD23– CD24– CD37– CD72– B-B4+ CD38+ cIg– sIg– | CD25– CD28+ B7+ CTLA-4– CD30+ CD40+ CD71+ HLA-I+ HLA-II– | CD32+ CD57– CD65– CD77+ | CD11a– CD11b– CD11c– CD18+ CD44+ CD49b– CD49d+ CD49e+ CD49f+ CD54+ CD56– CD58+ PNA+ | 9,17 18,72 |
| XG-9 | CD2– CD3– CD5– CD13– CD14– CD15– CD33– | CD10– CD19– CD20– CD23+ CD24– CD37– CD72– B-B4+ CD38+ cIg– sIg– | CD25– CD28+ B7– CTLA-4– CD30+ CD40– CD71+ HLA-I+ HLA-II– | CD32– CD57– CD65– CD77+ | CD11a+ CD11b– CD11c– CD18+ CD44+ CD49b– CD49d+ CD49e– CD49f– CD54+ CD56– CD58+ PNA+ | 9,17 18,72 |

[a] + ; strong, definite protein expression, mostly >10–20% cells positive; (+) = weak protein expression, qualitatively and quantitatively (commonly <10% cells positive); – = no protein expression; ± = contradictory results.
[b] EBV positive (Drexler, personal communication).

In summary, the EBV+ LCL type of non-neoplastic B cell line is generally recognizable by the surface marker profile CD19+, CD20+, CD138–, CD28–, CD11a+, CD49e+ and CD56–, while the phenotype of MM is characterized by the expression of CD19–, CD20–, CD138+, CD28+, CD11a–, CD49e– and CD56+/– (Table 1).

## 2. CLINICAL CHARACTERIZATION

Cell lines have been established from bone marrow, peripheral blood, plasmacytoma, pleural effusions and ascites of patients with advanced disease at a terminal stage (Table 2). During the course of the disease from the chronic to the acute phase, the proliferative compartment progressively increases [74]. In the late phases, MM cells frequently invade extramedullary sites, from which the majority of the cell lines are derived (62 characterized cell lines derived from 54 patients).

The data listed in Table 2 include a few cell lines derived from bone marrow of MM in the absence of extramedullary spread (DOBIL-6, HL407, ILKM2, ILKM3, LB 84-1, MEF-1, MM5.1, MM-C1, MM-Y1, TH, UCD-HL461, UMJF-2.). Interestingly, stage III disease was found in four of these patients. In the case of LB 84-1, MEF-1 and UMJF-2, Bence-Jones protein (BJP) was present in the MM patients and diffuse plasmacytosis was described in UMJF-2. The DOBIL-6 is derived from a nonsecretory MM. Three additional cell lines were established from nonsecretory MM, but these were from disseminated disease (FLAM-76 and KMS-12-PE and KMS-12-BM, the two latter lines being derived from different sites in the same patient). A few cell lines were derived from plasmacytomas (AMO1, delta-47, FR4 and its subline AD3, KMM-1, KPMM2, LOPRA-1, NOP-2).

## 3. IMMUNOPHENOTYPICAL CHARACTERIZATION

The established MM cell lines share common phenotypic characteristics with freshly isolated MM cells, including Ig gene rearrangements, a unique myeloma protein and the requirement for stromal cell-derived growth factors, including IL-6 and IGF-I. Also several surface markers are expressed on continuous MM cell lines and fresh MM cells, although some can differ in expression.

The CD38 marker of normal plasma cells is also expressed on MM cells. However, CD38 is not restricted to the plasma cell compartment, nor is it B-cell lineage specific. PCA-1 is expressed on malignant plasma cells of MM patients, with some specificity for disseminating tumor cells escaping the bone marrow compartment. CD28 is expressed on fresh MM cells and on many MM cell lines [17,72,75] (Table 3). The CD28 molecule expressed on MM cells was recently proved to be functional due to its binding to its receptor B7, inducing activation of PI-3 kinase in MM cells [75]. CD19 is expressed on normal plasma cells, while MM cells are usually negative.

The combined expression of phenotypic markers can be useful for the authentication of MM cell lines. Normal plasma cells are CD28– , but CD40+. Immunophenotypic studies have also demonstrated the expression of CD40 on primary cells of MM and on established cell lines [72]. The expression of CD40 is related to progression and was seen in 7/13 cell lines examined (XG-2, XG-3, XG-4, XG-6, XG-8, RPMI 8226, LP-1), although with the exception of XG-2 expression was weak, and negative in XG-1, XG-5, XG-7, U-266 and L363 (Table 3). All MM cell lines examined express CD28, a feature of advanced disease which is not expressed during the chronic phase of MM [72]. It is interesting that in the ANBL-6 cell line, the expression of CD40 and signaling via CD40 may induce autocrine IL-6 production and IL-6 autocrine growth stimulation [80]. Other reports also show enhanced secretion of IL-6 as a result of CD40L stimulation in MM cell lines [68].

CD138 (syndecan-1/B-B4) is a transmembrane heparan sulfate proteoglycan that binds ECM components and growth factors, such as fibroblast growth factor (FGF). It mainly mediates adhesion to collagen type I of ECM [81,82]. Similar to CD38, the expression of CD138 is a feature both of normal and malignant plasma cells derived from patients with MM. It has been suggested that the immature phenotype expressing CD45+, CD19– does not express CD138, while the more mature phenotype CD45-, CD19– does express CD138. Syndecan-1 may participate in the homing to or maintenance of tumor cells in the marrow, as well as in the cytokine-mediated growth regulation of MM cells. The fact that CD-138 is not presented on other hematopoietic, or lymphoid cells in patients with MM, while present on MM cells, emphasizes the usefulness of this marker in the purification and selection of primary cells for studies of the biology of MM [81].

The integrins are important cell adhesion molecules in MM. The very late antigens (VLA) are adhesion molecules belonging to the integrin $\beta$ family with a possible role in maintaining the tumor cells within the bone marrow [77]. Ligands of the ECM consist of collagen and laminin (VLA-2 (CD49b)), and fibronectin (VLA-4 (CD49d) and VLA-5 (CD49e)). Binding of MM cells to bone marrow stromal cells and osteoblasts through VLA-4 (CD49d)/VCAM-1 may induce IL-6 production in the stromal cells [83]. Immature subpopulations of MM are negative for VLA5 (CD49e), while mature cells are positive [84]. Normal and malignant plasma cells express VLA-4, but not other VLAs, including VLA-5 (CD49e) [77]. MM cell lines JIM-1, JIM-3, JJN-3, EJM, U-266 are negative for CD49e, and only RPMI 8226 was positive [77]. The cell lines were positive for VLA-4 (CD49d), while other VLAs (CD49a, CD49b) were generally not expressed. The results of Drew et al. [77] were confirmed by Kim et al. [85] comparing one authentic MM (U-266) with an EBV+ LCL (IM-9), and by Ohmori et al. [25] in the phenotypic characterization of the DOBIL-6 cell line. The expression of VLA-6 (CD49f) is heterogeneous in MM cells, while VLA3 (CD49c) does not appear to be expressed. However, the expression of CD49c and CD49f in MM cell lines seems to be highly variable [69,77].

The $\beta2$ family of integrins includes the lymphocyte function associated antigen (LFA-1) (CD18/CD11a), Mac1 (CD18/CD11b) and p150.95 (CD18/CD11c). The pattern of $\beta2$-integrin expression is heterogenous in normal and malignant plasma cells from MM patients. Authentic MM cell lines (L363, JJN-3, U-266, LP-1, Karpas 707, OPM-1, EJM, LB 84-1) are negative or weak for $\beta2$ integrins, while LCLs (ARH-77 and Fr/Fravel) are positive for CD18, CD11a, CD11b and CD11c [69] (Tables 2 and 3). The expression of the $\beta2$ integrins CD18/CD11a was determined in malignant plasma cells from MM patients [86]. In normal plasma cells and in MM cells from patients with inactive disease CD11a expression was absent, indicating that LFA expression correlated with proliferation [78,86]. However, these data were not confirmed by others [1,17,18], showing a decreased expression of CD11a in MM cells as compared to normal plasma cells. These data suggest the lack of expression of LFA-1 (CD11a) is associated with highly proliferative and highly malignant MM cells. The other $\beta2$ integrins, including CD11b and CD11c, are not expressed on fresh MM cells and cell lines derived from MM [69,77,85].

CD56 is an isoform of the neural adhesion molecule N-CAM. It belongs to a group of adhesion molecules (the Ig-superfamily) which also includes the intercellular adhesion molecule ICAM-1 (CD54). CD54 is one of the ligands of LFA-1 [69]. CD54 is invariably expressed by normal plasma cells, freshly isolated MM cells and established MM cell lines (L363, JJN-3, U-266, LP-1, Karpas 707, OPM-1, EJM, LB 84- 1) [69]. The expression of CD54 was also demonstrated in KMS-5 [78], DOBIL-6 [25], DP-6, KAS-6/1, KP-6, FLAM-6, KPMM2, UTMC-2 and the XG1- 9 cell lines (Table 3). CD54 was also detected in the LCLs ARH77 and Fr/Fravel (Table 2).

In line with the use of the combined expression pattern of CD28 and CD40, the expression of CD19 and CD56 can be used to determine the authenticity of MM lines. Normal plasma cells are frequently CD19+ and CD56–, while MM cells are almost invariably positive for CD56. The CD19+ CD56– phenotype was never detected in MM cells. CD56 (NKH- 1/Leu19) binds heparan sulfate of the ECM, which may be present in the BM environment, and this would be consistent with the finding that loss of CD56 may be associated with a more aggressive disseminating tumor [17,87,88]. The

CD56 was expressed in XG-1 and XG-7 and absent or weakly expressed in the XG-3, RPMI 8226, L363 and LP-1 MM cell lines [72] (Table 3). These data were confirmed by van Riet et al. [69], who showed the absence of CD56 expression in L363, JJN-3, U-266 and EJM cell lines. However, the LP-1, Karpas 707, OPM-1 and LB 84-1, KPMM2, KP-6, KAS-6/1, DP-6, JIM and DOBIL-6 lines stained positively for CD56. Interestingly, the CD56 expression has been suggested to be absent in cell lines derived from MM patients with plasma cell leukemia (PCL) [69,88,89]. CD56 was also absent in the LCLs Fr/Fravel and ARH-77, which express CD19 (Table 2).

## 4. CYTOKINE-RELATED CHARACTERIZATION

The proliferation and survival of MM cells in vivo is dependent on cytokines and a fine-tuned network of cell-cell and cell- extracellular matrix interactions [87,90,91]. In recent years IL-6 has been identified as an important factor regulating growth and survival of MM cells in vitro and in vivo. The study by Kawano et al., suggesting that IL-6 was an autocrine growth factor in MM, stimulated extensive studies of the role of this cytokine. IL-6 can induce growth of primary cells derived from MM patients in about 50% of cases [12,13]. The importance of IL-6 in the pathogenesis of MM was further emphasized by the anti-tumor effects of monoclonal IL-6 antibodies used in the treatment of MM patients [92]. The frequency of establishment of MM cell lines has considerably increased with the use of IL-6-producing feeder cells or recombinant growth factors (IL- 6, GM-CSF, and TNFα). Yet there is still no tissue culture technique that allows consistent success in the establishment of MM cell lines from the bone marrow of early stage MM patients [5,6,40,61,63].

It is difficult to provide a coherent picture of the capacity of MM cell lines to produce and secrete cytokines, and to express cytokine receptors, since there are few reports on the characterization of cytokine production and the response to cytokines in authentic MM cell lines. Also, the sensitivity of methods used varies (RT-PCR, northern blots, ELISA and bioassays). In addition, few studies take account of the changes in expression of cytokine and corresponding receptors that occur during progression in culture [6,93–95]. These difficulties are exemplified by the IL-6 dependent cell lines established with feeder cells (U-266, HL407) [6,95]. In these cell lines a weak expression of IL-6 mRNA, but no detectable IL-6 protein was found. During in vitro progression, the expression of IL-6mRNA increased and IL-6 secretion was induced in U-266 (–1984) and HL407 (late), in some cases coinciding with autonomous growth and survival in the absence of exogenous IL-6. Also the

IL-6R mRNA and the number of IL-6 receptors increased during a decade of continuous in vitro growth [94,95].

The relevance of an autocrine loop in vivo and in vitro has also been disputed [26]. The proposal for an autocrine loop in MM cells [12] was challenged by Portier et al. [96], who did not detect IL-6 mRNA in highly purified MM cells. However, recent studies using RT-PCR and in situ analysis have demonstrated the expression of IL-6mRNA in bone marrow plasma cells of MM patients [97,98].

In MM cell lines, IL-6mRNA was detectable in U-266, and antisense strategies and antibodies against IL-6R inhibited cell proliferation, indicating an autocrine growth stimulation in this cell line [19,94,95,99,100]. In addition to U-266 and HL407, further cell lines have been demonstrated to express IL-6mRNA and secrete IL-6 protein (OCI-My2, OCI-My3, DOBIL-6, DP-6, KP-6, KPMM2, MEF-1), while others express IL-6mRNA but no detectable protein (OCI-My1, OCI-My6, OCI- My7, ANBM-6, KAS-6/1, UTMC-2) (Table 4). In a few of these cell lines, autocrine IL-6 stimulation has been suggested [26,38,52,65]. This notion has also been supported in some cases by antisense strategies [129]. Most cell lines are negative for IL-6 mRNA, but do express IL-6R mRNA. The presence of functional receptors has in some cases been confirmed by binding assays and/or by flow cytometry analysis [9,26,38,65,95,109].

Numerous reports have established IL-6 as the major growth factor for primary cells and established cell lines (reviewed by Hawley and Berger [130]). Most MM cell lines can be stimulated by IL-6 (Table 4). Several IL-6 dependent cell lines have been described (ANBL-6, DP-6, FLAM-76, HL407 (early), ILKM2, ILKM3, JJN-1, KAS-6/1, KP-6, MM5.1, MM-A1, MM-C1, MM-Y1, MM-S1, OH-2, PCM6, U-1958, U-266-1970, XG-1, XG-2, XG-3, XG-4, XG-5, XG-6) (Table 4). As described above for HL407 and U-266, in DP-6 and KAS-6/1 the dependence on IL-6 decreased during long term in vitro culture.

Other members of the IL-6 type cytokine family, IL-11, leukemia inhibitory factor (LIF), oncostatin M (OM) and ciliary neurotrophic factor (CNTF) can induce proliferation in MM cells [122,131]. The IL-11, LIF, OM and CNTF share the same transducer chain (gp130) and induce similar biological effects as IL-6 (reviewed by Klein [92]). In two IL-6 dependent cell lines (XG-4, XG-6) proliferation was induced, while two other cell lines (XG-1 and XG-2) were unresponsive. This correlated with the expression of IL-11R and LIFR$\beta$ in the responsive cell lines [122]. In particular LIF-, Oncostatin M-, IL-11-, or CNTF- dependent MM cell lines have been established [92,122,125]. However, in a previous study, the absence of IL-11R, IL-11 and response to IL-11 was reported in two cell lines autonomous for growth (U-266, RPMI 8226) [118]. Established MM cell lines have been suggested

*Table 4.*  Multiple myeloma cell lines: cytokine-related characterization

| Cell line | Cytokine receptor expression[a] | Cytokine production[d] | Proliferation response to cytokines[b] | Growth arrest/apoptosis response to cytokines[c,d] | Dependency on cytokines | Ref. |
|---|---|---|---|---|---|---|
| ANBM-6/ ANBL-6 | Flow cytometry: IGF-IR+ | RT-PCR: IL-6 mRNA+ ELISA: IL-6– | 3H-TdR: IGF-I+, IL-6+ | IFNα+ (inh. of IL-6 ind. growth) | IL-6 | 101,102 |
| DOBIL-6 | RT-PCR: IL-6R/gp80 mRNA+ IL-6R mRNA+ | RT-PCR: IL-6 mRNA+ IL-6+ (5440 pg/100 000 cells/ml/96 h), IL-1α–, TNFα– RT-PCR: PTHrP mRNA+ | IL-6+ | neutr. anti-IL-6 ab+ (growth inh.) serum depletion + (apoptosis) | 25 | |
| DP-6 | Flow cytometry: IL-6R/gp80 (CD126)+ gp130 (CD130)+ IGF-IR+ | RT-PCR: IL-6 mRNA+ ELISA/bioassay: IL-6+ | 3H-TdR: IL6+, IL-1α+, IL-1β+, IL-3+, IGF-1+. IL-10+, TNFα+, IL-2(+), IL-11(+), IL-12–, IL-13(+) GM-CSF(+), LIF–, OSM–, IL-4(+), CNTF–, IFNγ(+) | TGFβ+ (growth inh.) IFNα + (growth inh.) anti-IL-6R ab+ (growth inh.), (<1 yr after initiation of cell line) anti-IL-6R ab(+) (growth inh.), (>1 yr after initiation of cell line) anti-IGF-IR + (growth inh.) | IL-6, later partly dependent | 26,103 101,102 |
| EJM | RT-PCR/binding assay: IGF-IR+ | RT-PCR/RIA: IGF-I+ Northern: IL-1β mRNA– Immunoassay: GM-CSF– | 3H-TdR: IL6+, IL-1(+) IL-3–, IL-7–, TNFα– IGF-I– IGF-I truncated analogue+ (truncated IGF-BP site) | IFNα+ (growth inh.) anti-IGF-IR ab+ (growth inh.) | | 27 8 104 105,106 |

*Continued on next page*

Table 4. (continued)

| Cell line | Cytokine receptor expression[a] | Cytokine production[d] | Proliferation response to cytokines[b] | Growth arrest/apoptosis response to cytokines[c,d] | Dependency on cytokines | Ref. |
|---|---|---|---|---|---|---|
| FLAM-76 | | bioassay: IL-6– | 3H-TdR: IL-6(+) | IFNα+ (growth inh.) | IL-6 | 28 |
| HL407 | RT-PCR: IL-6R mRNA+ | RT-PCR: IL-6 mRNA+ bioassay: IL-6– In HL407L, late subclone of IL-6 independent cell line: bioassay: IL-6+ >2 μg/ml IL-6 | IL-6+, IL-1–, IL-2–, IL-3–, IL-4–, IL-5–, IL-7–, G-CSF–, GM-CSF–, IGF-I+ In HL407L, late IL-6 independent subclone: IL-6–, IGF-I+ | anti-IGF-IR ab+ (growth inh.) | IL-6, later independent of stromal cells/ independent of IL-6 (HL407L) | 6, 105, 106 |
| ILKM2 | | IL-6– (in supernatant) | 3H-TdR: IL-1α+, IL-1β+ IL-2–, IL-3–, IL-4–, IL-6+ IFNα+, IFNβ–, IFNγ–, EGF–, TNFα(+), G-CSF–, GM-CSF–, MDF (Macrophage CM)+ | anti-IL-6 ab+ (inh. MDF induced growth) | IL-6 | 5 |
| ILKM3 | | IL-6– (in supermatant) | 3H-TdR: IL-1α–, IL-1β– IL-2–, IL-3–, IL-4–, IL-6+ IFNα–, IFNβ–, IFNγ–, TNFα–, G-CSF–, GM-CSF–, EGF–, MDF (Macrophage CM)+ | anti-IL6 ab+ (partly inh. MDF induced growth) | IL-6 | 5 |

*Continued on next page*

*Table 4.* (continued)

| Cell line | Cytokine receptor expression[a] | Cytokine production[d] | Proliferation response to cytokines[b] | Growth arrest/apoptosis response to cytokines[c,d] | Dependency on cytokines | Ref. |
|---|---|---|---|---|---|---|
| JJN-1 | IL-2R− | | 3H-TdR: IL-2−, IL-6+ | | IL-6 | 4 |
| JJN-2 | IL-2R− | | 3H-TdR: IL-2−, IL-6+ | | IL-6 | 4 |
| JJN-3 | | Northern: IL-1β mRNA− Immunoassay: GM-CSF− | IL-6+ | | | 27 |
| Karpas 707 | Northern: IL-6 mRNA+ RT-PCR: IL-6R mRNA+ RT-PCR/binding assay: IGF-IR+ | Northern: IL-6mRNA− RT-PCR/ RIA: IGF-I+ RT-PCR/RIA: IL-6− | IL-6− IGF-1+ | anti-IGFIR ab+ (growth inh.) | | 8,104; 14,107; 105–106, Georgii-Hemming per. comm. |
| KAS-6/1 | Flow cytometry: IL-6R/gp80(CD126) + gp130 (CD130)+, IGF-IR+ RT-PCR: LIFRβ mRNA+ | RT-PCR: IL-6 mRNA+ ELISA/bioassay: IL-6− RT-PCR: CNTF mRNA+, OSM mRNA+, IL-11 mRNA−, LIF mRNA− ELISA: OSM+ | 3H-TdR: IL-6+, IL-1α(+), IL-1β(+), IL-3+, IGF− I+, IL-10+, TNFα+, IL-2(+), IL-11+, IL-12−, IL-13(+) GM-CSF(+), LIF+, OSM+, IL-4(+), CNTF−, IFNγ+, IFNα+ | TGFβ(+) (growth inh.) | IL-6, later less dependent on exogenous IL-6 | 26,103 101,102 |
| KHM-1A | | Northern: IL-6 mRNA+ TNFα mRNA+ | TNFα+, IL-6+ | anti-TNF-α ab− (growth inh.) | | 108 |
| KHM-11 | | bioassay: IL-6+ (200 pg/ml) | IL-6+, IL-11−, GM-CSF− | | | 34 |

*Continued on next page*

Table 4. (continued)

| Cell line | Cytokine receptor expression[a] | Cytokine production[d] | Proliferation response to cytokines[b] | Growth arrest/apoptosis response to cytokines[c,d] | Dependency on cytokines | Ref. |
|---|---|---|---|---|---|---|
| KP-6 | Flow cytometry: IL-6R/ gp80(CD126)+, gp130(CD130)+, IGF-IR+ | RT-PCR: IL-6mRNA+ ELISA/bioassay: IL-6+ ELISA: OSM– | 3H-TdR: IL6+, IL-1α (+), IL-1β(+), IL-3+, IGF-I+, IL-10+, TNFα+, IL-2+, IL-11+, IL-12–, IL-13(+) GM-CSF(+), LIF–, OSM+, IL-4(+), CNTF– | IFNγ+ (growth inh.) IFNα+ (growth inh.) TGFβ+ (growth inh.) anti-IL-6R ab+ (growth inh. <1 yr after initiation of cell line) anti-IL-6R ab– (growth inh. >1 yr after initiation of cell line) | IL-6, later more dependent on exogenous IL-6 | 26,103 101,102 |
| KPMM2 | Flow cytometry: IL-6R+ RT-PCR: IL-6R mRNA+ | ELISA: IL-6+ ( 79.7 pg/ml) PCR: IL-6 mRNA+ | 3H-TdR: IL-6+, IL-11–, OSM–, IL-3–, SCF–, GM-CSF–, IL-1α–, IL-2–, IL-4–, IL-5–, IL-7–, IL-8–, IL-9–, IL-10–, EPO–, TNFα–, TGFβ– | IFNα+ (growth inh.) IFNγ+ (growth inh.) anti-IL-6R ab (PM1)+ (growth inh.) anti-IL-6 ab (SK2)+ (growth inh.) | | 38 |
| L363 | Northern: IL-6R mRNA– Northern: IL-6R/gp130 mRNA+ RT-PCR: IL-6R mRNA+ IL-6R+ (Flow cytometry/bind. assay) | Northern: IL-6 mRNA– | IL-6– 3H-TdR: IL-6/sIL-6R+ | | | 14,107 109 |
| LB-831 | EGFR– | Immunoassay: GM-CSF– | | | | 40,8 |

*Continued on next page*

*Table 4.* (continued)

| Cell line | Cytokine receptor expression[a] | Cytokine production[d] | Proliferation response to cytokines[b] | Growth arrest/apoptosis response to cytokines[c,d] | Dependency on cytokines | Ref. |
|---|---|---|---|---|---|---|
| LB-832 | EGFR– | | | | | 40 |
| LP-1 | RT-PCR/bind. assay: IGF-IR+ RT-PCR: IL-6R mRNA+ | RT-PCR/RIA: IGF-I+ RT-PCR/ELISA: IL-6– | IL-6+ (tyr phosph. of Shc) IGF-I+ | anti-IGF-IR ab+ (growth inh.) | | 110; 105,106, Georgii-Hemming per. commun. |
| MEF-1 | RT-PCR: IL-6R mRNA– | RT-PCR: IL-6 mRNA+ bioassay: IL-6, 18 pg/ml/48h | | | | 44 |
| MM5.1 | RT-PCR: LIFRβ mRNA+ In subclone MM5.2: LIFRβ mRNA+ | | 3H-TdR: OSM+, LIF–, SCF–, IL-6–, IL-10–, IL-11–, GM-CSF–, G-CSF–, IL-6/sIL-6R+ | | CM of BM stromal cells, later independent (subclone MM5.2) | 46 |
| MM-A1 | Flow cytometry: IL-6R+ | Northern: IL-6 mRNA– ELISA: IL-6– | 3H-TdR: IL-6+, IL-1α–, IL-2–, IL-3–, IL-4–, IL-5–, G-CSF–, GM-CSF– | | IL-6 | 7 |
| MM-C1 | Flow cytometry: IL-6R+ | ELISA: IL-6– | 3H-TdR: IL-6+, IL-1α–, IL-2–, IL-3–, IL-4–, IL-5–, G-CSF–, GM-CSF– | | IL-6 | 7 |

*Continued on next page*

*Table 4.* (continued)

| Cell line | Cytokine receptor expression[a] | Cytokine production[d] | Proliferation response to cytokines[b] | Growth arrest/apoptosis response to cytokines[c,d] | Dependency on cytokines | Ref. |
|---|---|---|---|---|---|---|
| MM-M1 | | Northern: IL-6 mRNA–, RT-PCR: IL-6 mRNA–, bioassay: IL-6– | 3H-TdR: IL-6+ | | | 47 |
| MM-S1 | EPO rec.+ Flow cytometry: IL-6R+ | Northern: IL-6 mRNA–, RT-PCR: IL-6 mRNA–, ELISA: IL-6– | EPO+, 3H-TdR: IL-6+, IL-1α–, IL-2–, IL-3–, IL-4–, G-CSF–, GM-CSF–, IL-5+ | anti-IL-6 ab– (growth inh.) | IL-6 (partly dependent) | 7,111 47 |
| MM-Y1 | Flow cytometry: IL-6R+ | Northern: IL-6 mRNA–, ELISA: IL-6– | 3H-TdR: IL-6+, IL-1α–, IL-2–, IL-3–, IL-4–, IL-5–, G-CSF–, GM-CSF– | | IL-6 | 7 |
| OPM-1 | | Northern: TGFβ1 mRNA+ | | Dexamethasone–, TGFβ1–, TGFβ2+ (growth inh.) | | 56 112 |
| OPM-2 | | Northern: TGFβ1 mRNA– | | Dexamethasone+ (apoptosis) TGFβ1+, TGFβ2+ (growth inh.) | | 56 112 |
| OCI-My1 | Northern: IL-6R mRNA+ | Bioassay/ELISA: IL-6–, RT-PCR: IL-6 mRNA+, ELISA: IL-1β–, Northern: IL-1β–, GM-CSF– | IL-6– (colony formation) | anti-IL-6 ab– (inh. of prol.) | | 52 |

*Continued on next page*

*Table 4.* (continued)

| Cell line | Cytokine receptor expression[a] | Cytokine production[d] | Proliferation response to cytokines[b] | Growth arrest/apoptosis response to cytokines[c,d] | Dependency on cytokines | Ref. |
|---|---|---|---|---|---|---|
| OCI-My2 | Northern: IL-6R mRNA+ | Bioassay/ELISA: IL-6+ RT-PCR: IL-6 mRNA+ ELISA: IL-1β– Northern: IL-1β–, GM-CSF– | IL-6– (colony formation) 3H-TdR: IL-6+ | | | 52 |
| OCI-My3 | Northern: IL-6R mRNA+ | Bioassay/ELISA: IL-6+ RT-PCR: IL-6 mRNA+ ELISA: IL-1β– Northern: IL-1β–, GM-CSF– | IL-6– (colony formation) | anti-IL-6 ab– (inh. of prol.) | | 52 |
| OCI-My4 | Northern: IL-6R mRNA+ | Bioassay/ELISA: IL-6– RT-PCR: IL-6 mRNA– ELISA: IL-1β– Northern: IL-1β–, GM-CSF– | IL-6+ (colony formation) 3H-TdR: IL-6+ | | | 52 |
| OCI-My5 | Northern: IL-6R mRNA+ | Bioassay/ELISA: IL-6– RT-PCR: IL-6 mRNA– ELISA: IL-1β– Northern: IL-1β–, GM-CSF– | IL-6– (colony formation) | anti-IL-6 ab– (inh. of prol.) | | 52 |
| OCI-My6 | Northern: IL-6R mRNA+ | Bioassay/ELISA: IL-6– RT-PCR: IL-6 mRNA+ ELISA: IL-1β– Northern: IL-1β–, GM-CSF– | IL-6+ (colony formation) 3H-TdR: IL-6+ | | | 52 |

*Continued on next page*

Table 4. (continued)

| Cell line | Cytokine receptor expression[a] | Cytokine production[d] | Proliferation response to cytokines[b] | Growth arrest/apoptosis response to cytokines[c,d] | Dependency on cytokines | Ref. |
|---|---|---|---|---|---|---|
| OCI-My7 | Northern: IL-6R mRNA+ | Bioassay/ELISA: IL-6–; RT-PCR: IL-6 mRNA+; ELISA: IL-1β–; Northern: IL-1β–, GM-CSF– | IL-6+ (colony formation); 3H-TdR: IL-6+ | | | 52 |
| OH-2 | Flow cytometry: p55/p75 TNFR + | | 3H-TdR: TNFα+, LTα(+); IL-12–, M-CSF–, GM-CSF–, IL-1β–, IL-3–, IL-7–, IL-6+ IL-6/TNFα+ | TGFβ+ (growth inh.); neutr. anti-TNF ab– (IL-6 ind. prolif.) neutr. anti-IL-6 ab– (TNF ind. prolif.) | IL-6 and/or TNFα | 55 |
| PCM6 | | 113; RIA: PDGF–, EGF–, IL-1β– | | | IL-6 | 57 |
| RPMI8226 | RT-PCR: IL-6R/gp80 mRNA– | RT-PCR: IL-6 mRNA– | 3H-TdR: IL-6– | TNF+ | | 114 |
| | IL-6R mRNA+ | RT-PCR/Northern: IL-6 mRNA– | 3H-TdR: GM-CSF+ | | | 66 |
| | RT-PCR/bind. assay: IL-6R+ | Northern: M-CSF mRNA+ (3.5 kb) | 3H- TdR: IL-11– | | | 25 |
| | bind. assay: TNFR+ IGF-IR+ | Northern: IL-1β mRNA– Immunoassay: GM-CSF– | | | | 47 |
| | | | | | | 115,116 |
| | bind. assay: IL-11R– | | | | | 117 |
| | RT-PCR: GM-CSFRα mRNA+, | Northern/bioassay: IL-11– | | | | 104 |
| | GM-CSFRβ mRNA+ | | | | | 8,19,118 |

*Continued on next page*

*Table 4.* (continued)

| Cell line | Cytokine receptor expression[a] | Cytokine production[d] | Proliferation response to cytokines[b] | Growth arrest/apoptosis response to cytokines[c,d] | Dependency on cytokines | Ref. |
|---|---|---|---|---|---|---|
| TH | | IL-1–, IL-6–, TNFα– | 3H-TdR: IL-1–, IL-2–, IL-4–, CD28–, IL-6+ IL-6+ (stim. Ig production) IL-6+ (increased expression of CD28 and CD38) | | | 60 |
| U-1958 | Northern:IL-6R mRNA+, IFNγR+ | Northern:IL-6 mRNA– | IL-6+, IL-1β–, IL-2–, TNFα–, GM-CSF–, IGF-I± | IFNα+ (growth inh.) IFNγ+ (growth inh.) | IL-6 | 14,61, 107 |
| | | | IL-3–, insulin– | anti-IGF-IR ab+ (growth inh.) | | 105,106 |
| U-1996 | Northern: IL-6R mRNA+, IFNγR+ | Northern: IL-6 mRNA– | IL-6+ | IFNα– (growth inh.) IFNγ– (growth inh) | | 14,61, 107 |
| U-266 | Northern: IL-6R mRNA+ IFNγR+ RT-PCR:IL-6R/gp80 mRNA– IL-6R+ (binding assay) | In IL-6 indep. U-266-1984: Northern: IL-6 mRNA+ RT-PCR: IL-6 mRNA+ prot/bioassay: IL-6+ | In IL-6 indep. U-266-1984: IL-6–, IL-1β–, TNFα–, IL-2– 3H-TdR:IL-6/sIL-6R– IGF-I– | In IL-6 indep. U-266-1984: anti-IL-6R ab(PM1)+, IFNα+ IFNγ– (growth inh) anti-IGF-IR ab+ (growth inh) | IL-6 (U-266-1970) later indep. of IL-6 (U-266-1984) | 8,19,25, 47,61, 66,95, 104– 107, 109,115, 116, 118– 121 |

*Continued on next page*

*Table 4.* (continued)

| Cell line | Cytokine receptor expression[a] | Cytokine production[d] | Proliferation response to cytokines[b] | Growth arrest/apoptosis response to cytokines[c,d] | Dependency on cytokines | Ref. |
|---|---|---|---|---|---|---|
|  | Northern: IL-6R/gp130 mRNA+ IGF-IR+ bind assay: IL-11R– Northern: IL-11R– | Northern: M-CSF mRNA+ (3.5 kb) /biol.assay M-CSF+ RT-PCR: hrIL-17 mRNA+ Northern: IL-1β mRNA– Northern/bioassay: IL-11– In IL-6 dep. U-266-1970: Northern: IL-6 mRNA+ prot/bioassay: IL-6– | 3H-TdR: IL-11– IL-1–, IL-2–, IL-3–, IL-4–, IL-5–, IL-7–, GM-CSF– In IL-6 dep. U-266-1970: IL-6+ 3H-TdR: IL-1β–, TNFα–, IGF-I– | In IL-6 dep. U-266-1970: anti-IL-6R ab(PM1)+, IFNα+, IFNγ+ (growth inh) anti-IGF-IR ab+ (growth inh) |  |  |
| UTMC-2 | Flow cytometry: IL-6R+ | ELISA: IL-6– RT-PCR: IL-6 mRNA+, TNFα mRNA–, IL-1β mRNA– | IL-6+, IL-1β–, IL-4–, TNFα–, TNFβ– | IL-6 antisense+ (growth inh) |  | 65 |
| XG-1 | Flow cytometry: IL-6R(gp80)+ (gp130)+ RT-PCR/Flow cytom.: LIFRβ–, IL-10R+, IL-11R–, CNTFRα– | Northern/bioassay: IL-6– RT-PCR/ELISA: OM +, IL-6+, LIF–, IL-11–, IL-10– Northern: IL-1β– Immunoassay: GM-CSF– | IL-6/GM-CSF+, 3H-TdR: CNTF–, LIF–, IL-11–, IL-6/IL-3+, OM– IFNα+ | IL-6 antagonist+ (growth inh., apoptosis) | IL-6 | 9,122 123 124,125 126 104 8,127 128 |

*Continued on next page*

Table 4.   (continued)

| Cell line | Cytokine receptor expression[a] | Cytokine production[d] | Proliferation response to cytokines[b] | Growth arrest/apoptosis response to cytokines[c,d] | Dependency on cytokines | Ref. |
|---|---|---|---|---|---|---|
| XG-2 | Flow cytom: IL-6R(gp80)+ (gp130)+ RT-PCR/Flow cytom.: LIFRβ–, IL-10R+, IL-11R–, CNTFRα– | Northern/bioassay: IL-6– RT-PCR/ELISA: OM+, IL-6–, LIF–, IL-11–, IL-10– Northern: IL-1β– | IL-6/GM- CSF+ 3H-TdR: CNTF–, LIF–, IL-11–, IL-6/IL-3+, OM– IFNα+ | | IL-6 | 9,122 8 126 104 127,128 124,125 |
| XG-3P | Flow cytom: IL-6R(gp80)+ (gp130)+ | Northern/bioassay: IL-6– Northern: IL-1β mRNA– | IL-6/GM-CSF+ IL-6/IL-3+ | | IL-6 | 9,122 104 |
| XG-3E | Flow cytom: IL-6R(gp80)+ (gp130)+ | Northern/bioassay: IL-6– Northern: IL-1β mRNA– | IL-6/GM-CSF+ IL-6/IL-3+ | | IL-6 | 9,122 104 |
| XG-4[e] | Flow cytom: IL-6R(gp80)+ (gp130)+ RT-PCR/Flow cytom: LIFRβ+, IL-10R+, IL-11R+, CNTFRα– | Northern/bioassay: IL-6– RT-PCR/ELISA: OM –, IL-6±, LIF–, IL-11–, IL-10+ Northern: IL-1β– | IL-6/GM-CSF+ 3H-TdR: CNTF+, LIF+, IL-11+, OM+ IFNα+ | | IL-6 | 9,122 124,125 126 104 127,128 |
| XG-5 | Flow cytom: IL-6R(gp80)+ (gp130)+ | Northern/bioassay: IL-6– | IL-6/GM-CSF+ | | IL-6 | 9,122 |

*Continued on next page*

*Table 4.* (continued)

| Cell line | Cytokine receptor expression[a] | Cytokine production[d] | Proliferation response to cytokines[b] | Growth arrest/apoptosis response to cytokines[c,d] | Dependency on cytokines | Ref. |
|---|---|---|---|---|---|---|
| | RT-PCR: CNTFRα mRNA− | Northern: IL-1β− | IFNα+<br>IL-6/IL-3+ | | | 126<br>104,125 |
| XG-6 | Flow cytom: IL-6R(gp80)+<br>RT-PCR/Flow cytom:<br>LIFRβ+,<br>IL-10R+, IL-11R+,<br>CNTFRα− | Northern/bioassay: IL-6−<br>RT-PCR/ELISA: OM −,<br>IL-6, LIF−,<br>IL-11−, IL-10+ | IL-6/GM-CSF+<br>IL-6/IL-3+<br>3H-TdR: CNTF−, LIF−,<br>IL-11+, OM+ | | IL-6 | 9,122<br>127,128<br>124,125 |
| XG-7 | Flow cytom: IL-6R(gp80)+<br>(gp130)+<br>RT-PCR: CNTFRα<br>mRNA+ | Northern/bioassay: IL-6−<br>RT-PCR/ELISA: OM+,<br>IL-6±,<br>LIF−, IL-11−, IL-10+ | IL-6/GM-CSF+<br>IL-6/IL-3+ | | IL-6 independent | 9,122<br>127,128<br>124,125 |
| XG-8 | Flow cytom: IL-6R(gp80)+<br>(gp130)+<br>RT-PCR: CNTFRα<br>mRNA+ | Northern/bioassay: IL-6− | IL-6/GM-CSF+<br>IL-6/IL-3+ | | | 9,122<br>125 |
| XG-9 | Flow cytom:IL-6R(gp80)+<br>(gp130)+<br>RT-PCR: CNTFRα<br>mRNA+ | Northern/bioassay: IL-6− | IL-6/GM-CSF+<br>IL-6/IL-3+ | | | 9,122<br>125 |

[a] Receptor expression at mRNA (RT-PCR , Northern) or at protein level; binding assay, or flow cytometry analysis. + = expression, − = no expression.

[b] Effects of cytokines on proliferation or growth as measured by thymidine incorporation (3H-TdR) or cell counting. + = induction of proliferation/growth, − = no effect.

[c] Effects of cytokines on growth inhibition or apoptosis. + = growth inhibitory or apoptosis inducing effect, (−); no effect.

[d] PTHrP; parathyroid hormone-related protein, MDF; Macrophage derived conditioned media (CM), OSM/OM; oncostatin M.

[e] Cell line responded by 3H-TdR incorporation to IL-6, LIF, OM and was by RT-PCR suggested to express CNTFRα [125,128].

to be classifiable into three different categories: (1) those dependent on IL-6 only, (2) those responding to some other member of the family of IL-6 type cytokines (OM, LIF), and (3) those independent of an IL-6 type cytokine [14,26,103,122,132].

The IL-6 unrelated growth factor IL-10 can induce proliferation in MM cells [127]. IL-10 has been suggested to induce proliferation via the up-regulation of IL-11R expression and the IL-11 signal transduction pathway [124,127,128]. Also, the expression of LIFR$\beta$ has been suggested to be in-duced by IL-10. Thus, relating back to the classification of MM cell lines, exogenous IL-10 will induce proliferation in MM cell lines dependent on IL-6 only by inducing functional LIFR$\beta$ expression. Furthermore, an autocrine IL-10 loop may be part of the growth regulation of MM cell lines stimulated by the IL-6 type family of cytokines [124]. IL-10 was produced by some MM cell lines (XG-4, XG-6, XG-7) but not in others (XG-1, XG-2). This finding was supported by the presence of both autocrine IL-6 and an autocrine OM loop in one cell line (XG-7) growing autonomously without the addition of exogenous IL-6. As IL-10 and/or OM is frequently produced in freshly isolated MM cells, these cell lines seem to be representative of MM cells in vivo. The primary cells were therefore suggested to be responsive to OM by upregulating LIFR expression in response to IL-10. However, the relevance of these findings to growth in vivo is not known [92,103,124].

Insulin-like growth factors (IGFs) produced by bone marrow stromal cells also influence growth and survival in MM. IGF-I has an anti-apoptotic effect on MM cells, via PI3-K activation of PKB/Akt and subsequent phos-phorylation of the *Bcl*-2 related Bad protein. Bad is thereby sequestered and cannot initiate apoptosis [133,134]. IGF-I may improve the survival of CD138 (syndecan-1/B-B4) positive MM cells in vitro, and a recent report suggests that IGF-I is an important chemoattractant in MM [105,135]. IGF-I was reported to be an important growth and/or survival factor in MM cell lines, possibly acting in some cell lines by an autocrine mechanism [105]. All the cell lines examined express functional IGF-IR (LP1, Karpas 707, EJM, HL407 (early), HL407 (late), U266-1970 and U-266-1984) and IGF-ImRNA and protein (LP-1, Karpas 707, EJM) [105,106] (Table 4). IGF-IR was also found to be expressed on U-266 and RPMI 8226 [115,116]. Using IL-6 dependent MM cell lines (DP-6, ANBL-6, KAS-6/1 and KP-6), IGF-I was also demonstrated to potentiate the IL-6 stimulated proliferative response, perhaps via autocrine IGF-I production [102].

Several other cytokines stimulating the growth of MM cells have been identified. In contrast, no consistent pattern of MM cell responsiveness to cy-tokines has been demonstrated. Together, such cytokines (G-CSF, GM-CSF, IL-5, TNF$\alpha$ TGF$\beta$) may act as indirect MM cell growth factors either by in-ducing IL-6 production or IL-6R expression (IL-3) [92,130,136]. In line with the need for feeder cells for growth and survival, cytokine production by MM of IL-1$\alpha$/IL-1$\beta$, TNF$\alpha$/$\beta$ and TGF$\beta$ has also been suggested to stimulate paracrine IL-6 production by bone marrow stromal cells [8,137,138].

GM-CSF was added to IL-6 to establish the XG (1-9) cell lines [8,9]. Increased DNA synthesis was reported in RPMI 8226 in response to GM-CSF [117]. However, using immunoassays, GM-CSF was not expressed in the MM cell lines (RPMI 8226, U-266, JJN-3, EJM, LB-831, XG-1) [8] (Table 4). A truncated form of monocyte-macrophage (M)-CSF that is functionally active has been identified in malignant plasma cells and in cell lines (RPMI 8226 and U-266) [66]. A synergistic growth promoting effect of IL-3 with IL-6 was demonstrated in several cell lines [9,139,140]. In some IL-6 dependent cell lines TNF$\alpha$ induces proliferation (DP-6, KAS- 6/1, KP-6) and in one case a TNF$\alpha$ dependent MM cell line has been described (OH-2) [55].

IL-1 is considered important in MM as it plays a role in osteoblast activation (OAF). More importantly, IL-1$\beta$ may control IL-6 production and the expression of adhesion molecules of stromal cells implicated in homing [104]. Some controversy exists concerning the synthesis of IL-1 by MM. IL-1 production was not detected in CD138 (syndecan-1) positive cells, while the expression of IL-1 and IL-6 was found in the CD138 negative population of non-myeloma cells [141]. However, using in situ analysis, weak expression was observed of IL-1$\beta$ and of both IL-1$\alpha$ and IL-1$\beta$ in 10/31 and 22/31 samples of freshly isolated MM cells [104]. In cell lines, IL-1$\beta$ transcript was not found in nine MM cell lines (including XG-1, XG-2, XG-3, XG-4, XG-5, EJM, JJN-3, RPMI 8226, U-266) (Table 4).

The use of IFN$\alpha$ in patients is controversial [130,142]. The heterogenous response of MM cells to IFN$\alpha$ in vitro may reflect the variable response in vivo. Some in vitro studies have shown a growth stimulatory effect of IFN$\alpha$ on primary MM cells and MM cell lines [126,143], but growth inhibition and cytotoxicity has also been described [144,145]. This effect seems to be dose dependent. Low concentrations of IFN$\alpha$ are stimulatory and high concentrations are inhibitory [144]. In the MM cell lines U-1958 and U-266-1970, a growth inhibitory effect of IFN$\alpha$ was seen at both low and high concentrations, while U-1996 was refractory to IFN$\alpha$ [107]. In U-266, the growth inhibition by IFN$\alpha$ was suggested to be due to the loss of surface IL-6R and a downregulation of gp130 disrupting the autocrine loop [121]. The IL-6 dependent cell lines KP-6 and ANBL-6 are also growth-inhibited by IFN$\alpha$ [101]. IFN$\alpha$ stimulated proliferation of 4/5 established cell lines (XG-1, XG-2, XG-4, XG-5) initially dependent on IL-6 for growth [126]. In these cell lines IFN$\alpha$ stimulation was suggested to lead to autonomous growth by autocrine IL-6 production [126]. The stimulatory effect of IFN$\alpha$ has been confirmed in another cell line, KP-6, that is growth stimulated by IFN$\alpha$ [26,101]. However, the mechanism of growth stimulation has been challenged, as upregulation of IL-6 or IL-6 R was not considered to be a pathway for IFNα stimulated growth [101].

IFNγ generally inhibits growth in primary cells derived from MM patients and in IL-6 dependent MM cell lines [107,146,147].


## 5. GENETIC CHARACTERIZATION

Cytogenetic studies show aneuploidy in MM. Extensive structural and numerical chromosomal alterations have been observed, including loss of chromosomes and chromosome fragments in 30–50% of the cases and the presence of marker chromosomes [148–150]. Fluorescence in situ hybridization has identified cytogenetic abnormalities in 80–90% of bone marrow cells from patients with MM [151,152].

The hallmark genetic lesion in many B cell tumors is a translocation, often involving a proto-oncogene, to the Ig heavy chain (IgH) locus (14q32), or less frequently involving one of the IgL loci [91]. A 14q32 translocation and/or illegitimate switch recombination fragments (ISRF) were seen in 19/21 cell lines analyzed, indicating that the translocation to the IgH locus is an almost universal event in MM cell lines [153].

Compared to other B cell tumors, MM is genetically very heterogenous. This heterogeneity is obvious both between and within MM clones and is reflected by the diversity of chromosomal partners to the IgH locus, including 1p13, 6p25, 8q24, 12q24, 16q23, 21q22, 11q23, 1q21, 3p11, 6p21, 7q11 and 18q21, as deduced from karyotypes of tumors and cell lines [91,153]. The translocation breakpoints cloned from eight MM lines were demonstrated to involve six different loci; 4p16 (JIM-3, KMS-11), 6 (SK-MM-1), 8q24 (KMM-1), 11q13 (SK-MM-2, KMS-12, U- 266), 16q23 (JJN-3), and 21q22 (KMM-1) (Table 5) [153,154].

The IgH translocation most frequently involves 11q13 (*cyclin D1*), 4p16 (*FGFR3*) and 16q23 (c-*maf*). In the established MM cell lines these three loci are involved in about 25% of the lines and are associated with ectopic expression of cyclin D1, fibroblast growth factor receptor 3 (FGFR3) gene and the c-*maf* oncogene, respectively. In 2/3 MM cell lines overexpressing cyclin D1, there is a translocation t(11;14)(q13;q32). The expression of cyclin D1 mRNA was found in MM-M1, KMS-12 and SK-MM-2. However, in the MM-M1 cell line the candidate translocation breakpoint 11q13 into the switch region of IgH loci was not identified [156]. FLAM-76 and U-266 overexpress cyclin D1 mRNA, and in U-266 the translocation involves the switch region of IgH loci [153]. The overexpression of cyclin D1/PRAD1 gene and an amplification of the cyclin D1 gene in the absence of t(11;14) was demonstrated in the KHM-11 cell line, while it was absent in cells of

*Table 5.* Multiple myeloma cell lines: genetic characterization

| Cell line | Cytogenetic karyotype | Unique translocations, translocations involving 14q32 and/or ISRF[a] | Unique gene alterations aberrant gene expression[b] | Ref. |
|---|---|---|---|---|
| ACB-885 | 46XY, −1, −2, −6, −7, −8, −10, −12, −13, +21, −22, +8mar t(1;10)(1qter-1q22::10p11.2-10qter), t(2;10)(2qter- −2p11.2::10p11.2-10pter), t(1;9)(1qter-1q11::9p13- 9p22), t(2;8)(2pter-2p11.2::8q24-8pter), t(1;9)(1pter- 1p32::9p21-9q22), t(3;12)(p?;q24) | | elevated c-*myc* mRNA (no RFLP) | 20 155 |
| ACB-1085 | 45XY, −1, −2, −6, −7, −8, −10, −12, −13, +21, −22, +8mar, no discernable difference from ACB-885 | | elevated c-*myc* mRNA (no RFLP) | 20 155 |
| AD3 | 109, XXXY, −Y, −1, −1, −4, +7, −8, −8, −8, −9, −9, −10, −11, +12 −12, −13, −13, −14, −14, −15, −16, +19, +20, +20, +22, +22, del(1)(q21), del(1)(q21), +der(1)t(1;7)(q11;p11.2), +der(1)t(1;7)(q11;p11.2), +der(1)t(?;1)(1;?) (?p32q22;?), +der(1)t(?;1)(1;?)(?;p32q22;?), der(2) t(2;?)(q23;?), del(5)(q11.2q35), der(5)t(5;?)(q11.2;?), +der(5;?)(q11.2;?), i(6q), der(6)t(6;?)(q21;?), der(8) dic(1;8)(8pter-8q24.1::1p13-1q42:), +der(8)dic(1;8) (8pter-8q24.1::1p13-1q42:), +der(8)t(1:8)(q25:q24), +der(8)t(8;?)(q24:?), +der(9)t(9;?)(p22;?), der(11) t(11;?)(q21;?), +der(13)t(1:13;?)(13qter-13p11::?:: 1p11-1q42:), +der(13)t(13;?)(p11;?), +der(14)t(8;14) q24.1;q32), +der(14)t(8;14)(q24.1;q32), +der(14)t(8;14) (q24.1;q32), +der(15)t(9;15)(q13;p11), +der(15)t(9;15) (q13;p11), i(21q), +7mar | 14q+, 8q+ | germline c-*myc* | 21 |

*Continued on next page*

*Table 5.* (continued)

| Cell line | Cytogenetic karyotype | Unique translocations, translocations involving 14q32 and/or ISRF[a] | Unique gene alterations aberrant gene expression[b] | Ref. |
|---|---|---|---|---|
| AMO1 | pseudodiploid 2p+, 8q+, 10q+, 12q+, 12p-, 14q+, 15p+ | | | 22 |
| ANBL-6 | near diploid population 44, X-X, -1, t(3;9)(p21;q12), der(6) t(1;6)(q11;q13), der(?8)t(8;?)(?p11.1;?), -9, der(10) t(9;10)(q13;q26), -13, -14x2, der(15)t(15;?)(q1?5;?), -17, -22.+7mar[cp3], near tetraploid population 87-88, XX, -Xx2, -1x2, -5, der(6) t(1;6)(q11;q13)x2, der(?8)t(8;?)(?p11.1;?)x2, -9x2, der(10)t(9;10)(q13;q26)x2, -13x2, -14x3, -15, der(15) t(15;?)(q1?5;?), -17, x2, -22x2, +15-17mar[cp4] | ISRF/translocation into IgH switch region involving 16q23 | *c-maf* mRNA overexpression | 23 153 170 |
| delta-47 | 46, X, -Y, -1, +der(1)t(1;?)(p34;?), -2, +der(2)t(2;?)(p23;?), -8, -10, -11, -11, -12, -14, +der(14)+t(14;?)(q34;?), +6mar | 14q+, no translocation into IgH switch region | | 24 153 |
| DOBIL-6 | 47, X, -X, der(1)t(1;3)(p36.1-p25)t(1;22)(q21;q13.1), der(2;10)(p11.2;p11.2)del(2)(q32.2q35), inv(3)(q25 q29), add(3)(p25), add(6)(p21.3), +9, der(10)t(2;10)(p11.2; p11.2), der(11)t(6;11)(p21.3;q13.1), t(12;16)(p11.2; q12)der(14)t(11;14)(q13;q32.1), der(17)t(14;17)(q24; q21), i(17)(q10), +18, add(19)(p13.3), der(19)t(17;19)(q21; q13.3), del(20)(q11.2q13.1), der(22)t(1;22)(q21;q13.1) | t(11;14)(q13;q32) | PRAD1-IgH (cyclin D1) mRNA not overexpressed | 25 |
| EJM | 48X, +1, del(3q), -4, +5, del(6)(q21), +9p, +9p, +11p, +11p, +12q, -13, -13, -14, iso(14q), -15, -15, -16, -16, -18, +18p, +20, -21, -22, +7mar | | *p53*mut (ex5) LOH expression of *bcl-2*+ | 27 |

*Continued on next page*

*Table 5.* (continued)

| Cell line | Cytogenetic karyotype | Unique translocations, translocations involving 14q32 and/or ISRF[a] | Unique gene alterations abberant gene expression[b] | Ref. |
|---|---|---|---|---|
| FLAM-76 | 43, X, Y, −8, −13, −14, −21, −22, t(11;14)/q13;q32), +2mar | t(11;14)(q13;q32), 14q+, 8q+ 11q13, no translocation into IgH switch region | cyclin D1mRNA overexpression, selective expression of one c-*myc* allele | 28 153 157 |
| FR4 | 67, X, −Y, +3, +7, +9, +11, +12, −13, −13, −14, −15, +16, +17, +18, +19, +20, +22, +22, del(1)(q21), del(1)(q21), +der(1)t(1;7)(q11;p11.2), der(1)t(?;1)(;?)(?;p32q22:?), del(5)(q22q35), +i(6q), der(8)dic(1;8)(p13;q24.1), +der(8)t(8:?)(q24:?), der(13)t(1;13)(q11;p11), +der(13) t(1;13)(q11;p11), der(14)t(8;14)(q24.1;q32), der(14) t(8;14)(q24.1;q32), der(15)t(9:15)(q13;p11), i(21q), 3mar | | germline c-*myc* | 21 |
| JIM-1 | no detected t(14:18) | | expression of *bcl-2*− | 30 |
| JIM-2 | hypotetraploid karyotype | | expression of *bcl-2*(+) | 30 |
| JIM-3 | | 14q+, translocation into IgH switch region involving 4p16.3 | expression of *bcl-2*+ FGFR3 mRNA− | 30 153,156 |
| JIM-4 | | | expression of *bcl-2*− | 30 |
| JJN-1 | 40XX, −7, +der(7)t(7;11)(q32;q13), −9, −10, −11, −12, −13, −14, −16, −20, −20, +mar, del(6)(q25), del(8)(p21), 14q+ | | | 4 |
| JJN-2 | 75XXX, del(Xq), −1, 1q+, Iq−+, −2, −3, del(3)(q23), 5p+, 5p+ del(6)(q15), del(6)(q15), −7, −7, +der(7)t(7;11)(q32;q12), del(8)(p21), del(8)(p21), −9, −9, −10, −10, −11, −11, −12, −12, −13, −13, −14, 14, +14q+, +14q+, −15, −16, −16, −17, −20, del(20)(p11), +4mar | | | 4 |

*Continued on next page*

*Table 5.* (continued)

| Cell line | Cytogenetic karyotype | Unique translocations, translocations involving 14q32 and/or ISRF[a] | Unique gene alterations aberrant gene expression[b] | Ref. |
|---|---|---|---|---|
| JJN-3 | | 14q+, translocation into IgH switch region involving 16q23.1 | *p53*wt | 4 |
| | | | selective expression of one | 157 |
| | | | c-*myc* allele | 153 |
| | | | c-*maf* mRNA overexpression | 170 |
| | | | expression of *bcl*-2+ | 158 |
| Karpas 620 | 76X, +7 mar der from chr 1, 8, 11, 13, 14, 17 t(1;14)(q11;q32.3), t(1;17)(q11;p13.1), t(1;11) (q32.1;q13.3), t(8;11)(q24.22;q13.3), t(8;14) (q24.1;q32.3), t(11;13)(q13.3;q14.3)+ numerical alter. incl. chr 2, 3, 4, 5, 6, 9, 12, 22 and 7, 15, 16, 18, 19, 20, 21 | | | 31 |
| Karpas 707 | hypodiploid 45 Ph+, monosomy 6, 12 and 16, 4p+, +6p, monosomy 5, 13, 17 and trisomy 9, +2mar, 45XY, Ph+, 4p+, +6p, -6, -12, -16, t(6q;7q) | Philadelphia chr Ph+ (q1.1) | germline c-*myc* | 32 |
| | | | germline *bcl*-2 | 120 |
| | | | expression of *bcl*- 2+ | 159 |
| KHM-11 | 81-95 add (X)(q22), add(X)(q12), der(1)t(1;19)(p11;q11) add (1)(q21)x2, der(1)t(1:8)(q10:q10)X2, t(2:5)(q35;q11) x2, dup(3)(q21:p25), i(11)(q10), der(15)t(15:21)(q10, q10) x2, mar1 and mar2 | no detectable t(11;14) or t(11;22) | cyclin D1/PRAD amplification and overexpression | 160 |
| | | | overexpression | 152 |
| KMM-1 | 47, X, Y, 1q+, -2, +t(1;2)(cen;cen), +7, 12q+, 14q+, +mar | translocation into IgH switch region involving 21q22.1 and 8q24.13 | c-*myc* mRNA overexpression | 35 |
| | | | | 161 |
| | | | | 153 |

*Continued on next page*

*Table 5.* (continued)

| Cell line | Cytogenetic karyotype | Unique translocations, translocations involving 14q32 and/or ISRF[a] | Unique gene alterations aberrant gene expression[b] | Ref. |
|---|---|---|---|---|
| KMS-5 | Triploid, no karyotype available | | c-*myc* mRNA overexpression | 161 |
| KMS-11 | Triploid, no karyotype available | missense mut in codon 373 of FGFR3 | c-*maf* overexpression | 37 |
| | | translocation into IgH switch region involving 4p16.3 and 16q23 | selective expression of one c-*myc* allele | 161 |
| | | | c-*myc* mRNA overexpression | 153,170 |
| | | | | 156 |
| | | | FGFR3 mRNA overexpression | 157,168 |
| KMS-12-PE | Hypodiploid 41, no karyotype available | t(11;14)(q13;q32) t(11;9)(q13;q34) | cyclin D1mRNA overexpression | 153 |
| | | 11q13, translocation into IgH switch region involving 11q13 and 14 | elevated c-*myc* mRNA | 161 |
| KMS-12-BM | Hypodiploid 41, no karyotype available | t(11;14)(q13;q32) t(11;9)(q13;q34) | elevated c-*myc* mRNA | 161 |
| KMS-18 | 42, add(1)(q32), add(10)(q24), add(17)(p11) | t(4;14)(p16.3;q32.3) | | 37 |
| KPMM2 | 46, XX, der(1;19)(q10;q10), t(3;14)(q21;q32)-4, t(6;11)(p12;p15), der(10)add(10)(p13)dic(9;10)(q10;q26), +16 | t(3;14)(q21;q32)(3q21/CDCL1/BM28) | | 38 |
| L363 | 49, X, +8M, −5, −6, +7, −8, −8, 14q+, −22 | | germline c-*myc* | 39 |
| | | | germline *bcl-2* | 120 |
| | | | expression of *bcl-2*+ | 159 |

*Continued on next page*

*Table 5.* (continued)

| Cell line | Cytogenetic karyotype | Unique translocations, translocations involving 14q32 and/or ISRF[a] | Unique gene alterations abberant gene expression[b] | Ref. |
|---|---|---|---|---|
| LB-831 | t(1;?)(q32;?), t(1;15)(p11;p11), del(3)(p21;p25), t(5;?)(q35;?), del(7)(p15;), t(7;?)(p22;?), ?HSR 7(p22), i(8q), inv(11)(p11;q13), t(13;13)(p12;p12), +3, +4, +4, +5, +6, +8, −9, +10, +11, +14, −16, −17, +18, +19, +20, −22 | | *p53*mut(ex5) | 40 |
| | | | | 158 |
| LB-832 | t(1;?)(q32;?), t(1;15)(p11;p11), del(3)(p21;p25), t(5;?)(q35;?), del(7)(p15;), t(7;?)(p22;?), ?HSR 7(p22), i(8q), inv(11)(p11;q13), t(13;13)(p12;p12), +3, +4, +4, +5, +6, +8, −9, +10, +11, +14, −16, −17, +18, +19, +20, −22 | | | 40 |
| LB 84-1 | Hyperdiploid -X, +2, −4, −14, +17, +18, +20, +21, +22, del(1)(p36), t(2;?)(37;?), del(3)(q22), dup(3)(q26:q29), dup(3), del(5)(p14), t(5;?)(q35), del(6)(q15), del(6)(q21), del(7)(q31) | | germline c-*myc*(ex 1) | 79 |
| LOPRA-1 | 3n+−, 70, XX, −X, −1, −4, −6, −8, −13, −16, +7, +18, +21, +i(1q), +i(1q), +6q−, +3mar | | | 41 |
| LP-1 | 73, XX, dup(1)(p13:p23), +3, +3, inv4(p12:p16), −4, der5 t(5;?)(q31;?), +5, del(6)(q23), +del(6)(q23), +7, +7, +8, +8, der(9)t(9;12)(q34:q22), +9, +10, +10, +11, der(12)t(12:13) (p13:q11), der(12), −13, −13, +15, +15, +16, +17, +17, +18, +19, +20, +21, +21, +22, +22, +3mar | t(4;14) | altered expr. of c-*myc* *13.2kb* c- *myc* transcr. (enz polymorph/RFLP) c-*maf* mRNA expression(+) | 42 |
| | | | | 170 |
| MEF-1 | 50, X, −X, +der(3)add(3)(p22)add(3)(q13), t(4;6)(q21:p23), +5, add(7)(q36), +8, t(11:14)(q13, q32), +der(14)t(11;14) (q13:q32), der(19)dic(1:19)(p13:p13.3) | t(11;14) | *p53* mut (ex7) cyclin D1 overexpression germline c-*myc* | 44 |

*Continued on next page*

*Table 5.* (continued)

| Cell line | Cytogenetic karyotype | Unique translocations, translocations involving 14q32 and/or ISRF[a] | Unique gene alterations abberant gene expression[b] | Ref. |
|---|---|---|---|---|
| MM.1 | 44, XX, −8, −13, −14, −16, −21, del(1)(p13p22), t(2:?)(q37;?), t(3:?)(p25:?), t(6:?)(q22:?), t(12;14)(q24.3;q32.3), +der(8)t(8;13) (q21;q22), +der(16)t(8;16)(p21.1;q12), +der(21)t(1;21) (q12;p13), t(2:?)(q37;?), del(8) (8pter−8q11;13q31−13qter), clonal additional markers | t(12;14)(q24.3;q32.3)  12q24, translocation into IgH switch region involving  16q23 | c-*maf* mRNA overexpression | 45  53  170 |
| MM-M1 | no available karyotype | 11q13, no translocation into IgH switch region | cyclin D1 mRNA overexpression selective expression of one c-*myc* allele | 153  157 |
| MT3 | ?74, 1q−, 2q+, 3q−, 4q+, 13q−, 14q+, 22q−, mar1, mar2, +DM, −2, −5, −14, −15, +1, +3, +5, +6, +7, +8, +11, +12, +13, +14, +15, +16, +17, +18, +19, +20, +21, +22 | t(?;22)(?;q1.1-q1.3) | germline c-*myc* | 49 |
| NCI-H929 | 90, XX 8q+ dup(1)(q11-25), t(10q, 12p), del(12p), +1mar | translocation into IgH switch region involving 4p16 | germline/rearranged c-*myc*  (no enz polymorph/RFLP) c-*maf* mRNA expression FGFR3 mRNA expression | 50  153  168 |
| NOP-2 | 47(47, X, −Y, inv dup del(1)(p13-q21 q21), +6, +7, t(8;22) (q24;q11), t(11;14)(q13;q32), −15, +der(15)t(15;1) (p11;p22) | t(11;14)(q13;q32) | | 51 |

*Continued on next page*

Table 5.  (continued)

| Cell line | Cytogenetic karyotype | Unique translocations, translocations involving 14q32 and/or ISRF[a] | Unique gene alterations aberrant gene expression[b] | Ref. |
|---|---|---|---|---|
| OCI-My5 | no karyotype published | t(14;16), no ISRF/Ig translocation to IgH switch region | p16 wt (PCR), p16 protein– MDM2 protein overexpression | 68 162 |
| | | | c-maf overexpression | 153 170 |
| OPM-1 | 76, 79 +del(1)(p32), +del(1)(p32), +der(3)t(1;3)(p22;p21), +der(3)t(3;7)(p13;p15), t(4;14)(p16;q13), +del(7)(p15), +der(8)t(1;8)(q12;q22), der(8)t(1;8)(q12;q22), +del(14)(q13), +der(22)t(1:22)(q12;p13) | | germline c-myc germline bcl-2 | 56 120 159 |
| OPM-2 | 65~75 +del(1)(p13), +del(1)(p13), +del(1)(q31), +der(1)t(1;2)(p13;p13), +del(5)(q13), +del(10)(q22), +del(10)(q22), +del(10)(q22), t(1:22)(q12;p13), der(22)t(1;22)(q12;p13) | translocation into Ig switch region involving 4p16.3 missense mutations of FGFR3 | germline c-myc germline bcl-2 selective expression of one c-myc allele FGFR3 mRNA overexpression | 56 120 157 168 159 |
| PCM6 | 45X, –X, 1p+, 1p–, +1q–, t(2;8)(q23;p23), 3p–, 6q–, 12q–, 14q+, –16, –17, +mar | | | 57 |
| RPMI 8226 | 58-67, t(1;14)(p13:q32), del(2)(q35), t(3;?)(q29), t(5;6)(p13;p12), del(6)(q15), del(6)(q11), t(9;?)(p24;?), del(11)(q23), t(11;?)(p11;?), del(11)(p11), t(17;?)(q25;?), HSR(21)(q22) | translocation into IgH switch region + (unidentified partner) 1p13, variant translocation t(16;22) | p53mut (ex8) LOH p16wt(PCR), p16 protein– c-maf mRNA overexpression K-ras mutation codon12 MDM2 protein overexpression | 2 163 58 162 153 170 158 164 |

Continued on next page

*Table 5.* (continued)

| Cell line | Cytogenetic karyotype | Unique translocations, translocations involving 14q32 and/or ISRF[a] | Unique gene alterations aberrant gene expression[b] | Ref. |
|---|---|---|---|---|
| SK-MM-1 | 33–45X, –X, 1p–, +2, t(4;19), 6p–, t(8q;10q), –9, –10, t(11q; 15q), +11p, t(13p;1q), 14q+, 14p+, –15, –18, –19, –20, +22 | 14q+, translocation into IgH switch region involving 6 and 6p25 | germline c-*myc* / IRF4 mRNA overexpression | 59 / 153 / 165 |
| SK-MM-2 | 69–84 +13-15mar, +1, +1, 1p–, 1p–, +2, 2q+, +3, +5, +6, +6, +6, (+6), 6q–, 6q–, (6q–), +9, +10, +11, +12, 14q+, –14, +15, –17, +19, +20, +21 | 11q13, translocation into IgH switch region involving 11q13.3 | germline c-*myc* / cyclin D1 mRNA overexpression | 59 / 153 |
| TH | 69–75 XX der(1)t((1;9)(1qter-1p11::9p21-qter), del(2)(p24) x2, der(5)t(5;18)(p15;q23), inv(8)(p23q13), der(10), t(10;?)(q26;?), der(13)t(13;21)(p11;p11)x2, 14q+, +14q+, 2–4 struct abn chr 18 with breakpoints in 18q21 and 18q23 | | germline *bcl-2* | 60 |
| U-1957 | 46, del(1)(p22), t(1;16)(q24;q24), t(6;13)(q21;q14), 10p+, del(10)(p12), 14q+, +7 | | | 166 |
| U-1958 | | | germline c-*myc* / germline *bcl-2* / expression of *bcl-2*+ | 120 / 159 |
| U-1996 | 82, 2xint del(1)(p22-p34), int del(2)(q21-q31), t(3;12), 2x4p+, t(6;?), del(6)(q21), del(8)(p21), del(12)(q24.1), t(13;?), 14q+, 2x14q+, 16q+, 14?, der(16)? | | germline c-*myc* / germline *bcl-2* / expression of *bcl-2*+ | 166 / 120 / 159 |
| U-2030 | aneuploid 60: t(1;12)(q21;p13)x2, t(1;?)(p34;?), del (1)(q11), 1q (cen-ter), int del (3)(q21-q25), t(5;16)(q11;q22), iso (6q), 8p-, t(9;?)(p22;?), 10p+, iso(11q), del(11)(q21), 16q+, 19p+, +12mar | | expression of *bcl-2*+ | 62,166 |
| U266B1 | 43–44, 1p+(1qter-p34:::?), t(2;?)(2qter-p25::?), 3q+, | translocation into IgH switch region involving 11q13.3 | *p53*mut (ex5) LOH | 166 |

*Continued on next page*

*Table 5.* (continued)

| Cell line | Cytogenetic karyotype | Unique translocations, translocations involving 14q32 and/or ISRF[a] | Unique gene alterations, aberrant gene expression[b] | Ref. |
|---|---|---|---|---|
| (U-266) | 3q−, 4p−, del(6q23), 7q+(q36), del(8)(p23), 9q+(9pter− q34 ::?), iso(10q), 11q+(11pter-q22::?), 12p+, del(14)(q24), 16-like, dic | | *p16*wt (PCR), *p16* prot− (meth.) | 163 |
| | | | germline *c-myc* (*c-myc* mRNA−) | 120 |
| | | | L-*myc* mRNA | 153 |
| | | | cyclin D1 mRNA overexpression | 158 |
| | | | biallelic loss of Rb1 Rb1 mRNA− | 167 |
| | | | germline *bcl-2* (1984) expression of *bcl-2*+ | 159 |
| | | | 4-fold ampl *bcl-2* (1970) overexpression of *bcl-2*+ | |
| UCD-HL461 | 80-83 X, multiple abnormalities includ. abn 1, der(11), der(13) +1mar | | | 63 |
| UMJF-2[c] | 46, XY/47XY, +12/92, XXYY | | | 64 |
| UTMC-2 | 43 t(1;5)t(2;15) no detailed karyotype | translocation into IgH switch region involving 4p16.3 | FGFR3 mRNA expression | 65 |
| | | | | 153 |
| XG-1 | hypodiploid, monosomy 13, t(11;14;?)(q13;q32;?) | | *p53* mut (ex5) | 9 |
| | | | N-*ras* mut codon12 | 164 |
| | | | | 158 |
| XG-2 | hyperdiploid, t(11;14;?)(q13;q32;?) | | *p53* mut(ex5) | 9 |
| | | | K-*ras* mut codon12 | 164 |
| | | | | 158 |
| XG-3P | hyperdiploid der14 t(14;?)(q32;?) | | *p53*wt | 9 |
| | | | | 158 |

*Continued on next page*

*Table 5.* (continued)

| Cell line | Cytogenetic karyotype | Unique translocations, translocations involving 14q32 and/or ISRF[a] | Unique gene alterations aberrant gene expression[b] | Ref. |
|---|---|---|---|---|
| XG-3E | hyperdiploid | | | 9 |
| XG-4 | hyperdiploid | | *p53* mut(ex5) | 9 |
| | | | | 158 |
| XG-5 | hypodiploid, t(11;14)(q13;q32), t(8;14)(q24;q32) | | *p53* mut(ex8)LOH | 9 |
| | | | | 158 |
| XG-6 | hypodiploid, 13q-, der14 t(11;14)(q13;q32) | | | 9 |
| XG-7 | hypodiploid, monosomy 13 | t(6;14)(q32) | IRF4 mRNA overexpression | 9 |
| | | translocation breakpoint 6p25 | | 165 |
| XG-8 | hyperdiploid, 13q-, der14 t(11;14)(q13;q32) | | | 9 |
| XG-9 | | | | 9 |

[a] ISRF – illegitimate switch recombination fragment (translocation into IgH switch region involving unknown or identifiable partners). By Southern blot analysis the translocation breakpoints were identified.
[b] RFLP – restriction fragment length polymorphism; mut – mutation; ampl – gene amplification; LOH – loss of the wild type allele.
[c] EBV positive (Drexler, personal communication).

the pleural effusion from which the cell line was established [160]. The t(11;14)(q13;q32) translocation is associated with aggressive disease and poor prognosis [91].

In four out of 21 MM cell lines the FGFR3 gene is dysregulated by t(4;14)(p16.3;q32.3), and the translocation breakpoint 4p16 was cloned from these cell lines (KMS-11, JIM-3, NCI-H929, OPM-2) (Table 5). With the exception of JIM-3, the FGFR3 was expressed in KMS-11, NCI-H929, OPM-2 and UTMC2, the latter cell line also harboring the t(4;14). Eight cell lines (RPMI 8226, ark, SK-MM-1, delta-47, H1112, KMM-1, TH and U-266) were negative for the translocation and displayed only weak expression of FGFR3 by RT-PCR (Tables 5 and 7). In ANBL-6, FLAM-76, SK-MM-2, JIM-3, JJN-3, KMS-12, MM-M1, MM.1 and OCI-My5, no FGFR3 mRNA was detected [168]. In the KHM-11 cell line, FGFR3 was translocated into the VH domain of the IgH locus and expression of the FGFR3 was confirmed [160,168]. An activating mutation of FGFR3 resulted in progression and ligand independent stimulation seems to occur frequently in MM with t(4;14). In two of the four MM cell lines examined (KMS-11, OPM-2), activation as a result of somatic mutation was found (Table 5). In the MM cell lines LP-1, UTMC-2, NCI-H929, JIM-3 and OPM-2, the IgH translocation simultaneously dysregulates two genes with oncogenic potential; FGFR3 on der(4) and Multiple Myeloma SET domain (MMSET) on der(4) [169].

The t(14;16)(q32.3;q23) was present in 5/21 MM cell lines (KMS-11, MM.1, JJN-3, ANBL-6). These cell lines also overexpress c-*maf*, as do OCI-My5 and RPMI 8226, but no candidate IgH switch translocation breakpoint fragment has been identified. FISH analysis revealed the presence of a t(14;16) in OCI-My5. In RPMI 8226, the c-*maf* was suggested to be involved in variant translocations e.g. t(16;22). Also in the NCI-H929 cell line c-*maf* mRNA was detected. Other cell lines (U-266, ark, SK-MM-1, KMM-1, OPM-2, FLAM-76, delta-47) were negative for expression of c-*maf* [170].

In a few cases the IgH translocation in MM may involve other partners, including 8q24 (c-*myc*) in less than 5% of the cases, 18q21 (*bcl*-2), 11q21-24 (mixed lineage leukemia *MLL*) and 6p21.1 (*IRF4*). The t(6;14)(p25;q32) translocation breakpoint was cloned from the SK-MM-1 cell line and was found to map to 6p25, near IRF4/(multiple myeloma oncogene MUM1)/ICSAT/LSIRF), a member of the interferon regulatory factor (IRF) family of transcription factors. This abnormality was also found in XG-7, but not in other cell lines (RPMI 8226, U-266, EJM, XG-1, XG-2, XG-4, XG-5, XG-6 (Table 5). IRF-4 mRNA was also overexpressed in these cell lines as compared to cell lines not carrying the alterations of the MUM/IRF4 [165]. However, in most of the cases the chromosome partner is not identified (14q+) [154].

This search for genetic alterations involved in the pathogenesis of MM has led to the identification of proto-oncogenes with a possible pathogenetic role. One of the candidates is c-*myc*, deregulated as a consequence of the reciprocal translocation of c-*myc* to one of the Ig loci in another human B cell tumor, Burkitts lymphoma. A similar translocation, resulting in a deregulated c-*myc*, has been implicated in the pathogenesis of rat immunocytoma and murine plasmacytoma [171]. Translocations leading to the juxtaposition of c-*myc* to one of the Ig loci are rare in MM [172–176]. In MM cell lines, only two cases (LP-1 and NCI-H929) of structural c-*myc* rearrangements have been reported [42,50,177]. In the absence of structural alterations, elevated expression of c-*myc* has been reported in freshly isolated MM cells [155,173,178,179]. In the cell lines ACB-885, ACB-985, ACB-1085, RPMI 8226 [155], elevated c-*myc* mRNA in the absence of restriction fragment length polymorphism (RFLP) and/or gene amplifications was reported (Tables 5 and 7). Also, an elevated expression of c-*myc* mRNA and protein was demonstrated in KMM-1, KMS-5, KMS-11, KMS-12PE and KMS-12BM in the absence of detectable translocations or hypomethylation [161]. In one MM cell line (KMM-1), a translocation involving the IgH locus and 8q24 was detected by Southern blotting [154]. In the absence of detectable rearrangements c-*myc* mRNA and protein are highly expressed in MM cell lines (U-1996, Karpas 707, L363 and OPM-1), compared to cell lines of other origin harboring *myc* amplification [120]. In the U-266 cell line and its IL-6 dependent subclone U-266-1970, c-*myc* mRNA was not detected, but L-*myc* mRNA and protein were seen at elevated levels.

Expression from only one c-*myc* allele was seen in 5/20 MM cell lines (JJN-3, KMS-11, FLAM-76, OPM-2, MM-M1), suggesting a cis-deregulation (possibly due to hypomethylation) [157]. However, these cell lines lack translocation of the c-*myc* gene or changes in the encoded c-*myc* protein. Mechanisms that may deregulate c-*myc* include rearrangements affecting the PVT1 locus or rearrangements of the MLVI4 region located 20kb downstream of c-*myc* [180,181], although this was not confirmed by others [176]. Rearrangements of the MLVI4 region were not identified in RPMI 8226, U-266, JJN-3, SK-MM-1, SK-MM-2, KMM-1, OPM-2, H1112, ark, TH, MM-M1, FLAM-76, KMS-11, MM.1, delta-47, JIM-3, UTMC-2, ANBL-6, OCI-My5 or in the KMS-12 MM cell lines [157]. Altered mRNA turnover resulting in the increased half-life of c-*myc* mRNA as a result of posttranscriptional events was not identified in MM [155]. Moreover, expression of the c-*myc* protein is often elevated independently of c-*myc* mRNA levels, pointing to a deregulated translational control of c-*myc* expression [182].

Rearrangements of *bcl*-2 by translocation t(14;18) and altered expression have been implicated in the pathogenesis of non-Hodgkin's lymphoma [183].

In MM these translocations are rare [176]. *Bcl*-2 is expressed at high but variable levels in MM cell lines and primary MM cells. The JJN-3, EJM, U-266, Karpas 707 and JIM-3 were demonstrated to express elevated levels of *bcl*-2 protein in the absence of a translocation t(14;18) [184]. Levels of *bcl*-2 mRNA and protein are elevated 4-fold in the U-266-1970 subline of U-266, while other lines (U-266, U-1996, U-1958, Karpas 707, L363) express elevated levels of *bcl*-2 in the absence of t(14;18) [159]. The high expression of *bcl*-2 by normal plasma cells and in MM cells in the absence of detectable translocations, suggests that this may be a marker of long-lived post-follicular cells rescued from apoptosis during germinal centre selection [159,184]. Overexpression of *bcl*-2 related genes may also result in the rescue of MM cells from apoptosis induced by dexamethasone or IL-6 deprivation and be implicated in the development of chemoresistance [185–187].

Deletion of the retinoblastoma gene Rb1 was found in more than 50% of bone marrow specimens obtained from patients with MM [188]. This is in line with the frequent occurrence of monosomy 13 in MM and deletions involving 13q24. However, nullisomy of chromosome 13 was not found in this study and is only rarely detected in MM [91]. In the U-266 cell line a biallelic loss and absence of Rb protein was reported [167]. As in the case of mutations of *ras* and p53, the Rb alterations seem to be associated with tumor progression rather than constituting an early genetic event [167].

Mutations have been identified in N- and K-*ras* oncogenes and occur in about 30% of newly diagnosed MM patients and in 70% of MM patients at relapse. Mutations in *ras* may be restricted to stage III MM and PCL, suggesting that they may be important during tumor progression [16,164,167,174,175]. A high frequency of activating mutations in codon 61 of the N-*ras* gene in MM was reported [174]. However, others found this mutation occurs in only about 25% of the activating *ras* mutations [164]. Of 10 MM cell lines (RPMI 8226, U-266, LB-831, EJM, JJN-3, XG-1, XG-2, XG-3, XG-4 and XG-5) examined, three (XG-1, XG-2 and RPMI 8226) contained a mutation in codon 12 in K- or N-*ras* [164]. The finding of Portier was confirmed in another study of U-266, EJM and LP-1 cell lines. In none of these cell lines could *ras* mutations be observed [167]. The introduction of a constitutively active N-*ras* into an IL-6 dependent MM cell line ANBL-6 (ANBL-6/Ras) resulted in significant IL-6 independent growth and reduced apoptosis suggesting that activation of *ras* oncogenes may result in growth factor independence and suppression of apoptosis [189].

Mutations of p53 are infrequent in MM (5–10% of patients), and are associated with progressive disease [158,164,190]. In established cell lines, however, the frequency of p53 mutations is 80% [158]. This is in line with previous reports, and suggests that the mutations may arise during tumor progression [191]. Among eight cell lines, five exhibited only the mutant form, indicative of loss of wild type sequences (U-266, EJM, XG-5, RPMI 8226, LB-831). In three cell lines (XG-1, XG-2 and XG-4), both mutated and wild-type alleles were present. The expression of p53 was found in all cell lines tested and the mutations were not correlated with the expression levels of p53 or with autonomous growth [158]. In the U-266 and EJM cell lines, a p53 mutation involving exon 5 was identified, while the LP1 cell line contained wild-type p53 [167].

Overexpression of the MDM2 gene, binding and inactivating p53, has been reported in MM and in the cell lines RPMI 8226 and U-266 [162].

p16$^{INK4}$ is frequently hypermethylated in MM. This occurs more frequently in advanced disease of MM and in cell lines. Inactivation of p16$^{INK4}$ by hypermethylation may be associated with disseminating disease and development of plasma cell leukemia [192]. In MM, a high expression of p16$^{INK4}$ correlated with the loss of cyclin D1 expression and IL-6 responsiveness and was suggested to be confined to a mature phenotype in MM (CD49e+, MPC-1+) [193].

## 6. FUNCTIONAL CHARACTERIZATION

The population doubling time ($T_{do}$) of continuous MM cell lines ranges from 16 h to up to 144 days (Table 6). All MM cell lines determined to be authentic were EBV negative. Some were tested and reported to be free of mycoplasma infection (31/81). The cytochemical staining profile of plasma cell and MM cell lines is characterized by $\beta$-glucuronidase (BGLU) and $\alpha$-naphthyl acetate esterase (ANAE) expression [11]. A typical profile can be described as follows; AcP+, NASDCAE–, ANAE/ANBE(+), BGLU+, while the cell lines tested also generally show weak or negative staining for Px and PAS (Table 6).

The staining of the LCLs qualitatively resembles that of normal B cells with reactivity only with acid phosphatase (AcP), naphthol-AS-D acetate esterase (NASDAE) and BGLU [11] (Table 2).

Most MM cells produce only light chains, either $\lambda$ or $\kappa$ (ILKM3, JJN-2, Karpas 620, Karpas 707, KMM-1, KMS-11, KMS-18, MEF-1, MM.1, MM5.1, MM5.2, MM-S1, NOP-2, OCI-My5, OPM-1, OPM-2, RPMI 8226, SK-MM-1, SK-MM-2, U-1996). A few cell lines produce IgG ($n = 17$), IgA ($n = 13$) and in one cell line each, IgD and IgE. The secretion rate of IgG per $10^6$ cells/24 h is >80 $\mu$g/ml in NCI-H929, 50 $\mu$g/ml in LP1, 40–50 $\mu$g/ml in MT3, 40 $\mu$g/ml in OH-2, 23 $\mu$g/ml in LB-831 and LB-832, about 20 $\mu$g/ml in 15 $\mu$g ($\lambda$-chains) in RPMI 8226, 5–10 $\mu$g/ml in EJM, 4–9 $\mu$g/ml ($\lambda$-chains) in MM.1, 4–9 $\mu$g/ml in U-266, 4 $\mu$g/ml in UTMC-2, about 3 $\mu$g/ml in KPMM2, 1.5 $\mu$g/ml in U-1958, about 1 $\mu$g/ml in SK-MM-2 cell line and <1

*Table 6.* Multiple myeloma cell lines: functional characterization

| Cell line | Doubling time $T_{do}$/generation time $T_c$ | EBV status[a,b] | Cytochemistry | Heterotransplantability into nude mice | Special functional features | Histopathology/cell size/growth pattern in tissue culture | Mycoplasma status | Ref. |
|---|---|---|---|---|---|---|---|---|
| ACB-885 | 30–35 h | EBV– | MGP+, PAS+ | | | ecc. nuclei, basophilic cytopl. | | 20 |
| ACB-1085 | 30–35 h | EBV– | MGP+, PAS+ | | | ecc. nuclei, basophilic cytopl. | | 20 |
| AD3 | 16 h | EBNA– | AcP+, ANBE+ | | amylase mRNA+ | blastoid nuclei, prominent nucleoli, ER, Golgi zone/ adherent islets, floating clusters | | 21 |
| AMO1 | | EBNA– | | | | plasmablast-plasmacyte | | 22 |
| ANBL-6 | | EBNA– | Px–, ANAE(+), AcP+ | | | plasmacell morph. | | 23 |
| delta-47 | 40 h | EBNA– | | | | ecc. nuclei, basophilic cytopl. bi-, multinucl. cells | | 24 |
| DOBIL-6 | 36 h | EBV– | | in nude mice 5/5 | | primitive plasmacytoid morph. | | 25 |
| DP-6 | | EBNA– | | | | basophilic cytopl., ecc. nuclei | | 26 |
| EJM | 72 h | EBV– | PAS–, SBB–, nonspecific esterase– | | | plasma cell morph. multinucl. cells/ single cells | neg | 27 |
| FLAM-76 | N D | EBNA– | AcP+, Px–, SBB– NASDCAE–, PAS–, AlkP–, ANBE (+) | | | ecc. nuclei, basophilic cytopl. multinucl. (5–10%) cells with eosinophilic cytopl. immature plasma cell morph. | | 28 |
| FR4 | 131 h | EBNA– | AcP+, ANBE+ | | amylase mRNA+ | blastoid nuclei, prominent nucleoli, ER, Golgi zone adh. paving stones floating clusters | | 21 |
| GM2132 (= RPMI 8226) | | EBNA– | | | | | | 29 |
| HL407 | | EBNA– | | | | plasma cell morph., ecc. nuclei prominent Golgi | | 6 |
| ILKM2 | 120 h | EBNA– | | | | plasmacytoid features | | 5 |
| ILKM3 | 96 h | EBNA– | | | | plasmacytoid features | | 5 |
| JIM-1 | | EBNA– | | | | plasmablast morph. | neg | 30 |

*Continued on next page*

*Table 6.* (continued)

| Cell line | Doubling time $T_{do}$/generation time $T_c$ | EBV status[a,b] | Cytochemistry | Heterotransplantability into nude mice | Special functional features | Histopathology/cell size/growth pattern in tissue culture | Mycoplasma status | Ref. |
|---|---|---|---|---|---|---|---|---|
| JIM-2 | | EBNA– | | | | plasmablast morph. | neg | 30 |
| JIM-3 | | EBNA– | | | | plasmablast morph. | neg | 30 |
| JIM-4 | | EBNA– | | | | plasmablast morph. | neg | 30 |
| JJN-1 | 10 d | EBNA– | | | | plasma cell/plasmablast morph. mono/bi nucl. cells, | | 4 |
| JJN-2 | 3–4 d | EBNA– | | | | plasma cell/plasmablast morph. | | 4 |
| JJN-3 | | EBV– | | | | large plasmacytoid blast cells | | 4, 27 |
| Karpas 620 | | EBNA– | AcP+, dual esterase–, SBB– | | | plasma cell morph. | neg | 31 |
| Karpas 707 | 48–72 h | EBNA– | AcP+, NASDCAE+ | | | prom. nucleoli and RER, highly dev. Golgi | | 32 |
| KAS 6/1 | | EBNA– | | | | plasmablasts basophilic cytopl. | | 26 |
| KHM-1α | 5 d | EBNA– | | | | multinucl. plasmablasts | | 33 |
| KHM-1B | 2 d | EBNA– | | | | basophilic cytopl. | | 33 |
| KHM-11 | | | | | amylase activity+ | plasmablast morph. basophilic cytopl. | | 34 |
| KMM-1 | 29 h/36–40 h | EBNA– | Px–, PAS–, MGP+, NASDAE– | nude mice– | 0.2% clon. eff (soft agar) | basophilic cytopl. ecc. nuclei immature plasma cell/single cells | neg | 36 |
| KMS-5 | 24 h | EBNA– | Px–, PAS–, MGP+, NASDAE– | nude mice (s.c.)+ | 16% clon. eff (soft agar) | basophilic cytopl. ecc. nuclei/single cells | neg | 36 |
| KMS-11 | 36 h | EBNA– | Px–, PAS–, MGP+, NASDAE– | nude mice– | 2.3% clon. eff (soft agar) | basophilic cytopl. ecc. nuclei | neg | 36 |
| KMS-12-PE | 62 h | EBNA– | Px–, PAS–, MGP+, NASDAE– | nude mice– | <0.1% clon.eff (soft agar) | basophilic cytopl. ecc. nuclei | neg | 36 |
| KMS-12-BM | 56 h | EBNA– | Px–, PAS–, MGP+, NASDAE– | nude mice– | <0.1% clon.eff (soft agar) | basophilic cytopl. ecc. nuclei | neg | 36 |
| KMS-18 | 72 h | EBV– | | | | plasmacytoid morph. | neg | 37 |
| KP-6 | | EBNA– | | | | ovoid ecc. nuclei | | 26 |

*Continued on next page*

Table 6.  (continued)

| Cell line | Doubling time $T_{do}$/generation time $T_c$ | EBV status[a,b] | Cytochemistry | Heterotransplantability into nude mice | Special functional features | Histopathology/cell size/growth pattern in tissue culture | Mycoplasma status | Ref. |
|---|---|---|---|---|---|---|---|---|
| KPMM2 | 48 h | EBV– | AcP+, Px–, NASDCAE–, PAS–, AlkP–, ANBE– | | | plasmablast/plasma cell morph. | neg | 38 |
| L363 | 65–75 h | EBNA– EA– VCA+ | PAS–, Px–, ANAE– AcP+, BGLU+ | in nude mice 0/18 | | large multinucl. cells/ 9–15 $\mu$m/ single cells | | 39 |
| LB-831 | $T_c$ 38.37 h | EBNA(–) (<5%EBNA+) EBV– | AcP+, BGLU+ | | clonable in plate assay | plasma cell morph. multinucl. cells size heterogeneity | neg | 40 91 |
| LB-832 | $T_c$ 26.4 h | EBNA(–) (<5%EBNA+) EBV– | AcP+, BGLU(+) | | clonable in plate assay | plasma cell morph. bi-, multinucl. cells | neg | 40 51 |
| LB 84-1 | $T_c$ 34.2 h | EBV– | ANBE+, NASDCAE+ | | | plasma cell morph. multinucl. cells double membr. bound mitochondr. | | 79 |
| LOPRA-1 | 30 h | EBNA– | AcP+, BGLU+, PAS(+), unspec esterase(+), Px–, AlkP– | | clonable on feeder cells | well- differentiated plasma cells, bi- or mononucl. cells highly developed Golgi | neg | 41 |
| LP-1 | 50 h | EBNA– | AcP+, PAS– | | ind. to diff by PWM or TPA | basophilic cytopl. multinucl. cells | neg | 43 |
| MEF-1 | 36 h | EBV– | | | | ecc. nuclei, basophilic cytopl. | | 44 |
| MM.1 | 72 h | EBNA– | PAS–, Px–, MGP+, AcP+, ANBE(+), ANAE+, AANAE+ | | not clonable in soft agar or limiting dilution | ecc. nuclei, bi- or multinucl. cells well developed RER, prom. Golgi, single cell/ small clusters | neg | 45 |
| MM5.1 | | EBNA– | | | | plasmacytic morph./ small aggregates | | 46 |
| MM5.2 | | EBNA– | | | | | | 46 |
| MM-A1 | 48 h | EBNA– | | | cloning eff. (methyl cellulose): 0 (–IL6), 1.8% (+IL6) | immature plasmablastic | | 46 7 |

Continued on next page

*Table 6.* (continued)

| Cell line | Doubling time $T_{do}$/generation time $T_c$ | EBV status[a,b] | Cytochemistry | Heterotransplantability into nude mice | Special functional features | Histopathology/cell size/growth pattern in tissue culture | Mycoplasma status | Ref. |
|---|---|---|---|---|---|---|---|---|
| MM-C1 | 96 h | EBNA– | | | cloning eff. (methyl cellulose): 0 (–IL6), 0 (+IL6) | | | 7 |
| MM-M1 | | EBNA– | Px– | | | plasma cell morph./well dev. RER abundant mitochondria | | 47 |
| MM-S1 | 24 h | EBNA– | | not transplantable, transplantable subclone S6B45 (+IL6 cDNA) | cloning eff. (methyl cellulose): 0.5% (–IL6), 21.5% (+IL6) | plasmacytoid morph. | | 111 |
| MM-Y1 | 49 h | EBNA– | | | cloning eff. (methyl cellulose): 0 (–IL6), 1.7% (+IL6) | | | 7 |
| MT3 | 40–50 h | EBNA– | AcP+, BGLU+, PAS– | | | basophilic cytopl., ecc. polynuclei cultured/ 17–40 $\mu$m cell size | neg | 49 51 |
| NCI-H929 | 50 h | EBNA– | MGP+, AcP+, ANAE+, BGLU+, PAS– | | | multinucl. immature plasma cells 20–50 $\mu$m cell size single cells/loose clusters | neg | 50 51 |
| NOP-2 | 48–72 h | EBNA– | | | | ecc. nuclei basophilic cytopl. | neg | 51 |
| OCI-My1 | 24 h | EBV– | | | | | | 52 |
| OCI-My2 | 24 h | EBV– | | | | | | 52 |
| OCI-My3 | 32 h | EBV– | | | | | | 52 |
| OCI-My4 | 130 h | EBV– | | | | | | 52 |
| OCI-My5 | 8 h | EBV– | | | clonable in methyl cellulose | | | 53 |
| OCI-My6 | 45 h | EBV– | | | | | | 52 |
| OCI-My7 | 57 h | EBV– | | | | | | 52 |
| OH-2 | | EBV– | | | | ecc. nuclei, bi- multinucl. cells plasma cell appearance | neg | 55 |

*Continued on next page*

Table 6. (continued)

| Cell line | Doubling time T_do/generation time T_c | EBV status[a,b] | Cytochemistry | Heterotransplantability into nude mice | Special functional features | Histopathology/cell size/growth pattern in tissue culture | Mycoplasma status | Ref. |
|---|---|---|---|---|---|---|---|---|
| OPM-1 | 36–42 h | EBNA– | Px–, PAS(+), AcP+ | | | blastoid, convoluted nuclei basophilic cytopl./ clumps | | 56 |
| OPM-2 | 30–36 H | EBNA– | Px–, PAS(+), AcP+ | | | blastoid nuclei/ single cells | | 56 |
| PCM6 | 40–50 h | EBV– (PCR) | | | | periph. nuclei, basophilic cytopl. | | 57 |
| RPMI8226 | 27.5 h | EBNA– | ANAE+, BGLU+, AcP(+) | | 8.08% cloning eff. (soft agar) | immature cellmorph., no mature plasma cells, swollen ER | | 2 51 58 |
| SK–MM–1 | 32 h | EBNA– | AcP+, MGP+, SBB– | | | central or ecc. nuclei basophilic. cytopl. Bi–, multi-nucl. cells/28–50 μm cell size | neg | 59 |
| SK–MM–2 | 60 h | EBNA– | AcP+, MGP+, SBB– | | | basophilic. cytopl. Large ecc. nuclei. Bi– or multinucl. cells/ 22–32 μm cell size | neg | 59 |
| TH | 24 h | EBNA– | | | | small lymphoid cells/ plasmacytoid cells/blastoid cells, basophilic cytopl. | | 60 |
| U–1957 | 48 h | EBNA– | AcP+, NASDCAE–, ANAE+ BGLU+ | | not clonable in agarose, clonable as single cells on feeder cells | plasmablast/cell morph. multinucl. cells ecc. nuclei/ single cells | | 61 |
| U–1958 | 60 h (92–136 h) | EBNA– | AcP+, NASDCAE–, ANAE+ BGLU+ | | not clonable in agarose, clonable as single cells on feeder cells | plasmablast/cell morph. multinucl. cells ecc. nuclei/ single cells | | 61 |
| U–1996 | 48 h | EBNA– | AcP+, NASDCAE–, ANAE+ BGLU+ | | not clonable in agarose, clonable as single cells on feeder cells | immature plasmablast/cell multinucl. cells ecc. nuclei/ single cells | | 61 |
| U–2030 U266B1 (U–266) | 36 h 108–144 h | EBNA– EBV– EBNA– | AcP+, ANAE–, BGLU+ ANAE+, BGLU+, AcP(+) | | clonable in agarose not clonable in agarose | plasmablasts basophilic cytopl., ecc. nuclei/ 6–16 μm cell size single cells | | 62 3,194 51 |

Continued on next page

*Table 6.* (continued)

| Cell line | Doubling time $T_{do}$/generation time $T_c$ | EBV status[a,b] | Cytochemistry | Heterotransplantability into nude mice | Special functional features | Histopathology/cell size/growth pattern in tissue culture | Mycoplasma status | Ref. |
|---|---|---|---|---|---|---|---|---|
| UCD-HL461 | 47 h | EBNA- | | | CM suppress PWM-stim. Ig production of normal PBL | ecc nucl. basoph cytopl./ single cells | | 63 |
| UMJF-2[c] | 67–90 h (24–72 h) | EBV- | | | UMJF-2 CM suppress PWM-stim. Ig production of normal PBMNC | centr. placed nuclei, basophilic cytopl., plasmablast morph./ 10–12 $\mu$m cell size | | 64 |
| UTMC-2 | 48 h | EBV- | | | | ecc. nuclei, basophilic cytopl./ single cells | | 65 |
| XG-1 | | | | | | plasmablastic immature | neg | 9 |
| XG-2 | | | | | | plasmablastic immature | neg | 124,125 |
| XG-3P | | | | | | plasmacytic differentiated | | 9 |
| XG-3E | | | | | | plasmacytic differentiated | neg | 9 |
| XG-4 | | | | | | plasmacytic differentiated | neg | 9 |
| XG-5 | | | | | | plasmablastic immature | neg | 9 |
| XG-6 | | | | | | plasmacytic differentiated | neg | 9 |
| XG-7 | | | | | | plasmablastic immature | neg | 9 |
| XG-8 | | | | | | plasmacytic differentiated | neg | 9 |
| XG-9 | | | | | | plasmablastic immature | neg | 9 |

a As for the KHM-11 and XG cell lines (1–9), the EBV status has not been reported. In each of these cell lines the authentication was demonstrated by Ig rearrangements consistent with fresh tumor cells of the patient from which the cell line was derived.

b In the OCI-My cell lines 1–7 cell lines, the characterization including EBV-status was made elsewhere [54] with no reference to the individual cell lines and/or designation.

c EBV positive (Drexler, personal communication).

Px – Peroxidase; PAS; periodic acid–Schiff; MGP – methyl green protein; ANAE – *alpha*–naphthyl acetate esterase; ANBE – *alpha*-naphtyl butyrate esterase; AcP – acid phosphatase; AlkP – alkaline phosphatase; NASDCAE – naphthol AS-D chloro acetate esterase; NASDAE – naphthol AS-D acetate esterase; SBB – Sudan Black B; ORO – Oil Red O; BGLU – $\beta$-glucuronidase.

$\mu$g/ml in UCD-HL461. Some MM cell lines are Ig non-producers ($n = 9$), reflecting the status (nonsecretory) in the MM patient in four cases. In other cases, the Ig production was lost during establishment (Table 2).

Transplantability in nude mice was reported in a few cases (DOBIL-6, KMS-5). Interestingly, heterotransplantability was reported in the subline S6B45 of MM-S1 with ectopic IL-6 cDNA and autonomous growth [48]. A number of the cell lines can be cloned and form colonies in soft agar or methyl cellulose (KMM-1, KMS-5, KMS-12, LB-831, LB-832, MM-S1, RPMI 8226) while MM-A1, MM- Y1, U-1957, U-1958 and U-1996 were only clonable in the presence of IL-6 or feeder cells. Phenotypic alterations (including improved growth rate, development of feeder cell independence, capacity for growth as colonies in soft agar, growth as subcutaneous tumors in nude mice) have also been reported to occur as a consequence of long term cultivation, as exemplified by the U-266(–1984) cell line [93–95].

In one of the cell lines, LP-1, PWM or TPA can induce differentiation. The expression of unique genes (amylase mRNA) associated with the occurrence of hyperamylasemia in the patient, was found in three cell lines (AD3, a subclone of FR4), FR4, and KHM-1B (Table 6).

## 7. PUTATIVE BUT UNCONFIRMED CELL LINES

As described, MM cell lines are phenotypically very heterogenous. However, they can be distinguished from LCLs (see Table 1). The following criteria should be used to distinguish MM from LCLs: (1) morphology; MM cells have a plasmablast/plasma cell morphology including eccentric nuclei with prominent nucleoli, a prominent RER and Golgi and perinuclear zone, (2) capacity for a high rate of Ig secretion, (3) surface antigen profile, (4) aneuploidy, with numerous structural and numerical aberrations and, most importantly (5) the lack of EBV. In some reports characterization by these criteria is insufficient. Taking these markers into account, this review distinguishes four categories of cell lines; (1) authentic MM cell lines, (2) EBV-positive lymphoblastoid cell lines (LCLs) (ARH-77, Fr/FRAVEL, GM1312, GM1500, HS-Sultan, IM-9 and MC/CAR), and (3) putative, but insufficiently characterized cell lines (ACB-985, ard- 1, ark, C23/11, col, H112, HGN-5, HSM-2, HSM-2.3, KMM 56, LA 49, mer, Oda, ram). In these cell lines the EBV status and/or the phenotypic characterization or structural abnormalities consistent with the fresh tumor cells were not reported. Without further characterization these are regarded as putative but unconfirmed cell lines (Table 7).

*Table 7.* Unconfirmed cell lines (not characterized, not verified, EBV status unknown, other)

| Cell line* | Patient | Treatment status/ specimen site | Culture medium/ other requirements for establishment | Ig production | Features | Ref. |
|---|---|---|---|---|---|---|
| ACB-985 | IgG/κ MM | T/BM | | | elevated *c-myc* mRNA (no RFLP) | 155 |
| ard-1 | MM | BM | RPMI1640+10%FCS | | B-B4+ CD19− CD45(+), requirement of IL-6 for growth | 82 |
| ark | MM | BM | RPMI1640+10%FCS | | B-B4+ CD19− CD45−, no requirement for IL-6 | 82 |
| ARP-1 | IgA/κ MM | BM | RPMI1640+10%FCS | IgA/κ | CD9+ CD10− CD19− CD38+ CD45+ CD56− cIg+ RT-PCR/Flow cyt: IL6R+ RT-PCR/bioassay IL-6+ (0.4 pg/mL) *bcl-2* (low expression) *p53-/-* | 195 196 |
| C23/11 | MM | PB | RPMI1640+10%FCS no requirement for feeder cells/growth factors | κ/λ | $T_{do}$ 12-15h eccentric nuclei, binucleated grow in clusters | 197 |
| col | MM | BM | RPMI1640+10%FCS | | B-B4− CD19+ no requirement for IL-6 | 82 |
| H112 | MM | | | | no available karyotype t(11;14), no ISRF/translocation into IgH switch region, cyclin D1 overexpression | 153 |
| HGN-5 | IgA/κ MM | | | | CD11a+ CD54+, *c-myc* rearrangement | 78 |

*Continued on next page*

*Table 7.* (continued)

| Cell line* | Patient | Treatment status/ specimen site | Culture medium/ other requirements for establishment | Ig production | Features | Ref. |
|---|---|---|---|---|---|---|
| HSM-2 | 90 M IgM/κ bi-phenotypic leukemia/ PCL | BM | IMDM+20%FCS adherent monolayers/ 10 U/ml IL-6 cloned by limiting dilution | IgM/κ | Px–, ANBE–, NASDCAE–, PAS+, SBB– Plasmablast/plasma cells/ 15-25 μm cell size Karyotype: 45, 1q+, 3q–, 4q–, 5q–, 7p+, 10p–, 10q+, 19p+, 22p+, –5, –9, –13, –15, –15, –22, –Y MTT assay: IL-6+, IL-1–, IL-2–, IL-3–, IL-4–, IL-5–, GM-CSF– Dependence of cytokine: IL-6+ CD38+ PCA-1+ sIg+ cIg+ | 139 |
| HSM-2.3 | (subclone of HSM-2) | | IMDM+5-20%FCS IL-3(100 U/ml)/ IL-6(0.5 U/ml) | IgM/κ | MTT assay: IL-6+, IL-3+ dependence of cytokine: IL-6+, IL-3+ CD10+ CD19+ CD38+ PCA-1+ | 139 |
| KMM 56 | 62 M IgD/λ BJP MM | PE | αMEM+20%FCS adherent to fibroblasts/ feeder cells, later feeder independent growth | λ | T$_{do}$: 36 h Px–, PAS–, E–, EA–, EAC–, sIg– immature plasmacytoid basophilic cytoplasm hyperdiploid | 198 |

*Continued on next page*

Table 7. (continued)

| Cell line* | Patient | Treatment status/ specimen site | Culture medium/ other requirements for establishment | Ig production | Features | Ref. |
|---|---|---|---|---|---|---|
| LA 49 | 59 F IgD/λ BJP | PE | RPMI 1640+40%FCS fibroblasts overlaid with agar/fibroblast CM | IgD/λ | immature blasts/mature plasmacytes basophilic cytoplasm large ecc. nucleus/15–20 μm cell size colony formation in agar overlay in the presence of fibroblasts polyploid | 199 |
| mer | MM | BM | RPMI1640+10%FCS | | B-B4– CD19+ no requirement for IL-6 | 82 |
| Oda | IgD/λ plasmacytoma | localized plasmacytoma | | | Ig production: IgD/λ | 200 |
| ram | MM | BM | RPMI1640+10%FCS | | B-B4(+) CD19+ no requirement for IL-6 | 82 |
| SIK | IgG/κ MM | BM | RPMI1640+10%FCS | | CD9– CD10– CD19– CD38+ CD45+ CD56– RT-PCR; IL-6R mRNA+, IL-6 mRNA+ Flow cyt: IL-6R–; Bioassay: IL-6+ | 195 |
| UCLA#1 | PCL | PB | RPMI1640+10%FCS | | CD38+ IL-6R+, Anti-apoptotic response to IL-6 | 201 |

* For all putative but unconfirmed cell lines the insufficient characterization concerns EBV negativity status and in some cell lines also relates to the clinical data, analysis of markers consistent with the fresh tumor cells, immunophenotypes and cytogenetics.

# References

1.  Helfrich, MH, Livingston, E, Franklin, IM, Soutar, RL. *Blood Rev* 11: 28–38, 1997.
2.  Matsuoka, Y, Moore, GE, Yagi, Y, Pressman, D. *Proc Soc Exp Biol Med* 125: 1246–1250, 1967.
3.  Nilsson, K, Bennich, H, Johansson, S, Pontén, J. *Clin Exp Immunol* 7: 477–489, 1970.
4.  Jackson, N, Lowe, J, Ball, J, Bromidge, E, Ling, NR, Larkins, S, Griffith, MJ, Franklin, IM. *Clin Exp Immunol* 75: 93–99, 1989.
5.  Shimizu, S, Yoshioka, R, Hirose, Y, Sugai, S, Tachibana, J, Konda, S. *J Exp Med* 169: 339–344, 1989.
6.  Scibienski, R, Paglieroni, T, Caggiano, V, Lemonello, D, Gumerlock, P, Mackenzie, M. *Leukemia* 6: 940–947, 1992.
7.  Okuno, Y, Takahashi, T, Suzuki, A, Ichiba, S, Nakamura, K, Fukumoto, M, Okada, T, Okada, H, Imura, H. *Leukemia* 5: 585–591, 1991.
8.  Zhang, XG, Bataille, R, Jourdan, M, Saeland, S, Banchereau, J, Mannoni, P, Klein, B. *Blood* 76: 2599–2605, 1990.
9.  Zhang, XG, Gaillard, JP, Robillard, N, Lu, ZY, Gu, ZJ, Jourdan, M, Boiron, JM, Bataille, R, Klein, B. *Blood* 83: 3654–3663, 1994a.
10. Nilsson, K, Ponten, J. *Int J Cancer* 15: 321–341, 1975.
11. Nilsson, K. *Hamatol Bluttransfus* 20: 253–264, 1977.
12. Kawano, M, Hirano, T, Matsuda, T, Taga, T, Horii, Y, Iwato, K, Asaoku, H, Tang, B, Tanabe, O, Tanaka, H, et al. *Nature* 332: 83–85, 1988.
13. Klein, B, Zhang, XG, Jourdan, M, Content, J, Houssiau, F, Aarden, L, Piechaczyk, M, Bataille, R. *Blood* 73: 517–526, 1989.
14. Jernberg, H, Pettersson, M, Kishimoto, T, Nilsson, K. *Leukemia* 5: 255–265, 1991a.
15. Nilsson, K, Klein, G. *Adv Cancer Res* 37: 319–380, 1982.
16. Ralph, P. *Cancer* 56: 2544–2545, 1985.
17. Pellat-Deceunynck, C, Barille, S, Puthier, D, Rapp, MJ, Harousseau, JL, Bataille, R, Amiot, M. *Cancer Res* 55: 3647–3653, 1995a.
18. Pellat-Deceunynck, C, Amiot, M, Bataille, R, van Riet, I, van Camp, B, Omede, P, Boccadoro, M. *Blood* 86: 4001–4002, 1995b.
19. Barut, BA, Zon, LI, Cochran, MK, Paul, SR, Chauhan, D, Mohrbacher, A, Fingeroth, J, Anderson, KC. *Leuk Res* 16: 951–959, 1992.
20. Brox, LW, Belch, A, Pollock, E, He, XX, De Braekeleer, M, Lin, CC. *Cancer Genet Cytogenet* 27: 135–144, 1987.
21. Tagawa, S, Doi, S, Taniwaki, M, Abe, T, Kanayama, Y, Nojima, J, Matsubara, K, Kitani, T. *Leukemia* 4: 600–605, 1990.
22. Shimizu, S, Takiguchi, T, Fukutoku, M, Yoshioka, R, Hirose, Y, Fukuhara, S, Ohno, H, Isobe, Y, Konda, S. *Leukemia* 7: 274–280, 1993.
23. Jelinek, DF, Ahmann, GJ, Greipp, PR, Jalal, SM, Westendorf, JJ, Katzmann, JA, Kyle, RA, Lust, JA. *Cancer Res* 53: 5320–5327, 1993.
24. Ishii, K, Yamato, K, Kubonishi, I, Taguchi, H, Ohtsuki, Y, Miyoshi, I. *Am J Hematol* 41: 218–224, 1992.
25. Ohmori, M, Nagai, M, Fujita, M, Dobashi, H, Tasaka, T, Yamaoka, G, Kawanishi, K, Taniwaki, M, Takahara, J. *Br J Haematol* 101: 688–693, 1998.
26. Westendorf, JJ, Ahmann, GJ, Greipp, PR, Witzig, TE, Lust, JA, Jelinek, DF. *Leukemia* 10: 866–876, 1996a.
27. Hamilton, M, Ball, J, Bromidge, E, Lowe, J, Franklin, I. *Br J Haematol* 75: 378–384, 1990.

28. Kubonishi, I, Seto, M, Shimamura, T, Enzan, H, Miyoshi, I. *Cancer* 70: 1528–1535, 1992.
29. Goldstein, M, Hoxie, J, Zembryki, D, Matthews, D, Levinson, AI. *Blood* 66: 444–446, 1985.
30. Barker, HF, Hamilton, MS, Ball, J, Drew, M & Franklin, IM. In: Radl, J, van Camp, B. (eds.), *EURAGE Monoclonal Gammopathies III: Clinical Significance and Basic Mechanisms*, EURAGE, Leiden, the Netherlands), pp 155–158, 1991.
31. Nacheva, E, Fischer, PE, Sherrington, PD, Labastide, W, Lawlor, E, Conneally, E, Blaney, C, Hayhoe, FG, Karpas, A. *Br J Haematol* 74: 70–76, 1990.
32. Karpas, A, Fischer, P, Swirsky, D. *Science* 216: 997–999, 1982.
33. Matsuzaki, H, Hata, H, Takeya, M, Takatsuki, K. *Blood* 72: 978–982, 1988.
34. Hata, H, Matsuzaki, H, Sonoki, T, Takemoto, S, Kuribayashi, N, Nagasaki, A, Takatsuki, K. *Leukemia* 8: 1768–1773, 1994.
35. Togawa, A, Inoue, N, Miyamoto, K, Hyodo, H, Namba, M. *Int J Cancer* 29: 495–500, 1982.
36. Namba, M, Ohtsuki, T, Mori, M, Togawa, A, Wada, H, Sugihara, T, Yawata, Y, Kimoto, T. *In Vitro Cell Dev Biol* 25: 723–729, 1989.
37. Otsuki, T, Nakazawa, N, Taniwaki, M, Yamada, O, Sakaguchi, H, Wada, H, Yawata, Y, Ueki, A. *Int J Oncol* 12: 545–552, 1998.
38. Goto, H, Shimazaki, C, Tatsumi, T, Yamagata, N, Fujita, N, Tsuchiya, M, Koishihara, Y, Ohsugi, Y, Nakagawa, M. *Leukemia* 9: 711–718, 1995.
39. Diehl, V, Schaadt, M, Kirchner, H, Hellriegel, KP, Gudat, F, Fonatsch, C, Laskewitz, E, Guggenheim, R. *Blut* 36: 331–338, 1978.
40. Durie, BG, Vela, E, Baum, V, Leibovitz, A, Payne, CM, Richter, LC, Grogan, TM, Trent, JM. *Blood* 66: 548–555, 1985.
41. Lohmeyer, J, Hadam, M, Santoso, S, Forster, W, Schulz, A, Pralle, H. *Br J Haematol* 69: 335–343, 1988.
42. Corradini, P, Ladetto, M, Voena, C, Palumbo, A, Inghirami, G, Knowles, DM, Boccadoro, M, Pileri, A. *Blood* 81: 2708–2713, 1993.
43. Pegoraro, L, Malavasi, F, Belloni, G, Massaia, M, Boccadoro, M, Saglio, G, Guerrasio, A, Benetton, G, Lombardi, L, Coda, R, Avanzi, G. *Blood* 73: 1020–1027, 1989.
44. Yufu, Y, Goto, T, Choi, I, Uike, N, Kozuru, M, Ohshima, K, Taniguchi, T, Motokura, T, Yatabe, Y, Nakamura, S. *Cancer* 85: 1750–1757, 1999.
45. Goldman-Leikin, RE, Salwen, HR, Herst, CV, Variakojis, D, Bian, ML, Le Beau, MM, Selvanayagan, P, Marder, R, Anderson, R, Weitzman, S, et al. *J Lab Clin Med* 113: 335–345, 1989.
46. Van Riet, I, De Greef, C, Aharchi, F, Woischwill, C, De Waele, M, Bakkus, M, Lacor, P, Schots, R, van Camp, B. *Leukemia* 11: 284–293, 1997.
47. Suzuki, A, Takahashi, T, Okuno, Y, Fukumoto, M, Fukui, H, Koishihara, Y, Ohsugi, Y, Ohno, Y, Imura, H. *Leuk Res* 15: 1043–1050, 1991.
48. Okuno, Y, Takahashi, T, Suzuki, A, Fukumoto, M, Nakamura, K, Fukui, H, Koishihara, Y, Ohsugi, Y, Imura, H. *Exp Hematol* 20: 395–400, 1992.
49. Donelli, A, Narni, F, Tabilio, A, Emilia, G, Selleri, L, Colo, A, Zucchini, P, Montagnani, G, Torelli, G, Torelli, U. *Int J Cancer* 40: 383–388, 1987.
50. Gazdar, AF, Oie, HK, Kirsch, IR, Hollis, GF. *Blood* 67: 1542–1549, 1986.
51. Nagai, T, Ogura, M, Morishita, Y, Okumura, M, Kato, Y, Morishima, Y, Hirabayashi, N, Ohno, R, Saito, H. *Int J Hematol* 54: 141–149, 1991.
52. Hitzler, JK, Martinez-Valdez, H, Bergsagel, DB, Minden, MD, Messner, HA. *Blood* 78: 1996–2004, 1991.

53. Wandl, U, Hoang, T, Minden, M, Messner, HA. *J Cell Physiol* 136: 384–388, 1988.
54. Takahashi, T, Lim, B, Jamal, N, Tritchler, D, Lockwood, G, McKinney, S, Bergsagel, DE, Messner, HA. *J Clin Oncol* 3: 1613–1623, 1985.
55. Borset, M, Waage, A, Brekke, OL, Helseth, E. *Eur J Haematol* 53: 31–37, 1994.
56. Katagiri, S, Yonezawa, T, Kuyama, J, Kanayama, Y, Nishida, K, Abe, T, Tamaki, T, Ohnishi, M, Tarui, S. *Int J Cancer* 36: 241–246, 1985.
57. Takahira, H, Kozuru, M, Hirata, J, Obama, K, Uike, N, Iguchi, H, Miyamura, T, Yamashita, S, Kono, A, Umemura, T. *Exp Hematol* 22: 261–266, 1994.
58. Bellamy, WT, Dalton, WS, Gleason, MC, Grogan, TM, Trent, JM. *Cancer Res* 51: 995–1002, 1991.
59. Eton, O, Scheinberg, DA, Houghton, AN. *Leukemia* 3: 729–735, 1989.
60. Weinreich, SS, von dem Borne, AE, van Lier, RA, Feltkamp, CA, Slater, RM, Wester, MR, Zeijlemaker, WP. *Br J Haematol* 79: 226–234, 1991.
61. Jernberg, H, Nilsson, K, Zech, L, Lutz, D, Nowotny, H, Scheirer, W. *Blood* 69: 1605–1612, 1987a.
62. Jernberg, H, Bjorklund, G, Nilsson, K. *Int J Cancer* 39: 745–751, 1987c.
63. Scibienski, RJ, Paglieroni, TG, MacKenzie, MR. *Leukemia* 4: 775–780, 1990.
64. Farnen, JP, Tyrkus, M, Hanson, CA, Cody, RL, Emerson, SG. *Leukemia* 5: 574–584, 1991.
65. Ozaki, S, Wolfenbarger, D, deBram-Hart, M, Kanangat, S, Weiss, DT, Solomon, A. *Leukemia* 8: 2207–2213, 1994.
66. Nakamura, M, Merchav, S, Carter, A, Ernst, TJ, Demetri, GD, Furukawa, Y, Anderson, K, Freedman, AS, Griffin, JD. *J Immunol* 143: 3543–3547, 1989.
67. Burk, KH, Drewinko, B, Trujillo, JM, Ahearn, MJ. *Cancer Res* 36: 2508–13, 1978.
68. Urashima, M, Chauhan, D, Uchiyama, H, Freeman, GJ, Anderson, KC. *Blood* 85: 1903–1912, 1995.
69. Van Riet, I, De Waele, M, Remels, L, Lacor, P, Schots, R, van Camp, B. *Br J Haematol* 79: 421–427, 1991.
70. Miller, CH, Carbonell, A, Peng, R, Paglieroni, T, MacKenzie, MR. *Cancer* 49: 2091–2096, 1982.
71. Duperray, C, Klein, B, Durie, BG, Zhang, X, Jourdan, M, Poncelet, P, Favier, F, Vincent, C, Brochier, J, Lenoir, G, et al. *Blood* 73: 566–572, 1989.
72. Pellat-Deceunynck, C, Bataille, R, Robillard, N, Harousseau, JL, Rapp, MJ, Juge-Morineau, N, Wijdenes, J, Amiot, M. *Blood* 84: 2597–2603, 1994.
73. Ritts, RE, Jr, Ruiz-Arguelles, A, Weyl, KG, Bradley, AL, Weihmeir, B, Jacobsen, DJ, Strehlo, BL. *Int J Cancer* 31: 133–141, 1983.
74. Van Riet, I. *Pathol Biol (Paris)* 47: 98–108, 1999.
75. Zhang, XG, Olive, D, Devos, J, Rebouissou, C, Ghiotto-Ragueneau, M, Ferlin, M, Klein, B. *Leukemia* 12: 610–618, 1998.
76. Westendorf, JJ, Ahmann, GJ, Lust, JA, Tschumper, RC, Greipp, PR, Katzmann, JA, Jelinek, DF. *Curr Top Microbiol Immunol* 194: 63–72, 1995.
77. Drew, M, Barker, HF, Ball, J, Pearson, C, Cook, G, Franklin, I. *Leuk Res* 20: 619–624, 1996.
78. Kawano, MM, Huang, N, Tanaka, H, Ishikawa, H, Sakai, A, Tanabe, O, Nobuyoshi, M, Kuramoto, A. *Br J Haematol* 79: 583–858, 1991.
79. Durie, BG, Grogan, TM, Spier, C, Vela, E, Baum, V, Rodriquez, MA, Frutiger, Y. *Blood* 73: 770–776, 1989.
80. Westendorf, JJ, Ahmann, GJ, Armitage, RJ, Spriggs, MK, Lust, JA, Greipp, PR, Katzmann, JA, Jelinek, DF. *J Immunol* 152 117–128, 1994.

81.  Jourdan, M, Ferlin, M, Legouffe, E, Horvathova, M, Liautard, J, Rossi, JF, Wijdenes, J, Brochier, J, Klein, B. *Br J Haematol* 100: 637–646, 1998.
82.  Ridley, RC, Xiao, H, Hata, H, Woodliff, J, Epstein, J, Sanderson, RD. *Blood* 81: 767–774, 1993.
83.  Uchiyama, H, Barut, BA, Mohrbacher, AF, Chauhan, D, Anderson, KC. *Blood* 82: 3712–3720, 1993.
84.  Kawano, MM, Huang, N, Harada, H, Harada, Y, Sakai, A, Tanaka, H, Iwato, K, Kuramoto, A. *Blood* 82: 564–570, 1993.
85.  Kim, I, Uchiyama, H, Chauhan, D, Anderson, KC. *Br J Haematol* 87: 483–493, 1994.
86.  Ahsmann, EJ, Lokhorst, HM, Dekker, AW, Bloem, AC. *Blood* 79: 2068–2075, 1992.
87.  Caligaris-Cappio, F, Gregoretti, MG, Nilsson, K. In: Caligaris-Cappio, F. (ed.), *Human B Cell Populations*, Vol 67, Karger, Basel, pp 102–113, 1997.
88.  De Greef, C, van Riet, I, Bakkus, MH, van Camp, B. *Leukemia* 12: 86–93, 1998.
89.  Barker, HF, Hamilton, MS, Ball, J, Drew, M, Franklin, IM. *Br J Haematol* 81: 331–335, 1992.
90.  Caligaris-Cappio, F, Gregoretti, MG, Merico, F, Gottardi, D, Ghia, P, Parvis, G, Bergui, L. *Leuk Lymphoma* 8: 15–22, 1992.
91.  Hallek, M, Bergsagel, P, Anderson, K. *Blood* 91: 3–21, 1998.
92.  Klein, B. *Curr Opin Hematol* 5: 186–191, 1998.
93.  Hellman, L, Josephson, S, Jernberg, H, Nilsson, K, Pettersson, U. *Eur J Immunol* 18: 905–910, 1988.
94.  Nilsson, K, Jernberg, H, Pettersson, M. *Curr Top Microbiol Immunol* 166: 3–12, 1990.
95.  Jernberg Wiklund, H, Pettersson, M, Carlsson, M, Nilsson, K. *Leukemia* 6: 310–318, 1992a.
96.  Portier, M, Rajzbaum, G, Zhang, XG, Attal, M, Rusalen, C, Wijdenes, J, Mannoni, P, Maraninchi, D, Piechaczyk, M, Bataille, R, et al. *Eur J Immunol* 21: 1759–1762, 1991.
97.  Hata, H, Xiao, H, Petrucci, MT, Woodliff, J, Chang, R, Epstein, J. *Blood* 81: 3357–3364, 1993.
98.  Sati, HI, Apperley, JF, Greaves, M, Lawry, J, Gooding, R, Russell, RG, Croucher, PI. *Br J Haematol* 101: 287–295, 1998.
99.  Freeman, GJ, Freedman, AS, Rabinowe, SN, Segil, JM, Horowitz, J, Rosen, K, Whitman, JF, Nadler, LM. *J Clin Invest* 83: 1512–1518, 1989.
100. Schwab, G, Siegall, CB, Aarden, LA, Neckers, LM, Nordan, RP. *Blood* 77: 587–593, 1991.
101. Jelinek, DF, Aagaard Tillery, KM, Arendt, BK, Arora, T, Tschumper, RC, Westendorf, JJ. *J Clin Invest* 99: 447–456, 1997a.
102. Jelinek, DF, Witzig, TE, Arendt, BK. *J Immunol* 159: 487–496, 1997b.
103. Westendorf, JJ, Jelinek, DF. *J Immunol* 157: 3081–3088, 1996b.
104. Costes, V, Portier, M, Lu, ZY, Rossi, JF, Bataille, R, Klein, B. *Br J Haematol* 103: 1152–1160, 1998.
105. Georgii-Hemming, P, Wiklund, HJ, Ljunggren, O, Nilsson, K. *Blood* 88: 2250–2258, 1996.
106. Georgii-Hemming, P, Stromberg, T, Janson, ET, Stridsberg, M, Wiklund, HJ, Nilsson, K. *Blood* 93: 1724–1731, 1999.
107. Jernberg-Wiklund, H, Pettersson, M, Nilsson, K. *Eur J Haematol* 46: 231–239, 1991b.
108. Hata, H, Matsuzaki, H, Takatsuki, K. *Acta Haematol* 83: 133–136, 1990.
109. Diamant, M, Hansen, MB, Rieneck, K, Svenson, M, Yasukawa, K, Bendtzen, K. *Leuk Res* 20: 291–301, 1996.

110. Neumann, C, Zehentmaier, G, Danhauser-Riedl, S, Emmerich, B, Hallek, M. *Eur J Immunol* 26: 379–384, 1996.

111. Okuno, Y, Takahashi, T, Suzuki, A, Ichiba, S, Nakamura, K, Hitomi, K, Sasaki, R, Imura, H. *Biochem Biophys Res Commun* 170: 1128–1134, 1990.

112. Johnson, BH, Gomi, M, Jakowlew, SB, Moriwaki, K, Thompson, EB. *Cell Growth Differ* 4: 25–30, 1993.

113. Borset, M, Medvedev, AE, Sundan, A, Espevik, T. *Cytokine* 8: 430–438, 1996.

114. Salmon, SE, Soehnlen, B, Dalton, WS, Meltzer, P, Scuderi, P. *Blood* 74: 1723–1727, 1989.

115. Freund, GG, Kulas, DT, Mooney, RA. *J Immunol* 151: 1811–1820, 1993.

116. Freund, GG, Kulas, DT, Way, BA, Mooney, RA. *Cancer Res* 54: 3179–3185, 1994.

117. Villunger, A, Egle, A, Kos, M, Egle, D, Tinhofer, I, Henn, T, Uberall, F, Maly, K, Greil, R. *Br J Haematol* 102, 1069–1080, 1998.

118. Paul, SR, Barut, BA, Bennett, F, Cochran, MA, Anderson, KC. *Leuk Res* 16: 247–252, 1992.

119. Zhou, L, Peng, S, Duan, J, Zhou, J, Wang, L, Wang, J. *Biochem Mol Biol Int* 45: 1113–1119, 1998.

120. Jernberg-Wiklund, H, Pettersson, M, Larsson, LG, Anton, R, Nilsson, K. *Int J Cancer* 51: 116–123, 1992b.

121. Schwabe, M, Brini, AT, Bosco, MC, Rubboli, F, Egawa, M, Zhao, J, Princler, GL, Kung, HF. *J Clin Invest* 94: 2317–2325, 1994.

122. Zhang, XG, Gu, JJ, Lu, ZY, Yasukawa, K, Yancopoulos, GD, Turner, K, Shoyab, M, Taga, T, Kishimoto, T, Bataille, R, et al. *J Exp Med* 179: 1337–1342, 1994b.

123. Petrucci, MT, Ricciardi, MR, Ariola, C, Gregorj, C, Ribersani, M, Savino, R, Ciliberto, G, Tafuri, A. *Ann Hematol* 78: 13–18, 1999.

124. Gu, ZJ, Costes, V, Lu, ZY, Zhang, XG, Pitard, V, Moreau, JF, Bataille, R, Wijdenes, J, Rossi, JF, Klein, B. *Blood* 88: 3972–3986, 1996a.

125. Gu, ZJ, Zhang, XG, Hallet, MM, Lu, ZY, Wijdenes, J, Rossi, JF, Klein, B. *Exp Hematol* 24: 1195–1200, 1996b.

126. Jourdan, M, Zhang, XG, Portier, M, Boiron, JM, Bataille, R, Klein, B. *J Immunol* 147: 4402–4407, 1991.

127. Lu, ZY, Zhang, XG, Rodriguez, C, Wijdenes, J, Gu, ZJ, Morel-Fournier, B, Harousseau, JL, Bataille, R, Rossi, JF, Klein, B. *Blood* 85: 2521–2527, 1995a.

128. Lu, ZY, Gu, ZJ, Zhang, XG, Wijdenes, J, Neddermann, P, Rossi, JF, Klein, B. *FEBS Lett* 377: 515–518, 1995b.

129. Levy, Y, Tsapis, A, Brouet, JC. *J Clin Invest* 88: 696–699, 1991.

130. Hawley, RG, Berger, LC. *Leuk Lymphoma* 29: 465–475, 1998.

131. Nishimoto, N, Ogata, A, Shima, Y, Tani, Y, Ogawa, H, Nakagawa, M, Sugiyama, H, Yoshizaki, K, Kishimoto, T. *J Exp Med* 179: 1343–1347, 1994.

132. Spets, H, Jernberg-Wiklund, H, Sambade, C, Soderberg, O, Nilsson, K. *Br J Haematol* 98: 126–133, 1997.

133. Harrington, EA, Bennett, MR, Fanidi, A, Evan, GI. *EMBO J* 13: 3286–3295, 1994.

134. Datta, SR, Dudek, H, Tao, X, Masters, S, Fu, H, Gotoh, Y, Greenberg, ME. *Cell* 91: 231–241, 1997.

135. Vanderkerken, K, Asosingh, K, Braet, F, van Riet, I, van Camp, B. *Blood* 93: 235–241, 1999.

136. Anderson, KC, Jones, RM, Morimoto, C, Leavitt, P, Barut, BA. *Blood* 73: 1915–1924, 1989.

137. Carter, A, Merchav, S, Silvian-Draxler, I, Tatarsky, I. *Br J Haematol* 74: 424–431, 1990.

138. Urashima, M, Ogata, A, Chauhan, D, Hatziyanni, M, Vidriales, MB, Dedera, DA, Schlossman, RL, Anderson, KC. *Blood* 87: 1928–1938, 1996.
139. Kobayashi, M, Miyagishima, T, Imamura, M, Maeda, S, Gotohda, Y, Iwasaki, H, Sakurada, K, Miyazaki, T. *Br J Haematol* 78: 217–221, 1991.
140. Kobayashi, M, Tanaka, J, Imamura, M, Maeda, S, Iwasaki, H, Tanaka, M, Tsudu, Y, Sakurada, K, Miyazaki, T. *Br J Haematol* 83: 535–538, 1993.
141. Borset, M, Helseth, E, Naume, B, Waage, A. *Br J Haematol* 85: 446–451, 1993.
142. Idestrom, K, Cantell, K, Killander, D, Nilsson, K, Strander, H, Willems, J. *Acta Med Scand* 205: 149–154, 1979.
143. Brenning, G. *Scand J Haematol* 35: 178–185, 1985.
144. Brenning, G, Jernberg, H, Gidlund, M, Sjoberg, O, Nilsson, K. *Scand J Haematol* 37: 280–288, 1986.
145. Einhorn, S, Fernberg, JO, Grander, D, Lewensohn, R. *Eur J Cancer Clin Oncol* 24: 1505–1510, 1988.
146. Portier, M, Zhang, XG, Caron, E, Lu, ZY, Bataille, R, Klein, B. *Blood* 81: 3076–3082, 1993.
147. Palumbo, A, Battaglio, S, Napoli, P, Omede, P, Fusaro, A, Bruno, B, Boccadoro, M, Pileri, A. *Br J Haematol* 86: 726–732, 1994.
148. Dewald, GW, Kyle, RA, Hicks, GA, Greipp, PR. *Blood* 66: 380–390, 1985.
149. Ferti, A, Panani, A, Arapakis, G, Raptis, S. *Cancer Genet Cytogenet* 12: 247–253, 1984.
150. Sawyer, JR, Waldron, JA, Jagannath, S, Barlogie, B. *Cancer Genet Cytogenet* 82: 41–49, 1995.
151. Drach, J, Angerler, J, Schuster, J, Rothermundt, C, Thalhammer, R, Haas, OA, Jager, U, Fiegl, M, Geissler, K, Ludwig, H, et al. *Blood* 86: 3915–3921, 1995.
152. Tabernero, D, San Miguel, JF, Garcia-Sanz, M, Najera, L, Garcia-Isidoro, M, Perez-Simon, JA, Gonzalez, M, Wiegant, J, Raap, AK, Orfao, A. *Am J Pathol* 149: 153–161, 1996.
153. Bergsagel, PL, Chesi, M, Nardini, E, Brents, LA, Kirby, SL, Kuehl, WM. *Proc Natl Acad Sci USA* 93: 13931–13936, 1996.
154. Bergsagel, PL, Nardini, E, Brents, L, Chesi, M, Kuehl, WM. *Curr Top Microbiol Immunol* 224: 283–287, 1997.
155. Fourney, R, Palmer, M, Ng, A, Dietrich, K, Belch, A, Paterson, M, Brox, L. *Dis Markers* 8: 117–124, 1990.
156. Chesi, M, Bergsagel, PL, Brents, LA, Smith, CM, Gerhard, DS, Kuehl, WM. *Blood* 88: 674–681, 1996.
157. Kuehl, WM, Brents, LA, Chesi, M, Huppi, K, Bergsagel, PL. *Curr Top Microbiol Immunol* 224: 277–282, 1997.
158. Mazars, GR, Portier, M, Zhang, XG, Jourdan, M, Bataille, R, Theillet, C, Klein, B. *Oncogene* 7: 1015–1018, 1992.
159. Pettersson, M, Jernberg-Wiklund, H, Larsson, LG, Sundstrom, C, Givol, I, Tsujimoto, Y, Nilsson, K. *Blood* 79: 495–502, 1992.
160. Sonoki, T, Nakazawa, N, Hata, H, Taniwaki, M, Nagasaki, A, Seto, M, Yoshida, M, Kuribayashi, N, Kimura, T, Harada, N, Mitsuya, H, Matsuzaki, H. *Int J Hematol* 68: 459–461, 1998.
161. Ohtsuki, T, Nishitani, K, Hatamochi, A, Yawata, Y, Namba, M. *Br J Haematol* 77: 172–179, 1991.
162. Teoh, G, Urashima, M, Ogata, A, Chauhan, D, DeCaprio, JA, Treon, SP, Schlossman, RL, Anderson, KC. *Blood* 90: 1982–1992, 1997.

163. Urashima, M, Teoh, G, Ogata, A, Chauhan, D, Treon, SP, Sugimoto, Y, Kaihara, C, Matsuzaki, M, Hoshi, Y, DeCaprio, JA, Anderson, KC. *Clin Cancer Res* 3: 2173–2179, 1997.

164. Portier, M, Moles, JP, Mazars, GR, Jeanteur, P, Bataille, R, Klein, B, Theillet, C. *Oncogene* 7: 2539–2543, 1992.

165. Iida, S, Rao, PH, Butler, M, Corradini, P, Boccadoro, M, Klein, B, Chaganti, RS, Dalla-Favera, R. *Nat Genet* 17: 226–230, 1997.

166. Jernberg, H, Zech, L, Nilsson, K. *Int J Cancer* 40: 811–817, 1987b.

167. Corradini, P, Inghirami, G, Astolfi, M, Ladetto, M, Voena, C, Ballerini, P, Gu, W, Nilsson, K, Knowles, DM, Boccadoro, M, et al. *Leukemia* 8: 758–767, 1994.

168. Chesi, M, Nardini, E, Brents, LA, Schrock, E, Ried, T, Kuehl, WM, Bergsagel, PL. *Nat Genet* 16: 260–264, 1997.

169. Chesi, M, Nardini, E, Lim, RS, Smith, KD, Kuehl, WM, Bergsagel, PL. *Blood* 92: 3025–3034, 1998b.

170. Chesi, M, Bergsagel, PL, Shonukan, OO, Martelli, ML, Brents, LA, Chen, T, Schrock, E, Ried, T, Kuehl, WM. *Blood* 91: 4457–4463, 1998a.

171. Klein, G. *Cell* 32: 311–315, 1983.

172. Sumegi, J, Hedberg, T, Bjorkholm, M, Godal, T, Mellstedt, H, Nilsson, MG, Perlman, C, Klein, G. *Int J Cancer* 36: 367–371, 1985.

173. Selvanayagam, P, Blick, M, Narni, F, van Tuinen, P, Ledbetter, DH, Alexanian, R, Saunders, GF, Barlogie, B. *Blood* 71: 30–35, 1988.

174. Neri, A, Murphy, JP, Cro, L, Ferrero, D, Tarella, C, Baldini, L, Dalla-Favera, R. *J Exp Med* 170: 1715–1725, 1989.

175. Paquette, RL, Berenson, J, Lichtenstein, A, McCormick, F, Koeffler, HP. *Oncogene* 5: 1659–1663, 1990.

176. Ladanyi, M, Wang, S, Niesvizky, R, Feiner, H, Michaeli, J. *Am J Pathol* 141: 949–953, 1992.

177. Hollis, GF, Gazdar, AF, Bertness, V, Kirsch, IR. *Mol Cell Biol* 8: 124–129, 1988.

178. Palumbo, AP, Pileri, A, Dianzani, U, Massaia, M, Boccadoro, M, Calabretta, B. *Cancer Res* 49: 4701–4704, 1989.

179. Nobuyoshi, M, Kawano, M, Tanaka, H, Ishikawa, H, Tanabe, O, Iwato, K, Asaoku, H, Sakai, A, Kuramoto, A. *Br J Haematol* 77: 523–528, 1991.

180. Bakkus, MH, Brakel-van Peer, KM, Michiels, JJ, van't Veer, MB, Benner, R. *Oncogene* 5: 1359–1364, 1990.

181. Palumbo, AP, Boccadoro, M, Battaglio, S, Corradini, P, Tsichlis, PN, Huebner, K, Pileri, A, Croce, CM. *Cancer Res* 50: 6478–6482, 1990.

182. Paulin, FE, West, MJ, Sullivan, NF, Whitney, RL, Lyne, L, Willis, AE. *Oncogene* 13: 505–513, 1996.

183. Tsujimoto, Y, Gorham, J, Cossman, J, Jaffe, E, Croce, CM. *Science* 229: 1390–1393, 1985.

184. Hamilton, MS, Barker, HF, Ball, J, Drew, M, Abbot, SD, Franklin, IM. *Leukemia* 5: 768–771, 1991.

185. Sangfelt, O, Osterborg, A, Grander, D, Anderbring, E, Ost, A, Mellstedt, H, Einhorn, S. *Int J Cancer* 63: 190–192, 1995.

186. Tu, Y, Xu, FH, Liu, J, Vescio, R, Berenson, J, Fady, C, Lichtenstein, A. *Blood* 88: 1805–1812, 1996.

187. Harada, N, Hata, H, Yoshida, M, Soniki, T, Nagasaki, A, Kuribayashi, N, Kimura, T, Matsuzaki, H, Mitsuya, H. *Leukemia* 12: 1817–1820, 1998.

188. Dao, DD, Sawyer, JR, Epstein, J, Hoover, RG, Barlogie, B, Tricot, G. *Leukemia* 8: 1280–1284, 1994.

189. Billadeau, D, Jelinek, DF, Shah, N, LeBien, TW, van Ness, B. *Cancer Res* 55: 3640–3646, 1995.

190. Preudhomme, C, Facon, T, Zandecki, M, Vanrumbeke, M, Lai, JL, Nataf, E, Loucheux-Lefebvre, MH, Kerckaert, JP, Fenaux, P. *Br J Haematol* 81: 440–443, 1992.

191. Ollikainen, H, Syrjanen, S, Koskela, K, Pelliniemi, TT, Pulkki, K. *Scand J Clin Lab Invest* 57: 281–289, 1997.

192. Tasaka, T, Asou, H, Munker, R, Said, JW, Berenson, J, Vescio, RA, Nagai, M, Takahara, J, Koeffler, HP. *Br J Haematol* 101: 558–564, 1998.

193. Kawano, MM, Mahmoud, MS, Ishikawa, H. *Br J Haematol* 99: 131–138, 1997.

194. Nilsson, K. *Int J Cancer* 7: 380–396, 1971.

195. Hardin, J, MacLeod, S, Grigorieva, I, Chang, R, Barlogie, B, Xiao, H, Epstein, J. *Blood* 84: 3063–3070, 1994.

196. Gazitt, Y, Fey, V, Thomas, C, Alvarez, R. *Int J Oncol* 13: 397–405, 1998.

197. Prosser, E, Carroll, K, O'Kennedy, R, Clark, K, Grogan, L, Otridge, B. *Biochem Soc Trans* 17: 1055–1056, 1989.

198. Niho, Y, Shibuya, T, Yamasaki, K, Kimura, N. *Int J Cell Cloning* 2: 161–172, 1984.

199. Jobin, ME, Fahey, JL, Price, Z. *J Exp Med* 140: 494–507, 1974.

200. Ishihara, N, Kiyofuzi, T, Oboshi, S. *Proc Jap Cancer Assoc, 36th Annual Meeting* 120, 1977.

201. Xu, FH, Sharma, S, Gardner, A, Tu, Y, Raitano, A, Sawyers, C, Lichtenstein, A. *Blood* 92: 241–251, 1998.

# Chapter 5

# T-Cell Acute Lymphoblastic Leukemia and Natural Killer Cell Lines

Harry W. Findley, Heather L. Johnson and Stephen D. Smith
*Department of Pediatrics, Emory University School of Medicine, 2040 Ridgewood Drive NE, Atlanta, GA 30322 and Department of Pediatrics, The University of Kansas Medical Center, 3901 Rainbow Boulevard, 3032 Delp Pavilion, Kansas City, KS 66160-7357, USA. Tel: +1-913-588-6340; Fax: +1-913-588-2245; E-mail:ssmith8@kumc.edu*

## 1. INTRODUCTION

This chapter reports the characterization of forty-four cell lines that were established from patients with T-cell acute lymphoblastic leukemia (T-ALL) and five natural killer (NK) cell lines established from patients with leukemia with features of NK cells. Several subclones (variant cell lines derived from a parental line) and sister cell lines (cell lines derived independently from the same patient but from a different specimen or at a different time during the clinical course) are also described.

Acute lymphoblastic leukemia with a T-lymphocyte phenotype (T-ALL) has unique clinical, immunophenotypic, biochemical, and karyotypic features. Approximately 15% of children with leukemia have a thymic (T) cell phenotype as characterized by the ability to rosette with sheep RBCs or react with monoclonal antibodies associated with the T-lymphocyte lineage. Children with T-ALL characteristically are male, commonly present during adolescence, possess an anterior mediastinal mass and have a high white count. Immunophenotypically, T-ALL cells are characterized as immature T lymphocytes that possess pan-T cell antigens as detected by monoclonal antibodies. Early in thymocyte development, there is rearrangement of the T-cell receptor (*TCR*) $\alpha$, $\beta$, $\gamma$ or $\delta$ genes associated with expression of CD7. Many T-ALL cells express CD2 and CD5 and cytoplasmic CD3. With T-cell maturation, *TCR* ($\alpha/\beta$, $\gamma/\delta$) is expressed on the cell surface.

T-ALL cells have a high adenosine deaminase (ADA) level and a low nucleoside phosphorylase (NP) level compared to normal T-lymphocytes. The T-ALL cell lines are generally terminal deoxynucleotidyl transferase (TdT) positive [165]. While the retrovirus human T-cell leukemia/lymphoma virus

*J.R.W. Masters and B.O. Palsson (eds.), Human Cell Culture Vol. III, 157–206.*
© 2000 *Kluwer Academic Publishers. Printed in the Netherlands.*

type 1 is commonly associated with adult T-cell leukemia/lymphoma, T-ALL is not associated with the human T-cell virus type 1 or 2.

T-ALL is characterized by a lack of hyperdiploidy (>50 chromosomes) and a high incidence of chromosome translocations involving the T-cell receptor *(TCR)* $\alpha/\delta$ or $\beta$ locus. About 40% of T-ALL cases have nonrandom breakpoints and translocations within the 14q11, 7q34, or 7p15 region, which contain the *TCR* $\alpha/\delta$, *TCR* $\beta$, and *TCR* $\gamma$ genes, respectively. At least ten recurring chromosome translocations have been observed in T-ALL. These translocations commonly result in fusion of genes that encode a chimeric protein [133]. The most frequent recurring abnormality in childhood T-ALL is t(11;14)(p13;q11), which accounts for about 7% of cases [138]. The t(11;14)(p15;q11) is found in about 1% of T-ALL cases. In both translocations, the breakpoint on 14q occurs within the *TCR* $\alpha/\delta$ locus and the breakpoint on 11p15 occurs within a rhombotin related gene *(RBTN1)*. The gene on 11p13 *(RBTN2)* has extensive homology with *RBTN1* and both genes are normally involved in erythroid differentiation. Ectopic production of RBTN is thought to be a major factor in T-cell transformation.

Newly established cell lines should be compared to the patient's leukemia cells by an analysis of morphology, karyotype, immunophenotype, Southern blot, Northern blot, and protein expression. Two T-ALL cell lines with NK activity (SPI 801, SPI 802) were demonstrated to be contaminated with the K-562 cell line by analysis of the gene rearrangement pattern on Southern blot analysis [48,90]. Using karyotype analysis, it was found that KE-37 was contaminated with SKW3 cells, because both shared the t(8:14)(q24:q11) which was lacking in the original KE-37 cells [93]. Evaluation of established cell lines by DNA fingerprinting, immunophenotyping, karyotyping, and isoenzyme analysis has helped to identify cell line cross contamination and mislabeling [61]. At least one stock of the cell line EU-7 has been shown to be cross-contaminated with CCRF-CEM, emphasizing the need to screen stocks of EU-7 by a method such as DNA profiling.

## 2. CLINICAL CHARACTERISTICS

The first leukemia cell line was established by Dr. George Foley in 1964 [45]. In vitro cell cultures were produced from the peripheral blood of a two-year-old girl with leukemia. The resultant CCRF-CEM cell line was subsequently found to have a T-cell phenotype. A series of T-ALL cell lines (the MOLT lines) has been established by Dr. Jun Minowada [106]. The availability of the CEM and MOLT lines has been critical to progress in understanding the cellular and molecular biology of T-ALL, and the pioneering work of these investigators in establishing and characterizing those cell lines has greatly contributed to leukemia research. The RPMI 8402 cell line [113], the Jurkat cell line [146], and the KARPAS 45 cell line [76] have also been widely distributed. The MOLT 1-4 lines were established simultaneously from the same patient (a 19-year-old male) at first relapse [106]. While numerous studies of MOLT 3 and 4 have been reported, MOLT 1 and 2 have been less well characterized.

The clinical characteristics of the 44 T-ALL cell lines are summarized in Table 1a. One cell line was established in the 1960s, 9 cell lines in the 1970s, 26 in the 1980s, and 8 in the 1990s. Corresponding to the low incidence of T-ALL in adults, most of the cell lines were derived from children (median age 11 years) and the sex distribution was predominantly male. Interestingly, most of the cell lines were established at relapse and only eleven cell lines were established at the time of diagnosis. Twenty cell lines were cultured from bone marrow samples, eighteen from peripheral blood and one from pleural effusion. The origin of the cell line from the patient was verified in a relatively high proportion of cases. Generally this was accomplished by a direct comparison between the patient's leukemia cells and the established cell line (growing continuously for >12 months) by comparing the mono-clonal antibody profile, karyotype or the Southern blot *TCR* rearrangement profile. The medium used was predominantly RPMI-1640 with 10–20% fetal calf serum. While McCoy's 5a medium (and a hypoxic environment) was used for the initial growth of SUP-T cells, the cells grew in RPMI-1640 with 10% fetal calf serum once established [154]. Many T-ALL cell lines (e.g. CCRF-CEM, Jurkat, MOLT-4, etc.) are available from the American Type Cell Culture (ATCC) Rockville, MD and international cell banks.

The clinical characteristics of the NK cell lines are summarized in Table 1b. Natural killer (NK) cells represent a small proportion of the peripheral blood lymphocytes that frequently display azurophilic granules and thus have been named large granular lymphocytes (LGLs). Detailed phenotypic and genotypic studies of LGLs demonstrate that there are two distinct cell pop-ulations: LGLs which are CD3–, CD16+, CD56+ and lack rearrangement of the *TCR* genes and LGLs which are CD3+, CD16+, CD56+ and possess rearrangement of the *TCR* genes [96]. Patients with CD3+ LGL leukemias often exhibit specific clinical manifestations such as rheumatoid arthritis, recurrent bacterial infections, chronic neutropenia and pure red cell aplasia. Patients with CD3– LGL leukemias commonly present with hepatospleno-megaly, lymphadenopathy, skin involvement, and bone marrow infiltration [96]. The first cell line with natural killer (NK) activity (YT cells) was estab-lished in 1983 [172]. Five NK cell lines have been reported during the past decade, established from peripheral blood from adults with features of leuk-emia. Interleukin-2 (or conditioned media) was added to the initial culture

*Table 1a.* T-ALL cell lines: clinical characteristics

| Cell line[a] | Patient age/sex[b] | Diagnosis[c] | Treatment status | Specimen site[d] | Authentication[e] | Year est. | Culture medium[f] | Availability[g] | Primary ref. |
|---|---|---|---|---|---|---|---|---|---|
| BE-13 | 11 F | T-ALL | Relapse | BM | | 1981 | RPMI 1640 + 20% FCS | DSMZ | 46 |
| CCRF-CEM[h] | 2 F | T-ALL | Relapse | PB | | 1964 | Eagle MEM + 10% FCS | ATCC, CCR, JCRB | 45 |
| DND-41 | 13 M | T-ALL | Not stated | Not stated | | 1977 | RPMI 1640 + 12% FCS | Author | 36,98 |
| DU.528 | 16 M | T-ALL | Relapse | PB | Yes | 1985 | RPMI 1640 + 10% HS + 10% FCS | Author | 91 |
| EU-7[q] | 16 F | T-ALL | Relapse | BM | Yes | 1991 | RPMI 1640 + 20% FBS | Author | 175 |
| EU-9 | 8 M | T-ALL | Relapse | BM | Yes | 1993 | RPMI 1640 + 20% FBS | Author | 175 |
| HPB-ALL | 14 M | T-ALL | Diagnosis | PB | | 1973 | RPMI 1640 + 20% FBS | Author | 114 |
| JM/JURKAT | 14 M | T-ALL | Relapse | PB | | 1974 | RPMI 1640 +10% FCS | ATCC, ECACC | 146 36 |
| KARPAS 45 | 2 M | T-ALL | Not stated | BM | Yes | 1977 | RPMI 1640 + 10% FCS | Author | 76 |
| KE-37 | 27 M | T-ALL | Not stated | Not stated | | 1985 | RPMI 1640 + 5% FCS | DSMZ | 36 |
| KH-1 | 9 M | T-ALL | Relapse | PB | | 1982 | RPMI 1640 + 20% FCS | Author | 115 |
| K-T1 | 16 M | T-ALL | Relapse | BM | Yes | 1982 | McCoy 5A + 15% FCS | Author | 152 |
| L-KAW | 6 M | T-ALL | Relapse | PB | Yes | 1990 | RPMI 1640 + 10% FCS | Author | 104 |
| Loucy | 38 F | T-ALL | Relapse | PB | Yes | 1990 | RPMI 1640 + 30% FBS | DSMZ | 11 |
| MOLT-1[i] | 19 M | T-ALL | Relapse | PB | | 1971 | RPMI 1640 + 10% FBS | DSMZ | 106 |

*Continued on next page*

*Table 1a.* (continued)

| Cell line[a] | Patient age/sex[b] | Diagnosis[c] | Treatment status | Specimen site[d] | Authentication[e] | Year est. | Culture medium[f] | Availability[g] | Primary ref. |
|---|---|---|---|---|---|---|---|---|---|
| MOLT-2-4 | | | | | | | | ATCC, DSMZ, ECACC, JCRB, RGB | |
| MOLT-12[j] | 2 F | T-ALL | Diagnosis | BM | Yes | 1983 | RPMI 1640 + 10% FBS | Author | 39,110 |
| MOLT-13 | 2 F | T-ALL | Relapse | BM | Yes | 1983 | RPMI 1640 + 10% FBS | Author | 39,110 |
| MOLT-14 | 2 F | T-ALL | Relapse | BM | Yes | 1983 | RPMI 1640 + 10% FBS | ATCC | 39,110 |
| MOLT-16[k] | 5 F | T-ALL | Relapse | PB | Yes | 1985 | RPMI 1640 +10% FCS | DSMZ | 39,110 |
| MOLT-17 | 5 F | T-ALL | Relapse | PB | Yes | 1985 | RPMI 1640 + 10% FCS | DSMZ | 39,110 |
| MT-ALL | 15 M | T-ALL | Relapse | PB | Yes | 1989 | IMDM with 20% FCS | Author | 56,57 |
| P12/Ichikawa | 7 | T-ALL | Not stated | PB | | 1978 | Nude Mice | DSMZ | 168 |
| P30/Ohkubo | 11 F | T-ALL | Relapse | BM | Yes | 1982 | RPMI 1640 + 10% FCS | JCRB | 66 |
| PEER | 4 F | T-ALL | Relapse | PB | Yes | 1977 | RPMI 1640 + 20% FCS | DSMZ, JCRB | 136 |
| PER 117 | 1 M | T-ALL | Relapse | BM | Yes | 1987 | RPMI 1640 + 20% FCS | Author | 77 |
| PER 255 | 5 M | T-ALL | Diagnosis | BM | Yes | 1989 | RPMI 1640 + Human Serum | Author | 79 |
| PER 423 | 5 M | T-ALL | Diagnosis | BM | Yes | 1993 | RPMI-1640 + FCS + IL-2 | Author | 80 |
| PF-382[l] | 6 F | T-ALL | Relapse | PE | Yes | 1985 | RPMI 1640 + 10% FCS | DSMZ | 130 |

*Continued on next page*

*Table 1a.* (continued)

| Cell line[a] | Patient age/sex[b] | Diagnosis[c] | Treatment status | Specimen site[d] | Authenti-cation[e] | Year est. | Culture medium[f] | Availability[g] | Primary ref. |
|---|---|---|---|---|---|---|---|---|---|
| RPMI 8402 | 16 F | T-ALL | Relapse | PB | | 1972 | RPMI 1640 + 20% FCS | DSMZ, CCR | 71,113 |
| SUP-T2 | 52 F | T-ALL | Diagnosis | BM | Yes | 1984 | McCoy 5A + 15% FCS | Author | 152 |
| SUP-T3 | 12 M | T-ALL | Relapse | PB | Yes | 1984 | McCoy 5A + 15% FCS | Author | 152 |
| SUP-T6 | 7 M | T-ALL | Diagnosis | BM | Yes | 1986 | McCoy 5A + 15% FCS | Author | 153 |
| SUP-T7 | 19 M | T-ALL | Diagnosis | BM | Yes | 1985 | McCoy 5A + 15% FCS | Author | 153 |
| SUP-T8 | 8 F | T-ALL | Relapse | BM | Yes | 1987 | McCoy 5A + 15% FCS | Author | 154 |
| SUP-T9 | 10 F | T-ALL | Relapse | BM | Yes | 1987 | McCoy 5A + 15% FCS | Author | 154 |
| SUP-T10 | 8 M | T-ALL | Relapse | BM | Yes | 1987 | McCoy 5A + 15% FCS | Author | 154 |
| SUP-T12 | 17 M | T-ALL | Diagnosis | PB | Yes | 1987 | McCoy 5A + 15% FCS | Author | 154 |
| SUP-T13 | 3 F | T-ALL[m] | Relapse | PB | Yes | 1987 | McCoy 5A + 15% FCS | Author | 154 |
| SUP-T14 | 6 M | T-ALL | Diagnosis | BM | Yes | 1987 | McCoy 5A + 15% FCS | Author | 154 |
| TALL-1 (a) | 28 M | T-ALL | Relapse | BM | | 1976 | RPMI 1640 + 20% FCS | Author | 111 |
| TALL-1 (b) | 4 M | T-ALL | Relapse | Not stated | Yes | 1987 | IMDM + 20% FBS | Author | 92 |
| TALL-101 | 9 M | T-ALL | Relapse | Not stated | Yes | 1987 | IMDM + 20% FBS + GM-CSF | Author | 92 |
| TALL-103/2[n] | 6 M | LL[o] | Relapse | BM | Yes | 1988 | IMDM + 10% FBS + IL-2 | Author | 121 |
| TALL-104 | 2 M | T-ALL | Relapse | PB | Yes | 1991 | IMDM + 10% FBS + IL-2 | ATCC | 122 |
| TALL-105 | 9 F | T-ALL | Diagnosis | BM | Yes | 1991 | IMDM + 10% FBS | Author | 122 |

*Continued on next page*

*Table 1a.* (continued)

| Cell line[a] | Patient age/sex[b] | Diagnosis[c] | Treatment status | Specimen site[d] | Authenti- cation[e] | Year est. | Culture medium[f] | Availability[g] | Primary ref. |
|---|---|---|---|---|---|---|---|---|---|
| TALL-106 | 15 M | T-ALL | Diagnosis | BM | Yes | 1991 | IMDM + 10% FBS + IL-3 | Author | 122 |
| UHKT-42 | 12 M | AUL[p] | Relapse | PB | Yes | 1988 | RPMI 1640 + 20% FCS | Author | 157 |

[a] Names of cell lines are indicated as given in the original literature. Subclones (variant cell lines derived from a parental cell line) and sister cell lines (derived independently from the same patient from different specimens or at different time points) are indicated for each cell line.

[b] Age at the time of establishment of cell line.

[c] Diagnosis is indicated as given in the original literature.

[d] BM – bone marrow; PB – peripheral blood; PE – pleural effusion.

[e] Evidence (eg. cytogenetic marker chromosomes, immunophenotype, others) that the cell line was derived from the patient indicated.

[f] Culture medium as indicated in the original literature; cell line may grow with other media and/or supplements.

[g] Available from cell banks (American Type Culture Collection ATCC; Coriell Cell Repository CCR; Deutsche Sammlung von Mikroorganismen und Zelkulturen DSMZ; European Collection of Animal Cell Cultures ECACC; Japanese Collection of Research Bioresources JCRB; and Riken Gene Bank RGB), or from the original investigator.

[h] CCRF-CEM C1 and CCRF-CEM-C7 are sublines of CCRF-CEM.

[i] MOLT 2-4 are subclones of MOLT 1.

[j] MOLT 13-14 are sister cell lines which were established from the same patient as MOLT 12.

[k] MOLT 17 is a sister cell line which was established from the same patient as MOLT 16.

[l] GI-CO-T-9 is a PF-382 subclone which was selected because of the elaboration of factors which suppress T-cell proliferation [131].

[m] Patient originally incorrectly reported as lymphoblastic lymphoma.

[n] TALL-103/2 is a T-lymphoid subclone of TALL 103/3. TALL 103/3 is an IL-3 dependent cell line which has lost T-cell specific markers and acquired a myeloid phenotype (CD15+, CD33+) [121].

[o] LL – lymphoblastic lymphoma, cell line established from BM in leukemic phase at relapse.

[p] AUL – acute undifferentiated leukemia.

[q] At least one stock of EU-7 cells has been found to be in fact CCRF-CEM.

*Table 1b.* NK cell lines: clinical characterization

| Cell line[a] | Patient age/sex[b] | Diagnosis[c] | Treatment status | Specimen Site[d] | Authenti- cation[e] | Year est. | Culture medium[f] | Availability[g] | Primary ref. |
|---|---|---|---|---|---|---|---|---|---|
| NK-92 | 50 M | NK-NHL | Not stated | PB | Yes | 1994 | αMEM + 15% FCS+ 12.5% HS + IL-2+ Hydrocortisone | Author | 51 |
| NKL | 62 M | Chronic leukemia LGL | Relapse | PB | Yes | 1994 | RPMI 1640 + 15% AB Serum + IL-2 | Author | 139 |
| NOI 90 | 45 M | NK-NHL | Diagnosis | PB | Yes | 1990 | RPMI 1640 + 10% FCS | Author | 142 |
| TKS-1 | 21 M | NK cell leukemia | Not stated | PB | Yes | 1992 | RPMI 1640 + 10% FCS + IL-2 | Author | 89 |
| YT[h] | 15 M | LL[i] | Not stated | Pericardial effusion | Yes | 1983 | RPMI 1640 + 10% FCS | Author | 172 |
| YT2C2 | | | | | | 1985 | RPMI 1640 + 10% FCS | Author | 160,173 |
| YT3C | | | | | | 1984 | RPMI 1640 + 10% FCS | Author | 172,173 |

[a] Names of cell lines are indicated as given in the original literature. Subclones (variant cell lines derived from a parental cell line) and sister cell lines (derived independently from the same patient from different specimens or at different time points) are indicated for each cell line.

[b] Age at the time of establishment of cell line.

[c] Diagnosis is indicated as given in the original literature.

[d] BM: bone marrow; PB: peripheral blood.

[e] Evidence (eg. cytogenetic marker chromosomes, immunophenotype, others) that the cell line was derived from the patient indicated.

[f] Culture medium as indicated in the original literature; cell line may grow with other media and/or supplements.

[g] Available from the original investigator.

[h] YT2C2 and YT3C are subclones of YT.

[i] LL – lymphoblastic leukemia.

media in the NK cell cultures and the NK cell line (NOI-90) was found to have an IL-2 autocrine loop.

## 3. IMMUNOPHENOTYPE

T-ALL cases have distinctive immunophenotypes that generally reflect their origin from cells at different stages of intrathymic differentiation. Reinherz and others have proposed models of thymic maturation based on expression of surface antigens, which have been useful for classification of T-ALL [137,107]. This classification was subsequently modified after the introduction of additional Leukocyte Workshop antibodies and Cluster Designation (CD) categories [30,134]. T-ALL/thymocyte-maturation stages include the early thymocyte phenotype (stage I), characterized by expression of CD2, 5 and 7 without CD1, 3, 4 or 8; the intermediate thymocyte phenotype (stage II), with expression of CD1, 2, 5 and 7 and variable expression of CD3, 4 and 8; and the mature thymocyte phenotype (stage III), expressing CD2, 3, 5 and 7 and in most cases either CD4 or 8 (the latter antigens generally being expressed singly rather than in combination). CD1 (the common thymocyte antigen) is negative in stage I and III cases, and therefore its presence is diagnostic for the stage II intermediate (or common) thymocyte group. CD3 may be weakly expressed on the cell surface in some stage II cases, or may be present only in the cytoplasm. Also, stage II cases may express both CD4 and CD8. However, some T-ALL cell lines and primary T-ALL cells do not fit into any one category due to their heterogeneity in antigen expression.

During T-cell differentiation, *TCR* genes rearrange to permit receptor expression and diversity. Two types of *TCR* molecules have been identified on the surface of T-ALL and normal T-cells, both of which consist of a polymorphic heterodimer of two polypeptide chains $\alpha\beta$ or $\gamma/\delta$. In normal T-cells, the *TCR* is linked to a monomorphic glycoprotein complex (CD3). In T-ALL, however, CD3 is commonly expressed in the cytoplasm but not on the cell surface, indicating transformation at an immature stage of development [167]. T-ALL cell lines lacking surface CD3 are generally also negative for surface *TCR* [16].

MOLT-3 and -4 belong to the intermediate thymocyte category (Stage II), based on expression of CD1, 2, 5 and 7 on >50% of the cells. CD4 and 8 are expressed on a minority of cells, and CD3 is negative [110]. MOLT-12 also has the intermediate thymocyte phenotype (Stage II thymocyte), characterized by expression of CD1, 2, 5, and 7 [39,110]. Unlike MOLT-3 and -4, MOLT-12 also expresses surface CD3 on >50% of the cells. MOLT-13 and -14 differ from MOLT-12 in having lost expression of CD1 and 2; since MOLT-13 and -14 were established from the same patient as MOLT-

12 after relapse, this antigen loss may represent the appearance of a less mature phenotype at relapse [39]. Based on the absence of CD1, 2, 4 and 8, MOLT-13 and -14 appear to have an immature thymocyte phenotype (Stage I). However, both MOLT-13 and -14 express surface CD3 on at least 50% of cells, which is associated with a more differentiated phenotype (Stage II or III); therefore, it is difficult to assign these lines to a precise category. MOLT-16 appears to represent a similar stage of differentiation (Stage I) based on absence of CD1, 4, and 8, and the presence of CD2, 5 and 7 [39]. However, as in the case of MOLT-13 and -14, the presence of surface CD3 on >50% of these cells complicates the assignment, and this line has features of Stage II (intermediate thymocyte), except for a lack of CD1.

The immunophenotypes of the T-ALL cell lines are summarized in Table 2a. Most of the T-ALL cell lines express both CD5 (31/37) and CD7 (37/39) and many express CD1 (19/43), CD3 (23/44), or CD4 (25/43). Several T-ALL cell lines express CD2 (29/41) and about 70% express CD8 (25/40). Of the 25 cell lines that express CD8, most also express CD2 (21/25) and dual expression of CD2/CD8 is associated with a mid-thymocyte stage of T-cell development [137]. When evaluated, T-ALL cell lines have rearrangements of the *TCR* genes ($\alpha/\beta/\gamma/\delta$) and many express the *TCR* $\alpha/\beta$ or $\gamma/\delta$ on the cell surface. While CD10 expression was detected on 20% (7/34) of the T-ALL cell lines tested, other B lymphocyte markers are not expressed, with the exception of MOLT-3 and -4 which express the B-lineage marker CD9. Myelomonocytic markers are uniformly negative.

T-ALL cells are generally TdT positive, a feature shared with normal cortical thymocytes but not mature lymphocytes [37]. Most of the cell lines express TdT (25/34) and are HLA-DR negative (31/33). Only 2 cell lines express the interleukin-2 receptor CD25 (TALL-101, TALL- 103/2).

The immunophenotypic characteristics of the NK cells are summarized in Table 2b. Two populations of normal LGLs have been described: CD3−, CD16+, CD56+, *TCR*− germline and CD3+, CD16+, CD56+, *TCR*− rearranged. The YT and NK92 cells are CD3−, CD16−, CD56+; NKL cells are CD3−, CD16−, CD56−; TKS1 cells are CD3−, CD16+, CD56− and NOI90 cells are CD3+, CD16−, CD56−. All of the NK cell lines (except YT subclones YT2C2) express CD25 and proliferate in response to IL-2 or conditioned media. The YT2C2 cell line was specifically subcloned to provide cells that lack CD25 for comparative analysis with CD25+ YT cells [173]. While YT cells possess EBV, it is interesting to note that the cells lack CD21 which is the putative EBV receptor. All the NK cell lines are CD4− but express the pan-T cell antibodies CD7, while lacking B-lymphoid and myelomonocytic markers. Only the TKS1 cells express CD16.

Table 2a. T-ALL cell lines: immunophenotype characteristics

| Cell line | T-/NK-cell marker[a] | T-cell receptor[b] | B-cell marker | Myelomonocytic marker | Progenitor/activation marker | Adhesion marker | Ref. |
|---|---|---|---|---|---|---|---|
| BE-13 | CD1+ CD8−<br>CD3+<br>CD4+ | | sIg− | | | | 46 |
| CCRF-CEM[c] | CD1− CD5+<br>CD2+ CD7+<br>CD3− CD8−<br>CD4+ | TCRα R/G<br>TCRβ G/R<br>TCRγ R/R<br>TCRδ D/D<br>TCRα, β,<br>γ expressed<br>on northern blot | CD9− cIg−<br>CD10+ sIg−<br>CD19−<br>CD20−<br>CD21−<br>CD24− | | CD18+<br>CD25−<br>CD28+<br>CD71+<br>HLA-DR−<br>TdT+ | | 112, 54, 94, 16 |
| DND-41 | CD1+ CD5+<br>CD2+ CD7+<br>CD3+ CD8−<br>CD4+ | TCRα G/G<br>TCRβ G/R<br>TCRγ G/R<br>TCRδ G/R<br>TCRβ, γ,<br>δ expressed<br>by northern blot | CD9−<br>CD10+<br>CD24− | | CD18+<br>CD25−<br>CD28+<br>HLA-DR−<br>TdT+ | | 4, 98, 94 |
| DU.528[d] | CD1− CD7+<br>CD2− CD8−<br>CD3−<br>CD4− | TCRα R<br>TCRβ R<br>TCRγ R<br>TCRδ R | | CD14− | CD25+<br>CD34(+)<br>CD71+<br>HLA-DR−<br>TdT− | | 91, 8, 112 |

*Continued on next page*

Table 2a. (continued)

| Cell line | T-/NK-cell marker[a] | T-cell receptor[b] | B-cell marker | Myelomonocytic marker | Progenitor/activation marker | Adhesion marker | Ref. |
|---|---|---|---|---|---|---|---|
| EU-7[f] | CD1+ CD2− CD5+ CD3+ CD7+ CD4+ CD8+ | | CD10+ CD19− CD20− CD24− | CD13− CD15+ CD33− | CD34− CD38+ HLA-DR− TdT+ | | 175 |
| EU-9 | CD1+ CD2+ CD5+ CD3− CD7+ CD4− CD8+ | | CD10− CD19− CD20− CD24− | CD13− CD15+ CD33− | CD34− CD38+ HLA-DR− TdT+ | | 175 |
| HPB-ALL | CD1+ CD5+ CD2+ CD7+ CD3+ CD8+ CD4+ | TCRγ R | CD9− cIg− CD10+ sIg− CD19− CD20− CD21− CD24− | | HLA-DR− TdT+ | | 84, 12 |
| JM/JURKAT | CD1+ CD5+ CD2+ CD7+ CD3+ CD8+ CD4+ | TCRγ R | CD9+ CD10− CD24− | | HLA-DR− TdT+ | | 36, 84, 94 |
| KARPAS 45 | CD1− CD5(+) CD3− CD8+ CD4+ | | CD10− sIg− | | | | 76, 26 |

*Continued on next page*

Table 2a. (continued)

| Cell line | T-/NK-cell marker[a] | T-cell receptor[b] | B-cell marker | Myelomonocytic marker | Progenitor/activation marker | Adhesion marker | Ref. |
|---|---|---|---|---|---|---|---|
| KE-37 | CD1– CD5+<br>CD2– CD7+<br>CD3– CD8–<br>CD4+ | TCRγ R | CD9–<br>CD10–<br>CD24– | | CD30+<br>HLA-DR–<br>TdT– | | 36, 84, 94, 58 |
| KH-1 | CD1+ CD5+<br>CD2+ CD8+<br>CD3+<br>CD4+ | | sIg– | | CD71– | | 115 |
| K-T1 | CD1– CD5–<br>CD2+ CD7+<br>CD3+ CD8+<br>CD44 | | cIg– sIg– | | CD38+<br>CD71+<br>HLA-DR–<br>TdT+ | | 152 |
| L-KAW | CD1– CD5+<br>CD2+ CD7+<br>CD3+ CD8+<br>CD4– CD56–<br>CD57– | TCRα E<br>TCRβ E | CD10–<br>CD19–<br>CD20–<br>CD21– | CD13–<br>CD14–<br>CD16–<br>CD33– | CD18+<br>CD25–<br>CD28+<br>CD34+<br>CD45 RA–<br>CD45 RO+<br>CD71+<br>HLA-DR– | CD11a+ | 104 |
| LOUCY | CD1– CD4+<br>CD2– CD8+<br>CD3+ | | CD10+ cIg–<br>CD19– sIg– | | CD25–<br>TdT– | CD11b– | 11 |

*Continued on next page*

*Table 2a.* (continued)

| Cell line | T-/NK-cell marker[a] | T-cell receptor[b] | B-cell marker | Myelomonocytic marker | Progenitor/activation marker | Adhesion marker | Ref. |
|---|---|---|---|---|---|---|---|
| MOLT-3[e] | CD1+ CD5+<br>CD2+ CD7+<br>CD3– CD8(+)<br>CD4+ CD57(+) | TCRβ R<br>TCRγ G | CD9+<br>CD10– sIg–<br>CD19– | CD13– | HLA-DR–<br>TdT+ | CD11a+<br>CD11b– | 55, 108, 109, 156, 16 |
| MOLT-4 | CD1+ CD5+<br>CD2– CD7+<br>CD3– CD8+<br>CD4+ CD57+ | TCRβ R<br>TCRγ G | CD9+<br>CD10– sIg–<br>CD19–<br>CD24– | CD13– | HLA-DR–<br>TdT+ | CD11a– | 55, 84, 108, 109, 156, 144 |
| MOLT-12 | CD1+ CD5+<br>CD2+ CD7+<br>CD3+ CD8+<br>CD4– | | CD10–<br>CD19– sIg–<br>CD20– | CD13–<br>CD15– | CD25–<br>CD38+<br>HLA-DR–<br>TdT+ | CD11b– | 39, 110 |
| MOLT-13 | CD1– CD5+<br>CD2– CD7+<br>CD3+ CD8<br>CD4– | TCRγ E<br>TCRδ E | CD10–<br>CD19– sIg–<br>CD20– | CD13–<br>CD15– | CD25–<br>CD38+<br>CD147+<br>HLA-DR–<br>TdT+ | CD11b– | 39, 110, 88, 85, 31, 32 |
| MOLT-14 | CD1– CD5+<br>CD2– CD7+<br>CD3+ CD8–<br>CD4– | TCRγ E<br>TCRδ E | CD10– sIg–<br>CD19–<br>CD20– | CD13–<br>CD15– | CD25–<br>CD38+<br>HLA-DR–<br>TdT+ | CD11b– | 39, 110, 88, 31, 32 |

*Continued on next page*

*Table 2a.* (continued)

| Cell line | T-/NK-cell marker[a] | T-cell receptor[b] | B-cell marker | Myelomonocytic marker | Progenitor/activation marker | Adhesion marker | Ref. |
|---|---|---|---|---|---|---|---|
| MOLT-16 | CD1− CD5+ <br> CD2+ CD6+ <br> CD3+ CD7+ <br> CD4− CD8− | TCRα E <br> TCRβ E | CD10− sIg− <br> CD19− <br> CD20− | CD13− <br> CD15− <br> CD16− | CD25− <br> CD28− <br> CD38− <br> HLA-DR− <br> TdT+ | CD11b− | 39, 110, 88, 16, 32 |
| MOLT-17 | CD1− CD5+ <br> CD2+ CD6+ <br> CD3+ CD7+ <br> CD4− CD8− | TCRα E <br> TCRβ E | CD10− <br> CD19− <br> CD20− | CD13− <br> CD15− | CD25− <br> CD38+ <br> HLA-DR− <br> TdT+ | CD11b− | 39, 88, 16 |
| MT-ALL | CD5− <br> CD2+ CD7+ <br> CD3+ CD8− <br> CD4− | TCRα E <br> TCRβ E | | CD13+ <br> CD14− <br> CD33− | CD25− | | 57, 56, 74 |
| P12/Ichikawa | CD1+ CD5+ <br> CD2+ CD7+ <br> CD3− CD8+ <br> CD4+ | | CD9− <br> CD10− <br> CD24− | | CD30+ <br> HLA-DR− <br> TdT+ | | 168, 58 |
| P30/Ohkubo | CD1− CD5+ <br> CD3− CD8− <br> CD4− | TCRβ G <br> TCRγ G | CD10+ cIg− <br> sIg− | | HLA-DR+ <br> TdT+ | | 66, 84 |

*Continued on next page*

*Table 2a.* (continued)

| Cell line | T-/NK-cell marker[a] | T-cell receptor[b] | B-cell marker | Myelomonocytic marker | Progenitor/activation marker | Adhesion marker | Ref. |
|---|---|---|---|---|---|---|---|
| PEER | CD1– CD5+<br>CD2– CD7+<br>CD3+ CD8–<br>CD4+ | TCRα G/G<br>TCRβ G/R<br>TCRγ G/R<br>TCRδ R/G<br>TCRβ, γ,<br>δ expressed<br>on northern blot | CD9+<br>CD10–<br>CD24– | | CD18+<br>CD25–<br>CD28+<br>HLA-DR+<br>TdT– | | 4, 94 |
| PER 117 | CD1– CD7+<br>CD2– CD8–<br>CD3–<br>CD4– | | | CD15– | CD38+<br>CD71–<br>HLA-DR–<br>TdT– | | 77 |
| PER 255 | CD1+ CD5–<br>CD2+ CD6–<br>CD3– CD7–<br>CD4+ CD8(+) | TCRβ– G/R | CD10–<br>CD19–<br>CD20– | | CD38+<br>CD71–<br>TdT+ | CD11b– | 79 |
| PER 423 | CD1– CD5–<br>CD2– CD6–<br>CD3– CD7+<br>CD4– CD8+<br>CD57+ | TCRβ– R | CD10–<br>CD19– | CD16– | CD25–<br>CD122+<br>HLA-DR– | CD11b(+) | 80 |

*Continued on next page*

*Table 2a.* (continued)

| Cell line | T-/NK-cell marker[a] | T-cell receptor[b] | B-cell marker | Myelomonocytic marker | Progenitor/activation marker | Adhesion marker | Ref. |
|---|---|---|---|---|---|---|---|
| PF-382 | CD1+ CD5+<br>CD2+ CD7+<br>CD3−<br>CD4− CD8+ | | CD9+<br>CD10−<br>CD24− | | TdT− | | 130 |
| RPMI 8402 | CD1− CD5+<br>CD2+ CD7+<br>CD3− CD8−<br>CD4+ | TCRγ R | CD10+ | | HLA-DR−<br>TdT+ | | 71, 84 |
| SUP-T2 | CD1− CD5+<br>CD2+ CD7+<br>CD3− CD8+<br>CD4− | | CD10− sIg− | CD13−<br>CD33− | CD38−<br>CD71+<br>HLA-DR−<br>TdT+ | | 151 |
| SUP-T3 | CD1+ CD5+<br>CD2+ CD7+<br>CD3− CD8+<br>CD4+ | | CD10− sIg− | CD13−<br>CD33− | CD25−<br>CD38+<br>CD71−<br>HLA-DR−<br>TdT+ | | 151 |
| SUP-T6 | CD1− CD5+<br>CD2+ CD7+<br>CD3− CD8+<br>CD4− | TCRβ R<br>CRγ R | CD10− sIg− | CD13−<br>CD33− | CD25−<br>HLA-DR−<br>TdT+ | | 154 |

*Continued on next page*

*Table 2a.* (continued)

| Cell line | T-/NK-cell marker[a] | T-cell receptor[b] | B-cell marker | Myelomonocytic marker | Progenitor/activation marker | Adhesion marker | Ref. |
|---|---|---|---|---|---|---|---|
| SUP-T7 | CD1– CD5+ CD2– CD7+ CD3+ CD4+ CD9+ | TCRβ R | CD10– sIg– | CD13– CD33– | CD25– CD38+ CD71+ HLA-DR– TdT+ | | 153 |
| SUP-T8 | CD1– CD5+ CD2– CD7+ CD3– CD8– CD4+ | TCRβ G/G TCRγ G/G | CD10– sIg– | CD13– CD33+ | CD25– HLA-DR– TdT– | | 154 |
| SUP-T9 | CD1+ CD5+ CD2+ CD7+ CD3– CD8+ CD4+ | TCRβ R TCRγ R | CD10– sIg– | CD13– CD33– | CD25– HLA-DR– TdT+ | | 154 |
| SUP-T10 | CD1– CD5– CD2– CD7+ CD3– CD8– CD4– | TCRβ G/G TCRγ G/G | CD10– sIg– | CD13– CD33– | CD25– HLA-DR– TdT+ | | 154 |
| SUP-T12 | CD1+ CD5+ CD2+ CD7+ CD3– CD8+ CD4+ | TCRβ R TCRγ R | CD10– sIg– | CD13– CD33– | CD25– HLA-DR– TdT+ | | 154 |

*Continued on next page*

Table 2a. (continued)

| Cell line | T-/NK-cell marker[a] | T-cell receptor[b] | B-cell marker | Myelomonocytic marker | Progenitor/activation marker | Adhesion marker | Ref. |
|---|---|---|---|---|---|---|---|
| SUP-T13 | CD1− CD5+<br>CD2+ CD7+<br>CD3+ CD8+<br>CD4+ | TCRβ G/G<br>TCRγ R | CD10− sIg− | CD13−<br>CD33− | CD25−<br>HLA-DR−<br>TdT+ | | 154 |
| SUP-T14 | CD1− CD5+<br>CD2+ CD7+<br>CD3+ CD8−<br>CD4+ | TCRβ G/G<br>TCRγ R | CD10− sIg− | CD13−<br>CD33− | CD25−<br>HLA-DR−<br>TdT+ | | 154 |
| TALL-1(a) | CD1+ CD5+<br>CD2+ CD7+<br>CD3+ CD8+<br>CD4+ | | CD9−<br>CD10−<br>CD24− | | HLA-DR−<br>TdT+ | | 36, 156, 94 |
| TALL-1(b) | CD1+ CD5+<br>CD2+ CD7+<br>CD3+ CD8+<br>CD4+ | | CD10− | | HLA-DR−<br>TdT+ | | 92 |
| TALL-101 | CD1− CD5−<br>CD2− CD7−<br>CD3− | | CD10−<br>CD19− | CD16+<br>CD38+ | CD25+<br>CD71−<br>TdT− | CD11b+ | 164 |
| TALL-103/2 | CD1− CD2+<br>CD7+<br>CD3+ CD8+<br>CD4− CD56+ | TCRγ E<br>TCRδ E | CD34−<br>CD38+ | CD14−<br>CD15−<br>CD33+ | CD71+<br>HLA-DR−<br>IL-2α+<br>IL-2β+ | CD11b−<br>CD16−<br>CD64− | 121, 122 |

Continued on next page

*Table 2a.* (continued)

| Cell line | T/NK-cell marker[a] | T-cell receptor[b] | B-cell marker | Myelomonocytic marker | Progenitor/activation marker | Adhesion marker | Ref. |
|---|---|---|---|---|---|---|---|
| TALL-104 | CD2+ CD7+<br>CD3+ CD8+<br>CD4– CD56+ | TCRα E<br>TCRβ E | | CD33–<br>CD45– | CD34– | CD11b– | 122 |
| TALL-105 | CD1+<br>CD2+ CD7+<br>CD3+ CD56–<br>CD4– | TCRα E<br>TCRβ E | | CD15–<br>CD33– | IL-2α–<br>IL-2β–<br>CD34+ | CD11b– | 122 |
| TALL-106 | CD1+ CD2+<br>CD7+<br>CD3+ CD56–<br>CD4(+) | TCRα E<br>TCRβ E | | CD15–<br>CD33– | IL-2α–<br>IL-2β–<br>CD34+ | CD11b+ | 122 |
| UHKT-42 | CD1a– CD5+<br>CD1b– CD7+<br>CD2– CD8–<br>CD3– CD56–<br>CD4– CD57– | TCRβ R<br>TCRδ R | CD10– sIg–<br>CD19–<br>CD20–<br>CD21–<br>CD24– | CD13–<br>CD14–<br>CD15–<br>CD16–<br>CD33–<br>CD41a–<br>CD61– | CD25–<br>CD28+<br>CD34+<br>CD38+<br>CD45 RO+<br>CD71+<br>HLA-DR–<br>TdT– | | 157 |

[a] + – strong, definite protein expression (more than 10% of cells positive); (+) – weak protein expression qualitatively and quantitatively, less than 10% cells positive; (–) – no protein expression; all immunophenotyping is for surface expression. [b] TCR – T cell receptor, R – rearranged, G – germ line, D – deleted, E – expressed on cell surface. [c] CCRF-CEM cells are cytoplasmic CD3+ and surface CD3– [16]. [d] DU.528 is a stem cell line with the capacity to produce cells of the T-lymphoid, granulocyte/monocyte and erythroid lineage. [e] MOLT 2-4 are subclones of MOLT-1. Comprehensive immunophenotype and other data have not been reported for MOLT 1 and 2. [f] At least one stock of EU-7 cells has been found to be in fact CCRF-CEM, and this cell line should not be used unless the stocks available are shown to be distinct from CCRF-CEM.

*Table 2b.* NK cell lines immunophenotypic characteristics

| Cell line | T-/NK-cell marker[a] | T-cell receptor[b] | B-cell marker | Myelo-monocytic marker | Progenitor/ activation marker | Adhesion marker | Ref. |
|---|---|---|---|---|---|---|---|
| NK92 | CD1– CD7+<br>CD2+ CD8–<br>CD3– CD28+<br>CD4– CD56+<br>CD5– | TCRβ G<br>TCRγ G | CD10–<br>CD19–<br>CD20–<br>CD23– | CD14–<br>CD16– | CD25+<br>CD34–<br>CD45+<br>CD54+<br>CD122+<br>HLA-DR– | CD11a+ | 51 |
| NKL | CD1– CD7–<br>CD2+ CD8–<br>CD3– CD26+<br>CD4– CD27+<br>CD5– CD28–<br>CD6+ CD56–<br>CD57– | TCRα/β<br>not expressed<br>TCRγ/δ<br>not expressed | CD19–<br>CD20– | CD14–<br>CD16–<br>CD29+ | CD25+<br>CD38+<br>CD69–<br>CD71+<br>CD81+<br>CD94+<br>CD95+<br>HLA-DR+ | CD11a+<br>CD11b–<br>CD43+ | 139 |
| NOI 90 | CD1 (+) CD7+<br>CD2– CD8–<br>CD3+ CD56+<br>CD4 (+) CD57+<br>CD5– | TCRβ G<br>TCRγ G | CD10– | CD13–<br>CD14–<br>CD16– | CD25 (+)<br>CD34– | | 142 |
| TKS1 | CD2+ CD5–<br>CD3– CD8–<br>CD4– CD56–<br>CD57– | TCRβ G<br>TCRγ G | CD10–<br>CD19–<br>CD20– | CD16+ | CD25+<br>CD122– | | 89 |
| YT | CD1– CD5–<br>CD2– CD6–<br>CD3– CD7+<br>CD4– CD8–<br>CD5– CD28+<br>CD56+<br>CD57– | TCRβ G<br>TCRγ G<br>TCRδ G | CD10–<br>CD19–<br>CD20–<br>CD21–<br>CD23–<br>CD24– | CD13–<br>CD15–<br>CD16– | CD25+<br>CD30+<br>CD33–<br>CD38–<br>CD45RA–<br>CD45RO+<br>HLA-DR+ | CD11b– | 172,173 |
| YT2C2 | CD1– CD2–<br>CD6– CD3–<br>CD7+ CD4–<br>CD8– CD5–<br>CD28+ CD56+<br>CD57– | TCRβ G<br>TCRγ G<br>TCRδ G | CD10–<br>CD19–<br>CD20–<br>CD21–<br>CD23–<br>CD24– | CD13–<br>CD15–<br>CD16– | CD25–<br>CD30+<br>CD33–<br>CD38–<br>CD45RA–<br>CD45RO+<br>HLA-DR+ | CD11b– | 173,160 |
| YT3C | CD1–<br>CD2– CD6–<br>CD3– CD7+<br>CD4– CD8–<br>CD5– CD28+<br>CD56+<br>CD57– | TCRβ G<br>TCRγ G<br>TCRδ G | CD10–<br>CD19–<br>CD20–<br>CD21–<br>CD23–<br>CD24– | CD13–<br>CD15–<br>CD16– | CD25+<br>CD30+<br>CD33–<br>CD38–<br>CD45RA–<br>CD45RO+<br>HLA-DR+ | CD11b– | 173,172,160 |

[a] + – strong definite protein expression (more than 10% of cell positive); (+) – weak protein expression qualitatively and quantitatively, less than 10% cells positive; − – no protein expression.
[b] TCR = T cell receptor, R – rearranged, G – germ line, D – deleted, E = expressed on cell surface.

## 4. CYTOKINE-RELATED CHARACTERIZATION

Cytokine receptor expression, cytokine production, cytokine induced proliferation and differentiation and cytokine dependency of T-ALL cell lines are
summarized in Table 3a. Of note is that 39 of the 44 T-ALL cell lines were
established without the addition of exogenous cytokines. Five T-ALL cell
lines show increased proliferation in response to IL-2 and three cell lines are
dependent on IL-2 for cellular proliferation. Phorbol ester induces IL-2R in
some T-ALL cell lines and supplemental IL-2 could induce NK-like activity
against K562 cells. PER 117, PER 255 and PER 423 express c-kit receptors
and PER 423 cells proliferate synergistically when cultured with human stem
cell factor (SCF) and IL-2 [82].

MOLT-3 expresses the hemopexin receptor and proliferation is stimulated
by heme-hemopexin [155]. PHA induces MOLT-4 cells to secrete IL-2 and
IL-4 [31]. MOLT-13, -14, and -16 constitutively produce and secrete IL-
2 at low levels (<1 U/ml), and production is increased after stimulation
with PHA, interferon (IFN)$\gamma$ (MOLT-13, -14), IFN$\alpha$ (MOLT-16), and TNF$\alpha$
(MOLT-13, -16) [31,32]). Interestingly, IFN$\alpha$ and $\gamma$ have reciprocal effects
on MOLT-13, -14 compared to MOLT-16: IFN$\alpha$ inhibits IL-2 production
by MOLT-13 and -14, while IFN$\gamma$ augments IL-2 production by MOLT-16.
Like MOLT-4, MOLT-13, -14, and -16 also produce low levels (<1 ng/ml) of
IL-4 upon stimulation with PHA [32]. Furthermore, anti-CD7 but not PHA
induces expression of GM-CSF by MOLT-13 [18].

MOLT-16 cells express IL-2R, and since these cells also secrete IL-2, the
possibility for autocrine growth exists, although this has not been demonstrated [4]. MOLT-14 and -16 also express mRNA for the cholecystokinin
(CCK)-B/gastrin receptor (specific for hormones CCK and gastrin) as well
as gastrin mRNA. Although the presence of the respective proteins was
not examined, proliferation of MOLT-16 is inhibited by a receptor-specific
antagonist in the absence of added ligand, suggesting a possible gastrin-
mediated autocrine loop [73]. Although differentiation of MOLT cell lines
in response to cytokines has not been reported, MOLT-13 acquires NK-like
activity against K562 cells following incubation with IL-2 or phorbol ester
(PMA) [4].

Dr. Lange and colleagues established several T-ALL cell lines that are
dependent on cytokines for continuous cell proliferation. One T-ALL cell
line (TALL-101) responds to GM-CSF and IL-3 and is GM-CSF dependent for *in vitro* growth. TALL-103/2 is IL-2 dependent and responds to the
cytokines IL-2 and IL-1. Under the influence of IL-2, TALL-103 cells have a
T-lymphoid monoclonal antibody profile while TALL-103 cells cultured with
IL-3 have a myeloid immunophenotype [20,145]. TALL-104 cells proliferate
in response to IL-2 and are IL-2 dependent. TALL-105 cells proliferate in

Table 3a. T-ALL cell lines: cytokine-related characterization

| Cell line | Cytokine receptor expression | Cytokine production | Proliferation response to cytokines | Differentiation response to cytokines | Dependency on cytokines | Ref. |
|---|---|---|---|---|---|---|
| DND-41 | | | | IL-2 induced NK like activity to K562 cells. | | 4 |
| HPB-ALL | | | | Forskolin augmented PMA induced expression of IL-2R | | 120 |
| MOLT-13 | IL-2R induced by treatment with IL-2 or PMA. | IL-2 production in unstimulated and PHA stimulated cultures; increased by addition hIFN-γ and hTNF-α, inhibited by IFNα. Anti CD7 induced transcription of GM-CSF gene. | | IL-2 induced NK like activity against K562 cells and induced expression of the β sub-unit of IL-2R. | | 4,31,32,18 |
| MOLT-14 | IL-2R induced by treatment with IL-2 or PMA. Cholecystokinin-B/gastrin receptor expressed; cells demonstrated ligand dependent proliferation in serum-free media. | IL-2 production in unstimulated and PHA stimulated cultures increased by addition of IFNγ, inhibited by IFNα. Cholecystokinin-B/gastrin may act as autocrine factor. IL-4 production followed PHA stimulation. | | Treatment with IL-2 or PMA induced NK like activity against K562 cells, expression of IL-2R. | | 4,73,31,32,69 |

*Continued on next page*

Table 3a. (continued)

| Cell line | Cytokine receptor expression | Cytokine production | Proliferation response to cytokines | Differentiation response to cytokines | Dependency on cytokines | Ref. |
|---|---|---|---|---|---|---|
| MOLT-16 | IL-2R induced by treatment with IL-2 or PMA. Cholecystokinin-B/gastrin receptor expressed. 68-kD GTP-binding protein immunoprecipitated with TcR/CD-3 complex. | IL-2 production following stimulation with PHA was increased by IL-1, INFα, or TNFα. INFα increased apoptosis following PHA stimulation. Cholecystokinin-B/gastrin may act as an autocrine factor. | | Treatment with IL-2 or PMA induced NK-like activity against K562 cells, expression of IL-2 receptor. | | 4,32,33,68,69, 73,127 |
| PEER | | | | PHA and PMA increased expression of IL-4R, IL-2Rβ and IL-2Rα. | | 59 |
| PER 117 | Expressed low level of c-kit receptor. PMA, IL-1 induced expression of IL-2R. | IL-2. IL-1 required for optimal IL-2 secretion. | IL-2 | | | 82,143,78,81 |
| PER 255 | Expressed low level of c-kit receptor. | | | | | 82 |
| PER 423 | Tac-Mikβ1+ Expresses c-kit receptor. | | Cells proliferated in response to IL-2. Human steel factor acts in synergy with IL-2 to promote cellular proliferation. | | IL-2 | 82,80 |

*Continued on next page*

Table 3a. (continued)

| Cell line | Cytokine receptor expression | Cytokine production | Proliferation response to cytokines | Differentiation response to cytokines | Dependency on cytokines | Ref. |
|---|---|---|---|---|---|---|
| TALL-101 | | | GM-CSF, IL-3 | | GM-CSF | 92 |
| TALL-103/2 | | Activation signals induced production of IFNγ, TNF-α, and GM-CSF. | IL-2, IL-1 | | IL-2 | 145,20 |
| TALL-104 | | Activation signals induced production of IFNγ, TNF-α, and GM-CSF. | IL-2 | | IL-2 | 20,122 |
| TALL-105 | | | IL-1, IL-7 | | None | 122 |
| TALL-106 | | | IL-2, IL-3, IL-4, IL-6 | | IL-3 | 122 |
| UHKT-42 | CD25– IL-2Rβ– | PHA and TPA induced IL-2 production. | | TPA induced CD-25 expression. | | 157 |

IFN – interferon; TNF – tumor necrosis factor; GM-CSF – granulocyte- macrophage colony stimulating factor.

response to IL-1 and IL-7 but continuous cell line growth is growth factor in-dependent. TALL-106 cells respond to IL-2, IL-3, IL-4, and IL-6 and require IL-3 for continuous growth *in vitro* [122].

Cytokine receptor expression, cytokine production, cytokine-induced pro-liferation and differentiation, and cytokine dependency of NK cell lines are summarized in Table 3b. The NK cell lines generally possess IL-2R and respond to exogenous IL-2. NK92 cells proliferate in response to IL-2 and IL-7 and are dependent on IL-2 for continuous cell proliferation *in vitro* [51]. NOI-90 cells have an autocrine loop with IL-2, and anti-IL-2 antibodies block cell proliferation [142]. The NK92 and TKS1 cells are IL-2 dependent, and YT cells are dependent on conditioned media for continuous cell growth. The YT2C2 cells were subcloned from YT to be IL-2 independent, and YT2C2 cells do not respond to exogenous IL-2.

## 5. GENETIC CHARACTERIZATION

The genetic characteristics of the T-ALL cell lines are summarized in Table 4a. No established T-ALL cell line has a normal karyotype. While most T-ALL cell lines have complex karyotypes with multiple abnormalities, a few have a single chromosome translocation (EU-9, KH-1, SUP-T7, TALL-101, TALL-104, TALL-105) or certain chromosome deletions (Jurkat, PF-382, SUP-T2, SUP-T14). Many of the 40 karyotyped T-ALL cell lines have com-mon recurring chromosome breakpoints which occur at or near the *TCR* $\beta$ gene (7q34, N = 6) or the *TCR* $\alpha/\delta$ gene (14q11.2, N = 7) or both (N = 1). The t(8;14)(q24;q11) is observed in approximately 2% of T-ALL cases and breakpoints occur within the MYC protooncogene and the *TCR* $\alpha$ gene. The t(8;14)(q24;q11) was demonstrated in MOLT-16, TALL-101, TALL-103, TALL-105, and TALL-106 cell lines. In MOLT-16 cells, c-myc is juxtaposed with the *TCR* $\alpha$ loci [44,101]. Specifically, the MOLT-16 break-point at 14(q11) occurs within the joining (J) region of the *TCR* $\alpha$ gene. The *TCR* $\alpha$ constant region and part of the J region are then translocated to the 3' side of c-myc on chromosome 8 [149]. Cytogenetic analysis has not been reported for MOLT-12, -13 and -14.

The most frequently recurring abnormality in childhood T-ALL is t(11;14)(p13;q11), which accounts for about 7% of cases and this translo-cation was observed in the TALL-104 cell line [138]. The t(1;14)(q32;q11) occurs in about 3% of T-ALL cases. The breakpoint at 1p32 occurs in TAL1 whereas the breakpoint on 14q11 occurs at a TCR $\alpha/\delta$ site, as has been reported in the DU.528 cell line [8,12,44]. The Karpas 45 cell line has a t(X;11)(q13;q23.3) which results in a fusion of the *AFX* gene on chromosome X band q13 to the *MLL* gene [100].

*Table 3b.* NK cell lines: cytokine-related characterization

| Cell line | Cytokine receptor expression | Cytokine production | Proliferation response to cytokines | Differentiation response to cytokines | Dependency on cytokines | Ref. |
|---|---|---|---|---|---|---|
| NK92 | IL-2 p55 positive IL-2 p75 positive | | IL-2, IL-7 induced cellular proliferation. No proliferation with IL-1, IL-6, TNFα, IFNα, IFNγ | | IL-2 | 51 |
| NKL | IL-2Rα expressed | | IL-2 | | IL-2 | 139 |
| NOI 90 | Low number of high affinity IL-2R | IL-2 | No response to exogenous IL-2 | | Autocrine IL-2 Expressed a low number of high affinity IL-2R with autocrine IL-2 loop. | 142 |
| TKS-1 | IL-2Rα positive IL-2Rβ initially positive but negative after 6 months in culture | | IL-2 | | IL-2 | 89 |
| YT | Inducible high affinity IL-2R IL-2Rβ expressed | | | PHA/PMA reduced IL-2Rβ expression and induced IL-2Rα. IL-1α promoted expression of IL-2Rα. Forskolin induced both high and low affinity IL-2R | Conditioned media. | 172,59,120 |
| YT2C2 | IL-2Rα not expressed IL-2Rβ expressed | | No response to exogenous IL-2 | | | 173,160 |

*Table 4a.* T-ALL cell lines: genetic characterization

| Cell line | Cytogenetic karyotype | Unique translocations ($\rightarrow$ fusion genes) | Unique gene alterations, receptor gene rearrangements[a] | Ref. |
|---|---|---|---|---|
| BE-13 | Not reported | | No p53 protein, p53 wt, Rb protein expressed | 23–25 |
| CCRF-CEM | 90–91, XXY, −5, −6, −9, −9, iso (9q) x2, +14, +14, −16, −17, −18, −19, +20, +20, +22, +22, +4 mar | | p53M, Rb protein expressed, p15 P, p16D/D, Rb-wt | 23, 65, 176, 40, 14, 99 |
| | 46, X, −X, −9, −18, +20, + der(18)t(X;18)(q12:p11.3), del(8)(p12), del(9)(p21), + der(9)t(9:?)(p23:?) | | | |
| DND-41 | Not reported | | p15G/D, p16D/M | 14 |
| DU.528 | 46, XY, del(1)(p33), t(1;14)(p33;q11), del(13)(q14), der(14), 19q+ | t(1;14)(p32:q11) results in truncated SCL gene SCL – TCR-$\delta$ | Rb protein expressed | 23, 12, 64, 8, 44 |
| EU-7 | 47, XX, +20, t(8;9)(p11.2;p24), del(9)(p22), inv(9) | | p16D, p15D, p53M | 176, 43, 177 |
| EU-9 | 46, XY, t(1;7)(p34; q34) | $\beta$-TCR-B constant region and LCK gene | p16D, p15P, p53 not expressed | 176, 43, 177 |
| HPB-ALL | 94-96, XXYYY, t(1;5)(q23;q25)x2, del(2)(p22)x2, −3, −3, −9, add(14)(q32), −16, −16, −18, +19, +del(20)(q11)x2, +del(22)(q21)x2, +4mar | | p15P, p16P | 65, 14 |

*Continued on next page*

*Table 4a.* (continued)

| Cell line | Cytogenetic karyotype | Unique translocations (→ fusion genes) | Unique gene alterations, receptor gene rearrangements[a] | Ref. |
|---|---|---|---|---|
| JM/JURKAT | 45, X, −Y, del(2)(p23), del(8)(q23) | | p53M, p15D/D p16D/D, Rb protein expressed | 23, 65, 40, 14 |
| KARPAS 45 | 84, −Y, −Y, t(X;11)(q13;q23.3), der(X)t(X;11)(q13;q23.3)t(1:5)(q25:q13.1) x2, −2, −3, −4, del(4)(q21.1q31.1), +6, −9, der(11)t(14;11)(11;X)(q11:p13q23.3:q13), −13, −14, +19, −20, −21 | AFX-MLL fusion | | 87, 100, 13 |
| KE-37FN | 46, XY, der(9)t(9;?)(p24:?)/45, X, −Y, der(9)t(9;?)(p24;?), +variation | | | 93 |
| KH-1 | 46, XY, t(8q+;15q−)(8pter→8qter::15q15→15qter) | | | 115 |
| K-T1 | 46, XY, t(1;11)(p36.2;p13) del(2)(p16.3p22.1), del(9)(p12p21.1) | | p16D, p15D, p53 − wt | 152, 176 |
| L-KAW | 46, XY, −1, −2, +6, −7, −9, −14, −18, del(3)(p23), + der(2)t(2;?)(p23;?), + der(7)t(7;7)(q12;q35), + der(9)t(9;?)(q34:?), + der(14)t(14:?)(q11:?), + mar | | | 104 |
| LOUCY | 45, X, −X, del(5)(pter→q15::q35→qter), t(16;20)(p12:q13) | | p53M | 11, 132 |

*Continued on next page*

*Table 4a.*  (continued)

| Cell line | Cytogenetic karyotype | Unique translocations (→ fusion genes) | Unique gene alterations, receptor gene rearrangements[a] | Ref. |
|---|---|---|---|---|
| MOLT-3 | Hypertetraploid (modal n = 97) XXYY, +4, +6, −7, −7, +8, +8, −9, +11, −14, +15, −18, −18, +20, del(1)(q21;q42), der(2)t(2;2)(p15;q11), del(6)(q13;q21)x2, i(17q), +der(7)x2, t(7;7)(p15;q11) | | Expressed alternately spliced P53, p15DG/GG, p16DD/DR, Rb-wt | 55, 27, 63, 176, 40 |
| MOLT-4 | Hypertetraploid (modal n = 98) XXYY, +4, −7, −7, +8, +11, −14, +15, −18, −18, +20, +20, del(1)(q21;q42), del(6)(q13q21), i(17q), +der(7)x2, t(7;7)(p15;q11) | | p16D/R, Expressed alternately spliced p53, Rb protein expressed | 23, 55, 123, 27, 63 |
| MOLT-13 | | | p15D, p16D | 14 |
| MOLT-14 | | | p16D/R, p15D/R, p53M | 123, 23, 14 |
| MOLT-16 | 45, XX, −7, −9, t(3;11)(q26;p14), t(8;14)(q24;q11), dic(9;15)(p11;p11), + der(7)t(7;7)(7qter→7p15::7q11→7qter), + der(9)t(9;9)(p24;p11) | MYC-TCRα | p53 MRb, protein expressed, p16D/R, p15D/R | 14, 39, 149, 101, 123, 23, 75 |
| MOLT-17 | | | p15D, p16D | 14 |
| MT-ALL | 47, XY, +19, del(6)(q15q25)t(1;10;12)(q25;p13;p13) | | | 56, 57 |
| P12/Ichikawa | Hypotetraploid with an increased number of chromosomes in groups A, C, D, F, and G. | | p16D, p15P | 168, 14 |

*Continued on next page*

*Table 4a.* (continued)

| Cell line | Cytogenetic karyotype | Unique translocations (→ fusion genes) | Unique gene alterations, receptor gene rearrangements[a] | Ref. |
|---|---|---|---|---|
| P30/Ohkubo | 46 (45), X (X), del(2)(qter→p23), del(9)(p12→q31), t(11;12)(11pter→11q25::12q13→12qter; 12pter→ 12q13::11q25→11qter) | | | 66 |
| PEER | 46, XX, −4, + der(4) rea (4), del(5)(q21q23), del(6)(q14q22), del(9)(p12p21), i(9p) | | p16 D, p15 P | 99, 14 |
| PER 117 | 46, XY, t(1;11)(p13; p15), −16, +18 | | | 77 |
| PER 255 | 46, XY, t(7;10)(q32-34;q24), t(9;12)(p22;p12–13) | HOX11 – TCRβ | | 79, 83 |
| PF-382 | 46, X, Xq−, 15p+ | | | 130 |
| RPMI 8402 | 100% of the cells possessed 1 or 2 minute chromosomes and 1 subtelocentric chromosome | | P53G, Rb protein expressed | 23, 71 |
| SUP-T2 | 46, XX, del(6)(q21q27) | | | 152 |
| SUP-T3 | 46, XY, del(4)(q31q35), t(7;9)(q34;q32) | | | 153,170, 169 |
| SUP-T6 | 46, XY, t(7;9)(q34;q32), del(6)(q21) | | | 154 |
| SUP-T7 | 46, XY, t(7;19)(q34;p13.1) | LYL1 – TCRβ | | 154, 102, 28 |

*Continued on next page*

*Table 4a.* (continued)

| Cell line | Cytogenetic karyotype | Unique translocations ($\rightarrow$ fusion genes) | Unique gene alterations, receptor gene rearrangements[a] | Ref. |
|---|---|---|---|---|
| SUP-T8 | 44, XX, del(1)(p32p35), 1p+, +1p+, 2q+, −4, t(4;19)(q21:p13), 4q−, 7p+, +8, −9, 10p+, del(11)(q21q25), −12, −16, −17, der(17)t(12;17)(p13;p13), 19p+, −22, +mar | | | 154 |
| SUP-T9 | 46, XX, t(6;14;21)(q23;q11.2;q22), del(11)(q23q25), t(15;21)(q15q22) | | | 154 |
| SUP-T10 | 47, XY, del(5)(q31), t(7;11)(p13;p13), t(8;12)(q13;p13), t(9;16)(p22;p13), t(17;18)(q11.2;q23), +mar | | | 154 |
| SUP-T12 | 60, XY, +Y, t(1;7)(p34;q34), +17, +der(1)t(1;7)(p34;q34), +4, +6, +7, +8, +8, +10 +13, +16, +18, +19, +19 | LCK – TCRβ | | 154, 162, 17 |
| SUP-T13 | 46, XX, t(1;8)(q32;q24), t(1;5)(q41;p11), del(9)(q24q34), t(11;19)(q24;p13) | | | 154 |
| SUP-T14 | 46, XY, del(6)(q15), del(9)(p22) | | | 154 |
| TALL-1(a) | Hypertetraploid with chromosome numbers ranging from 95 to 101 | | p15 P, p16 P | 111, 14 |
| TALL-1(b) | 46, XY, del(6)(q23), t(1;8)(q32;q24) | | | 92 |

*Continued on next page*

*Table 4a.* (continued)

| Cell line | Cytogenetic karyotype | Unique translocations (→ fusion genes) | Unique gene alterations, receptor gene rearrangements[a] | Ref. |
|---|---|---|---|---|
| TALL-101 | 46, XY, t(8;14)(q24;q11), inv(9) | | | 92 |
| TALL-103 | 47, XY, −9, −13, t(8;14)(q24;q11) der(12)t(12;13)(p11-13;q11-14), +12, +17, +mar | | | 122 |
| TALL104 | 46, XY, t(11;14)(p13;q11) | | | 122 |
| TALL105 | 46, XX, t(8;14)(q24;q11) | | | 122 |
| TALL-106 | 46, t(8;14)(q24;q11), inv(14)(q11q32) | | | 122 |

[a] G – germ line, M – mutated, D – deleted, wt – wild type, P – DNA present on Southern blot, R – rearranged.
[b] At least one stock of EU-7 cells has been found to be in fact CCRF-CEM, and this cell line should not be used unless the stocks available are shown to be distinct from CCRF-CEM.

*Table 4b.* NK cell lines: genetic characterization

| Cell line | Cytogenetic karyotype | Unique translocations ($\rightarrow$ fusion genes) | Unique gene alterations, receptor gene rearrangements[a] | Ref. |
|---|---|---|---|---|
| NK92 | Near-tetraploid with multiple aneuploidies and structural rearrangements. | | | 51 |
| NKL | 47, XY, add(1)(q42), +6, del(6)(q15q23), del(17)(p11) | | | 139 |
| NOI 90 | 47, XY, +8, 17p– | | | 142 |
| TKS1 | 46, XY, 17p+, 21p+ | | | 89 |
| YT | Tetraploid (83–95 with mode = 92), duplicated marker of 4q+ consistently found. | | | 172 |

T-ALL cases often have translocations with a breakpoint involving the $\beta$ T-cell receptor (*TCR*) at chromosome 7 band q34-35 and several other partner genes [133]. The t(7;19)(q34;p13) as seen in the SUP-T7 cell line involves a fusion of LYL (19p13) to *TCR* $\beta$ [102]. The t(7;9)(q34;q32) as seen in the SUP-T3 and SUP-T6 cell lines involves fusion of the TAL2 gene (9q32) and *TCR* $\beta$ [170], and the t(1;7)(p34;q34) as seen in the EU-9 and SUP-T12 cell lines involves the LCK gene (1p34) and *TCR* $\beta$ [17,162].

Both MOLT-3 and -4 are hypertetraploid (modal 97 and 98 chromosomes, respectively) with shared chromosomal gains, losses and deletions. In particular, both lines have an intrachromosomal translocation t(7:7)(p15;q11) and a rearrangement of *TCRγ* gene. However, the breakpoint at 7p15 does not involve the *TCR* γ joining or constant regions [55].

T-ALL is commonly associated with a deletion in chromosome 9p21, which occurs in approximately 60% of patients. The 9p21 region is the locus for the Ink4 family of Cdk4 inhibitors, including p15$^{ink}$4b and p16$^{ink}$4a, and T-ALL cells frequently have a deletion of one or both of these tumor-suppressor genes [6,124,135]. Deletion of p15/p16 is thought to deregulate cell-cycle in tumor cells by abrogating inhibition of cyclin-dependent kinase 4 (Cdk4). The 9p21 region also contains the methylthioadenosine phosphorylase (*MTAP*) gene, and T-ALL cells with 9p21 deletions are typically deficient in *MTAP*, an enzyme important in the salvage pathway for adenine and methionine [5]. Five cell lines have a deletion of the short arm of chromosome 9 with the resultant loss of p15, p16 and *MTAP*.

Another important tumor suppressor gene and regulator of cell cycle progression, the p53 gene, is frequently mutated in T-ALL cells at relapse, with an incidence of approximately 25% [23,171]. MOLT-3, -4, -14, and -16 as well as EU-7 express a mutant or alternately-spliced p53; EU-9 does not express p53.

The karyotype of the NK cell lines is summarized in Table 4b. No established NK cell line has a normal karyotype. Two cell lines (NK92, YT) have near tetraploidy and three NK cell lines have abnormalities involving the short arm of chromosome 17. No evaluation of p53, p16 or p15 has been reported in the NK cell lines.

## 6. SPECIAL FUNCTIONS AND CYTOCHEMISTRY

The functional characterizations of the T-ALL cell lines are listed in Table 5a. Biochemically, T-ALL cells have aberrant expression of certain enzymes involved in purine biosynthesis. Specifically, T-ALL cells have high levels of adenosine deaminase and low levels of purine nucleoside phosphorylase and 5′-nucleotidase relative to normal T-cells and non-T-ALL lymphoid malignancies. Additionally, T-ALL cells may show a deficiency in methyl-thioadenosine phosphorylase (*MTAP*) activity due to a 9p21 deletion, as discussed earlier.

Leukemia cells often appear to be altered in their responses to apoptotic stimuli, and this may be due to aberrant expression or regulation of certain anti-apoptosis genes. The MDM2 (Murine-Double-Minute 2) gene codes for a protein that inhibits p53 function, thereby making the cell more resistant to factors that activate p53-mediated apoptosis. MDM2 is frequently over-expressed in ALL cells that express a wild-type p53 [15,175]. Similarly, members of the Bcl-2 family having anti-apoptotic activity (Bcl-2 and Bcl-Xl) may be expressed at high levels (or show abnormal regulation following exposure to DNA-damaging agents) in both T-lineage and B-lineage ALL [29,43]. T-ALL cells may also express high levels of Fas (CD95), a member of the tumor-necrosis-factor receptor superfamily which is capable of indu-cing apoptosis upon binding of its ligand (FasL). However, the sensitivity of Fas+ T-ALL cells to FasL is heterogeneous and may be affected by other anti-apoptosis proteins [34,177].

Several of the MOLT lines show abnormal regulation of purine and pyrimidine pathway enzymes commonly associated with T-ALL, and these lines have been widely used in studies of the effects of antimetabolites on T-ALL cells [5,166,174]. The HPB-ALL and PEER lines have been used to evaluate the effects of deoxyadenosine and deoxycoformycin on T-ALL cells expressing high levels of adenosine deaminase [41].

Several T-ALL cell lines, including MOLT-3, Jurkat and CCRF-CEM, have been shown to be induced by phorbol esters (such as TPA) to differenti-ate into cells with more mature phenotypes. TPA induces MOLT-3 and Jurkat cells to acquire increased sheep RBC rosette-forming ability (corresponding to increased expression of CD2) and lose TdT activity. Jurkat cells respond to TPA and differentiate but CCRF-CEM cells do not [116]. When MOLT-3 and Jurkat cells are treated with TPA, CD3 cells increase but the proportions of CD4, 1, and 8 cells decrease [117]. Phorbol dibutyrate (PDB) induces a rapid and reversible loss of the expression of CD4 antigen and a slower increase in CD2 antigen on the Jurkat cells [19]. When treated with thymosin fraction 5, MOLT-3 cells increase $5'$-ectonucleotidase activity, but decrease adenosine deaminase activity, proliferating activity, and percent of cells that are pos-itive for CD1. TPA induces purine nucleoside phosphorylase and increases the expression of CD2 with suppression of TdT activity but does not affect $5'$-ectonucleotidase or adenosine deaminase activity [67]. CEM cells were induced with TPA to express a cell surface antigen pattern of both suppressor and cytotoxic T-lymphocytes [140]. HPB-ALL cells, when exposed to TPA, differentiate into a more mature T-cell phenotype as judged by morphologic changes, loss of TdT activity, and increased sheep red blood cell rosetting

*Table 5a.* T-cell lines: functional characterization

| Cell line | Doubling time | EBV status | Cytochemistry | Differentiation | Heterotransplant-ability into mice | Special functional features | Ref. |
|---|---|---|---|---|---|---|---|
| BE-13 | | | | | | Dexamethasone sensitive Inhibited by haptoglobulin related antisense cDNA | 47,147 |
| CCRF-CEM | | | Oil Red O– PAS+ ALP– ACP– ANAE+ | TPA induced ↑CD3/CD8 | | Cyclosporin induced apoptosis Ectopic p16 expression resulted in growth arrest TPA induced carboxylic esterase, LDH and acid phosphatase. Subclone has defective methotrexate polyglutamate synthesis. | 140,45,72 40,38,105 |
| DU. 528 | 48–96 hrs | | NBT(+), Perox–, NSE(+) | PMA induced myeloid differentiation with ↑NBT, ↑MPO, ↑NSE | | | 91 |
| EU-7* | 45–55 hrs | Neg | NSE+ SB– SE– Perox– | | SCID mice | Colony formation in methylcellulose Bcl-2 family gene expression: Bcl-2–, Bcl-xl+, Bax+ (by W/Blot) MDM2+ by W/Blot Fas+ (high level expression by FC) | 175,43,177 |

*Continued on next page*

*Table 5a.* (continued)

| Cell line | Doubling time | EBV status | Cytochemistry | Differentiation | Heterotransplant-ability into mice | Special functional features | Ref. |
|---|---|---|---|---|---|---|---|
| EU-9 | 45–55 hrs | Neg | NSE+ Perox– SB– SE– | | SCID mice | Colony formation in methylcellulose Bcl-2 family gene expression: Bcl-2–, Bcl-xl+, Bax+ (by W/Blot) MDM2– by W/Blot Fas+ (low level expression by FC) | 175,43,177 |
| HPB-ALL | | EBNA– | | TPA induced ↓TdT, ↓CD1, ↑SRBC | | HPB-ALL cells are inhibited by deoxyadenosine and deoxycoformycin HPB-ALL cells are infectable by EBV. 1, 25 dihydroxy vitamin D3 did not inhibit HPB-ALL cells | 118,114,41, 128,119 |
| JM/JURKAT | | EBNA– | PAS– ACP+ NSE– Perox– | TPA induced ↑CD2, ↑CD3, ↓TdT PDB induced ↓CD4 | | Ectopic p16 expression resulted in growth arrest TPA induced acid phosphatase | 117,19,146, 40,38 |
| KARPAS 45 | | EBNA– | SB–, Perox–, ANAE–, ANBE–, PAS(+) ACP+ ALP– | | | | 76 |
| K-T1 | 68 hrs | | | | | | 152 |

*Continued on next page*

Table 5a. (continued)

| Cell line | Doubling time | EBV status | Cytochemistry | Differentiation | Heterotransplant-ability into mice | Special functional features | Ref. |
|---|---|---|---|---|---|---|---|
| L-KAW | 36 hrs | | NSE(+)<br>PAS+<br>Perox−<br>N AS-D CE− | | | PHA induced growth inhibition and apoptosis of L-KAW cells. | 104 |
| LOUCY | 36–48 hrs | | ACP+<br>PAS−<br>Perox−<br>SB−<br>Esterase− | | | | 11 |
| MOLT-3 | 3–5 days | Neg | AP+,<br>Perox-<br>SB− | In SCID mice, ↑CD5, ↑CD8, ↓CD45RA<br>TPA induced ↑CD2, ↑CD3, ↑PNP, ↓TdT<br>TF5 induced ↓ADA and ↓CD1<br>PMA and A23187 induced calpain secretion. | Nude and SCID mice | Colony formation in agar<br>Overexpressed H-ras c-myc, v-myb, v-Hras, N-Ras, v-vef, v-fos by N Blot<br>Sublines resistant to MTX overexpressed DHFR; some with ↓ expression of folate carrier gene.<br>Sensitive to anti-CD-7-PAP in SCID mice.<br>Subline resistant to trimetrexate with increased DHFR and MDR1.<br>Increased CTP synthetase activity. | 106,60,163, 125,126,95, 49,50,3, 117,166,67, 35,86,53 |

*Continued on next page*

*Table 5a.* (continued)

| Cell line | Doubling time | EBV status | Cytochemistry | Differentiation | Heterotransplantability into mice | Special functional features | Ref. |
|---|---|---|---|---|---|---|---|
| MOLT-4 | 3–5 days | Neg | AP+ Perox− SB− | In SCID mice, ↑CD5, ↑CD8, ↓CD45RA TPA induced selective loss of CD-4 TPA induced resistance to deoxyguanosine. | Nude and SCID mice | Overexpressed H-ras c-myc Expressed TALLA-1 Ly6 gene overexpressed Expressed methylthioadenosine phosphorylase Cyclosporin induced apoptosis Expressed specific 1, 25 dihydroxy vitamin $D_3$ receptors and vitamin $D_3$ inhibits cellular proliferation. Cells overexpressed cyclin D3 TPA induced carboxylic esterase and acid phosphatase | 106,174,163, 70,49,50, 158,159,141, 72,161,5, 53,38 |
| MOLT-13 | | Neg | | | | Anti-CD7 induced increase in cytoplasmic free calcium. | 39,18 |
| MOLT-16 | Neg | Neg | | | Growth in SCID mice inhibited by anti-CD-7 antibody. | Expressed methylthioadenosine phosphorylase and is resistant to AMP-synthesis inhibitor alanosine. | 39,7,5 |
| MT ALL | | | | | Growth in SCID mice inhibited by anti-CD-7 antibody. | | 74 |
| P12/Ichikawa | | | | | | 1, 25 dihydroxy vitamin D3 did not inhibit P12/Ichikawa cells. | 119 |

*Continued on next page*

*Table 5a.* (continued)

| Cell line | Doubling time | EBV status | Cytochemistry | Differentiation | Heterotransplant-ability into mice | Special functional features | Ref. |
|---|---|---|---|---|---|---|---|
| P30/Ohkubo | 30 hrs | Neg | Perox− PAS+ ACP+ ALP− ANAE− | | | | 66 |
| PEER | 36 hrs | Neg | SB− Oil Red O− Perox− NS+ PAS+ Muramidase− ACP+ | | | PEER cells strongly inhibited by deoxyadenosine plus deoxycoformycin | 136,41 |
| PER 117 | 55–65 hrs | | ACP+ PAS+ Perox− NSE− | | | | 77 |
| PER 255 | | Neg | ACP+ PAS+ Perox− NSE− SB− | | | | 79 |

*Continued on next page*

Table 5a. (continued)

| Cell line | Doubling time | EBV status | Cytochemistry | Differentiation | Heterotransplant-ability into mice | Special functional features | Ref. |
|---|---|---|---|---|---|---|---|
| PF-382 | | Neg | ACP+ PAS− Perox− ANAE− | | | Cell line released a factor which inhibits CFU-GM and BFU-E growth *in vitro*. Cell line released potent enhancers of monocyte effect on BFU-E growth | 9,10,130 |
| RPMI 8402 | | | | TPA induced a more mature phenotype. TPA induced TdT loss and resistance to deoxyguanosine. | | Alpha and beta interferon inhibits cellular proliferation. Cells expressed 1, 25 dihydroxy vitamin D3 receptors. | 141,148,150, 161 |
| SUP-T2 | 130 hrs | Neg | | | | | 152 |
| SUP-T3 | 61 hrs | Neg | | | | | 152 |
| SUP-T6 | | Neg | | | | | 152 |
| SUP-T7 | | Neg | | | | | 153 |
| SUP-T8 | | Neg | | | | | 154 |
| SUP-T9 | | Neg | | | | | 154 |
| SUP-T10 | | Neg | | | | | 154 |
| SUP-T12 | | Neg | | | | | 154 |
| SUP-T13 | | Neg | | | | Anti-CD3 inhibits DNA synthesis and triggers cell death. | 154,178 |
| SUP-T14 | | Neg | | | | | 154 |
| TALL-1 (a) | 60–90 hrs | | | | | | 111 |
| TALL-103/2 | | | | | | Lyses only NK targets. | 21 |

*Continued on next page*

*Table 5a.* (continued)

| Cell line | Doubling time | EBV status | Cytochemistry | Differentiation | Heterotransplant-ability into mice | Special functional features | Ref. |
|---|---|---|---|---|---|---|---|
| TALL-104 | | | | | | Lyses a broad range of tumor targets. | 21,22 |
| UHKT-42 | | | ACP+ PAS+ Perox− SB− NSE− | TPA induced IL-2R expression. | | | 157 |

ACP – acid phosphatase; ALP – alkaline phosphatase; ANAE – alpha naphthol acetate esterase; ANBE – alpha naphthol butyrate esterase; FC – flow cytometry; N AS-D CE – naphthol AS-D Chloroacetate Esterase; NBT – nitro blue tetrazolium; NSE – non-specific esterase; PAS – periodic acid-schiff; Perox – peroxidase; PMA – phorbol myristate acetate; SB – sudan black; TF5 – thymosin fraction 5; TPA – 12-O-tetradecamoylphorbol 13 – acetate.
\* At least one stock of EU-7 cells has been found to be in fact CCRF-CEM, and this cell line should not be used unless the stocks available are shown to be distinct from CCRF-CEM.

*Table 5b.* NK cell lines: functional characterization

| Cell line | Doubling time | EBV status | Cytochemistry | Differentiation | Heterotransplant-ability into mice | Special functional features | Ref. |
|-----------|---------------|------------|---------------|-----------------|-----------------------------------|-----------------------------|------|
| NK92 | 24 hrs | EBV– | | | | NK activity against both K562 and Daudi | 51 |
| NKL | 24–48 hrs +(IL2) | | | | | NKL cells lysed K562, JURKAT, Daudi cells and mediated antibody-dependent cellular cytotoxicity. | 139 |
| NOI90 | 48 hrs | | Perox– NSE– ACP+ | | | Lacks NK or LAK activity Expressed a low number of high affinity IL-2R with autocrine IL-2 loop. | 142 |
| TKS-1 | | Not stated | ACP+ NSE+ (inhibited by sodium fluoride) | | | High cytotoxic activity against K562 Raji and Daudi target cells. | 89 |
| YT | | EBV infected | Perox– PAS+ | | | YT cells exhibit NK and antibody-dependent cellular cytotoxicity | 173 |
| YT2C2 | | EBV infected | | | | YT2C2 cells do not exhibit NK or antibody-dependent cellular cytotoxicity. | 173 |
| YTC3 | | EBV infected | | | | YT3C cells do not exhibit NK or antibody-dependent cellular cytotoxicity. | 173 |

ACP – acid phosphatase; NSE – non-specific esterase; PAS – periodic acid- schiff; Perox – peroxidase.

*Table 6.* Lymphoblast cell lines: unconfirmed cell lines (not fully characterized, not verified, other)

| Cell line | Patient | Features | Ref. | Remarks |
|---|---|---|---|---|
| ALL-Sil | | t(10;14)(q24;q11) HOX 11 – TCR delta | 62,97 | Provided by Dr. Jun Minowada, Fujisaki Cell Center |
| K3P | T-cell lymphoma | t(10;14)(q24;q11) HOX 11 – disregulated | 42 | |
| MOLT-10 | | T-ALL cell lines established by Dr. Jun Minowada, Fujisaki Cell Center | 110,109 | Full characterization not reported. |
| MOLT-11 | | T-ALL cell lines established by Dr. Jun Minowada, Fujisaki Cell Center | 110,109 | Full characterization not reported. |
| REX | From a patient with lymphoblastic leukemia | CD1– CD2+ CD3+ CD4+ CD5+ CD6+ CD8+ | 1,2 | Used to dissect human T-Cell receptor. |
| SUP-T11 | From BM from 74 y/o M with leukemia | CD1– CD2+ CD3+ CD4– CD5+ CD7+ CD8– 40, X, –Y, t(7;15;?)(q34;q22?)-1, 1q–, 2p+, –4, t(5;9;?), der(5q), 6q+, –9, 9q–, –10, 12p+, –13, t(14;14)(q11.2;q32), –15, –16, 16q+, –17, 19p+, –21, 21q+, –22, +5mar | 154 | While the patient was originally reported to have T-ALL, the patient's age and the presence of t(14;14) supports the diagnosis of T-CLL. |

ACP – acid phosphatase; NSE – non-specific esterase; PAS – periodic acid- schiff; Perox – peroxidase.

(CD2) activity and the disappearance of CD1 [118]. RPMI 8402 cells, when treated with TPA, acquire a more differentiated phenotype [141].

The functional characterization of the NK cell lines are listed in Table 5b. The NK cell lines YT and the sub-clones YT2C2 and YTC3 are EBNA positive and are the first reported EBV+ NK cell lines. While NOI-90 cells lack NK or LAK activity, cell lines NK-92, YT, and TSK were toxic against NK, ADCC, K562, Daudi, and Raji targets.

Cell lines which have not been fully characterized are reported in Table 6. The ALL-Sil and K3P cell lines carry the t(10;14)(q24;q11) and have been used to evaluate the HOX 11 gene. The T-ALL cell lines MOLT-10 and MOLT-11 have not been fully reported. The Rex cell line, a T-ALL cell line, has been used for identification of $\alpha$ and $\beta$ subunits of the human T-cell receptor. The Rex cells have also been used to identify and study ion channels which are activated by Ti or T3 antibodies [1,2,129]. While the SUP-T11 cell line was originally reported as T-ALL, a review of the clinical and laboratory characterization revealed that this patient most likely had T-CLL. This impression was supported by the presence of the t(14;14)(q11;q32) which is a characteristic marker of T-CLL [103,154].

## References

1.  Acuto O, Fabbi M, Smart J et al. *Proc Natl Acad Sci USA* 81: 3851–3855, 1984.
2.  Acuto O, Hussey RE, Fitzgerald KA et al. *Cell* 34: 717–726, 1983.
3.  Arkin H, Ohnuma T, Kamen BA et al. *Cancer Res* 42: 1655–1660, 1989.
4.  Ariyasu T, Kimura N, Kikuchi M et al. *Leukemia* 5: 807–812, 1991.
5.  Batova A, Diccianni MB, Nobori T et al. *Blood* 88: 3083–3090, 1996.
6.  Batova A, Diccianni MB, Yu JC et al. *Cancer Res* 57: 832–836, 1997.
7.  Baum W, Steininger H, Bair JH et al. *Br J Haematol* 95: 327–338, 1996.
8.  Begley CG, Aplan PD, Davey MP et al. *Proc Natl Acad Sci USA* 86: 2031–2035, 1989.
9.  Bellone G, Foa R, Fierro MT et al. *Leukemia* 1: 603–608, 1987.
10. Bellone G, Avanzi GC, Lista P et al. *J Cell Phys* 135: 127–132, 1988.
11. Ben-Bassat H, Shlomai Z, Kohn G et al. *Cancer Genet Cytogenet* 49: 241–248, 1990.
12. Bernard O, Guglielmi P, Jonveaux P et al. *Genes Chromosomes & Cancer* 1: 194–208, 1990.
13. Borkhardt A, Repp R, Haas OA et al. *Oncogene* 14: 195–202, 1997.
14. Brandter LB, Heyman M, Rasool O et al. *Eur J Haematol* 56: 313–318, 1996.
15. Bueso-Ramos CE, Yang Y, deLeon E et al. *Blood* 82: 2617–2623, 1993.
16. Burger R, Hansen-Hagge T, Drexler HG et al. *Leukemia* Research 23: 19–27, 1999.
17. Burnett RC, Thirman MJ, Rowley JD et al. *Blood* 84: 1232–1236, 1994.
18. Carrel S, Salvi S, Rafti F et al. *Eur J Immunol* 21: 1195–1200, 1991.
19. Cassel DL, Hoxie JA, Cooper RA. *Cancer Res* 43: 4582–4586, 1983.
20. Cesano A, Santoli D. In Vitro *Cell Dev Biol* 28a: 657–662, 1992a.
21. Cesano A, Santoli D. In Vitro *Cell Dev Biol* 28a: 648–656, 1992b.
22. Cesano A, Visonneau S, Santoli D. *Anticancer Res* 18: 2289–2295, 1998.
23. Cheng J, Haas M. *Mol Cell Biol* 10: 5502–5509, 1990a.

24. Cheng J, Scully P, Shew J-Y et al. *Blood* 75: 730–735, 1990b.
25. Cheng J, Yee JK, Yeangin J et al. *Cancer Res* 52: 222–226, 1992.
26. Chou JL, Barker CR, Prospero TD. *Leukemia Res* 10: 211–220, 1986.
27. Chow VT, Quek HH, Tock EP. *Cancer Lett* 73: 141–148, 1993.
28. Cleary ML, Mellentin JD, Spies J et al. *J Exp Med* 167: 682–687, 1988.
29. Conroy LA, Alexander DR. *Leukemia* 10: 1422–1435, 1996.
30. Crist WM, Shuster JJ, Falletta J et al. *Blood* 72: 1891–1897, 1988.
31. Dao T, Holan V, Minowada J. *Cell Immunol* 151: 451–459, 1993a.
32. Dao T, Holan V, Minowada J. *Int. J. Hematol* 57: 139–146, 1993b.
33. Dao T, Ariyasu T, Holan V. *Cell Immunol* 155: 304–311, 1994.
34. Debatin KM, Krammer PH. *Leukemia* 9: 815–820, 1995.
35. Deshpande RV, Goust JM, Hogan EL et al. *J Neurosci Res* 42: 259–265, 1995.
36. Drexler HG, Gaedicke G, Minowada J. *Leukemia Res* 9: 549–559, 1985.
37. Drexler H, Messmore H, Menon M et al. *Acta Haematol* 75: 12–17, 1986.
38. Drexler HG. *Blut* 54: 79–87, 1987.
39. Drexler HG, Minowada J. *Hematol Oncol* 7: 115–125, 1989.
40. Drexler HG. *Leukemia* 12: 845–859, 1998.
41. Duan DS, Smith W, Sadee W et al. *Thymus* 19: 1–10, 1992.
42. Dube ID, Kamel-Reid S, Yang CC et al. *Blood* 78: 2996–3003, 1991.
43. Findley HW, Gu L, Yeager AM et al. *Blood* 89: 2986–2993, 1997.
44. Finger LR, Kaga J, Christopher G et al. *Proc Natl Acad Sci USA* 86: 5039–5043, 1989.
45. Foley GE, Lazarus H, Farber S et al. *Cancer* 18: 522–529, 1965.
46. Galili N, Galili U, Ravid Z et al. *Human Lymph Diff* 1: 123–130, 1981.
47. Galili U, Peleg A, Milner Y, Galili N. *Cancer Res* 44: 4594–4601, 1984.
48. Gignac SM, Steube K, Schleithoff L et al. *Leukemia and Lymphoma* 10: 359–368, 1993.
49. Goldman J, McGuire WA. *Pediatric Hematol Oncol* 9: 357–363, 1992a.
50. Goldman J, McGuire WA. *Pediatric Hematol Oncol* 9: 309–316, 1992b.
51. Gong J-H, Maki G, Klingemann H-G. *Leukemia* 8: 652–658, 1994.
52. Gong J, Bhatia U, Traganos F et al. *Leukemia* 9: 893–899, 1995.
53. Gong M, Yess J, Connolly T et al. *Blood* 89: 2494–2499, 1997.
54. Greaves MF, Rao J, Hariri G, Verbi W et al. *Leukemia Res* 5: 281–299, 1981.
55. Greenberg JM, Gonzalez-Sarmiento R, Arthur D et al. *Blood* 1755–1760, 1988.
56. Griesinger F, Arthur DC, Drunning R et al. *J Exp Med* 169: 1101–1120, 1989a.
57. Griesinger F, Greenberg JM, Arthur DC et al. Hem *Blut* 32: 214–219, 1989b.
58. Gruss H-J, DaSilva N, Hu Z-B et al. *Leukemia* 8: 2083–2094, 1994.
59. Guizani L, Lang P, Stancou R, Bertoglio J. *Lymphokine and Cytokine Res* 12: 25–31, 1993.
60. Gunther R, Chelstrom LM, Finnegan D et al. *Leukemia* 7: 298–309, 1993.
61. Hane B, Tummler M, Jager K et al. *Leukemia* 6: 1129–1133, 1992.
62. Hatano M, Roberts CWM, Minden M et al. *Science* 253: 79–82, 1991.
63. Hayata M, Oshimura J, Minowada J et al. *In Vitro* 11: 361–368, 1975.
64. Hershfield MS, Kurtzberg J, Harden E et al. *Proc Natl Acad Sci USA* 81: 253–257 1984.
65. Heyman M, Grander D, Brondum-Nielsen K et al. *Leukemia* 8: 425–434, 1994.
66. Hirose M, Minato K, Tobinai K et al. *Gann* 73: 600–605, 1982.
67. Ho AD, Ma DDF, Price G et al. *Leukemia* Res 7: 779–86, 1983.
68. Holan V, Minowada J. *Folia Biol* 39: 1–13, 1993.
69. Holan V, Dao T, Kosarova M. *Cell Immunol* 157: 549–555, 1994.
70. Horie M, Okutomi K, Taniguchi Y. *Genomics* 53: 365–368, 1998.
71. Huang CC, Hou Y, Woods LK et al. *JNCI* 53: 655–660 1974.

72.  Ito C, Ribeiro RC, Behm FG et al. *Blood* 91: 1001–1007, 1998.
73.  Iwata, N Murayama T, Matsumori Y et al. *Blood* 88: 2863–2869, 1996.
74.  Jansen B, Vallera DA, Jaszcz WB et al. *Cancer Res* 52: 1314–1321, 1992.
75.  Kawamura M, Ohnishi H, Guo S et al. *Leukemia Res* 23: 115–126, 1999.
76.  Karpas A, Hayhoe FGJ, Greenberger JS et al. *Leukemia Res* 1: 35–49, 1977.
77.  Kees UR, Ford J, Price PJ et al. *Leukemia Res* 11: 489–498, 1987.
78.  Kees UR. *Blood* 72: 1524–1529, 1988.
79.  Kees UR, Lukeis R, Ford J et al. *Blood* 74: 369–373, 1989.
80.  Kees UR, Ford J, Peroni SE et al. *Leukemia* 17: 51–59, 1993.
81.  Kees UR, Peroni SE, Ranford PR et al. *J Immunol* 168: 1–8, 1994.
82.  Kees UR, Ashman LK. *Leukemia* 9: 1046–1050, 1995.
83.  Kennedy MA, Gonzalez-Sarmiento R, Kees UR et al. *Proc Natl Acad Sci USA* 88: 8900–8904, 1991.
84.  Kimura N, Du T-P, Mak TW. European *J Immunol* 17: 1653–1656, 1987.
85.  Kirsch AH, Diaz LA, Bonish B et al. *Tissue Antigens* 50: 147–152, 1997.
86.  Kobayashi H, Dorai T, Holland JF et al. *Cancer Res* 54: 1271–1275, 1994.
87.  Kobayashi, H Espinosa III, R Thirman J et al. *Blood* 81: 3027–3033, 1993.
88.  Kohno K, Shibata Y, Matsuo Y et al. *Cellular Immunol* 131: 1–10, 1990.
89.  Kojima H, Suzukawa K, Yatabe Y et al. *Leukemia* 8: 1999–2004, 1994.
90.  Komiyama A, Kawai H, Miyagawa Y et al. *Blood* 6: 1429–1436, 1982.
91.  Kurtzberg J, Bigner SH, Hershfield MS. *J Exp Med* 162: 1561–1578, 1985.
92.  Lange BM, Valtieri D, Santoli D et al. *Blood* 70: 192–199, 1987.
93.  Larsen CJ, Mathieu-Mahul D, Bernard O et al. *Leukemia* 2: 247–248, 1988.
94.  LeBien T, Kersey J, Nakazawa S et al. *Leukemia Res* 6: 299–305, 1982.
95.  Li XX, Kobayashi H, Holland JF et al. *Leukemia Res* 17: 483–490, 1993.
96.  Loughran TP. *Blood* 82: 1–14, 1993.
97.  Lu M, Gong Z, Shen W et al. *EMBO J* 10: 2905–2910, 1991.
98.  Martin PJ, Giblett ER, Hansen JA. *Immunogenetics* 15: 385–398, 1982.
99.  Massaad L, Venuat A, Luccioni, C et al. *Cancer Genet Cytogenet* 53: 23–24, 1991.
100. McCabe NK, Kipiniak M, Kobayashi H et al. *Genes, Chromosomes & Cancer* 9: 211–224, 1994.
101. McKeithan TW, Shima-Rich EA, Le Beau MM et al. *Proc Natl Acad Sci USA* 83: 6636–6640,1986.
102. Mellentin JD, Smith SD, Cleary ML. *Cell* 58: 77–83, 1989.
103. Mengle-Gaw L, Willard HF, Smith CIE et al. *EMBO J* 6: 2273–2280, 1987.
104. Minegishi M, Minegishi N, Yanagisawa T et al. *Leukemia Res* 19: 433–442, 1995.
105. Mini E, Stanyon R, Coronnello M et al. *Tumori* 77: 95–99, 1991.
106. Minowada J, Ohnuma T, Moore GE. *J Natl Cancer Inst* 49: 891–895, 1972.
107. Minowada J, Koshiba H, Sagawa K et al. *J Cancer Res Clin Oncol* 101: 91–100, 1981.
108. Minowada J, Minato K, Srivastava BIS et al. In: Serrou B and Rosenfield H. (eds.) 1982a *Current Concepts in Human Imm & Cancer Imm*, North Holland, Amsterdam, pp 75–83.
109. Minowada J, Sagawa K, Trowbridge IS et al. 1982b In: Rosenberg SA and Kaplan HS (eds.), *Malignant Lymphomas, Etiology, Immunology, Pathology, Treatment*, Academic Press, New York, pp 53–74.
110. Minowada J, Drexler HG, Menon M. *Hematol-Bluttransfus* 29: 426–429, 1985.
111. Miyoshi I, Hiraki S, Tsubota T et al. *Nature* 267: 843–845, 1977.
112. Mohammad RM, Vistisen K, Al-Katib A. *Electrophoresis* 15: 1218–1224, 1994.
113. Moore GE, Woods LK, Minowada J et al. *In Vitro* 8: 434–439, 1973.

114. Morikawa S, Tatsumi E, Baba M et al. *Int J Cancer* 21: 166–170, 1978.
115. Nagasaka M, Maeda S, Mabuchi O et al. *Int J Cancer* 30: 173–180, 1982.
116. Nagasawa K, Howatson A, Mak TW. *J Cell Physiol* 109: 181–192, 1981.
117. Nagasawa K, Mak TW. *Cell Immunol* 71: 396–403, 1982.
118. Nakao Y, Matsuda S, Fujita T et al. *Cancer Res* 42: 3843–3850, 1982.
119. Nakao Y, Koizumi T, Matsui T et al. *JNCI* 78: 1079–1086, 1987.
120. Narumiya S, Hirata M, Nanba T et al. *Biochem & Biophys Res Comm* 143: 753–760, 1987.
121. O'Connor R, Cesano A, Kreider BL et al. *J Immunol* 145: 3779–3787, 1990.
122. O'Connor R, Cesano A, Lange B et al. *Blood* 77: 1534–1545, 1991.
123. Ogawa S, Hirano N, Sato N et al. *Blood* 84: 2431–2435, 1994.
124. Ohnishi H, Kawamura M, Ida K et al. *Blood* 86: 1269–1275, 1995.
125. Ohnoshi T, Ohnuma T, Takahashi I et al. *Cancer Res* 42: 1655–1660, 1982.
126. Ohnuma T, Arkin H, Takahashi I et al. *Mol Inter Nut & Cancer* 105–121, 1985.
127. Ohmura T, Sakata A, Onoue K. *J Exp Med* 176: 887–891, 1992.
128. Paterson RLK, Kelleher C, Amankonah TD et al. *Blood* 85: 456–464, 1995.
129. Pecht I, Corcia A, Pia M et al. *EMBO J* 6: 1935–1939, 1987.
130. Pegoraro L, Fierro MT, Lusso P et al. *J Natl Cancer Inst* 75: 285–290, 1985.
131. Ponzoni M, Montaldo PG, Lanciotti M et al. *Biochem and Biophys Res Comm* 150: 702–710, 1988.
132. Prokocimer M, Peller S, Ben-Bassat H. *Leukemia Lymphoma* 29: 607–611, 1998.
133. Rabbitts TH. Nature 372: 143–149, 1994.
134. Raimondi SC, Behm FG, Roberson PK et al. *Blood* 72: 1560–1566, 1988.
135. Rasool O, Heyman M, Brandter LB et al. *Blood* 85: 3431–3436, 1995.
136. Ravid Z, Goldblum N, Zaizou R et al. *Int J Cancer* 25: 705–710, 1980.
137. Reinherz EL, Kung PC, Goldstein G et al. *Proc Natl Acad Sci USA* 77: 1588–1592, 1980.
138. Ribeiro RC, Raimondi SC, Behm FG et al. *Blood* 78: 466–471, 1991.
139. Robertson MJ, Cochran KJ, Cameron C et al. *Exp Hematol* 24: 406–415, 1996.
140. Ryffel B, Henning CB, Huberman E. *Proc Natl Acad Sci USA* 79: 7336–7340, 1982.
141. Sacchi N, Fiorini G, Plevani P et al. *J Immunol* 130: 1622–1626, 1983.
142. Sahraoui Y, Perraki M, Theodoropoulou M et al. *Leukemia* 11: 245–252, 1997.
143. Salvati MD, Peroni SE, Ranford PR et al. *Immunology* 78: 449–454, 1993.
144. San Jose E, Sahuquillo AG, Bragado R et al. *Eur J Immunol* 28: 12–21, 1998.
145. Santoli D, O'Connor R, Cesano A et al. *J Immunol* 144: 4703–4711, 1990.
146. Schneider U, Schwenk H-U, Bornkamm G. *Int J Cancer* 19: 621–626, 1977.
147. Shalitin C, Valansi C, Lev A et al. *Anticancer Res* 18: 145–150, 1998.
148. Shibata T, Shimada Y, Shimoyama M. *Jpn J Clin Oncol* 15: 67–75, 1985.
149. Shima-Rich EA, Harden AM, McKeithan TW et al. *Genes Chromosomes and Cancer* 10: 363–371, 1997.
150. Shimada Y, Shimoyama M *Gann* 75: 1116–1124, 1984.
151. Smith SD, Shatsky M, Cohen PS et al. *Cancer Res* 44: 5657–5660, 1984.
152. Smith SD, Morgan R, Link MP et al. *Blood* 67: 650–656, 1986.
153. Smith SD, Morgan R, Gemmell R et al. *Blood* 71: 395–402, 1988.
154. Smith SD, McFall P, Morgan R et al. *Blood* 73: 2182–2187, 1989.
155. Smith A, Eskew JD, Borza CM et al. Exp *Cell* Res 232: 246–254, 1997.
156. Srivastava BI, Minowada J. *Leukemia Res* 7: 331–338, 1983.
157. Stockbauer P, Ariyasu T, Stary J et al. *Human Cell* 7: 40–46, 1994.
158. Takagi S, Fujikawa K, Imai T et al. *Int J Cancer* 61: 706–715, 1995.

159. Tatsumi E, Piontek C, Sugimoto T et al. *Am J Hematol* 17: 287–294, 1984.
160. Teshigawara K, Wang HM, Kato K et al. *J Exp Med* 165: 223–238, 1987.
161. Thomas GA, Simpson RU. *Cancer Biochem Biophys* 8: 221–234, 1986.
162. Tycko B, Smith SD, Sklar J. *J Exp Med* 174: 867–873, 1991.
163. Uittenbogaart CH, Anisman DJ, Tary-Lehman M et al. *Int J Cancer* 56: 546–551, 1994.
164. Valtieri M, Santoli D, Caracciolo BL. *J Immunol* 138: 4042–4050, 1987.
165. van de Griend RJ, van der Reijden HJ, Bolhuis RLH et al. *Blood* 62: 669–674, 1983.
166. van den Berg AA, van Lenthe H, Kipp JB et al. Eur J *Cancer* 31A: 108–112, 1995.
167. van-Dongen JJ, Krissansen GW, Wolvers-Tettero IL et al. *Blood* 71: 603–612, 1988.
168. Watanabe S, Shimosato Y, Kameya T et al. *Cancer Res* 38: 3494–3498, 1978.
169. Westbrook CA, Rubin CM, Le Beau MM et al. *Proc Natl Acad Sci USA* 84: 251–255, 1987.
170. Xia Y, Brown L, Yang CY et al. *Proc Natl Acad Sci USA* 88: 11416–11420, 1991.
171. Yeargin J, Cheng J, Haas M. *Leukemia* 6(Suppl): 85–91, 1992.
172. Yodoi J, Teshigawara K, Nikaido T et al. *J Immunol* 134: 1623–1630, 1985.
173. Yoneda N, Tatsumi E, Kawano S et al. *Leukemia* 6: 136–141, 1992.
174. Yu J, Batova A, Shao L et al. Clin *Cancer Res* 3: 433–438, 1997.
175. Zhou M, Gu L, James CD et al. *Blood* 85: 1608–1614, 1995a.
176. Zhou M, Gu L, James CD et al. *Leukemia* 9: 1159–1161, 1995b.
177. Zhou M, Yeager AM, Findley HW. *Leukemia* 12: 1756–1763, 1998.
178. Zhu L, Anasetti C. *J Immunol* 154: 192–200, 1995.

# Chapter 6

# Myelocytic Cell Lines

Hans G. Drexler
*DSMZ-German Collection of Microorganisms & Cell Cultures, Dept. of Human and Animal Cell Cultures, Braunschweig, Germany. Tel: +49-531-2616-160; Fax: +49-531-2616-150; E-mail: hdr@dsmz.de*

## 1. INTRODUCTION

The fully differentiated end stage cell of the myelocytic/granulocytic series is the mature granulocyte. The mature granulocyte has a polylobed nucleus, hence the designation "polymorphonuclear leukocyte" [1]. In the normal adult, granulocytes are produced in the bone marrow. Granulocytes pass through three compartments during their life span: maturation in the bone marrow, transit in the blood, and function in the tissues [1]. All cells of this group play an essential role in acute inflammation in protection against microorganisms, although they do not show any inherent specificity for antigens. Their predominant function is phagocytosis [2]. The three types of granulocytes (neutrophils, eosinophils and basophils) have a parallel development [3].

The earliest morphologically recognizable cell of the granulocytic series is the myeloblast, which gives rise sequentially to the promyelocyte, myelocyte, metamyelocyte, stab cell and polymorph [4]. From promyelocyte onwards, specific granules (neutrophilic, eosinophilic or basophilic) become increasingly conspicuous in the cytoplasm and they distinguish the common neutrophilic granulocytes from the much less common eosinophils and the normally rare basophils [5]. Mitoses may occur in immature cells up to the metamyelocyte stage. Between the pluripotent stem cell and the first morphologically recognizable granulocyte precursor are committed stem cells and progenitor cells of lesser or greater restriction, with an inverse relationship between commitment to differentiation and self-renewal capacity [6]. Normal clonogenic progenitors common to granulocytes, erythrocytes, monocytes and megakaryocytes are denoted as CFU-GEMM. These progenitors give rise to granulocyte- monocyte (CFU-GM) colonies.

 *J.R.W. Masters and B.O. Palsson (eds.), Human Cell Culture Vol. III, 207–236.*
© 2000 *Kluwer Academic Publishers. Printed in the Netherlands.*

Neutrophils constitute over 90% of the circulating polymorphs and possess two main types of granules: the primary (azurophilic) granules are lysosomes containing acid hydrolases, myeloperoxidase and lysozyme; the secondary (specific) granules contain lactoferrin in addition to lysozyme [2,5]. Eosinophils comprise between 2–5% of blood leukocytes [7,8]. The granules of eosinophils contain specific products, eventually released by degranulation, which are not produced by neutrophils or basophils [9]. Morphocytochemically, the eosinophilic and basophilic granules can be distinguished by their typical size and color. Eosinophilic granules are stained with Luxol fast blue which is specific for eosinophils, but are not metachromatic when stained with toluidine blue, a characteristic which is commonly used to identify basophils.

Basophils are found in small numbers in the circulation (commonly <1% of leukocytes). A "relative" of the basophil, the mast cell, is not seen at all in the peripheral blood and is only resident in body tissues. The relationship of the mast cell to the basophil is controversial. There are similarities, but also distinct differences between these two types of cells, and the interrelationship of the basophil-mast cell system requires further study [10–13]. While the basophil clearly arises in the bone marrow, the origin of mast cells has been more difficult to establish [14]. There may be a common progenitor for mast cells and basophils, because mast cell and basophil precursors have been found among the blast cells in CML. Various analyses of single mixed hematopoietic colonies (CFU-GEMM) also show human "basophil/mast cells" to have a clonal origin from circulating multipotent hematopoietic progenitors. Similarly, "basophils/mast cells" can be grown in suspension culture from umbilical cord mononuclear cells or the peripheral blood of individuals with allergies [15].

Malignancies involving the myelocytic cells are of acute or chronic nature. Clinicians distinguish the acute myeloid leukemias (AML) from the myeloproliferative disorders (most prominently CML) and the "pre-leukemic" myelodysplastic syndromes (MDS). While monocytic, erythroid and megakaryocytic acute leukemias have been assigned to the morphological subtypes FAB M4/M5, M6 and M7, respectively, the myelocytic subtypes of AML have been designated FAB M0, M1, M2 and M3, depending on the degree of apparent morphological maturation and cytochemical differentiation [16,17]. With regard to diagnosis, prognosis and treatment, the unique subtypes AML M2 and AML M3 are of particular interest [18–20]. Cytogenetically, AML M2 is associated with the t(8;21) [21,22], whereas the t(15;17) is specific for AML M3 [23]. Cell lines are available for these clinically and scientifically interesting AMLs (Table 1b). Equally important is the availability of cell lines as model systems for the rare basophilic, mast cell and eosinophilic leukemias [10] (Table 1b). MDS with its different entities may terminate in various subtypes of AML [24–26]. The most common chronic form of myelocytic leukemia is CML [27–29], which ends invariably in a blast crisis. The predominant cell type during this blast crisis has myelocytic features.

The establishment of human myelocytic cell lines has always been difficult [30,31]. However, the number of myelocytic cell lines has increased dramatically [32–34], and now more than 40 well-characterized and mostly authenticated myelocytic cell lines have been described (Tables 1–5). The first continuous human myelocytic leukemia cell line was HL-60, which was established in 1976 [47]. Due to its widespread availability, this cell line has become the quintessential *in vitro* paradigm for myeloid leukemia cells [111]. Another widely used, early myelocytic cell line is KG-1 [62]. Myelocytic cell lines have also been established from specific AML subtypes, CML in myeloid blast crisis, MDS and rare myeloid leukemias. Of particular importance are several unique promyelocytic, basophilic, mast cell and eosinophilic cell lines [10,100] (Table 1b).

The data listed in Tables 1–6 provide detailed characteristics of individual myelocytic cell lines. The lines differ greatly, despite their origin from the same or similar types of leukemia. This multiplicity of markers reflects the heterogeneity inherent in human myeloid leukemias. A number of the myelocytic cell lines described below are available from major cell banks (Appendix 1, p. 284).

## 2. CLINICAL CHARACTERIZATION

Fifty-one well-characterized myelocytic cell lines are listed in Table 1a. These lines were established from 34 patients with various forms of AML (1 AML M0, 3 AML M1, 14 AML M2, 5 AML M3, 6 AML M4, 1 eosinophilic AML, 4 unspecified AML), 13 patients with CML in blast crisis, and one patient each with CMML, MDS, mast cell leukemia or T-ALL. In the latter case, a lymphoid cell line was not established, but myelocytic cells grew in culture. Corresponding to the relatively low incidence of AML and CML in childhood, 46 myelocytic cell lines were derived from adults and only five from children; the sex distribution is 29 from male and 22 from female patients. Only 15 (29%) lines were established at the diagnosis of the disease (i.e. prior to therapy), whereas 36 (71%) lines were derived from cases at relapse, in blast crisis or resistant to therapy. Thirty-three (64%) and fifteen (29%) cell lines were developed from samples taken from peripheral blood and bone marrow, respectively. The origin of the cell lines from the presumed patient was confirmed in a relatively high percentage of cell lines; 29/51 (57%). Apart from Iscove's MEM (n = 3) and $\alpha$-MEM (n = 6), the

*Table 1a.* Myelocytic cell lines: clinical characterization

| Cell line[a] | Patient[b] age/sex | Diagnosis[c] | Treatment status[d] | Specimen site[e] | Authentication[f] | Year of est. | Culture medium[g] | Availability[h] | Primary ref. |
|---|---|---|---|---|---|---|---|---|---|
| AML14[i] | 68 M | AML M2 | refractory | PB | no | 1991 | RPMI 1640 + FBS | | 35,36 |
| AR230 | 52 F | CML-my BC | BC | PB | yes | 1991 | RPMI 1640 + FBS | Author | 37 |
| CML-C-1[j] | 42 F | CML-BC | BC | PB | yes | 1990 | RPMI 1640 + FBS + mouse BM stromal feeder layer | | 38 |
| CTS | 13 F | AML M1 (post BMT) | R | PB | no | 1992 | RPMI 1640 + FBS | Author | 39 |
| Ei501[k] | 17 F | AML M3 | D | BM | no | 1994 | RPMI 1640 + FBS | Author | 40 |
| EM-2[l] | 5 F | CML-my BC (post BMT) | 2nd R | BM | yes | 1982 | RPMI 1640 + FBS | DSMZ | 41 |
| EoL-1[m] | 33 M | hypereosinophilic syndrome → eosinophilic leukemia | D | PB | yes | 1984 | RPMI 1640 + FBS | DSMZ, RIKEN | 42 |
| FKH-1 | 61 M | Ph-neg CML with trilineage myelodysplasia → AML M4 | refractory | PB | yes | 1993 | RPMI 1640 + FBS + GM-CSF | | 43 |
| GDM-1 | 66 F | MPD (CML-like) → AML M4 | refractory | PB | yes | 1979 | RPMI 1640 + FBS | DSMZ | 44 |
| GF-D8 | 82 M | AML M1 | D | PB | no | 1989 | RPMI 1640 + FBS (or serum-free) + GM-CSF | Author | 45 |
| GM/SO | 50 F | CML-my BC | BC | BM | yes | 1988 | RPMI 1640 + FBS + GM-CSF | Author | 46 |
| HL-60 | 35 F | initially AML M3 – later changed to AML M2 | D | PB | yes | 1976 | RPMI 1640 + FBS | ATCC, DSMZ, IFO, JCRB, RIKEN | 47–49 |
| HMC-1 | 52 F | mast cell leukemia | refractory | PB | yes | 1994 | IMDM + FBS | Author | 15 |
| HNT-34 | 47 F | MDS (CMML) → AML M4 | refractory | PB | yes | 1994 | RPMI 1640 + FBS | Riken | 50 |
| HT93 | 66 M | AML M3 | R | PB | no | 1993 | RPMI 1640 + FBS | | 51 |
| IRTA17[n] | 25 F | AML M2 | R | BM | yes | 1993 | IMDM + FBS + G-CSF, GM-CSF or SCF | | 52 |
| K051[o] | 46 M | AML M2 | D | BM | yes | 1991 | α-MEM + FBS | Author | 53 |
| Kasumi-1 | 7 M | AML M2 (post-BMT) | 2nd R | PB | yes | 1989 | RPMI 1640 + FBS | DSMZ | 54 |
| Kasumi-3 | 57 M | AML M0 | D | BM | yes | 1990 | RPMI 1640 + FBS | Author | 55 |
| Kasumi-4 | 6 F | CML-my BC | BC | PB | yes | 1993 | RPMI 1640 + FBS | Author | 56 |
| KBM-7[p] | 39 M | CML-my BC (post BMT) | 2nd BC | BM | yes | 1984 | IMDM + FBS | Author | 57,58 |
| KCL-22 | 32 F | CML-BC | BC | PE | no | 1981 | RPMI 1640 + FBS | Author | 59,60 |
| KF-19[q] | 35 M | AML M2 | R | Pericardial effusion | no | 1981 | α-MEM + FBS | | 61 |

*Continued on next page*

*Table 1a.* (continued)

| Cell line[a] | Patient[b] age/sex | Diagnosis[c] | Treatment status[d] | Specimen site[e] | Authentication[f] | Year of est. | Culture medium[g] | Availability[h] | Primary ref. |
|---|---|---|---|---|---|---|---|---|---|
| KG-1[r] | 59 M | erythroleukemia → AML | R | BM | yes | 1977 | α-MEM or RPMI 1640 + FBS | ATCC, DSMZ, JCRB, RIKEN | 62,63 |
| KOPM-28 | 64 F | CML-my BC | BC | PB | no | 1982 | RPMI 1640 + FBS | Author | 64 |
| KT-1 | 32 M | CML-BC | BC | PB | yes | 1991 | RPMI 1640 + FBS | Author | 65 |
| KU-812[s] | 38 M | CML-BC | BC | PB | no | 1981 | McCoy's 5A or RPMI 1640 + FBS | DSMZ, IFO, JCRB, RIKEN | 66,67 |
| KY821[t] | 28 M | AML M2 | R | meninges | no | 1982 | RPMI 1640 + FBS | JCRB | 68 |
| KYO-1 | 22 M | CML-my BC | BC | PB | no | 1981 | RPMI 1640 + FBS | Author | 69 |
| LW/SO | 60 F | AML M4 → AML M2 | R | BM | no | | RPMI 1640 + FBS | | 70 |
| M20 | 10 F | AML M2 | D | PB | no | | RPMI 1640 + FBS | | 71 |
| Marimo | 68 F | essential thrombocythemia → t-AML M2 | terminal | BM | no | 1993 | RPMI 1640 + FBS | Author | 72 |
| MDS92 | 52 M | MDS (RARS/RAEB) (prior leukemic transformation) | D | BM | no | 1991 | RPMI 1640 + FBS + IL-3 | Author | 73 |
| MOLM-6[u] | 44 M | CML-my BC | BC | PB | no | 1992 | RPMI 1640 + FBS | Author | 74 |
| MUTZ-2 | 62 M | AML M2 | D | PB | yes | 1993 | α-MEM + FBS + SCF or 5637 CM | DSMZ | 75 |
| NB4[v] | 23 F | AML M3 | 2nd R | BM | yes | 1989 | RPMI 1640 + FBS | DSMZ | 76–78 |
| NKM-1 | 33 M | AML M4 | D | PB | no | 1981 | RPMI 1640 + FBS or serum-free | | 79 |
| OCI/AML-1[w] | 73 F | AML M4 | D | PB | no | 1987 | α-MEM + FBS + G-CSF or 5637 CM | Author | 80,81 |
| OCI/AML-4 | 35 F | Hodgkin's disease + AML M4 | D | PB | yes | 1987 | α-MEM + FBS (+ 5637 CM) | Author | 82 |
| OCI/AML-5 | 77 M | AML M4 | R | PB | yes | 1990 | α-MEM + FBS + GM-CSF or 5637 CM | DSMZ | 83 |
| OHN-GM | 60 M | Hodgkin's disease → t-MDS (RA/RAEB) (t-AML) | refractory | BM | yes | 1995 | RPMI 1640 + FBS + GM-CSF | Author | 84 |
| OIH-1 | 72 M | MDS (RAEB) → AML | D | PB | yes | 1993 | RPMI 1640 + FBS + G-CSF or GM-CSF | | 85 |
| PL-21 | 24 M | (mediastinal) granulocytic sarcoma → AML M3 | refractory | PB | no | 1981 | RPMI 1640 + FBS | | 59,86 |
| SKNO-1 | 22 M | AML M2 | 2nd R | BM | yes | 1990 | RPMI 1640 + FBS + GM-CSF | Author | 87 |
| SR-91 | 22 M | T-ALL (post-BMT) | R | PB | yes | 1991 | RPMI 1640 + FBS | Author | 88 |
| TI-1 | 42 M | AML M2 | D | PB | no | 1988 | RPMI 1640 + FBS | Author | 89 |
| UCSD/AML1 | 73 F | AML | R | BM | yes | 1989 | RPMI 1640 + FBS + GM-CSF | Author | 90 |

*Continued on next page*

*Table 1a.* (continued)

| Cell line[a] | Patient[b] age/sex | Diagnosis[c] | Treatment status[d] | Specimen site[e] | Authenti- cation[f] | Year of est. | Culture medium[g] | Availability[h] | Primary ref. |
|---|---|---|---|---|---|---|---|---|---|
| UF-1 | 33 F | AML M3 | 2nd R | PB | no | 1994 | RPMI 1640 + FBS | Author | 91 |
| YJ | 69 M | CMML (with eosinophilia) | BC | PB | no | 1994 | RPMI 1640 + FBS | Author | 92 |
| YNH-1 | 46 M | AML M1 | D | PB | yes | 1994 | RPMI 1640 + FBS + GM-CSF | Riken | 93 |
| YOS-M[x] | 77 M | CML-my BC | BC | PB | yes | 1990 | RPMI 1640 + FBS | Author | 94 |

[a] Names of cell lines are as given in the original literature; subclones (variant cell lines derived from a parental cell line) and sister cell lines (derived independently from the same patient from different specimens or at different time points) are indicated. [b] Age at the time of establishment of cell line. [c] Diagnoses are indicated as given in the original literature; →: disease progressed from a pre-malignancy/first malignancy into the final malignancy. [d] BC: at blast crisis; CP: in the chronic phase; D: at diagnosis (prior to therapy); R: at relapse; T: during therapy. [e] BM: bone marrow; PB: peripheral blood; PE: pleural effusion; Tu: tumor. [f] Evidence (e.g. cytogenetic marker chromosomes, immunoprofile, others) that this cell line was derived from the patient indicated. [g] Culture media as indicated in the original literature; cell line might also grow with other medium and/or supplements. [h] Availability from cell banks (ATCC; DSMZ; IFO; JCRB; RIKEN) or from the original investigator (author). [i] Subclone AML14.3D10 differentiates spontaneously to eosinophils. [j] CML-C-1 is the *in vitro* variant of cell line CML-N-1 (which was established in a nude mice and is serially transplantable *in vivo*, but failed to proliferate *in vitro* under standard liquid culture conditions). [k] Ei501 seems to be lost entirely due to yeast contamination. [l] The sister cell line EM-3 was established independently (EM-2 at day 28 and EM-3 at day 47 post- BMT) showing similar features. [m] The two sister subclones EoL-2 and EoL-3 show similar immunophenotypical and functional features; Eo-B is an EBV+ B-LCL from the same patient. [n] Sister cell line IRTA21 is immunophenotypically and cytogenetically different from IRTA17; both IRTA17 and IRTA21 are lost. [o] Sister cell line K052 was established at relapse showing different immunophenotypic, cytogenetic and oncogene point mutation features. [p] The subclone KBM-7/B5 is near-haploid, but has similar immunological and functional features. [q] Subclones KF-10AraC, KF-10ADR and KF-19VCR are constitutively resistant to cytosine arabinoside, adriamycin and vincristine. [r] Subclone KG-1a established between passages 15 and 35 of the original KG-1 is morphologically, cytochemically, immunologically and functionally less mature than KG-1. [s] Subclones KU-812-E and KU-812-F were isolated by semi-solid culture; these subclones possess similar properties as KU-812. [t] Subclone KY821A3 is methotrexate-resistant. [u] Seven cell lines from a single sample from the same patient were established: MOLM-7 and MOLM-11 represent the megakaryocytic lineage; MOLM-6, -8, -9, -10, -12 represent the myelocytic lineage. [v] Retinoic acid-resistant subclones NB4.306 and NB4-RAr were established. [w] Subclone OCI/AML-1a has similar features as the parental line OCI/AML-1. [x] YOS-B is a sister cell line, an EBV+ B-LCL carrying also the Philadelphia chromosome.

*Table 1b.* Myelocytic cell lines: cell lines with promyelocytic, eosinophil, basophil or mast cell features

| Promyelocytic | Eosinophil | Basophil | Mast cell |
|---|---|---|---|
| Ei501[a] | AML14[b] | GRW[b] | HMC-1 |
| HL-60[c] | EoL-1 | HL-60[b] | |
| HT93[a] | HL-60[b] | KU-812 | |
| NB4[a] | ME-1[b] | LAMA-84[b] | |
| UF-1[a] | OMA-AML-1[b] | | |
| | YJ | | |

[a] Cell line carries the AML M3-specific t(15;17)(q22;q21) translocation with the PML-RARA fusion gene.

[b] Only weak constitutive expression of eosinophil or basophil features which, however, can be up-regulated by differentiation-inducing agents.

[c] HL-60 does not carry the specific t(15;17)(q22;q21) translocation, but has promyelocytic morphology and responds to retinoic acid.

Cell lines ME-1, OMA-AML-1, GRW, and LAMA-84 are described in the chapters on monocytic and erythroid-megakaryocytic cell lines, respectively.

preferred choice of medium is RPMI 1640 (n = 41) supplemented with FBS; only two cell lines (GF-D8, NKM-1) were reported to also grow under serum-free conditions (Appendix 2, p. 286). Thirteen of these cell lines (25%) can be obtained from cell banks (Appendix 1, p. 284).

Of note are the sister cell lines which were established from the same patient (simultaneously or at different time points: EM-3, K052, MOLM 7–12), subclones with more or less significant phenotypic differences (derived from AML14, EoL-1, KBM-7, KF-10, KG-1, KU-812, KY821, NB4, OCI/AML-1) and complementary EBV+ B-LCLs from the same patient (Eo-B, YOS-B).

Several cell lines deserve special mention as they represent the *in vitro* counterparts of rare primary leukemia subtypes (Table 1b). While HL-60 has been the *in vitro* model cell line for AML M3 (or acute promyelocytic leukemia), these cells do not carry the AML M3-specific t(15;17) resulting in the *PML-RARA* fusion gene (Table 4). Nevertheless, HL-60 can be induced by a variety of reagents to undergo differentiation to a promyelocytic stage and beyond [31,114]. Several other authentic AML M3 promyelocytic cell lines have now been established: Ei501, HT93, NB4, and UF-1 (Table 1b). Some cell lines display features of eosinophils constitutively or upon induced differentiation. The best-known and most widely used cell line is EoL-1 (and its subclones EoL-2/-3). The KU-812 cell line was generated from a CML patient. In line with the biology of CML, KU-812 exhibits a multilineage differentiation capacity. However, the basophil differentiation potential is pre-

*Table 2.*  Myelocytic cell lines: immunophenotypical characterization

| Cell line | T-/NK-cell marker[a] | B-cell marker | Myelomonocytic marker | Erythroid-megakaryocytic marker | Progenitor/activation marker | Adhesion marker | Ref. |
|---|---|---|---|---|---|---|---|
| AML14 | CD2− | CD20− | CD13+ CD14(+) CD33+ | | CD34− | CD11b+ | 35 |
| AR230 | CD2− CD3− CD4+ CD5− CD7− CD8− | CD10− CD19− CD20− CD22− CD24− | CD13+ CD33+ CD35+ | | CD34− HLA-DR− | CD11b(+) | 37 |
| CML-C-1 | CD3− | CD19− | CD13+ CD33− | CD41a− | | CD11b− | 38 |
| CTS | CD1− CD2− CD3− CD4− CD5− CD7+ CD8− | CD10− CD19− CD20− | CD13+ CD14− CD33+ | CD41− GlyA− | CD34+ CD38− HLA-DR+ CD34+ HLA-DR+ | | 39 |
| Ei501 | | CD10− CD19− | CD13+ CD33+ | | CD34− CD38+ HLA-DR+ | CD11a+ CD18+ CD54+ | 40 |
| EM-2 | CD2− CD3− CD4+ CD5− CD7− CD8− CD28− CD57− | CD9− CD10− CD19− | CD13+ CD14+ CD15+ CD33+ | CD41− CD42b− vWF− | CD34− HLA-DR− TdT− | | 41,95,96 |
| EoL-1 | CD1b− CD2− CD3− CD4+ CD5− CD7− CD8− CD28− TCRαβ− | CD20− CD21− CD23− cy/sIg− | CD13− CD14− CD15+ CD16− CD32+ CD33+ CD64− | | CD34− CD71− HLA-DR+ TdT− | CD11a+ CD11b− CD54(+) | 42,95,97 |
| FKH-1 | CD2− CD3− CD4− CD5− CD7− CD8− | CD9+ CD10− CD19− CD20− CD21− cy/sIg− | CD13+ CD33+ | CD41a− CD61− | CD34− HLA-DR+ | CD11a+ CD11b+ CD11c+ | 43 |
| GDM-1 | CD3− CD7− | CD10− CD19− CD20− CD21− CD22− | CD13− CD14− CD15+ CD33+ | | CD34+ CD71+ HLA-DR+ TdT− | CD11b+ | 44,95 |
| GF-D8 | CD1− CD2− CD3− CD4+ CD57− TCRαβ− TCRδ− | CD10− CD20− | CD13+ CD14− CD15− CD33+ | CD41− | CD34− HLA-DR− TdT− | CD11a+ CD11b+CD11c− CD49d− CD54− | 45 |
| GM/SO | CD1− CD2+ CD3− CD4− CD5− CD7− CD8− CD56− CD57− | CD9+ CD10− CD19− CD20− CD22− CD24+ CD80− CD86− sIg− | CD13+ CD14− CD16− | CD36+ CD42b− | CD34+ HLA-DR+ | CD11b− | 46 |
| HL-60 | CD1− CD2− CD3− CD4+ CD5− CD7− CD8− | CD9+ CD10− CD19− CD20− CD22− CD24+ CD80− CD86− sIg− | CD13+ CD14− CD15+ CD33+ | CD41a− CD42b− CD61− vWF− | CD34− CD38+ CD71+ HLA-DR− TdT− | CD11b− CD44+ CD54− | 48,95 |
| HMC-1[b] | CD1− CD2− CD3− CD4 CD7− CD8− CD56+ | CD9− CD10− CD19− CD20− CD21− CD22− CD23− CD24− CD37+ CD40+ CD74− | CD12− CD13+ CD14− CD15− CD16− CD17+ CD32+ CD33− CD35− CD65− CD68− CD88+ | CD31− CD41− CD61+ CD63+ | CD34− CD38− CD69− CD71− HLA-DR− TdT− | CD11a+ CD11b− CD11c(+) CD18+ CD43+ CD44+ CD54+ | 15,98 |
| HNT-34 | CD2− CD3+ CD4+ CD7(+) CD8− | CD10− CD19− CD20− | CD13+ CD14− CD33+ | CD41a− CD42b− GlyA− | CD34+ HLA-DR− | | 50 |
| HT93 | | CD10− CD19− CD20− CD21− CD22− CD23+ | CD13− CD14− CD15(+) CD33+ | CD36− CD41− GlyA− | CD34+ CD38− HLA-DR− | CD11b− | 51,99 |
| IRTA17 | | CD10− CD19− CD20− | CD13+ CD14+ CD33− | CD36− | CD34+ HLA-DR(+) | | 52 |
| K051 | CD2− CD3− CD5− CD7− | CD10− CD19− CD20− | CD13+ CD14− CD33+ | CD41+ GlyA+ | CD34− HLA-DR− | | 53 |
| Kasumi-1 | CD2− CD3− CD4+ CD5− CD7− CD8− | CD10− CD19− CD20− CD21− | CD13+ CD14− CD15+ CD33+ CD68− | | CD34+ CD38+ CD71+ HLA-DR+ | CD11b− CD11c− | 54,95 |

*Continued on next page*

*Table 2.* (continued)

| Cell line | T-/NK-cell marker[a] | B-cell marker | Myelomonocytic marker | Erythroid-megakaryocytic marker | Progenitor/ activation marker | Adhesion marker | Ref. |
|---|---|---|---|---|---|---|---|
| Kasumi-3 | CD2– CD3– CD4+ CD5– CD7+ CD8– | CD10– CD19– CD20– | CD13+ CD14– CD15– CD33+ | CD36– CD41– CD42– | CD34+ HLA-DR+ | CD11a+ CD11b(+) CD11c(+) CD54+ | 55 |
| Kasumi-4 | CD2– CD3– CD4– CD7– CD8– | CD10– CD19– CD20– | CD13+ CD14– CD33+ | CD36– CD41a– CD42b– GlyA– PPO– | CD34+ HLA-DR+ | CD11b– | 56 |
| KBM-7 | CD4+ | CD20(+) | CD13+ CD14(+) CD33+ CD33+ CD65+ | CD9+ CD41a+ CD61+ GlyA– | CD34+ HLA-DR(+) TdT– | CD11b(+) | 58 |
| KCL-22 | CD1b– CD2– CD4(+) CD28– | CD20– CD80– CD86– cy/sIg– | CD13+ CD14– CD15+ CD33+ CD65+ | CD9+ CD41a+ CD61+ GlyA– | CD34+ HLA-DR– TdT– | CD11b– CD54+ | 60,100 |
| KF-19 | CD3– CD5– | CD19– | CD13+ CD14– CD33+ | | | CD11b– | 61 |
| KG-1 | CD1– CD2– CD3– CD4– CD5– CD7– CD8– CD28– CD56– sIg– CD57– TCRαβ– TCRδ– | CD9– CD10– CD19– CD20– CD21– CD22– CD24+ CD80+ CD86– sIg– | CD13+ CD14– CD15+ CD33+ CD65+ | CD36– CD41– CD42b– CD61+ GlyA– vWF– | CD34+ CD38+ CD71+ HLA-DR+ TdT– | CD11b+ CD44+ CD54+ | 30,95,99 |
| KOPM-28 | CD1a– CD2– CD3– CD4+ CD8– | CD10– CD19– CD20– CD22– cy/sIg– | CD13+ CD15– CD33+ | CD41+ PPO– | CD38+ CD71+ HLA-DR+ TdT– | CD11b+ | 64,101 |
| KT-1 | CD2– CD3– CD4+ CD5– CD7– CD8– | CD10– CD19– CD20– | CD13– CD14– CD33+ | CD41a– | CD34– HLA-DR– TdT– | CD11b– | 65 |
| KU-812[b] | CD2– CD3– CD5– CD7– | CD9+ CD10– CD19– CD20– CD21– CD22– CD23– CD24+ CD37– CD39– CD40+ CD80– CD86+ cy/sIg– | CD13+ CD14– CD15– CD16– CD17+CD32+ CD33+ CD35+ CD65– CD68+ CD88+ | CD31+ CD41+ CD42+ CD61+ CD63+ GlyA(+) | CD34– CD38– CD71+ HLA-DR– TdT– | CD11a– CD11b(+) CD11c– CD18(+) CD43+ CD44+ CD54+ CD69– | 66,95,98 |
| KY821 | CD2– CD3– | CD19– | CD13+ CD14– CD15+ CD33+ | | CD34– HLA-DR– HLA-DR– TdT– | | 68 |
| KYO-1 | CD2– | sIg– | | GlyA– | CD34(+) CD38(+) CD71+ HLA-DR– TdT– | CD11b+ | 69 |
| LW/SO | CD2– CD3(+) CD4+ CD5+ CD7+ CD8– CD56– | CD10– | CD15+ | CD41– CD42– | HLA-DR– | | 70 |
| M20 | CD7+ | sIg– | | | HLA-DR– | | 71 |
| Marimo | T-marker neg | B-marker neg | CD13+ CD14– CD15+ CD33– | ery-meg marker neg | CD34– HLA-DR– | | 72 |
| MDS92 | CD2– CD3– CD7– | CD19– CD20(+) | CD13+ CD14(+) CD33+ | CD41(+) CD61– GlyA– | CD34+ HLA-DR+ | CD11b+ | 73 |
| MOLM-6 | CD3– CD4+ | CD9(+) CD10– CD19– | CD13+ CD14– CD15+ CD33+ | CD41a– CD61– | CD34(+) CD71+ HLA-DR+ TdT– | CD11b+ | 74 |
| MUTZ-2 | CD1– CD3– CD4+ CD5– CD7– CD8– CD56– | CD10– CD19– CD20– | CD13+ CD14– CD15+ CD16– CD33+ CD65+ CD68+ | CD41a– CD42b– CD61– GlyA– | CD30+ CD34+ CD38+ CD71+ HLA-DR+ TdT– | CD11b– | 75,95 |

*Continued on next page*

*Table 2.*   (continued)

| Cell line | T-/NK-cell marker[a] | B-cell marker | Myelomonocytic marker | Erythroid-megakaryocytic marker | Progenitor/activation marker | Adhesion marker | Ref. |
|---|---|---|---|---|---|---|---|
| NB4 | CD2+ CD3- CD4+ CD5- CD6- CD7- CD8- TCRαβ- TCRγδ- | CD9+ CD10- CD19- CD20- CD23- | CD13+ CD14- CD15+ CD33+ CD68- | CD36- CD41- CD42- GlyA- | CD34- CD38+ HLA-DR- | CD11b+ CD11c- | 76,91,95 |
| NKM-1 | CD2- CD3- CD5- CD7- | CD10- CD19- CD20- | CD13- CD14- CD15+ CD16- CD32+ CD33- | CD41- GlyA- | HLA-DR+ | CD11b- CD18+ | 79 |
| OCI/AML-1 | CD7- CD8- | CD19- CD20- | CD13+ CD14(+) CD33+ | CD36+ CD41a- GlyA- | CD34+ CD38+ CD71+ HLA-DR+ | CD11b+ | 81,83,102 |
| OCI/AML-4 | CD1- CD2- CD3- CD7+ | CD10- CD19+ CD20- | CD13+ CD14+ CD33+ | CD41- GlyA- | CD34- HLA-DR+ TdT+ | CD11+ | 82 |
| OCI/AML-5 | CD3- CD7- CD8+ | CD19- | CD13+ CD14- CD15+ CD33+ CD68+ | | CD34+ HLA-DR+ TdT+ | | 83,95 |
| OHN-GM | CD3- CD4- CD5- CD7+ CD8- | CD10- CD19- CD20- | CD13+ CD33+ | CD41- GlyA- | CD34+ CD38+ CD71+ HLA-DR+ | | 84 |
| OIH-1 | CD1- CD2- CD3- CD4+ CD5- CD7+ CD8- | CD10- CD19(+) CD20- | CD13+ CD14- CD33+ | CD41a- CD42b- GlyA- | CD34+ HLA-DR+ | | 85 |
| PL-21 | CD2- CD4+ CD4+ | CD10- CD20- sIg- CD19+ | CD13+ CD14- CD15+ CD33+ CD13+ CD33+ | CD41a+ CD61+ | HLA-DR+ TdT- CD34+ | | 86,100 |
| SKNO-1 | CD4+ | CD19+ | | | | | 87 |
| SR-91 | CD1- CD2- CD3- CD4- CD5- CD7- CD8- CD56+ | CD10- CD19- CD20- | CD13- CD14- CD33+ | CD41- | CD34+ HLA-DR- | CD54+ | 88 |
| TI-1 | CD1- CD2- CD3- CD4- CD5- CD7- CD8- | CD19- CD20- | CD13+ CD14- CD33+ | CD41a- | CD34+ | | 89 |
| UCSD/AML1 | CD2- CD4- CD5- CD7+ CD8- | CD19- CD20- | CD13+ CD14- CD15+ CD33+ | CD36+ CD41a- CD42b+ GlyA- | CD34+ CD38- CD71+ HLA-DR+ TdT+ | CD49b+ | 90,102 |
| UF-1 | CD3- CD4- CD5- CD7+ CD8- | CD10- CD19- CD20- | CD13+ CD14+ CD33+ | CD41- | CD34- CD38+ HLA-DR- | CD11b- | 91 |
| YJ | | CD19- CD20- CD23- | CD14(+) CD16- CD33+ | | CD34- | | 92 |
| YNH-1 | CD1- CD2+ CD3- CD4- CD5- CD7- CD8- | CD10+ CD19- CD20- | CD13+ CD14- CD33+ | CD41a+ CD42b- HLA-DR- | CD34+ HLA-DR- | CD11a+ CD11b+ CD11c+ | 93 |
| YOS-M | CD2- CD3- CD4+ CD5- CD7- CD8- | CD9- CD10- CD19- CD20- CD21- | CD13(+) CD14+ CD33+ | | CD34+ HLA-DR+ TdT- | | 94 |

a +, strong, definite protein expression (mostly more than 10–20% cells positive); (+), weak protein expression, qualitatively and quantitatively (commonly <10% cells positive); –, no protein expression.
b Extensive immunophenotype in ref. 98.

*Table 3.* Myelocytic cell lines: cytokine-related characterization

| Cell line | Cytokine receptor expression[a] | Cytokine production[a] | Proliferation response to cytokines[b] | Differentiation response to cytokines[b] | Dependency on cytokines[c] | Ref. |
|---|---|---|---|---|---|---|
| AML14 | mRNA: GM-CSFRα+, IL-3Rα+, IL-3Rβ+, IL-5Rα+ | | GM-CSF+, IL-3+, IL-5+ | GM-CSF+, IL-3+, IL-5+ | | 35 |
| Ei501 | RT-PCR: IL-2Rγ+<br>protein: IL-2Rα−, IL-2Rβ+, IL-4Rα+ | RT-PCR: TGFβ+ | | | | 40 |
| EoL-1 | IL-2Rα+, Kit− | | | G-CSF+, TNFα+ | | 97 |
| FKH-1 | | | G-CSF+, GM-CSF+, IL-3+, SCF+ | | G-CSF or GM-CSF | 43 |
| GF-D8 | mRNA: G-CSFR+, GM-CSFRα+, GM-CSFRβ+, IL-2Rα+, IL-2Rβ+, IL-2Rγ+, IL-3Rα+, IL-4R(+), IL-7Rα(+), IL-9Rα(+)<br>protein: IL-2Rα−, IL-2Rγ− | IL-1β+ | GM-CSF+, IFNγ+, IL-3+, PIXY-321+, SCF+<br>inhibition: TGFβ1+ | | GM-CSF or IL-3 | 45,103 |
| GM/SO | | | GM-CSF+, IFNγ+, IL-1α(+), IL-4+, IL-13+, PIXY-321+, SCF+<br>inhibition: TGFβ1+ | | GM-CSF | 46,103 |
| HL-60 | mRNA: IGF-1R+, IGF-2R+, IL-2Rα(+), IL-4Rα+, IL-7Rα(+), IL-9Rα+<br>protein: IL-2Rα−, IL-2Rβ+, IL-2Rγ+, Kit− | | | | | 104 |
| HMC-1 | GM-CSFRα+, IL-1Rα(+), IL-2Rα(+), IL-2Rβ(+), IL-3Rα−, IL-3Rβ+, IL-6Rα−, IL-7Rα−, Kit+ | | | | | 10,98 |
| HT93 | IL-2Rα−, IL-2Rβ− | | G-CSF+, GM-CSF+ | | | 51 |
| IRTA17 | IL-2Rα(+) | | G-CSF+, GM-CSF+, IL-3+, SCF+ | | G-CSF, GM-CSF or SCF | 52 |
| Kasumi-1 | Kit+ | | G-CSF+, GM-CSF+, IL-3+, IL-6+ | | | 54 |
| Kasumi-3 | IL-2Rα+, Kit+ | | GM-CSF+, IL-2+, IL-3+, IL-4+, SCF+ | | | 55 |
| Kasumi-4 | | | GM-CSF+, IL-3+, IL-6+, SCF+ | | | 56 |
| KF-19 | | | inhibition: IFNγ+, TNFα+ | | | 61 |
| KG-1 | mRNA: IL-4Rα−, IL-7Rα−, IL-9Rα−<br>protein: IL-2Rα+, IL-2Rβ+, IL-2Rγ+, Kit− | | | | | 99,104 |
| KU-812 | EPO-R+, GM-CSFRα+, IL-1Rα+, IL-2Rα−, IL-2Rβ−, IL-3Rα+, IL-3Rβ+, IL-6Rα−, IL-7Rα−, Kit+ | | | | | 98 |
| LW/SO | IL-2Rα+, IL-2Rβ−, Kit−, TNFRI+, TNFRII+ | | | | | 70 |
| MDS92 | | | GM-CSF+, IL-3+, SCF+, TPO(+) | G-CSF+, GM-CSF+, IL-3+, SCF+ | IL-3 | 73,105 |
| MUTZ-2 | GM-CSFRα+, IL-2Rα+, IL-3Rα+, Kit+, M-CSFR+ | | bFGF+, FL+, G-CSF+, IFNβ+, IFN-γ+, IGF-1+, IL-6+, M-CSF+, SCF+, TNFα+ | | SCF | 75,103 |

*Continued on next page*

*Table 3.* (continued)

| Cell line | Cytokine receptor expression[a] | Cytokine production[a] | Proliferation response to cytokines[b] | Differentiation response to cytokines[b] | Dependency on cytokines[c] | Ref. |
|---|---|---|---|---|---|---|
| NKM-1 | G-CSFR+, M-CSFR+ | | G-CSF+, M-CSF+ | | | 79 |
| OCI/AML-1 | mRNA: Kit− | mRNA: GM-CSF+ | G-CSF+, GM-CSF(+), IFNβ+, IGF-1(+), IL-3+, IL-4(+), IL-6(+), M-CSF(+), PIXY-321+, SCF(+) inhibition: TGFβ1+, TNFα+, TNFβ+ | | G-CSF | 80,81,103 |
| OCI/AML-4 | mRNA: Kit+, M-CSFR+ | | G-CSF(+), GM-CSF+, IL-3+, SCF+ | | | 82 |
| OCI/AML-5 | Kit mRNA− | | FL+, G-CSF+, GM-CSF+, IFNβ+, IFNγ+, IL-3+, IL-6+, M-CSF+, PIXY-321+, SCF− inhibition: TGFβ1+, TNFα+ | | GM-CSF, IL-3 | 83,103 |
| OHN-GM | | | GM-CSF+ | | GM-CSF | 84 |
| OIH-1 | | | G-CSF+, GM-CSF+, IL-3+, SCF+ | | G-CSF, GM-CSF | 85 |
| SKNO-1 | | | G-CSF+, GM-CSF+, IFNα+, IFNβ+, IFNγ+, IL-3+, IL-5+, IL-6+, IL-13+, M-CSF+, PIXY-321+, SCF+ inhibition: TNFα+ | | GM-CSF | 87,103 |
| SR-91 | IL-2Rα− | | GM-CSF+, IL-6(+) | | | 88 |
| UCSD/AML1 | | | GM-CSF+, IL-3+, IL-4+, IL-5+, IL-6+, M-CSF+, PIXY-321+, SCF+ inhibition: TGFβ1+ | IL-6+ | GM-CSF | 90,103 |
| UF-1 | | | GM-CSF+, IL-3+, SCF+ inhibition: TGFβ1+ | | | 91 |
| YNH-1 | | | G-CSF+, GM-CSF+, IL-3+ | | G-CSF, GM-CSF or IL-3 | 93 |

[a] Receptor expression or cytokine production at the protein level (ELISA, McAb, RIA), unless otherwise indicated, e.g. at the mRNA level (by RT-PCR, Northern); +, strong, definite expression; (+), weak expression; −, no expression.

[b] Effects of cytokine exposure on proliferation or differentiation (−, no effect; +, positive effect).

[c] Upon growth factor withdrawal these cell lines will die by apoptosis.

*Table 4.* Myelocytic cell lines: genetic characterization

| Cell line | Cytogenetic karyotype | Unique translocations (→ fusion genes) | Unique gene alterations, receptor gene rearrangements* | Ref. |
|---|---|---|---|---|
| AML14 | 45, XY, −5, +8, −9, +13, −14, −18, −21, +der(3), +der(5q), +mar | | | 35 |
| AR230 | 44, XX, −2, −11, −14, −17, −17, inv(1)(p31−32p36) or add(1)(p31−32), add(2)(p13), del(9)(q22) or der(9)t(9:?)(q11−13:?), der(9)t(9:?)(p13:?)t(9:22)(q34·q11), add(15)(p11), add(18)(q21), der(11:21)(q10;p10), der(22)t(9:22)(q34·q11), +der(22)t(9:22)(q34·q11), +mar | t(9:22)(q34·q11) → *BCR-ABL* (e19/e18-a2) fusion gene | | 37 |
| CML-C-1 | 49, XX, +8, +8, +21, del(3)(q21q23), t(9:22)(q34·q11) | t(9:22)(q34·q11) → *BCR-ABL* fusion gene | | 38 |
| CTS | 46, XX, −17, −22, +2mar, t(6:11)(q27:q23) | t(6:11)(q27:q23) → *MLL-AF6* fusion gene | *IGH* R, *IGK* R, *IGL* G, *TCRB* G, *TCRD* R, *TCRG* G | 39 |
| Ei501 | 46, XX, −7, t(7:8)(q32:q13), t(15:17)(q22:q12) | t(15:17)(q22:q21) → *PML-RARA* fusion gene | | 40 |
| EM-2 | 74(70−86)<3n>X, −X, −X, +3, +4, +6, +6, +8, −9, +11, −14, −14, +15, +17, −19, +21, +22, +mar, der(5)t(5:?)(q13/15:?), der(9)t(9)t(9:22)(q34·q11), i(17q)x2, in some cells up to three copies of der(9) | t(9:22)(q34·q11) → *BCR-ABL* (b3-a2) fusion gene | *IGH* G, *P53* D/PM | 95,96, 106 |
| EoL-1 | 50(48−51)<2n>XY, +4, +6, +8, +19, del(9)(q22) | | | 42,95 |
| FKH-1 | 46, XY, −7, t(6:9)(p23:q34) | t(6:9)(p23:q34) → *DEK-CAN* fusion gene | | 43 |
| GDM-1 | 48(46−48)<2n>XX, +8, +13, −16, +mar, t(2:11)(q36:q13), del(6)(q21), t(7:?)(q35:?), del(12)(p13) | | *IGH* G, *TCRD* G; *MYC* amplified | 95 |
| GF-D8 | 45, XY, −5, 8q+, +8q+, 11q+, 12p−, −15, −17, +mar, del(7)(q32qter), inv(7)(q31.2q36) | | | 45 |
| GM/SO | 45, XX, −9, −17, −19, −22, 7p−, 9q+, +3mar, +der(9)t(9:22)(q34·q11), der(13q) | t(9:22)(q34·q11) → *BCR-ABL* (b3-a2) fusion gene | | 46 |
| HL-60 | 82(78−88)<4n>XX, −X, −X, −2, −3, −4, −5, −8, −9, −10, −14, −16, −17, −17, +3m, der(6)t(6:?)(q25:?)/dupt(6)(q23:qter)x2, del(9)(p22), del(11)(q22/23), der(16)t(16:17)(q22/23:q21−22)x2 | no t(15:17)(q22:q11) *PML-RARA* fusion gene | *P15INK4B* GD or GG, *P16INK4A* GD or GG; *P53* D; *MYC, NEU, NRAS* amplified | 95, 107 |
| HMC-1 | 46, XX, dir ins(10:16)(q2?2:q13q22) | | *KIT* PM | 15 |
| HNT-34 | 46, XX, t(3:3)(q21:q26), t(9:22)(q34·q11), 20q− | t(9:22)(q34·q11) → *BCR-ABL* (M/m-bcr) fusion genes t(3:3)(q21:q26) → *EVI1* overexpression | | 50 |
| HT93 | 46, XY, t(1:12)(q25:p13), 2q+, t(4:6)(q12:q13), t(15:17)(q22:q11) | t(15:17)(q22:q11) → *PML-RARA* bcr3 fusion gene t(1:12)(q25:p13) → *ETV6-ARG* fusion gene | | 51 |

*Continued on next page*

*Table 4.* (continued)

| Cell line | Cytogenetic karyotype | Unique translocations (→ fusion genes) | Unique gene alterations, receptor gene rearrangements* | Ref. |
|---|---|---|---|---|
| IRTA17 | 46, XX, t(16;21)(p11;q22), add(22)(p11) | t(16;21)(p11;q22) → TLS/FUS-ERG fusion gene | | 52 |
| K051 | 48, XY, +1, +2, 7q−, 17p−, 21q− | | P53 PM | 53 |
| Kasumi-1 | 45<2n>X, −Y, −9, −13, −16, +3mar, t(8;21)(q22;q22), der(9)t(9;?)(p22;?), der(15)t(?9;15)(?q11;?p11) | t(8;21)(q22;q22) → AML1-ETO fusion gene | IGH G, TCRB G | 54,95, 108 |
| Kasumi-3 | 46, XY, −8, t(2;5)(p13;q33), t(3;7)(q27;q22), del(5)(q15), add(16)(p11), add(16)(q13), +mar | t(3;7)(q27;q22) → EVI1 overexpression | IGH G, TCRB G, TCRD G, TCRG D | 55 |
| Kasumi-4 | 46, XX, inv(3)(q21q26), t(9;22;11)(q34;q11;q13) | t(9;22)(q34;q11) → BCR-ABL (b2-a2) fusion gene inv(3)(q21q26) → EVI1 overexpression | BCL1 G | 56 |
| KBM-7 | 48, X, −Y, +8, +8, −9, +15, −22, +22q−, del(6)t(6;?)(p23;?), t(9;22)(q34;q11), +t(9;22)(q34;q11), del(12p) | t(9;22)(q34;q11) → BCR-ABL (b2-a2) fusion gene | P53 PM | 58 |
| KCL-22 | 52, XX, +1p−, +6, +8, +8, +8, t(9;22)(q+;q−), +22q− | t(9;22)(q34;q11) → BCR-ABL (b2-a2) fusion gene | P53 D+PM | 59 |
| KG-1 | 45(42−47)<2n>X, −Y, −7, −12, +2mar, dup(7)(q12q33), der(8)t(8;?)(p12;?), der(11)t(1;11)(q11;p11), der(17)t(17;?)(p13;?) | t(9;22)(q34;q11) → BCR-ABL (b2-a2) fusion gene | P53 PM, RB1 R, NRAS mutation | 95 |
| KOPM-28 | 44, XX, −5, −17, −19, t(9;22)(q34;q11), +der(19)t(19;?)(p13;?) | | | 64 |
| KT-1 | 51, XXYY, +X, +Y, −5, −6, +8, +8, +19, +19, +mar, t(9;22)(q34;q11), t(9;22)(q34;q11) | t(9;22)(q34;q11) → BCR-ABL (b3-a2) fusion gene | IGH G, TCR G | 65 |
| KU-812 | 50-60(58). XY, +1, +4, +6, +6, +8, −9, +14, +15, +19, +21, +21, t(9;22)(q34;q11), +der(9)t(9;22)(q34;q11)x2, +der(11)(q11), ii(17q) | t(9;22)(q34;q11) → BCR-ABL (b2-a2) fusion gene | P53 PM | 66,95, 107 |
| KY821 | 86, X, X, −Y, +1, +1, +3, +3, +4, +4, +5, +5, +6, +6, +7, −7, +8, +8, +10, +11, +11, +12, +13, +13, +13, +14, +14, +15, +15, +16, +16, +18, +19, +19, +20, +20, +21, +22, +22, +der(2)t(2;?)(p25;?), +der(2)t(2;?)(p25;?), del(9)(p22p24), del(9)(p22p24), +mar | t(9;22)(q34;q11) → BCR-ABL (b3-a2) fusion gene | P53 PM | 68 |
| KYO-1 | 47, XY, +15, −16, −18, 6q−, 8q+, 12p+, 15p+, 17p+, 22p+, 22q−(Ph), +22q−(Ph) | t(9;22)(q34;q11) → BCR-ABL (b2-a2) fusion gene | P53 PM | 69, 109 |
| LW/SO | 45, X, −X, der(9)inv(9)(p12q13), del(9)(p22?) | | | 70 |
| Marimo | 44, X, X, −5, +18, ins(1;?)(q21;?), del(8)(q22), t(10;14;11)(q22;q32;q13), der(14)t(10;14;11)(q22;q32;q13), add(17)(p11), psu dic(18;9)(q23;p21)x2 | | MYC amplified | 72 |
| MDS92 | 44, XY, −7, −12, −13, del(5)(q13q35), add(14)(p11), add(22)(q13), +mar | | NRAS PM | 73 |
| MUTZ-2 | 48(46−50)<2n>XY, +8, +10 | | P53 PM | 75,95 |

*Continued on next page*

Table 4. (continued)

| Cell line | Cytogenetic karyotype | Unique translocations (→ fusion genes) | Unique gene alterations, receptor gene rearrangements* | Ref. |
|---|---|---|---|---|
| NB4 | 78(71–81)<3n>XX, −X, +2, +6, +7, +7, +11, +12, +13, +14, +17, −19, +20, +4mar, der(8)t(8:?)(q24:?), der(11)t(11:?)(?::11p1511q22.1::11q13−>22.1:), der(12)t(12:?)(p11:?), 14p+, t(15:17)(q22:q11−12.1), der(19)t(10:19)(q21.1:p13.3)x2 | t(15:17)(q22:q21) → *PML-RARA* (bcr1-2) fusion gene | *P15INK4B* DD, *P16INK4A* DD | 76, 95, 110 |
| NKM-1 | 47, XY, −6, +8, der(2)t(2:?)(q37:?), +der(6)t(6:?)(p23:?) | | | 79 |
| OCI/AML-1 | 46, XX, −6, +der(6)t(6:8)(p25:q22) | | | 81 |
| OCI/AML-5 | 48(44–48)<2n>XX, +1, +8, der(1)t(1:19)(p13:p13) | | | 95 |
| OHN-GM | 48, XY, −7, +8, +22, del(5)(q11.2q31), t(10:13)(q24:q14), +der(13)t(10:13)(q24:q14) | | *P53* PM, *RB1* D | 84 |
| OIH-1 | 49, X, −Y, −5, −11, +18, +20, +21, add(7)(q11), add(7)(p21), +del(7)(q22), add(8)(q22), +add(8)(q22), inv(9), add(15)(p11) | | *DCC* alteration | 85 |
| PL-21 | 46, XY, −11, +der(11)t(11:?)(11pter−11q23::?) | no t(15:17)(q22:q11) *PML-RARA* fusion gene | | 86 |
| SKNO-1 | 44, X, −Y, −16, +der(16q17q), −17.2q, 6q+, t(8:21)(q22:q22), 11p+, 19p+ | t(8:21)(q22:q22) → *AML1-ETO* fusion gene | *P53* PM | 87 |
| SR-91 | 50, X, −Y, +3, +6, +7, +13, −14, +18, dic(8:14)(p11:p11), i(17q) | | *IGH* G, *TCRB* G; *BCL2* G | 88 |
| TI-1 | 69 (65–71), XXY, +1, +4, +7, +8, +11, +12, +19, 2q+, 6p+, 9p+, 9p−, 13p+ | t(3:3)(q21:q26) → *EV11* overexpression | *IGH* G, *TCRB* G | 89, 75, 90, 108, 111 |
| UCSD/AML1 | 45, XX, −7, t(3:3)(q21:q26), t(12:22)(p13:q12) | t(12:22)(p13:q12) → *ETV6/TEL-MN1* fusion gene | | |
| UF-1 | 46, XX, add(1)(q44), add(6)(q12), add(7)(q36), t(15:17)(q21:q21) | t(15:17)(q22:q21) → *PML-RARA* fusion gene | | 91 |
| YJ | 45, X, −Y, −2, −5, −9, −14, −14, −15, −16, −17, +8mar, add(3)(q27), del(3)(p13), add(4)(q34), add(6)(q25), del(10)(p12), add(11)(p15), add(13)(q32), add(16)(q23), del(18)(q22) | | | 92 |
| YNH-1 | 46, XY, der(16)t(16:21)(p11:q22)t(1:16)(q12:q13), der(21)t(16:21)(p11:q22), der(6)t(6:12)(q13:q13), der(12)t(6:12)(q21:q13) | t(16:21)(p11:q22) → *TLS/FUS-ERG* fusion gene | *P53* PM | 93 |
| YOS-M | 46, XY, −20, del(7)(q22q32), t(9:22)(q34:q11), +mar | t(9:22)(q34:q11) → *BCR-ABL* (b2-a2) fusion gene | *IGH* G | 94 |

* Receptor gene arrangements: D – deleted; G – germline; PM – point mutation; R – rearranged; wt – wild type.

Table 5. Myelocytic cell lines: functional characterization

| Cell line | Doubling time | EBV status | Cytochemistry | Inducibility of differentiation | Heterotransplantability into mice | Special functional features | Ref. |
|---|---|---|---|---|---|---|---|
| AML14 | 24–36 h | | | TPA, vit. D3 → mono/macro differentiation; GM-CSF+IL-3+IL-5 → eosino differentiation; retinoic acid → neutro differentiation | | spontaneous/induced expression of specific eosino-granular proteins: Charcot-Leyden crystal protein, eosino-cationic protein, eosino-derived neurotoxin, eosino-lysophospholipase, eosino-peroxidase, major basic protein | 35, 112 |
| AR230 | | | | | into nude mice | | 37 |
| CML-C-1 | | | ACP+, Alcian Blue−, ANBE−, CAE−, MPO−, PAS+, SBB−, Toluidine Blue− | | | produces p230 BCR-ABL fusion protein | 38 |
| CTS | 72 h | | ACP−, ALP−, ANAE−, ANBE−, CAE−, MPO−, PAS− | (DMSO, hemin, retinoic acid, TPA no effect) | | | 39 |
| Ei501 | 3–4 d | | ANAE−, MPO+, PAS− | | | | 40 |
| EM-2 | 24–48 h | EBV− HTLV-I− HIV− | MSE− | DMSO, retinoic acid → neutro differentiation; TPA → mono/macro differentiation | into SCID mice | PO: BCL2 mRNA+, WT1 mRNA+; target for cytotoxicity assay | 95, 96, 113 |
| EoL-1 | 48–60 h | EBV− HTLV-I− HIV− | ACP+, ANAE+, ANBE(+), CAE−, Luxoll(+), MPO+, MSE+, PAS(+), SBB+, Toluidine Blue− | cAMP, DMSO, G-CSF, TNFα → eosino differentiation (retinoic acid no effect) | | colony formation in agar; no phagocytosis | 42, 95, 97 |
| FKH-1 | 54 h | EBV− | ANBE+, CAE−, MPO−, Toluidine Blue− | | | granules positive for sulphate glycoproteins | 43 |
| GDM-1 | 96–120 h | EBV− HTLV-I− HIV− | ACP+, ALP+, ANBE+, CAE−, GLC+, Lysozyme+, MPO(+), MSE−, PAS−, SBB+ | TPA → macro differentiation | | no colony formation in agar; phagocytosis+ | 44, 95 |
| GF-D8 | 48–72 h | EBV− | ACP+, ALP−, ANAE+ (NaF inhibitable), ANBE−, MPO(+), MSE−, SBB+ | TPA → mono/macro differentiation | | clonable/colony formation in agar | 45 |
| GM/SO | | | ANAE−, MPO−, PAS− | TPA → mono/macro differentiation | | | 46 |
| HL-60 | 25–40 h | EBV− HTLV-I− HIV− | ACP(+), Alcian Blue−, ALP−, ANAE−, ANBE(+), CAE+, Methyl Green−, MPO+, MSE−, NBT−, Oil Red O+, PAS+, Pyronin+, SBB+ | various reagents → neutro, mono, macro, eosino, baso differentiation[a] | into nude or SCID mice | colony formation in agar/methylcellulose; lysozyme production+; no phagocytosis; no chemotaxis; no adherence; mRNA: (α-1-antitrypsin+, azurocidin+, cathepsin G+, Charcot Leyden crystals+, defensin+, lactoferrin−, major basic protein+, myeloblastin+, N-elastase+, tryptase−; TF: mRNA: GATA1+, GATA2, SCL− | 48, 95, 115, 116 |

*Continued on next page*

Table 5. (continued)

| Cell line | Doubling time | EBV status | Cytochemistry | Inducibility of differentiation | Heterotrans-plantability into mice | Special functional features | Ref. |
|---|---|---|---|---|---|---|---|
| HMC-1 | 6–8 d | EBNA– | Alcian Blue+, CAE+, Luxol+, Lysozyme–, MPO–, MSE–, Toluidine Blue+ | spontaneous ery differentiation | into nude mice | clonable in methylcellulose; no histamine production; no IgE binding; protein: aminocaproate esterase+, cathepsin G(+), chymotrypsin–, CLC protein–, Elastase–, eosino major basic protein+, eosino peroxidase+, eosino- derived neurotoxin–, eosino cationic protein–, tryptase+; mRNA: α-tryptase+, carboxypeptidase A(+), Charcot Leyden crystal protein+, defensin–, granzyme A –, IgE-Rα–/β–/γ+, lactoferrin–, lysozyme(+), mast cell chymase–, mast cell tryptase+, N-elastase– | 10, 15, 98 |
| HNT-34 | 26–27 h | | ALP–, ANBE(+), CAE–, MPO–, PAS– | (DMSO, TPA no effect) | | | 50 |
| HT93 | 48 h | | ALP–, ANAE–, MPO(+) | ATRA → neutro/eosino differentiation; ATRA + G-CSF → neutro differentiation; ATRA + GM-CSF, IL-3, IL-5 → eosino/baso differentiation | | colony formation in methylcellulose | 51 |
| IRTA17 | 79 h | | ANBE+, CAE–, Lysozyme–, MPO– | | | | 52 |
| K051 | 48 h | EBNA– | ANAE–, MPO+, MSE–, PAS–, SBB+ | retinoic acid → ery differentiation | | MDR-1 mRNA+ | 53 |
| Kasumi-1 | 40–72 h | EBV– HTLV-1– HIV– | ALP–, ANBE–, CAE+, MPO+, MSE– | TPA → macro differentiation (DMSO no effect) | | TF: mRNA: GATA1–, GATA2+, SCL– | 54, 95, 116 |
| Kasumi-3 | 55–60 h | | ACP+, ALP–, ANBE–, CAE–, MPO–, PAS– | TPA → mono differentiation (DMSO, retinoic acid no effect) | | | 55 |
| Kasumi-4 | 55 h | | ACP+, ALP–, ANBE–, CAE–, MPO–, PAS– | (cytokines, DMSO, retinoic acid, TPA no effect) | | no α-granules, no demarcation membranes | 56 |
| KBM-7 | 22–24 h | EBNA– | ACP+, ANAE+, ANBE–, CAE+, MPO–, MSE–, PAS+ | | into nude mice | clonable/colony formation in agar | 57, 58 |
| KCL-22 | 24 h | EBNA– | ACP+, ALP–, ANBE–, Benzidine–, CAE–, MPO–, MSE–, PAS+, SBB– | (DMSO no effect) | into newborn hamsters | no phagocytosis | 60 |
| KF-19 | 27 h | | ANAE+, ANBE–, CAE+, MPO+, PAS+ | TPA → macro differentiation; DMSO, retinoic acid → neutro differentiation | into nude or SCID mice | no phagocytosis; resistant to AraC, adriamycin, vincristine | 61 |
| KG-1 | 40–50 h | EBV– HTLV-1– HIV– | ACP+, CAE+, MPO+, MSE–, PAS+ | TPA → mono/macro differentiation (DMSO, Na butyrate no effect) | | no ADCC; no phagocytosis; no chemotaxis; PHA-LCM → colony formation in agar; TF: mRNA: GATA1+, GATA2+, SCL– | 63, 95, 113, 116, 117 |

*Continued on next page*

*Table 5.* (continued)

| Cell line | Doubling time | EBV status | Cytochemistry | Inducibility of differentiation | Heterotransplantability into mice | Special functional features | Ref. |
|---|---|---|---|---|---|---|---|
| KOPM-28 | 22–24 h | | ACP+, ALP−, ANBE+, CAE+, MPO−, MSE−, PAS− | TPA → meg differentiation (DMSO, retinoic acid no effect) | | clonable/colony formation in agar; no phagocytosis | 64, 118 |
| KT-1 | 18–24 h | EBNA− | ANBE−, CAE−, MPO+, PAS(+) | | | colony formation in methylcellulose | 65 |
| KU-812 | 80–100 h | EBV− HTLV-I− HIV− | ACP+, Alcian Blue(+), ALP−, ANAE+, ANBE(+), Astra Blue+, CAE−, Lysozyme−, MPO−, MSE−, PAS(+), SBB−, Toluidine Blue+ | spontaneous ery differentiation; serum-free culture → baso differentiation; TPA → macro differentiation | | cloning/colony formation in agar; histamine production; no IgE binding; protein: cathepsin G−, chymase−, elastase−, tryptase−; mRNA: α-tryptase−, carboxypeptidase A+, Charcot Leyden crystal+, chymase−, defensin−, eosino peroxidase−, eosino-derived neurotoxin−, eosino cationic protein−, glycophorin+, granzyme A−, heparin core protein+, IgE-Rα+/β+/γ+, lactoferrin−, lysozyme+, mast cell chymase−, mast cell tryptase+, N-elastase−; α-, β-, γ-, δ-globin mRNA+; δ-aminolevulin synthase mRNA+; HbA+; HbF+, HB Bart's inducible; TF: mRNA: GATA1+, GATA2+, PU1+, SCL+ | 66, 95, 98, 116, 119 |
| KY821 | | | ANBE(+), CAE+, MPO+, MSE− | | | colony formation in methylcellulose; resistant to methotrexate | 68 |
| KYO-1 | 22 h | EBNA− | ANBE−, CAE−, MPO−, MSE− | | | colony formation in agar | 69 |
| LW/SO | | | ACP(+), ANAE−, PAS(+), SBB− | | | not clonable | 70 |
| M20 | | EBNA− | ACP+, ANAE(+), CAE+, MPO+, SBB+ | TPA → macro differentiation (DMSO, retinoic acid no effect) | | colony formation in methylcellulose; no phagocytosis; secretes lysozyme, prostaglandin E2 | 71 |
| Marimo | 25–28 h | | ANAE−, MPO− | DMSO, retinoic acid → neutro differentiation | | | 72 |
| MDS92 | 80–90 h | | MPO+, MSE− | G-CSF, GM-CSF, IL-3, SCF → neutro differentiation | | colony formation in agar | 73 |
| MOLM-6 | 24–60 h | | MPO+, MSE− | | | | 74 |
| MUTZ-2 | 48 h | EBV− HTLV-I− HIV− | MSE−, TRAP− | TPA → mono differentiation (retinoic acid no effect) | | clonable/colony formation in methylcellulose; PO: BCL2 mRNA+ | 75, 95 |
| NB4 | 36–48 h | EBV− HTLV-I− HIV− | ALP−, ANBE−, CAE−, MPO+, MSE+ | retinoic acid → neutro differentiation | | | 76, 95 |
| NKM-1 | 36–48 h | EBV− | ANBE−, CAE+, MPO+, NBT− | (DMSO, Na butyrate, retinoic acid, TPA no effect) | | colony formation in agar; no phagocytosis | 79 |

*Continued on next page*

*Table 5.* (continued)

| Cell line | Doubling time | EBV status | Cytochemistry | Inducibility of differentiation | Heterotransplantability into mice | Special functional features | Ref. |
|---|---|---|---|---|---|---|---|
| OCI/AML-1 | 32 h | | ANBE(+), CAE+, MPO+, MSE−, NBT− | responsive to retinoic acid (but no differentiation) | | colony formation in methylcellulose; RARA mRNA+ | 80 |
| OCI/AML-4 | 7 d (3 d)[b] | | ANBE(+), MPO(+), SBB(+) | sensitive to retinoic acid | | colony formation in methylcellulose | 82 |
| OCI/AML-5 | 30–50 h | EBV−, HTLV-I−, HIV− | MSE− | | | colony formation in methylcellulose | 83, 95 |
| OHN-GM | 60 h | EBV− | ANAE(+), MPO+ | | | | 84 |
| OIH-1 | 48 h | EBV− | ANBE+, CAE−, MPO−, PAS(+) | | | | 85 |
| PL-21 | 48–64 h | EBNA− | ACP+, ALP−, ANBE(+), CAE+, Lysozyme+, MPO+, MSE+, PAS(+), SBB+ | DMSO → macro differentiation | | phagocytosis+ | 86 |
| SKNO-1 | 48–72 h | EBNA− | ANBE−, CAE+, MPO+ | TPA → mono/macro differentiation | | PO: p53 overexpressed | 87, 120 |
| SR-91 | 28–32 h | EBV− | ACP−, ANAE(+), CAE−, MSE−, PAS−, SBB− | | | colony formation in methylcellulose; resistant to NK activity | 88 |
| TI-1 | 14 h | | ANBE−, CAE−, MPO−, MSE− | TPA → mono differentiation; hemin → ery differentiation | | HbF+ upon induction | 89 |
| UCSD/AML1 | 48–72 h | | ACP+, ANAE+, CAE−, MSE− | TPA → macro differentiation; IL-6 → meg differentiation | | | 90 |
| UF-1 | 72 h | | MPO+, NBT− | resistant to retinoic acids TPA | | mRNA: Charcot-Leyden crystal protein+, eosino cationic protein+, eosino-derived neurotoxin+, eosino peroxidase+, major basic protein+; TF: mRNA: C/EBPα+, EGR1(+), GATA1(+), GATA2+, MZF1+, PU1+ | 91, 121 |
| YJ | 24 h | | Fast Green+, neutral+ | → mono differentiation; retinoic acid → neutro differentiation | | | 92 |
| YNH-1 | 82 h | EBV− | ANBE−, CAE−, MPO+, PAS− | | | | 93 |
| YOS-M | 5–6 d | EBNA− | ANBE−, CAE(+), MPO+, MSE− | | | | 94 |

[a] Reviewed in [31,114].
[b] Doubling time is shortened by incubation with GM-CSF or IL-3 from 7 d to 3 d.
PHA-LCM – phytohemagglutinin-stimulated peripheral leukocytes conditioned medium; PO – (proto)-oncogenes; TF – transcription factors.

dominant. Thus, KU-812 is currently the prototype cell line for studies on basophil biology. The cell line HMC-1 has been developed from a patient with mast cell leukemia. These cells express a number of characteristic "mast cell-related antigens" and other typical features [10]. The HMC-1 line is widely used as a model for investigating the biology of human mast cells.

## 3. IMMUNOPHENOTYPE

Myelocytic cell lines generally express the pan-myelomonocytic surface antigens CD13 (41/47, 87%), CD15 (18/23, 78%) and CD33 (42/47, 82%), while they only rarely display the monocyte-associated marker CD14 (9/40, 22%) (Table 2). Other antigens physiologically expressed by myelocytic cells, such as CD16, CD32, CD64, CD65, CD68 [144], are mostly positive, but have not been tested on many cell lines. Apart from the relatively frequent expression of T-cell antigens CD4 and CD7 which are also seen on normal cells committed to myelocytic differentiation [145], the majority of these cell lines are negative for surface markers associated with T-cells, B-cells and erythroid-megakaryocytic cells (Table 2). Concerning the progenitor and activation markers, the cell lines were predominantly positive for CD34 (27/45, 60%), CD38 (11/15, 73%) and HLA-DR (26/46, 56%), but negative for the lymphoid nuclear enzyme marker TdT (3/20, 15%). The following profile of myelocytic cell lines emerges: CD4± CD7± CD13+ CD14– CD15+ CD33+ CD34+ CD38+ HLA-DR+, T-antigen negative, B-antigen negative, ery-meg-antigen negative.

## 4. CYTOKINE-RELATED CHARACTERIZATION

Normal myelocytic cells respond to signals provided by the cytokines G-CSF, GM-CSF and IL-3 regulating cellular proliferation and differentiation [146,147]. Cytokine receptor expression, cytokine production, and proliferative and differentiative responses to cytokines of the myelocytic cell lines are summarized in Table 3. Several cell lines were reported to be constitutively dependent on the addition of specific cytokines to the culture medium (see Appendix 3): FKH-1, GF-D8, GM/SO, IRTA17, MDS92, MUTZ-2, OC1/AML1, OC1/AML5, OHN-GM, OIH-1, SKNO-1, UCSD/AML1, YNH-1 (Table 3). These cells die within a few days when deprived of growth factors [103,148]. The necessary cytokines are usually G-CSF, GM-CSF, IL-3, or SCF. These growth factor-dependent cell lines also respond proliferatively to a variety of other cytokines (reviewed in ref. [103]) (Table 3).

*Table 6.* Myelocytic cell lines: unconfirmed cell lines (not immortalized, not characterized, not verified, other)

| Cell | Patient | Features | Ref. | Remarks* |
|---|---|---|---|---|
| 2L1 | from BM of 6 M with t-AML M5 (post-nasopharyngeal embryonal rhabdomyosarcoma and ganglioneuroma) | CD2+ CD10+ CD13+ CD15+ CD33+ CD38+ CD56+; *P53* PM; *MLL* amplified | 122 | insufficiently characterized |
| 8261 | from PB of patient with AML | ALP− CAE− NBT− PAS− SBB−; no phagocytosis; differentiation to monos/macros | 123 | insufficiently characterized |
| AML-CL | from PB of 27 F with AML M1 (at diagnosis) | CD13+ CD19− CD33+ CD45+; secretion of IL-1β; 46, XX; established, maintained and serially passaged in SCID mice, but no long-term *in vitro* growth | 124 | insufficiently characterized; no long-term *in vitro* growth |
| AML-PS | from BM of 61 M with AML M1 (3rd relapse) (post-sideroblastic anemia MDS) | CD13+ CD19− CD33+ CD45+; secretion of IL-1β; 46, XY/49, XY, −8, −9, +11, +12, +21, +21, +21; established, maintained and serially passaged in SCID mice, but no long-term *in vitro* growth | 124 | insufficiently characterized; no long-term *in vitro* growth |
| AML5q− | from patient with AML M2 | mRNA: GATA1−, GATA2+, GATA3−; EPO-R+ | 125,126 | insufficiently characterized |
| AP-1060 | from BM of 45 M with AML M3 (at relapse) | 46, XY, t(3;14)(p21;q11.2), t(15;17)(q22;q11.2); *PML-RARA* fusion gene; ATRA → terminal differentiation; A₃O₂ → apoptosis | 127 | insufficiently characterized |
| CC-AML | from PB of patient with AML M1 (refractory) | CD2− CD3− CD5− CD7− CD10− CD11+ CD13+ CD14+ CD15+ CD16+ CD19− CD20− CD21− CD33+ CD34+ CD45+ CD56− HLA-DR+; PIXY-321, SCF → proliferation; TPA → macro differentiation; DMSO → neutro differentiation | 128 | insufficiently characterized |
| EU-4 | from child with AML (relapse) | CD2− CD7− CD10− CD13+ CD15+ CD19− CD33+ HLA-DR−; *P15INK4B* D, *P16INK4A* D; mRNA: BAX+, BCL-XL+, MDM2+ | 129,130 | insufficiently characterized |
| HSM-911 | from PB of patient with acute mixed leukemia | CD7+ CD10− CD13+ CD14− CD19− CD33+ CD34+ CD41− GlyA− HLA-DR+; ANBE−, CAE−, MPO−; dependent on GM-CSF, IL-3 or SCF | 131 | insufficiently characterized |

*Continued on next page*

*Table 6.* (continued)

| Cell | Patient | Features | Ref. | Remarks* |
|---|---|---|---|---|
| KH-143 | from patient with NK-myeloid leukemia | CD2− CD11b− CD13− CD18− CD33+ CD34− CD41+ CD54− CD56+; IL-3Rα+, Kit+; 46−49, XY, Yp+, 1p−, 2p+, +3q−−, −8, 9q−, 11q−, 17p+, −18; *IGH* G, *IGK* G, *IGL* G, *TCRB* G, *TCRG* G; responsive to IL-3(+), SCF(+); inhibition by TGFβ+ | 132 | insufficiently characterized |
| KM-MDS | from BM of patient with MDS | CD2− CD11b+ CD13+ CD18+ CD33+ CD34+ CD41− CD54+ CD56−; G-CSFR+, GM-CSFRα+,IL-3Rα+, Kit+, c-mpl+; 46, XX, −4, 5q−, −7, 13q+, 20q−; responsive to G-CSF(+), IL-3(+), SCF(+), TPO(+); *IGH* G, *IGK* G, *IGL* G, *TCRB* G, *TCRG* G; phagocytosis+; H₂O₂ production+ | 132 | insufficiently characterized |
| KOPM 30 | from peripheral blood of patient with AML | EBV−; *P53* PM; t(9;22)(q34;q11) → *BCR-ABL* (e1-a2) fusion gene | 133 | insufficiently characterized |
| KOPM 55 | from patient with AML | CD2− CD3− CD10− CD13+ CD19− CD22− CD33+ CD41+ HLA-DR+; ANBE− MPO+; t(9;22)(q34;q11) | 101 | insufficiently characterized |
| M24 | from PB of patient with AML | HLA-DR− sIg−; EBNA−; ACP+ ANAE+ CAE+ MPO+ SBB+; TPA → macro differentiation; no phagocytosis | 134 | insufficiently characterized |
| M26 | from PB of patient with AML | HLA-DR− sIg−; EBNA−; ACP+ ANAE− CAE+ MPO+ SBB+; TPA → macro differentiation; no phagocytosis | 134 | insufficiently characterized |
| MDS-KZ | from patient with MDS (RAEBT) | dt 2 weeks; complex chromosome abnormalities, −4, 5q−, −7, 13q+, 20q−; responsive to G-CSF; vit. K2 → apoptosis | 135 | insufficiently characterized |
| MHH-203 | from PB of adult patient with AML | responsive to FL, G-CSF, IGF-I, IL-3, TPO | 103 | insufficiently characterized |
| MML-1 | from BM of patient with AML M1 | biphenotypic: CD3− CD4− CD5− CD8− CD13+ CD14+ CD15+ CD19+ CD20− CD33+ CD37− CD68+ HLA-DR+ cy/sIg−; 46<2n>X, del(2)(p22p23), +8, del(9)(p21p24) | 136 | insufficiently characterized |
| MO-91 | from patient with AML M0 | mRNA: GATA1+, GATA2+, GATA3+; EPO-R+ | 125,126 | insufficiently characterized |
| MPD | from patient with non-CML MPD | dt 72 h; CD11b+ CD11c+ CD14+ CD15+ CD16+ CD33+ CD54+; r(18); spontaneous neutro, mono, eosino, baso differentiation; mRNA Charcot-Leyden crystal protein+, lactoferrin+, major basic protein+, transcobalamin-1+ | 137 | insufficiently characterized |

*Continued on next page*

Table 6. (continued)

| Cell | Patient | Features | Ref. | Remarks* |
|------|---------|----------|------|----------|
| MTO-94 | from BM of patient with MDS | CD7+ CD13+ CD33+ CD34− HLA-DR+; 46, XY, i(17q); responsive to G-CSF, GM-CSF, IL-3, SCF | 138 | insufficiently characterized |
| MZ93 | from patient with CML-BC | CD1− CD2− CD3− CD4(+) CD7+ CD8(+) CD10− CD13+ CD14− CD15− CD19− CD20− CD21− CD22− CD25+ CD33+ CD34+ CD36− CD38(+) CD41− CD56+ CD122− GlyA− HLA-DR− | 99 | insufficiently characterized |
| OCI/AML-6 | from PB of 68 F with MDS → AML M4 (at diagnosis) in 1991 | responsive to G-CSF, GM-CSF, IFNγ, IL-3, IL-4, IL-13, M-CSF, PIXY-321, SCF | 103,139 | cross-contaminated with OCI/AML-2 |
| OU-AML-1 | from PB of 26 F with AML M4 (at diagnosis) | CD3− CD7− CD13+ CD14− CD19− CD33+ CD34− CD61− CD117− GlyA− MPO−; [hyperdiploid]; *P53* wild-type | 140 | cross-contaminated with OCI/AML-2 |
| OU-AML-2 | from PB of 52 F with AML M2 (at diagnosis) | CD3− CD7− CD13+ CD14− CD19− CD33+ CD34− CD61− CD117− GlyA− MPO−; [diploid]; *P53* wild-type | 140 | cross-contaminated with OCI/AML-2 |
| OU-AML-3 | from PB of 48 M with AML M4 (at diagnosis) | CD3− CD7+ CD13+ CD14− CD19− CD33+ CD34− CD61− CD117− GlyA− MPO−; [diploid]; *P53* wild-type; responsive to ATRA | 140 | cross-contaminated with OCI/AML-2 |
| OU-AML-4 | from PB of 39 M with AML M2 (at 1st relapse) | CD3− CD7+ CD13+ CD14− CD19− CD33+ CD34− CD61− CD117− GlyA− MPO−; [hyperdiploid]; *P53* wild-type | 140 | cross-contaminated with OCI/AML-2 |
| OU-AML-5 | from PB of 70 M with AML M5 (at 2nd relapse) | CD3− CD7+ CD13+ CD14− CD19− CD33+ CD34− CD61− CD117− GlyA− MPO−; [hyperdiploid]; *P53* wild-type | 140 | cross-contaminated with OCI/AML-2 |
| OU-AML-6 | from PB of 47 F with AML M1 (at 1st relapse) | CD3− CD7+ CD13+ CD14− CD19− CD33+ CD34− CD61− CD117− GlyA− MPO−; [hyperdiploid]; *P53* wild-type | 140 | cross-contaminated with OCI/AML-2 |
| OU-AML-7 | from PB of 63 F with AML M4 (at 1st relapse) | CD3− CD7+ CD13+ CD14− CD19− CD33+ CD34− CD61− CD117− GlyA− MPO−; [hyperdiploid]; *P53* wild-type | 140 | cross-contaminated with OCI/AML-2 |

*Continued on next page*

*Table 6.* (continued)

| Cell | Patient | Features | Ref. | Remarks* |
|------|---------|----------|------|----------|
| OU-AML-8 | from PB of 63 F with AML M4 (at diagnosis) | CD3− CD7+ CD13+ CD14− CD19− CD33+ CD34− CD61− CD117− GlyA− MPO−; [hyperdiploid]; *P53* wild-type | 140 | cross-contaminated with OCI/AML-2 |
| RDFD-2 | from PB of 56 M with AML M1 | ANBE− MPO+; dt 56 h; DMSO, retinoic acid → macro differentiation | 141 | insufficiently characterized |
| RED-3 | from PB of 24 M with T-ALL → AML (2nd relapse) | CD2+ CD4+ CD13+ CD15+ TdT− ; ACP+ ANAE+ MPO+ PAS+ SBB+; DMSO, retinoic acid → neutro differentiation; TPA → macro differentiation; *NRAS* point mutation; c-*MYC* amplification | 142 | insufficiently characterized; needs to be excluded, cross-contaminated with HL-60 |
| TMM | from PB of 62 M with CML-BC (post-MDS) in 1985 | CD3− CD10− CD13− CD14− CD15− CD19+ CD20+ CD33− CD37+ cy/sIg+ HLA-DR+; EBNA+; 46(42−46)<2n>XY | 95, 143 | not leukemia cell line, but normal EBV+ B-LCL |
| YS-1 | from patient with CML-BC | CD1− CD2− CD3− CD4− CD7− CD8− CD10+ CD13+ CD14− CD15(+) CD19+ CD20+ CD21(+) CD22+ CD25− CD33(+) CD34− CD36− CD38+ CD41− CD56− CD122− GlyA− HLA-DR+ | 99 | insufficiently characterized |

* For most cell lines, the insufficient characterization concerns the description of clinical data, authentication, immunoprofile and/or cytogenetics. Present status unknown: no data on this cell line have been published since its original description.

## 5. GENETIC CHARACTERIZATION

In common with other types of leukemia-lymphoma cell lines, the myelocytic cell lines are characterized by highly complex numerical and structural chromosomal aberrations (Table 4). Of particular note are the various translocations leading to the occurrence of specific fusion genes (Appendix 4, p. 288). The most common balanced translocation among these cell lines is the t(9;22) with the hybrid gene BCR-ABL which is associated in vivo with CML [28]. Fourteen such myelocytic Philadelphia+ cell lines have been developed [96,149,150]. Other unique translocations/fusion genes seen in myelocytic cell lines are t(6;9) *DEK-CAN*, t(8;21) *AML1-ETO*, t(12;22) *ETV6/TEL-MN1*, t(15;17) *PML-RARA*, t(16;21) *TLS/FUS-ERG* and alterations involving bands 11q23 (gene *MLL*) and 3q26 (gene *EVI1*). The cell lines provide models for these translocations [19,22,28,108,151–156].

Some cell lines have deletions, rearrangements, point mutations or other alterations of the tumor suppressor genes *P53* (EM-2, HL-60, K051, KBM-7, KCL-22, KG-1, KU-812, KY821, KYO-1, MUTZ-2, OHN-GM, SKNO-1, YNH-1), *RB1* (KG-1, OHN-GM) and *P15INK4B/P16INK4A* (HL-60, NB4); and the oncogenes *NRAS* (HL-60, KG-1, MDS92) and *MYC* (GF-D8, HL-60, Marimo).

## 6. FUNCTIONAL CHARACTERIZATION

The doubling times of myelocytic cell lines range from 20 hours to 6–8 days (Table 5). All 27 cell lines tested were negative for EBV infection. The following results on the cytochemical staining of myelocytic cell lines were seen: 89% ACP+, 0% ALP, 46% ANAE, 40% ANBE, 48% CAE, 50% lysozyme, 61% MPO, 33% MSE, 48% PAS, 64% SBB (Table 5). A typical cytochemical profile is: ACP+, ALP–, ANAE/ANBE±, CAE±, MPO+, MSE–, PAS±, SBB+. Extensive cytochemical investigations have been reported for cell lines CML-C-1, EoL-1, HL-60, HMC-1 and KU-812. The eosinophilic EoL-1 (Luxol fast blue+, toluidine blue negative), the mast cell line HMC-1 (toluidine blue+) and the basophilic KU-812 (toluidine blue+) showed the staining patterns expected for their cell types (Table 5).

Cellular differentiation induced by various agents has been reported for over 20 cell lines (Table 5). The principal inducers used were retinoic acid and DMSO, leading to neutrophil differentiation; and TPA, causing monocytic/macrophage maturation. The HL-60 cell line is the most widely and best studied target cell line in this regard [31,34,114]. HL-60 cells can be induced to differentiate into neutrophils, monocytes/macrophages, eosinophils or basophils by a multitude of specific inducers [157–159]. Of special interest are the various retinoic acid-resistant and acid-sensitive parental lines and their subclones (see also Tables 1a and 1b). Heterotransplantation into nude or SCID mice or hamster was reported for only seven cell lines (Table 5, Appendix 5, p. 289). Most cell lines can be cloned and form colonies in methylcellulose. Phagocytosis, adherence and chemotaxis are not properties of myelocytic cell lines. Expression of unique proteins specifically associated with certain subtypes of granulocytic cells (neutrophils, basophils, mast cells) was studied extensively in HL-60, HMC-1 and KU-812.

## 7. UNCONFIRMED CELL LINES

There is also a panel of myelocytic cell lines which, due to insufficient clinical, immunophenotypic, functional or other characterization were assigned to the category "Unconfirmed Cell Lines" (Table 6). These cell lines were established from patients with AML (M1, M2, M3, M4), acute mixed leukemia, NK-myeloid leukemia, MDS (RAEBT), MPD, and CML in blast crisis. The present status of some older cell lines (e.g. 8261, RDFD-2) is unknown as no further data have been published. The cells might be lost or a continuous cell line was not established. Of interest are two cell lines that can only be propagated by continuous serial passage in SCID mice (AML-CL, AML-PS). Cell lines RED-3 and TMM might be a cross-contaminated culture and a normal EBV+ B-LCL, respectively.

### References

1. Bessis M (ed.) *Blood Smears Reinterpreted*. Springer, Berlin, 1977.
2. Roitt I, Brostoff J, Male D (eds.) *Immunology*, 3rd edn., Mosby, London, 1993.
3. Hoffbrand AV, Pettit JE (eds.) *Sandoz Atlas Clinical Haematology*, Gower, London, 1988.
4. Hayhoe FGJ, Flemans RJ (eds.) *A Colour Atlas of Haematological Cytology*, 3rd edn., Wolfe, London, 1992.
5. Borregaard N, Cowland JB. *Blood* 89: 3503–3521, 1997.
6. Marmont AM, Damasio E, Zucker-Franklin D. In: Zucker-Franklin D, Greaves MF, Grossi CE, Marmont AM (eds.), *Atlas of Blood Cells. Function and Pathology*, 2nd edn., Fischer, Stuttgart, pp 157–253, 1988.
7. Gleich GJ, Adolphson CR. In: Dixon FJ (ed.), *Advances in Immunology 39*, Academic Press, pp. 177–253, 1986.
8. Spry CJF (ed.) *Eosinophils: A Comprehensive Review and Guide to the Scientific and Medical Literature*, Oxford, Oxford University Press, 1988.
9. Zucker-Franklin D. In: Zucker-Franklin D, Greaves MF, Grossi CE, Marmont AM (eds.), *Atlas of Blood Cells. Function and Pathology*, 2nd edn., Fischer, Stuttgart, pp 255–284, 1988.
10. Agis H, Beil WJ, Bankl HC et al. *Leukemia Lymphoma* 22: 187–204, 1996.

11. Weber S, Krüger-Krasagakes S, Grabbe J et al. *Int J Dermatol* 34: 1–10, 1995.
12. Denburg JA. *Blood* 79: 846–860, 1992.
13. Valent P, Ashman LK, Hinterberger W et al. *Blood* 73: 1778–1785, 1989.
14. Zucker-Franklin D. In: Zucker-Franklin D, Greaves MF, Grossi CE, Marmont AM (eds.), *Atlas of Blood Cells. Function and Pathology*, 2nd edn., Fischer, Stuttgart, pp 285–320, 1988.
15. Butterfield JH, Weiler D, Dewald G et al. *Leukemia Res* 12: 345–355, 1988.
16. Bennett JM, Catovsky D, Daniel MT et al. *Br J Haematol* 33: 451–458, 1976.
17. Bennett JM, Catovsky D, Daniel MT et al. *Br J Haematol* 78: 325–329, 1991.
18. Grignani F, Fagioli M, Ferrucci PF et al. *Blood Rev* 7: 87–93, 1993.
19. Grignani F, Fagioli M, Alcalay M et al. *Blood* 83: 10–25, 1994.
20. Warrell RP, de Thé H, Wang ZY et al. *N Engl J Med* 329: 177–189, 1993.
21. Kita K, Shirakawa S, Kamada N et al. *Leukemia Lymphoma* 13: 229–234, 1994.
22. Nucifora G, Rowley JD. *Leukemia Lymphoma* 14: 353–362, 1994.
23. Chen Z, Chen SJ. *Leukemia Lymphoma* 8: 253–260, 1992.
24. Bennett JM, Catovsky D, Daniel MT et al. *Br J Haematol* 51: 189–192, 1982.
25. Beris P. *Semin Hematol* 26: 216–233, 1989.
26. Koeffler HP. *Semin Hematol* 23: 223–236, 1986.
27. Bennett JM, Catovsky D, Daniel MT et al. *Br J Haematol* 87: 746–754, 1994.
28. Melo JV. *Blood* 88: 2375–2384, 1996.
29. Kantarjian HM, Deisseroth A, Kurzrock R et al. *Blood* 82: 691–703, 1993.
30. Koeffler HP, Golde DW. *Blood* 56: 344–350, 1980.
31. Koeffler HP. *Blood* 62: 709–721, 1983.
32. Lübbert M, Koeffler HP. *Blood Rev* 2: 121–133, 1988.
33. Lübbert M, Koeffler HP. *Cancer Rev* 10: 33–62, 1990.
34. Lübbert M, Herrmann F, Koeffler HP. *Blood* 77: 909–924, 1991.
35. Paul CC, Tolbert M, Mahrer S et al. *Blood* 81: 1193–1199, 1993.
36. Paul CC, Mahrer S, Tolbert M et al. *Blood* 86: 3737–3744, 1995.
37. Wada H, Mizutani S, Nishimura J et al. *Cancer Res* 55: 3192–3196, 1995.
38. Kubonishi I, Takeuchi S, Uemura Y et al. *Jpn J Cancer Res* 86: 451–459, 1995.
39. Kakuda H, Sato T, Hayashi Y et al. *Br J Haematol* 95: 306–318, 1996.
40. Weidmann E, Brieger J, Karakas T et al. *Leukemia* 11: 709–713, 1997.
41. Keating A, Martin PJ, Bernstein ID et al. In: Golde DW, Marks PA (eds.), *Normal and Neoplastic Hematopoiesis*, Liss, New York, pp 513–520, 1983.
42. Saito H, Bourinbaiar A, Ginsburg M et al. *Blood* 66: 1233–1240, 1985.
43. Hamaguchi H, Nagata K, Yamamoto K et al. *Br J Haematol* 102: 1249–1256, 1998.
44. Ben-Bassat H, Korkesh A, Voss R et al. *Int J Cancer* 6: 743–752, 1982.
45. Rambaldi A, Bettoni S, Tosi S et al. *Blood* 81: 1376–1383, 1993.
46. Oez S, Tittelbach H, Fahsold R et al. *Blood* 76: 578–582, 1990.
47. Collins SJ, Gallo RC, Gallagher RE. *Nature* 270: 347–349, 1977.
48. Gallagher R, Collins S, Trujillo J et al. *Blood* 54: 713–733, 1979.
49. Dalton WT Jr, Ahearn MJ, McCredie KB et al. *Blood* 71: 242–247, 1988.
50. Hamaguchi H, Suzukawa K, Nagata K et al. *Br J Haematol* 98: 399–407, 1997.
51. Kishi K, Toba K, Azegami T et al. *Exp Hematol* 26: 135–142, 1998.
52. Hiyoshi M, Yamane T, Hirai M et al. *Br J Haematol* 90: 417–424, 1995.
53. Abo J, Inokuchi K, Dan K et al. *Blood* 82: 2829–2836, 1993.
54. Asou H, Tashiro S, Hamamoto K et al. *Blood* 77: 2031–2036, 1991.
55. Asou H, Suzukawa K, Kita K et al. *Jpn J Cancer Res* 87: 269–274, 1996.
56. Asou H, Eguchi M, Suzukawa K et al. *Br J Haematol* 93: 68–74, 1996.

57. Andersson B, Beran M, Pathak S et al. *Cancer Genet Cytogenet* 24: 335–343, 1987.
58. Andersson B, Collins VP, Kurzrock R et al. *Leukemia* 9: 2100–2108, 1995.
59. Kubonishi I, Machida KI, Sonobe H et al. *Gann* 74: 319–322, 1983.
60. Kubonishi I, Miyoshi I. *Int J Cell Cloning* 1: 105–117, 1983.
61. Fukuda T, Kamishima T, Kakihara T et al. *Leukemia Res* 20: 931–939, 1996.
62. Koeffler HP, Golde DW. *Science* 200: 1153–1154, 1978.
63. Koeffler HP, Billing R, Lusis AJ et al. *Blood* 56: 265–273, 1980.
64. Mori T, Nakazawa S, Nishino K et al. *Leukemia Res* 11: 241–249, 1987.
65. Yanagisawa K, Yamauchi H, Kaneko M et al. *Blood* 91: 641–648, 1998.
66. Kishi K. *Leukemia Res* 9: 381–390, 1985.
67. Fukuda T, Kishi K, Ohnishi Y et al. *Blood* 70: 612–619, 1987.
68. Saito H, Kishi K, Narita M et al. *Leukemia Res* 16: 217–226, 1992.
69. Ohkubo T, Kamamoto T, Kita K et al. *Leukemia Res* 9: 921–926, 1985.
70. Oez S, Trautmann U, Smetak M et al. *Ann Hematol* 72: 307–316, 1996.
71. Treves AJ, Halperin M, Barak V et al. *Exp Hematol* 13: 281–288, 1985.
72. Yoshida H, Kondo M, Ichihashi T et al. *Cancer Genet Cytogenet* 100: 21–24, 1998.
73. Tohyama K, Tsutani H, Ueda T et al. *Br J Haematol* 87: 235–242, 1994.
74. Tsuji-Takayama K, Kamiya T, Nakamura S et al. *Human Cell* 7: 167–172, 1994.
75. Hu ZB, Ma W, Zaborski M et al. *Leukemia* 10: 1025–1040, 1996.
76. Lanotte M, Martin-Thouvenin V, Najman S et al. *Blood* 77: 1080–1086, 1991.
77. Dermime S, Grignani F, Clerici M et al. *Blood* 82: 1573–1577, 1993.
78. Duprez E, Ruchaud S, Houge G et al. *Leukemia* 6: 1281–1287, 1992.
79. Kataoka T, Morishita Y, Ogura M et al. *Cancer Res* 50: 7703–7709, 1990.
80. Wang C, Curtis JE, Minden MD et al. *Leukemia* 3: 264–269, 1989.
81. Nara N, Suzuki T, Nagata K et al. *Jpn J Cancer Res* 81: 625–631, 1990.
82. Koistinen P, Wang C, Yang GS et al. *Leukemia* 5: 704–711, 1991.
83. Wang C, Koistinen P, Yang GS et al. *Leukemia* 5: 493–499, 1991.
84. Nagai M, Fujita M, Ikeda T et al. *Br J Haematol* 98: 392–398, 1997.
85. Hamaguchi H, Inokuchi K, Nara N et al. *Int J Hematol* 67: 153–164, 1998.
86. Kubonishi I, Machida KI, Niiya K et al. *Blood* 63: 254–259, 1984.
87. Matozaki S, Nakagawa T, Kawaguchi R et al. *Br J Haematol* 89: 805–811, 1995.
88. Klingemann HG, Gong HJ, Maki G et al. *Leukemia Lymphoma* 12: 463–470, 1994.
89. Taoka T, Tasaka T, Tanaka T et al. *Blood* 80: 46–52, 1992.
90. Oval J, Jones OW, Montoya M et al. *Blood* 76: 1369–1374, 1990.
91. Kizaki M, Matsushita H, Takayama N et al. *Blood* 88: 1824–1833, 1996.
92. Yamaguchi Y, Nishio H, Kasahara T et al. *Leukemia* 12: 1430–1439, 1998.
93. Yamamoto K, Hamaguchi H, Nagata K et al. *Leukemia* 11: 599–608, 1997.
94. Yasukawa M, Yanagisawa K, Kohno H et al. *Br J Haematol* 82: 515–521, 1992.
95. Drexler HG, Dirks W, MacLeod RAF et al. (eds.) *DSMZ Catalogue of Human and Animal Cell Cultures*, 7th edn., DSMZ, Braunschweig, 1999.
96. Keating A. *Clin Haematol* 2: 1021–1029, 1987.
97. Mayumi M. *Leukemia Lymphoma* 7: 243–250, 1992.
98. Nilsson G, Blom T, Kusche-Gullberg M et al. *Scand J Immunol* 39: 489–498, 1994.
99. Toba K, Kishi K, Koike T et al. *Exp Hematol* 24: 894–901, 1996.
100. Drexler HG, Quentmeier H, MacLeod RAF et al. *Leukemia Res* 19: 681–691, 1995.
101. Kojika S, Sugita K, Inukai T et al. *Leukemia* 10: 994–999, 1996.
102. Taetle R, Payne C, Dos Santos B et al. *Cancer Res* 53: 3386–3393, 1993.
103. Drexler HG, Zaborski M, Quentmeier H. *Leukemia* 11: 701–708, 1997.
104. Schumann RR, Nakarai T, Gruss HJ et al. *Blood* 87: 2419–2427, 1996.

105. Tohyama K, Tohyama Y, Nakayama T et al. *Br J Haematol* 91: 795–799, 1995.
106. Feinstein E, Cimino G, Gale RP et al. *Proc Natl Acad Sci USA* 88: 6293–6297, 1991.
107. Drexler HG. *Leukemia* 12: 845–859, 1998.
108. Drexler HG, MacLeod RAF, Borkhardt A et al. *Leukemia* 9: 480–500, 1995.
109. Bi S, Lanza F, Goldman JM. *Leukemia* 7: 1840–1845, 1993.
110. de Thé H, Chomienne C, Lanotte M et al. *Nature* 347: 558–561, 1990.
111. Oval J, Smedsrud M, Taetle R. *Leukemia* 6: 446–451, 1992.
112. Paul CC Ackerman SJ, Mahrer S et al. *J Leuk Biol* 56: 74–79, 1994.
113. Uckun FM. *Blood* 88: 1135–1146, 1996.
114. Collins SJ. *Blood* 70: 1233–1244, 1987.
115. Abrink M, Gobl AE, Huang R et al. *Leukemia* 8: 1579–1584, 1994.
116. Shimamoto T, Ohyashiki K, Ohyashiki JH et al. *Blood* 86: 3173–3180, 1995.
117. Machado EA, Gerard DA, Lozzio CB et al. *Blood* 63: 1015–1022, 1984.
118. Tsuda H, Sakaguchi M, Kawakita M et al. *Int J Cell Cloning* 6: 209–220, 1988.
119. Blom T, Huang R, Aveskogh M et al. *Eur J Immunol* 22: 2025–2032, 1992.
120. Kawaguchi R, Yoshioka JI, Hikiji K et al. *Cancer Genet Cytogenet* 67: 157, 1993.
121. Kizaki M, Ueno H, Matsushita H et al. *Leukemia Lymphoma* 25: 427–434, 1997.
122. Felix CA, Megonigal MD, Chervinsky DS et al. *Blood* 91: 4451–4456, 1998.
123. Barak Y, Shore NA, Higgins GR et al. *Br J Haematol* 27: 543–549, 1974.
124. Giavazzi R, Di Berardino C, Garofalo A et al. *Int J Cancer* 61: 280–285, 1995.
125. Okabe M, Kunieda Y, Shoji M et al. *Leukemia Res* 19: 933–943, 1995.
126. Okabe M, Kunieda Y, Nakane S et al. *Leukemia Lymphoma* 16: 493–503, 1995.
127. Kim SH, Ding W, Cannizzaro L et al. *Blood* 92: 609a, 1998.
128. Srikanth S, Sokol L, Canfield M et al. *Blood* 82: 23a, 1993.
129. Zhou M, Gu L, James CD et al. *Leukemia* 9: 1159–1161, 1995.
130. Zhou M, Yeager AM, Smith SD et al. *Blood* 85: 1608–1614, 1995.
131. Takahashi M, Shigeno N, Takahashi H et al. *Leukemia Res* 21: 1115–1123, 1997.
132. Takiguchi T, Sakuma M, Kim C et al. *Br J Haematol* 102: 24, 1998.
133. Wada H, Asada M, Nakazawa S et al. *Leukemia* 8: 53–59, 1994.
134. Treves AJ, Barak V, Halperin M et al. *Immunol Lett* 12: 225–230, 1986.
135. Miyazawa K, Kawanishi Y, Yaguchi M et al. *Blood* 92: 253b, 1998.
136. Komada Y, Zhou YW, Zhang XL et al. *Blood* 86: 3848–3860, 1995.
137. Paul CC, Aly E, Baumann MA. *Blood* 92: 147b, 1998.
138. Mizobuchi N, Takahashi I, Horimi T et al. *Acta Medica Okayama* 51: 227–232, 1997.
139. Tohda S, Yang GS, Ashman LK et al. *J Cell Physiol* 154: 410–418, 1993.
140. Zheng A, Castren K, Saily M et al. *Br J Cancer* 79: 407–415, 1999.
141. Lucas DL, Knight RD, Khan EC et al. *Blood* 62: 152a, 1983.
142. Mallet M, Mane SM, Meltzer SJ et al. *Leukemia* 3: 511–515, 1989.
143. McCarty TM, Rajamaran S, Elder FFB et al. *Blood* 70: 1665–1672, 1987.
144. Drexler HG, Minowada J. *Leukemia Res* 10: 279–290, 1986.
145. Drexler HG, Thiel E, Ludwig WG. *Leukemia* 7: 489–498, 1993.
146. Callard R, Gearing A (eds.) *The Cytokine FactsBook*, Academic Press, London, 1994.
147. Nicola NA (ed.) *Guidebook to Cytokines and their Receptors*, Sambrook, Oxford, 1994.
148. Oval J, Taetle R. *Blood Rev* 4: 270–279, 1990.
149. Drexler HG. *Leukemia Res* 18: 919–927, 1994.
150. Drexler HG, MacLeod RAF, Uphoff CC. *Leukemia Res* 23: 207–215, 1999.
151. Nucifora G, Rowley JD. *Blood* 86: 1–14, 1995.
152. Rubnitz JE, Behm FG, Downing JR. *Leukemia* 10: 74–82, 1996.

153. Ohyashiki K, Ohyashiki JH, Shimamoto T et al. *Leukemia Lymphoma* 23: 431–436, 1996.

154. Golub TR, McLean T, Stegmaier K et al. *Biochim Biophys Acta* 1288: M7–M10, 1996.

155. Rabbitts TH. *Nature* 372: 143–149, 1994.

156. Drexler HG, Borkhardt A, Janssen JWG. *Leukemia Lymphoma* 19: 359–380, 1995.

157. Fischkoff SA, Pollak A, Gleich GJ et al. *J Exp Med* 160: 179–196, 1984.

158. Hutt-Taylor SR, Harnish D, Richardson M et al. *Blood* 71: 209–215, 1988.

159. Tomonaga M, Gasson JC, Quan SG et al. *Blood* 67: 1433–1441, 1986.

# Chapter 7

# Monocytic Cell Lines

Hans G. Drexler

*DSMZ-German Collection of Microorganisms and Cell Cultures, Department of Human and Animal Cell Cultures, Braunschweig, Germany. Tel: +49-531-2616-160; Fax: +49-531-2616-150; E-mail: hdr@dsmz.de*

## 1. INTRODUCTION

The cells comprising the "mononuclear phagocyte system" include promono-cytes and their precursors in the bone marrow, monocytes in the circulation and macrophages in tissues. The notion that these cells are parts of a system is derived from their common origin, functions and similar morphology. An older, but now revised concept was the so-called "reticuloendothelial system". The ideas and definitions of the monocytic series (including its precursors and progeny) have undergone drastic changes in recent decades [1]. Matters are further complicated by different denominations for the same type of cell, an example being "histiocyte" (used mainly by pathologists) and "macrophage" (applied predominantly by immunologists and hematologists). The classical theory according to which monocytes are derived from histiocytes (called reticulum cells by early investigators – hence the term "reticuloendothelial system") was found to be incorrect and was replaced by the reverse hypothesis.

The mononuclear phagocyte system forms a network which is found in many organs. It includes circulating cells, monocytes in the peripheral blood, and cells resident in tissues or fixed to the endothelial layer of blood capillaries, such as Kupffer cells in the liver, intraglomerular mesangial cells in the kidney, alveolar and serosal macrophages in the lung and pleura/peritoneum, microglia cells in the brain, and the sinus macrophages in the spleen and lymph node [2].

The monocytic cell lineage originates from the stem cell in the bone marrow as a common committed progenitor cell for the granulocyte (myelocytic) and monocyte-macrophage pathways; the so-called colony-forming unit granulocyte-macrophage (CFU-GM). This cell typically expresses cytokine receptors for GM-CSF and M-CSF, which stimulate further differentiation to

 *J.R.W. Masters and B.O. Palsson (eds.), Human Cell Culture Vol. III, 237–257.*
© 2000 *Kluwer Academic Publishers. Printed in the Netherlands.*

the promonocyte, the earliest morphologically identifiable cell of the series [3]. The promonocyte is capable of endocytosis and adherence to glass, plastic or other substrates. The subsequent stage, the monocyte, is phagocytic and microbiocidal. Some monocytes migrate into the organs and tissue systems to become macrophages. Once in the tissues, monocytes do not re-enter the circulation. Rather, they undergo transformation into tissue-specific macrophages with morphological and functional properties that are characteristic of the tissue in which they reside.

The mononuclear phagocyte system has two functions, performed by two different types of cells. First, the removal of particulate antigens by the "professional" phagocytic monocyte-macrophage; and second, the presentation of antigens to lymphocytes by the "antigen-presenting cell" [2].

There are various forms of acute and chronic leukemia in which a cell committed to monocytic differentiation appears to be the target of leukemogenesis. Morphologists discern the myelomonocytic subgroup AML FAB M4 (with its variant M4eo in which an increased percentage of eosinophils is seen) and the monocytic subgroup AML FAB M5 (with further distinction of the immature subtype M5a and the mature subtype M5b) [4]. Despite its high percentage of monocytic blasts, chronic myelomonocytic leukemia (CMML) has been assigned to the myelodysplastic syndromes (MDS) [5]. Finally, CML can occasionally lead to monoblastic blast crisis (instead of the more common myelocytic blast crisis).

More than 30 leukemia cell lines have been established since 1974 (Tables 1 and 6). Thirty-one monocytic cell lines have been described adequately (Table 1), while the remaining cell lines require further characterization, have been lost or their current status is not known (Table 6). The oldest monocytic cell line is U-937 [38]. Despite the original diagnosis of "generalized diffuse histiocytic lymphoma" (this nomenclature is no longer used), the U-937 cells do have clear-cut monocyte-associated features. U-937 and the next cell line to be established, THP-1 [35] are the prototype monocytic leukemia cell lines. Both lines are widely used and are available through several cell line banks (Appendix 1, p. 284). The lines listed in Table 1 are useful models for cells representing the different steps of monocytic maturation [84,85].

## 2.  CLINICAL CHARACTERIZATION

Thirty-one cell lines with monocytic characteristics are listed in Table 1. These cell lines were derived mainly from patients with either AML M4 (n = 10, 32%) or AML M5 (n = 16, 51%); two cell lines were established from patients with CML in blast crisis. The age of the patients ranged from 1 to 76 years (among them 8 children). Specimens were obtained predominantly

Table 1. Monocytic cell lines: clinical characterization

| Cell line[a] | Patient age/sex[b] | Diagnosis[c] | Treatment status[d] | Specimen site[e] | Authentication[f] | Year est. | Culture medium[g] | Availability[h] | Primary ref. |
|---|---|---|---|---|---|---|---|---|---|
| AML-1 | 12 M | AML M4 | R | | yes | | IMDM + FBS | ATCC | 6 |
| AML-193 | 13 F | AML M5 | R | | yes | | IMDM + GM-CSF or IL-3 | | 6 |
| CTV-1 | 40 F | AML M5 | R | PB | no | 1982 | RPMI 1640 + FBS | DSMZ | 7 |
| DOP-M1 | 1 F | AML M5a | T | cerebro-spinal fluid | yes | 1989 | RPMI 1640 + FBS | | 8 |
| FLG29.1 | 38 F | AML M5a | D | BM | no | 1987 | RPMI 1640 + FBS | | 9 |
| IMS-M1 | 33 M | AML M5a | D | BM | no | 1988 | IMDM + FBS | | 10,11 |
| KBM-3 | 32 F | AML M4 | resistant | PB | yes | | IMDM + FBS | Author | 12 |
| KBM-5 | 67 F | CML-mono BC | BC | PB | yes | | IMDM + FBS | Author | 13 |
| KP-1 | 2 F | AML M5 | D | PB | no | 1986 | RPMI 1640 + FBS | | 14 |
| KP-MO-TS | 12 M | AML M5b | D | PB | yes | 1987 | RPMI 1640 + FBS | Author | 15 |
| ME-1[i] | 40 M | AML M4eo | 2nd R | PB | yes | 1988 | RPMI 1640 + FBS | Author | 16,17 |
| ML-2[j] | 26 M | T-NHL → T-ALL → AML M4 | D | PB | yes | 1978 | RPMI 1640 + FBS | DSMZ | 18 |
| MOLM-13[k] | 20 M | MDS (RAEB) → AML M5a | R | PB | yes | 1995 | RPMI 1640 + FBS | Author | 19 |
| Mono Mac 6[l] | 64 M | myeloid metaplasia → AML M5 | R | PB | no | 1985 | RPMI 1640 + FBS | DSMZ | 20 |
| MUTZ-3 | 29 M | AML M4 | D | PB | yes | 1993 | α-MEM + FBS + 5637 CM or GM-CSF, IL-3, PIXY-321 | DSMZ | 21 |
| MV4-11 | 10 M | AML M5 | D | | yes | | IMDM or RPMI 1640 + FBS or IMDM + GM-CSF | ATCC, DSMZ | 6 |

*Continued on next page*

*Table 1.* (continued)

| Cell line[a] | Patient age/sex[b] | Diagnosis[c] | Treatment status[d] | Specimen site[e] | Authentication[f] | Year est. | Culture medium[g] | Availability[h] | Primary ref. |
|---|---|---|---|---|---|---|---|---|---|
| NOMO-1 | 31 F | AML M5a | 2nd R | BM | no | 1985 | RPMI 1640 + FBS or serum-free | Author | 22 |
| OCI/AML-2 | 65 M | AML M4 | D | PB | no | 1986 | $\alpha$-MEM + FBS | DSMZ | 23,24 |
| OMA-AML-1 | ? M | AML M4 | 5th R | PB | no | 1989 | RPMI 1640 + FBS + HS | Author | 25,26 |
| P31/Fujioka | 7 M | AML M5 | R | PB | no | 1980 | RPMI 1640 + FBS | JCRB | 27 |
| P39/Tsugane | 69 M | MDS (CMML) → AML M2 | D | PB | no | 1983 | RPMI 1640 + FBS | JCRB | 28 |
| PLB-985 | 38 F | AML M4 | R | PB | yes | 1985 | RPMI 1640 + FBS or serum-free | DSMZ | 29 |
| RC-2A[m] | adult M | AML M4 | refractory | PB | no | | RPMI 1640 + FBS | | 30 |
| RWLeu-4 | | CML-BC | BC | PB | no | | RPMI 1640 + FBS | Author | 31,32 |
| SCC-3 | 20 F | NHL (diffuse large cell – stage IV-A) | R | PE | no | 1985 | RPMI 1640 + FBS | JCRB | 33 |
| SKM-1 | 76 M | MDS (RAEBT) → AML M5 | refractory | PB | yes | 1989 | RPMI 1640 + FBS | JCRB | 34 |
| THP-1 | 1 M | AML M5 | R | PB | no | 1978 | RPMI 1640 + FBS | ATCC, DSMZ, JCRB, RIKEN | 35 |
| TK-1[n] | 22 M | T-lymphoblastic lymphoma → biclonal leukemia (T-ALL L2 + AML M4) | D | PB | no | 1984 | RPMI 1640 + FBS | Author | 36,37 |
| U-937 | 37 M | generalized, diffuse histiocytic lymphoma | refractory | PE | no | 1974 | RPMI 1640 + FBS (originally Ham's F10 + NCS) | ATCC, DSMZ, IFO, JCRB, RIKEN | 38 |

*Continued on next page*

*Table 1.* (continued)

| Cell line[a] | Patient age/sex[b] | Diagnosis[c] | Treatment status[d] | Specimen site[e] | Authentication[f] | Year est. | Culture medium[g] | Availability[h] | Primary ref. |
|---|---|---|---|---|---|---|---|---|---|
| UG3 | 26 F | AML M5 | T | PB | no | 1997 | IMDM + FBS + GM-CSF or IL-3 | Author | 39 |
| YK-M2 | 31 M | AML M5 | T | PB | yes | 1985 | RPMI 1640 + FBS | Author | 40 |

a Names of cell lines are as given in the original literature; subclones (variant cell lines derived from a parental cell line) and sister cell lines (derived independently from the same patient from different specimens or at different time points) are indicated for each cell line.

b Age at the time of establishment of cell line.

c Diagnoses are indicated as given in the original literature; →: disease progressed from a pre-malignancy/first malignancy into the final malignancy.

d BC – at blast crisis; CP – in the chronic phase; D – at diagnosis (prior to therapy); R – at relapse; T – during therapy.

e BM – bone marrow; PB – peripheral blood; PE – pleural effusion; Tu – tumor.

f Evidence (e.g. cytogenetic marker chromosomes, immunoprofile, others) that this cell line was derived from the patient indicated.

g Culture media as indicated in the original literature; cell line might also grow with other medium and/or supplements.

h Availability from cell banks (ATCC; DSMZ; IFO; JCRB; RIKEN) or from the original investigator (author).

i Five subclones (ME-2, ME-3, ME-F$_1$, ME-F$_2$, ME-F$_3$) with some different morphological, cytochemical and cytogenetic features were established.

j Sister cell lines ML-1 and ML-3 with different cytogenetic aberrations were established from the same specimen of the same patient.

k Sister cell line MOLM-14 with some different immunophenotypic and minor cytogenetic features established from the same specimen.

l Sister cell line Mono Mac 1 was established from the same specimen and the same patient.

m Subclone of original cell line RC-2 which was growth factor-dependent; sister cell line CESS-B is an EBV+ LCL and was established from the same specimen.

n Subclones TK-1B (pseudodiploid with complex chromosomal aberrations) and TK-1D (diploid normal karyotype) were established by cloning from the parental line TK-1.

Table 2. Monocytic cell lines: immunophenotypic characterization

| Cell line | T/NK-cell marker* | B-cell marker | Myelomonocytic marker | Erythroid-megakaryocytic marker | Progenitor/activation marker | Adhesion marker | Ref. |
|---|---|---|---|---|---|---|---|
| AML-1 | CD2− CD3− CD5− CD7+ | CD10− | CD13+ CD14+ CD15+ CD33+ | | HLA-DR+ | | 6 |
| AML-193 | CD2− CD3− CD4− CD5− CD7− CD8− | CD10− CD19− | CD13+ CD14− CD15+ CD16− CD33+ | CD41a− GlyA− | CD34− CD38− CD71+ HLA-DR+ | | 6,41, 42 |
| CTV-1 | CD2− CD3− CD4− CD5− CD6+ CD7+ CD8− CD28− CD57+ TCRαβ− TCRδ− | CD10− CD19− CD20− CD21− cy/sIg− | CD13− CD14(+) CD15+ CD33− CD68− | CD41a− CD61− | CD34− HLA-DR− TdT− | CD11b− | 7 |
| DOP-M1 | CD2− CD3− CD4− CD7− CD8− | CD10− CD19− | CD13− CD14− CD15+ CD33+ CD65+ | CD41− | HLA-DR− | CD11b− | 8 |
| FLG29.1 | CD1a− CD2− CD3− CD4− CD5− CD7− CD8− | CD19− CD20− CD21− CD22− CD23− CD24− | CD13+ CD14(+) CD15(+) CD32+ CD33(+) CD35(+) CD68+ | CD9+ CD31− CD36− CD41a− CD42b(+) CD61+ | CD34+ CD71+ HLA-DR− | CD11a− CD11b− CD18− CD44+ CD51(+) CD54+ CD55+ | 9 |
| IMS-M1 | CD4+ | | CD14+ CD33+ | | CD34+ CD38+ HLA-DR+ | | 11 |
| KBM-3 | CD2+ CD3− CD4+ CD5− CD8− | sIg− | CD14+ CD33+ | | CD34− HLA-DR+ TdT− | CD11b− | 12 |
| KBM-5 | CD2− CD4+ | CD19− CD20− | CD13+ CD14− CD33+ | | CD34− HLA-DR− TdT− | CD11a+ CD11c+ | 13 |
| KP-1 | CD2− CD3− CD4+ CD8− | | CD13+ CD14+ CD33+ | | HLA-DR+ | CD18+ | 14 |
| KP-MO-TS | CD1a− CD2− CD3− CD4+ CD5− CD8− | CD10− CD19− CD20− sIg− | CD13− CD14− CD15+ CD33+ CD35− | | HLA-DR+ TdT+ | CD11b+ | 15 |
| ME-1 | CD2− CD7− | CD10− CD20− | CD13+ CD14+ CD33+ | | CD34+ HLA-DR+ | CD11b(+) | 16 |
| ML-2 | CD1− CD2− CD3− CD4+ CD5− CD7+ CD8− CD28− CD57− TCRαβ− TCRδ− | CD10(+) CD19− CD20− CD21− | CD13+ CD14(+) CD15+ CD33+ | CD9− CD41a− CD61(+) | CD34− CD38+ CD71+ HLA-DR+ TdT− | CD11b+ | 24 |
| MOLM-13 | CD1− CD2− CD3− CD4(+) CD5− CD7− CD8− CD57− | CD10− CD19− CD20− CD21− CD22− CD23− CD40− | CD13− CD14− CD15+ CD32+ CD33+ CD64− CD65− CD68− CD87(+) CD91− CD92(+) CD93+ CD155− | CD9− CD41a− CD42b− CD61− CD62P− | CD34− HLA-DR− TdT− | CD11a+ | 19 |
| Mono Mac 6 | CD2− CD3− CD4(+) | CD19− CD20− | CD13+ CD14(+) CD15+ CD33+ CD68+ | GlyA− | CD34− HLA-DR+ TdT− | CD11b+ | 20,24 |

Continued on next page

Table 2. (continued)

| | | | | | | | |
|---|---|---|---|---|---|---|---|
| MUTZ-3 | CD1- CD3- CD4+ CD5- CD7- CD8- CD56- | CD10- CD19- CD20- CD23- | CD13+ CD14+ CD15+ CD16- CD32+ CD33+ CD64(+) CD65+ CD68+ | CD41a- CD42b- CD61- GlyA- | CD30+ CD34(+) CD38+ CD71+ HLA-DR+ TdT- | CD11b+ | 21 |
| MV4-11 | CD2- CD3- CD4+ CD5- CD7- CD8- | CD10- CD19- CD21- CD37- CD138+ | CD13+ CD14(+) CD15+ CD16- CD33+ | | CD34- CD38- HLA-DR+ | | 6,24, 41 |
| NOMO-1 | CD2- | cy/sIg- | | | | | 22 |
| OCI/AML-2 | CD3- | CD19- | CD13+ CD14- CD15+ CD33+ | | HLA-DR+ | | 24 |
| OMA-AML-1 | | | CD13+ CD14(+) CD15+ CD33+ | | CD34+ HLA-DR- | CD11c+ | 25 |
| P31/Fujioka | CD1- CD2- CD3- CD4- CD5- CD8- | CD10- CD20- CD80- CD86+ cy/sIg- | CD14+ CD15+ | | HLA- DR+ TdT- | CD11b+ CD54+ | 27 |
| P39/Tsugane | CD1- CD2- CD3- CD4+ CD5- CD8- | CD10- CD20- CD80- CD86- cy/sIg- | CD13+ CD14+ CD15+ | | HLA-DR- TdT- | CD11b- CD54+ | 28 |
| PLB-985 | CD3- | CD19- | CD13+ CD14- CD15+ CD33+ | | HLA-DR- | CD11b+ | 24,29 |
| RC-2A | | | CD14- | | HLA-DR+ | CD11b+ | 43 |
| RWLeu-4 | CD8- | CD21+ sIg- | CD14- CD15+ | | HLA-DR(+) | CD11b+ | 31 |
| SCC-3 | CD1- CD2- CD3- CD4- CD5- CD8- CD57- | CD10+ CD19- CD20- CD21- CD24- PCA-1+ | CD13+ CD14+ CD15- CD16- CD33- | CD36- | CD38+ CD71+ HLA-DR- | CD11b- | 33 |
| SKM-1 | CD2- CD3- CD4+ CD5- CD7- CD8- | | CD13+ CD14- CD33+ | | HLA-DR+ | CD11b+ | 34 |
| THP-1 | CD1- CD2- CD3- CD4+ CD7+ CD28- CD57- TCRαβ- | CD19- CD20- CD23(+) CD24+ CD80- CD86+ cy/sIg- | CD13+ CD14(+) CD15+ CD33+ CD68- | CD9- CD41a+ CD61+ | CD34- CD38+ CD71+ HLA-DR+ | CD11b+ CD54+ | 24,35 |
| TK-1 | CD1- CD2- CD3- CD4- CD5- CD7- CD8- TCRαβ- | CD10- CD20- cy/sIg- | CD13- CD14- CD15+ | | HLA-DR+ TdT- | CD11b- | 36 |
| U-937 | CD1- CD2- CD3- cyCD3- CD4+ CD7+ CD8- CD28- CD57- TCRαβ- TCRδ- | CD10- CD19- CD20- CD21- CD23+ CD24(+) CD75- CD80- CD86- cy/sIg- | CD13+ CD14(+) CD15+ CD17- CD33+ CD35+ CD68+ CD88(+) | CD9- CD41a- CD42b- vWF- | CD34- CD38(+) CD71+ HLA-DR- TdT- | CD11b+ CD44+ CD54+ | 24 |
| UG3 | | | CD13+ CD14(+) CD16- CD33+ CD64+ CD68+ | CD36+ | CD71+ HLA-DR- | CD11b+ CD11c- CD54+ | 39 |
| YK-M2 | CD1a- CD2- CD3- CD4- CD5- CD8- TCRαβ- | CD10- CD20- CD21- sIg- | CD13+ CD14+ CD15+ | | HLA- DR+ TdT- | CD11b- | 40 |

*, +, strong, definite protein expression (mostly more than 10-20% cells positive); (+), weak protein expression, qualitatively and quantitatively (commonly <10% cells positive); -, no protein expression.

*Table 3.*   Monocytic cell lines: cytokine-related characterization

| Cell line | Cytokine receptor expression[a] | Cytokine production[a] | Proliferation response to cytokines[b] | Differentiation response to cytokines[b] | Dependency on cytokines[c] | Ref. |
|---|---|---|---|---|---|---|
| AML-193 | IL-2Rα− | | G-CSF+, GM-CSF+, IGF-1+, IL-3+, PIXY-321+ <br> inhibitory: TGF-β1+ | | GM-CSF or IL-3 | 6,41,44 |
| FLG29.1 | G-CSFRα−, GM-CSFRα−, gp80+, IFN-γR+, IL-2Rα−, IL-3Rα+, IL-4Rα−, IL-6Rα−, Kit+, TNF-RI/II+ | | | | | 9 |
| IMS-M1 | Kit− | | | | | 11 |
| KBM-3 | Northern: M-CSFR+ | Northern: GM-CSF+ | G-CSF+, GM-CSF+, M-CSF+ | | | 12 |
| KBM-5 | | RT-PCR: TNF-α+ | inhibitory: TNF-α+ | | | 13 |
| ME-1 | | | GM-CSF+, IL-3+ | GM-CSF+, IL-3+, IL-4+ | | 16 |
| MOLM-13 | GM-CSFRα+, IFN-γR+, M-CSFR−, TNFRI/II− | | | IFN-γ+, IFN-γ+, TNF-α+ | | 19 |
| Mono Mac 6 | mRNA: IL-2Rα+, IL-2Rβ+, IL-2Rγ(+), IL-4Rα(+), IL-7Rα(+), IL-9Rα+ <br> protein: IL-2Rγ− | IL-1+ | | IFN-γ+ | | 45 |
| MUTZ-3[d] | GM-CSFRα+, IL-2Rα−, IL-3Rα+, Kit+, M-CSFR+ | | G-CSF+, GM-CSF+, IFN-β+, IFN-γ+, IGF-1+, IL-3+, IL-4+, IL-6+, IL-7+, insulin+, M-CSF+, MIP-1α+, PIXY-321+, SCF+ | | GM-CSF, IL-3, PIXY-321 | 21,44 |
| MV4-11 | RT-PCR: IGF-1R+, IGF-2R+, IL-2Rα− | RT-PCR: IGF-1+ | G-CSF+, GM-CSF+, IL-3+ | | | 6,41 |
| OCI/AML-2 | | | G-CSF+, GM-CSF+, M-CSF+ | | | 23 |
| OMA-AML-1 | | | G-CSF+, GM-CSF+, IL-3+, IL-6+, LIF+ | GM-CSF+, IL-3+, LIF+ | | 25,26 |
| SKM-1 | | | GM-CSF+ | | | 34 |
| U-937 | RT-PCR: IL-2Rα+, IL-4Rα+, IL-7Rα+, IL-9Rα+ <br> protein: GM-CSFRα+, IL-1Rα (+), IL-2Rα−, IL-2Rβ+, IL-2Rγ+, IL-3Rα(+), IL-3Rβ+, IL-6Rα+, IL-7Rα−, Kit− | PDGF− | | IFN-γ+ | | 45 |
| UG3 | G-CSFRα+ | | | | | 39 |

[a] Receptor expression or cytokine production at the protein level (ELISA, McAb, RIA), unless otherwise indicated, e.g. at the mRNA level (by RT-PCR, Northern); +, strong, definite expression; (+), weak expression; —, no expression. [b] Effects of cytokine exposure on proliferation or differentiation (−, no effect; +, positive effect). [c] Upon growth factor withdrawal these cell lines will die by apoptosis. [d] Detailed report on proliferative response to cytokines in ref. [21].

*Table 4.* Monocytic cell lines: genetic characterization

| Cell line | Cytogenetic karyotype | Unique translocations (→ fusion genes) | Unique gene alterations, receptor gene rearrangements[a] | Ref. |
|---|---|---|---|---|
| AML-1 | 51, XYY, +1p−, +6, +8, +8, +19, 2p−, 12p+ | | IGH G, IGK G, IGL G, TCRB G | 6 |
| AML-193 | 49, X, +3, +6, +8, −17, +der(17)t(17;17)(p13.1;q21.3) | | IGH G, IGK G, IGL G, TCRB G | 6 |
| CTV-1 | 92(89−94)<4n> XX/XXY/XXYY, −1, +6, +6, −17, t(1;7)(p34.2;q34)x2, del(3)(p21), ii(6q)x2, t(12;16)(q24.32;q11)x2 | | P53 PM | 7, 24 |
| DOP-M1 | 47, X, −X, −13, +19, +20, +mar | | | 8 |
| FLG29.1 | 45−69, 3p+ | | | 9 |
| IMS-M1 | 46, XY, −7, +8, −10, +12, 1p−, t(9;11)(p22;q23) | t(9;11)(p22;q23) → MLL-AF9 fusion gene | IGH G, TCRB G, TCRG G | 10, 11 |
| KBM-3 | 48(41−51)<2n>, −4, −12, +13, del(12)ins(12q+?), del(5)(p12 → ter), del(9)(p11(ter), del(16)t(16q+?), del(7)ins(7p+HSR) | | P53 PM | 12 |
| KBM-5 | 69−87, XX, +6, +7, +8, +8, +8, −9, −9, 9q+, −11, −11, −13, 14q−, +15, 17p+, +20, +22q−, +22q−, − multiple Ph, +frag, +DMs | Ph+ t(9;22)(q34;q11) → BCR-ABL (b3−a2) fusion gene | BCR, ABL, amplified; P53 PM | 13, 46 |
| KP-1 | 47(46−48)XX, 1p−, 10p+, 16q+, 19p+, +mar, t(11q+;19p−) | | | 14 |
| KP-MO-TS | 46, XY, −17, t(10;11)(p13;q21), +mar | | IGH G, IGK R, TCRB G, TCRG G | 15 |
| ME-1 | 47, XY, +8, inv(16)(p13q22), del(17)(p12p13) | inv(16)(p13q22) → CBFB-MYH11 fusion gene | | 16, 47 |
| ML-2 | 92(84−94)<4n>XX, −Y, −Y, −7, −9, −10, −10, +11, +12, +12, +13, +13, −15, −16, −17, −17, +18, +18, −20, −20, +4mar, der(1)t(1;?)(p21;?)x2, del(6)(q23)x2, der(6)t(6;11)(q27;?q23)x2, ?der(11)t(6;11)(q27;?q23)/del(11)(q23)x2, der(11)t(11;?)(? → 11p11 → 11q23)x2, der(11)t(11;?)(q11−13;?), dup(13)(q32 → qter)x2, der(18)t(15;?;18)(q21;?;q11)x2 | t(6;11)(q27;q23) → MLL-AF6 fusion gene | P16INK4A D; P53 PM; RB1 R[b]; IGH GR, TCRA GG, TCRB GR, TCRG GR[c] | 24, 48−51 |
| MOLM-13 | 49<2n>XY, +6, +8, +13, ins(11;9)(q23;p22p23), del(14)(q23.3q31.3) | ins(11;9)(q23;p22p23) → MLL-AF9 fusion gene | P15INK4B D, P16INK4A D | 19 |
| Mono Mac 6[d] | 84−90<4n>XX/XXX, −Y, +6, +7, −12, −13, −13, −16, −16, +2mar, t(9;11)(p22;q23)x2, add(10)(p11)x2, add(12)(q21), del(13)(q13q14) der(13)t(13;14)(p11;q12)x2, der(17)t(13;17)(q21;p11)x2 | t(9;11)(p21;q23) → MLL-AF9 fusion gene | | 24, 52 |
| MUTZ-3 | 46(44−48)<2n>XY, t(1;3)(q43;q13)inv(3)(q21;q26),t(2;7)(q36;q36)inv(7)(p15q36), t(12;22)(p13;q12) | t(12;22)(p13;q12) → ETV6/TEL-MN1 fusion gene? | | 21, 24, 53 |
| MV4-11 | 48(46−48)<2n>XY, +8, +18, +19, −21, t(4;11)(q21;q23) | t(4;11)(q21;q23) → MLL-AF4 fusion gene | IGH G, IGK RG, IGL G, TCRB G | 6, 24 |
| NOMO-1 | 47, XX, 1p+, 7q+, 13p+ | | | 22 |
| OCI/AML-2 | 48(43−49)<2n>XY, +6, +8, der(1)inv(1)(p36q31)t(1;6)(q13;p12), der(2)t(2;17)(q23;q24.1)del(2)(q14.2q36),der(3)t(1;3)(p36;p25),ins(3;2)(q21;q14.2q36), t(5;8)(q11.2;q24),der(6)t(6;16)(q31;p12)t(3;6)(q26;q24),inv(12)(p13.3q13.2), t(13;14)(q32/33;q24.2), der(17)t(2;17)(p23;q24.1) | | IGH G, IGK RG, IGL G, TCRB G | 24 |

*Continued on next page*

*Table 4.* (continued)

| Cell line | Cytogenetic karyotype | Unique translocations (→ fusion genes) | Unique gene alterations, receptor gene rearrangements[a] | Ref. |
|---|---|---|---|---|
| OMA-AML-1 | 46, XY, t(2;7;11)(p21;p12;p15), +mar | | | 25 |
| P31/Fujioka | 47, XY, −10, +15, +19, t(7;11)(7qter → 7p13::11q22 → 11qter;11pter → 11q22:), inv(9)(pter → p11::q13 → q11:), del(17)(qter → p11:) | | *NRAS* mutation | 27 |
| P39/Tsugane | 45(39–46)XXY,−7,−8,−16,−17,+22,+del(6)(pter → q15:), t(9:?)(q34:?), t(14:16)(14pter → 14q24::16q21 → 16qter; 16pter → 16q21::14q24 → 14qter) | | *NRAS* mutation | 28 |
| PLB-985 | 46<44–50><2n>X,−X,+6, dm, dup(1)(q21q41), del(3)(p21), del(6)(q21)x2 | | | 24 |
| RC-2A | 47, XY, +11, t(11;12) | | | 30 |
| RWLeu-4 | 56, 7p+, 12p+, 17p+, Ph | Ph+ t(9;22)(q34;q11) → *BCR-ABL* (b2-a2) fusion gene | | 31 |
| SCC-3 | 50, XX, −2, −4, −7, −15, +16, −20, +1p−, 1q−, 3q−, 6q−, 7p+, 8q−, 10q+, +11q+, +12q+, 12p−, 13p+, 17p+, +5 mar | | *P53* PM | 33 |
| SKM-1 | 46, XY, del(9)(q13q22), der(17)t(17:?)(p13:?) | | *P53* PM | 34 |
| THP-1 | 94(88–96)<4n>XY/XXY,−Y,+1,+3,+6,+6,−8,−13,−19,−22,−22,+2mar, add(1)(p11), del(1)(q42.2), i(2q), del(6)(p21)x2-4, i(7p), der(9)t(9;11)(p22:q23)i(9)(p10)x2, der(11)t(9;11)(p22;q23)x2, add(12)(q24)x1-2, der(13)t(8;13)(p11;p12), add(?18)(q21) | t(9;11)(p21;q23) → *MLL-AF9* fusion gene | *P15INK4B* DD, *P16INK4A* DD, *P53* D; *RB1* R | 24, 49, 54 |
| TK-1 | 46, XY, −14, −17, +der(14)t(14;17)(14pter(14q22::17q23(17q4ter). +der(17)t(11;14:17)t17pter(17q23::14q22(14qater::11q13(11qter) | | *IGH* G, *TCRB* G, *TCRG* R | 37, 40 |
| U-937 | 63(58–69)<3n>XXY,−2,−4,−6,+7,−9,−20,−21,+3mar, t(1;12)(q21;p13), der(5)t(1:5)(p22:q35), add(9)(p22), t(10;11)(p14;q23),i(11q),i(12p).add(16)(q22), add(19)(q13) | t(10;11)(p14;q23) → *MLL-CALM* fusion gene | *P53* PM | 24, 55 |
| UG3 | 46, XX, −7, +8, t(9;11)(p22;q23) | t(9;11)(p22;q23) → *MLL-AF9* fusion gene | | 39 |
| YK-M2 | 68<3n>X,−Y,+3,−4,−5,+6,+7,−10,−11,−16, del(17)(p11),+4mar | | | 40 |

a Receptor gene arrangements: D – deleted; G – germline; PM – point mutation; R – rearranged.

b Data from sister cell line ML-1.

c ML-2 is negative for the *IGH, TCRA, TCRB, TCRG* mRNAs.

d Mono Mac 1 is diploid sister cell line with fewer *in vitro* rearrangements [24].

*Table 5.* Monocytic cell lines: functional characterization

| Cell line | Doubling time | EBV status | Cytochemistry | Inducibility of differentiation | Heterotransplant-ability into mice | Special functional features | Ref. |
|---|---|---|---|---|---|---|---|
| AML-1 | | EBNA– HTLV-II– | ANAE+, CAE+, MPO–, PAS– | | | | 6 |
| AML-193 | | EBNA– HTLV-II– | ANAE(+), CAE–, MPO–, MSE–, PAS(+) | | | | 6 |
| CTV-1 | 30–40 h | EBV– HTLV-I– HIV– | ACP–, ANAE+ (NaF inhibitable), ANBE+, MPO–, MSE–, PAS+, SBB–, TRAP– | (AraC, DMSO, retinoic acid, TPA → no effect) | | | 7,24 |
| DOP-M1 | 48 h | | ANBE+ (NaF inhibitable), CAE+, MPO+ | (IFN-γ no effect) | | no phagocytosis | 8 |
| FLG29.1 | | EBV– | ACP+, ANBE+ (NaF inhibitable), CAE–, MSE+, NBT+, PAS–, SBB–, TRAP+ | TPA → osteoclast differentiation (vit. D3 no effect) | | responsive to calcitonin; expression of calcitonin and estrogen receptors | 9 |
| IMS-M1 | 36–48 h | | ANBE+, CAE–, MPO– | vit. D3 → mono/macro differentiation | | retinoic acids induce apoptosis | 11 |
| KBM-3 | 23 h | | ANBE+, CAE+, MPO+, MSE+, PAS– | | into nude mice or SCID mice | colony formation in agar | 12 |
| KBM-5 | 24–60 h | EBNA– | ACP+, ANBE+, CAE+, MPO(+), MSE+, PAS– | (DMSO, hemin, retinoic acid, TPA no effect) | into SCID mice | colony formation in agar/ methylcellulose; resistant to NK activity, IFN-α, IFN-γ; no phagocytosis | 13 |
| KP-1 | 96 h | | ANBE+ (NaF inhibitable), CAE–, MPO–, PAS– | TPA → macro differentiation | | produces lysozyme; phagocytosis+; scavenger receptor+ | 14 |
| KP-MO-TS | 48 h | EBNA– | ACP–, ANBE+, CAE(+), MPO+, PAS– | retinoic acid, TPA → mono/macro differentiation (DMSO, IFN-γ no effect) | | α1-antitrypsin+; phagocytosis(+); lysozyme production | 15 |
| ME-1 | 4–5 d | EBNA– | ANBE(+), CAE+, Luxol(+), MPO+, MSE+, Toluidine Blue(+) | GM-CSF, IL-3, IL-4, PHA-LCM → macro differentiation; serum-free culture → neutro differentiation; PHA-LCM → eosino differentiation; (DMSO, retinoic acid, vit. D3 no effect) | | produces lysozyme; phagocytosis+; weak colony formation in methylcellulose | 16,17,56 |
| ML-2 | 60 h | EBV– HTLV-I– HIV– | MSE+ | AraC, DMSO, TPA → mono/macro differentiation | | TF: GATA-1 mRNA–, GATA-2 mRNA+, SCL mRNA– | 24,57,58 |
| MOLM-13 | 3–4 d | MPO–, MSE+ | | IFN-γ (+TNF-α) → macro differentiation (GM-CSF, M-CSF no effect) | | | 19 |
| Mono Mac 6 | 50–60 h | EBV– HTLV-I– HIV– | ACP–, ANAE+ (NaF inhibitable), CAE–, MPO–, MSE+, PAS– | IFN-γ → mono differentiation | | phagocytosis+; lysozyme+; mRNA: α1-antitrypsin–, azurocidin+, C3 complement+, cathepsin G+, defensin–, N-elastase+, lactoferrin–, myeloblastin+, tryptase+ | 20,24,59 |

*Continued on next page*

*Table 5.* (continued)

| Cell line | Doubling time | EBV status | Cytochemistry | Inducibility of differentiation | Heterotransplant-ability into mice | Special functional features | Ref. |
|---|---|---|---|---|---|---|---|
| MUTZ-3 | 90–110 h | EBV– HTLV-I– HIV– | MSE+, MPO+, TRAP+ | retinoic acid, TPA → mono/macro differentiation | | colony formation in methylcellulose | 21,24 |
| MV4-11 | 50 h | EBV– HTLV-I– HIV– | ANAE+, CAE(+), MPO–, MSE+, PAS– | | | colony formation in agar | 6,24 |
| NOMO-1 | 24 h | EBNA– | ACP+, ANBE+, CAE–, MPO+, MSE+, NBT+ | TPA → macro differentiation | | phagocytosis+; lysozyme production+ | 22,60 |
| OCI/AML-2 | | EBV– HTLV-I– HIV– | MSE+, NBT– | | | colony formation in methylcellulose; retinoic acid receptor+ | 23,24 |
| OMA-AML-1 | | | ANAE+ (NaF inhibitable), NBT– | GM-CSF, IL-3 → eosino differentiation | into nude or SCID mice | phagocytosis+; clonable in suspension; forming cobblestone areas | 25,26 |
| P31/Fujioka | 80 h | EBNA– | ACP+, ALP–, ANBE+ (NaF inhibitable), CAE+, MPO–, PAS+ | | | phagocytosis+ | 27 |
| P39/Tsugane | | | ACP+, ANBE+ (NaF inhibitable), CAE+, MPO–, PAS+ | | into nude mice | phagocytosis+ | 28 |
| PLB-985 | 24–48 h | EBV– HTLV-I– HIV– | ALP–, ANAE+, MPO+, MSE+, PAS(+), SBB+ | Bt-cAMP, DMSO, retinoic acid → neutro differentiation; TPA → mono/macro differentiation | into nude mice | colony formation in agar; phagocytosis inducible | 24,29 |
| RC-2A | 48 h | EBNA– | ACP(+), ANAE+, MSE– | PHA-LCM → mono/macro differentiation | | cloning/colony formation in agar; weak antigen presentation | 30,43 |
| RWLeu-4 | 25–26 h | EBNA– | CAE+, MPO+, MSE+, NBT–, PAS+ | DMSO, TPA, Vit. D3 → macro differentiation | into nude mice | PO: c-MYC+ | 31,32 |
| SCC-3 | 48 h | EBNA– HTLV-I– | ANBE+ (NaF inhibitable), CAE–, MPO–, MSE–, PAS+ | TPA → macro differentiation | | phagocytosis(+) | 33 |
| SKM-1 | 48 h | EBNA– | ANBE+, MPO+, MSE+ | | | cloning/colony formation in agar; secretion of MPO; p53 overexpression | 34,61 |
| THP-1 | 35–40 h | EBV– HTLV-I– HIV– | ALP–, ANBE+ (NaF inhibitable), CAE–, MPO–, MSE+, PAS+, SBB– | TPA → macro differentiation | | antigen presentation+; ADCC–; phagocytosis+; lysozyme production+; mRNA: $\alpha$1-antitrypsin–, azurocidin+, C3 complement+, cathepsin G–, defensin–, N-elastase+, lactoferrin–, myeloblastin+, tryptase–; TF: GATA-1 mRNA–, GATA-2 mRNA–, SCL mRNA– | 24,35,58,59,62 |
| TK-1 | 36–48 h | EBNA– | ANAE+ (NaF inhibitable), CAE+, MPO+, MSE+, NBT– | TPA, Vit. D3 → mono/macro differentiation | | clonable; no phagocytosis | 36 |

*Continued on next page*

*Table 5.* (continued)

| Cell line | Doubling time | EBV status | Cytochemistry | Inducibility of differentiation | Heterotransplantability into mice | Special functional features | Ref. |
|---|---|---|---|---|---|---|---|
| U-937 | 30–40 h | EBV– HTLV-I– HIV– | ACP+, Alcian Blue(+), ALP–, ANAE+ (NaF inhibitable), ANBE+, CAE+, GLC+, MPO–, MSE(+), Oil Red O–, PAS+, SBB+, Toluidine Blue–, TRAP– | IFN-γ, retinoic acid, TPA, Vit. D3 → mono/macro differentiation | | ADCC+; phagocytosis+; no colony formation in agar; mRNA: α1-antitrypsin+, azurocidin–, C3 complement+, cathepsin G–, defensin–, N-elastase–, lactoferrin–, myeloblastin–, tryptase(+); lysozyme production | 24,38,59 |
| UG3 | 60–70 h | | ALP–, ANBE+, CAE+, MPO– | G-CSF → neutro differentiation; GM-CSF, M-CSF → macro differentiation; M-CSF + IL-4/IL-13 → osteoclast differentiation | | colony formation in methylcellulose (with cytokines) | 39,63 |
| YK-M2 | 60 h | EBNA– | ANBE+ (NaF inhibitable), CAE(+), MPO+, NBT– | Vit. D3 → mono/macro differentiation | | phagocytosis+ | 40 |

PHA-LCM – phytohemagglutinin-stimulated peripheral leukocytes conditioned medium; PO – (proto)-oncogenes; TF – transcription factors.

*Table 6.*  Monocytic cell lines: unconfirmed cell lines (not immortalized, not characterized, not verified, other)

| Cell line | Patient | Features | Ref. | Remarks[a] |
|---|---|---|---|---|
| 230 | from PB of 35 M with AML M4 in 1976 | CD2– HLA-DR+ sIg– TdT–; ACP+ ANBE(+) ALP– CAE– EBNA– MPO– PAS+ SBB– | 64 | insufficiently characterized; present status unknown |
| 2MAC[b] | normal donor | well-characterized normal macrophage cell line (immunoprofile, cytokines) | 65 | not leukemia cell line |
| AMoL I | from BM of child with AML M5 (at diagnosis) | CD3– CD11b– cy/sIg– HLA-DR+; ANAE+ (NaF inhibitable); no antigen presentation; no phagocytosis; produces IL-1 | 66 | insufficiently characterized; present status unknown |
| AMoL II | from BM of child with AML M5 (at diagnosis) | CD3– CD11b– cy/sIg– HLA-DR+; ANAE+ (NaF inhibitable); no antigen presentation; no phagocytosis; produces IL-1 | 66 | insufficiently characterized; present status unknown |
| DD | from PB of patient with malignant histiocytosis in leukemic phase | CD2– CD11a(+) CD14(+) CD16– CD19– CD32+ CD64– HLA-DR–; ACP– ANAE(+); LPS, TPA → mono/macro differentiation | 67 | insufficiently characterized |
| HBM-MI-1 and -2[b] | from BM of 1 M with diffuse cutaneous mastocytosis | established by SV40 transformation of parental, not immortal line HBM-M: CD11b(+) CD11c– CD13(+) HLA-DR–; ACP+ Alcian blue(+) ALP+ ANAE+ CAE+ MPO+ PAS+ Toluidine blue(+) | 68 | not leukemia cell lines |
| HL-92 | from PB of patient with AML M4 | CD2– CD11b+ HLA-DR+ sIg– TdT–; ACP(+) ANAE+ CAE+ EBNA– MPO– NBT+ SBB+; phagocytosis+; DMSO, retinoic acid → neutro differentiation; TPA → mono/macro differentiation | 69 | cell line apparently lost |
| J6-1 | | CD15+; M-CSFR mRNA+, TGF-$\beta$ mRNA+; produces M-CSF; GM-CSF, IL-3, M-CSF stimulate proliferation | 70 | insufficiently characterized |
| J-111 | from PB of 25 F with AML M5 | immunophenotypically, cytogenetically, etc. well characterized | 71 | cross-contaminated: in reality cell line HELA (ref. 72,73) |
| JOSK-I | from PB of 72 F with AML M4 in 1983 | immunophenotypically, cytogenetically, etc. well characterized | 74 | cross-contaminated: in reality cell line U-937 (ref. 24) |
| JOSK-K | from PB of 54 M with AML M5 in 1984 | immunophenotypically, cytogenetically, etc. well characterized | 74 | cross-contaminated: in reality cell line U-937 (ref. 24) |
| JOSK-M | from PB of 37 M with CML-mono BC in 1984 | immunophenotypically, cytogenetically, etc. well characterized | 74 | cross-contaminated: in reality cell line U-937 (ref. 24) |
| JOSK-S | from PB of 66 F with AML M5 in 1983 | immunophenotypically, cytogenetically, etc. well characterized | 74 | cross-contaminated: in reality cell line U-937 (ref. 24) |
| K1m[b] | from PB | well-characterized normal macrophage cell line (immunoprofile); phagocytosis+ | 75 | not leukemia cell line |
| KMT-2[b] | from normal umbilical cord blood | CD13+ CD15+ CD33+ CD34+ GlyA– HLA-DR+; ACP+ ANAE+ (NaF inhibitable) CAE– MPO– PAS+ SBB–; GM-CSF, IL-3, M-CSF stimulate proliferation | 76 | not leukemia cell line |
| KOCL-48 | from 0.5 F with first ALL L2, then AML M4 | CD2+ CD3– CD5+ CD7+ CD10+ CD13+ CD14+ CD15+ CD19+ CD22– CD33+ CD41b– HLA-DR+ cyIgM+ sIg–; ANBE+, MPO–; 50, X, t(4;11)(q21;q23) → *MLL-AF4* fusion gene; *IGH* RD, *IGK* G, *TCRB* G, *TCRG* RG | 77,78 | insufficiently characterized |
| MOBS-1 | | MSE+ | | cross-contaminated: in reality cell line U-937 |
| Na | from BM of 69 F with AML M2 → AML M4 | CD11c+ CD13+ CD14+; ANBE+ lysozyme+; t(8;17); *MYC* rearranged, amplified | 79 | insufficiently characterized |

*Continued on next page*

*Table 6.* (continued)

| Cell line | Patient | Features | Ref. | Remarks[a] |
|---|---|---|---|---|
| OCI-M3 | from PB of patient with AML M4 in 1986 | | 80 | insufficiently characterized; present status unknown |
| OCI/AML-3 | from PB of 57 M with AML M4 in 1987 | CD7+ CD8+ CD13+ CD14+ CD19+ CD33+ CD34(+) TdT+; retinoic acid receptor mRNA+; MSE+ | 81 | insufficiently characterized |
| OTC-4 | from PB of patient with AML M4 | G-CSF, GM-CSF inhibit proliferation, induce macro differentiation; TPA, vit. D3 → macro differentiation | 82 | insufficiently characterized |
| YAP[b] | from PB of 42 M with psoriasis vulgaris in 1993 | CD2– CD3– CD4+ CD8– CD11b+ CD14– CD16– CD19– CD25+ CD33+ CD56+ CD68–; ANBE+ (NaF inhibitable) MPO– PAS+; complex, highly rearranged karyotype; EBNA– SV40–; responsive to TPA | 83 | not leukemia cell line |

[a] For most cell lines, the insufficient characterization concerns the description of essential features, including clinical data, authentication, immunoprofile and/or cytogenetics. Present status unknown: no data on this cell line have been published since its original description.
[b] These cell lines were described to have monocytic/macrophage features, but were not derived from patients with leukemias or other hematological neoplasms.

at diagnosis (n = 9) or at relapse/refractory stage (n = 16). Patient material was obtained from the peripheral blood (n = 22), bone marrow (n = 3) or extra-hematological sites (n = 3). Cell lines were established (or when this was not indicated, at least described) in the 1970s (n = 3), 1980s (n = 23) and 1990s (n = 5). Authentication was provided for 14/31 (45%) cell lines. Three types of media are used to culture these cell lines, namely RPMI 1640 (n = 22), IMDM (n = 7) and $\alpha$-MEM (n = 2); most lines require addition of 10–20% FBS. Four cell lines could be cultured under serum-free conditions (AML-193, MV4-11, NOMO-1, PLB-985) (Appendix 2, p. 286). Cell lines AML-193, MUTZ-3 and UG3 are absolutely growth factor-dependent and will die in the absence of growth factors (Appendix 3, p. 287). Fourteen cell lines can be obtained from major cell banks (Appendix 1, p. 284). Subclones with variant immunological or cytogenetic features have been derived from the parental lines ME-1 and TK-1. Sister cell lines of ML-2, MOLM-13, and Mono Mac 6 were established from the same patient, but either at different time points, or the original material was split prior to or very early in the cell culture.

## 3. IMMUNOPHENOTYPE

Monocytic cell lines usually express the pan-myelomonocytic cell surface markers CD13 (20/25, 80%), CD15 (21/22, 95%), CD33 (22/24, 92%), and CD68 (5/7, 62%) (Table 2). A typical monocyte-associated cell surface antigen is CD14, found on 19/30 (63%) cell lines. It appears that expression of this marker is lost during *in vitro* culture of the originally CD14-positive primary AML cells [59]. The cell lines are commonly negative for typical B-cell, T-cell and erythroid-megakaryocytic-associated antigens, except for CD4 and CD7 which are also known to be expressed by normal monocytic precursor cells [86]. With regard to the progenitor cell markers CD34 and HLA-DR, 33% (5/15) and 69% (20/29) of lines, respectively, were described as positive. Finally, most cell lines display adhesion molecules CD11b and CD54. A typical immunoprofile of a monocytic leukemia cell line is: CD4± CD7± CD11b+ CD13+ CD14± CD15+ CD33+ CD34± CD54+ HLA-DR+, T-antigen negative, B-antigen negative, ery-meg-antigen negative.

## 4. CYTOKINE-RELATED CHARACTERIZATION

The cytokine which is uniquely effective on cells committed to monocyte differentiation is M-CSF. Several cell lines were described as responding proliferatively to this growth factor (IMS-M1, MUTZ-3, OCI/AML-2, UG3).

The related cytokines GM-CSF and IL-3 are effective on normal monocytic precursor cells and also induce the growth of various monocytic leukemia cell lines (AML-193, IMS-M1, ME-1, MUTZ-3, MV4-11, OCI/AML-2, OMA-AML-1, SKM-1, UG3) [44]. IFN-$\gamma$, known as the typical *in vitro* inducer of monocytic activation and differentiation, was also effective on monocytic cell lines. Three cell lines (AML-193, MUTZ-3, UG3) are absolutely growth factor-dependent on GM-CSF, IL-3 or PIXY-321 and will die within about one week in the absence of growth factor (Appendix 3, p. 287).

## 5. GENETIC CHARACTERIZATION

The monocytic cell lines display rather complex karyotypes (Table 4). Non-random unique translocations were found in several cell lines. Two cell lines (KBM-5, RWLeu-4) have the Philadelphia chromosome t(9;22)(q34;q11) which is typical for CML, the breakpoints lying in the major breakpoint cluster region (M-bcr) (Appendix 4, p. 288). Inversion (16)(p13q22), which is specific for the AML subtype M4eo, occurs in the cell line ME-1. Apart from these latter two alterations and the rare t(12;22) seen in cell line MUTZ-3, all other cell lines with specific translocations have a breakpoint at chromosome 11q23 involving the gene *MLL*: IMS-M1, ML-2, MOLM-13, Mono Mac 6, MV4-11, THP-1, U-937, UG3 (Appendix 4, p. 288). While 11q23-involving translocations were detected in various types of ALL and AML, monocytic AML commonly has this cytogenetic aberration [54]. Of note also are the deletions of the *P15INK4B*, *P16INK4A* and *P53* genes, the *NRAS* mutations, and the rearrangement of the *RB1* gene in various cell lines [87].

## 6. FUNCTIONAL CHARACTERIZATION

The monocytic cell lines have somewhat longer doubling times than other cell types, ranging from 1 to 4–5 days, mostly in the 24–48 h range (15/25, 60%) (Table 5). All cell lines tested were negative for EBV, HTLV-I/II and HIV (Table 5). With regard to the cytochemical profile of monocytic leukemia cell lines, the following results were reported: 8/10 (80%) ACP+, 0/5 ALP+, 26/26 ANAE/ANBE+, 13/21 (62%) CAE+, 11/25 (44%) MPO+, 17/21 (81%) MSE+, 2/7 (29%) NBT+, 9/17 (53%) PAS+, 2/5 (40%) SBB+, 2/4 (50%) TRAP+. The most prevalent monocytic enzyme marker is $\alpha$-naphthyl acetate or butyrate esterase [88], which is positive in all cell lines. The majority of this enzymatic activity stems from one unique isoenzyme, the so-called monocyte-specific esterase (MSE), detected in 80% of the cell lines tested. Importantly, this monocytic-specific activity can be inhibited by sodium fluoride [89–91].

Various agents can induce differentiation in human monocytic leukemia cells: including DMSO, IFN-$\gamma$, retinoic acid, TPA and vitamin D3 [92]. These biomodulators induce monocytic-macrophage differentiation in most cell lines examined (Table 5). U-937 and THP-1 cell lines provide two useful models for investigating the mechanisms of action of these agents [92–94]. Several cell lines can be heterotransplanted into nude or SCID mice (Table 5).

Functional features specific for monocytic cell lines, in concordance with their normal physiological counterparts [2], are phagocytosis (seen in 12 out of 15 cell lines tested), lysozyme production (in all six cell lines tested), and antigen presentation and antigen-dependent cell-mediated cytotoxicity. Unfortunately, the last two features, unique to monocytic cells, have rarely been examined.


## 7. UNCONFIRMED CELL LINES

The panel of cell lines requiring further characterization is large and diverse (Table 6). There are cell lines which are cross-contaminated with U-937 or the cervix carcinoma cell line HELA (the JOSK-series, J-111, MOBS-1). Several cell lines have interesting features and might be useful additions to the spectrum of available, well-characterized monocytic cell lines (e.g. OCI-M3, OCI/AML-3, OTC-4). The cell lines remain insufficiently characterized and in many cases their availability is not known and they might have been lost, including 230, AMOL I, AMOL II, DD, HL-92, J6-1, KOCL-48, Na, OCI-M3, OCI/AML-3. There are some cell lines with specific monocytic or macrophage properties which are not derived from patients with monocytic leukemias or hematological neoplasms, but were reported to be established from normal monocytic cells or macrophages (see references in Table 6). They seem to be representative *in vitro* models for this type of hematopoietic cell: 2MAC, HBM-MI-1/-2, K1m, KMT-2, YAP.

The osteoclast originates from an early precursor common to the mono-cytic and granulocytic lineages. However, tissue macrophages and monocytes isolated from the circulation significantly differ from osteoclasts in mor-phology, antigenic properties and other biological features. The cell line FLG29.1 (listed in Table 1 under monocytic leukemia cell lines) is quite unique in presenting monocyte-macrophage and osteoclast features [9]. Fol-lowing induction of differentiation with TPA, these cells lose monocytic char-acteristics and acquire more osteoclast-associated markers. Similarly, UG3 cells were described as acquiring osteoclast-like features upon treatment with cytokines [39,63].

# References

1. Bessis M (ed.) *Blood Smears Reinterpreted*, Springer, Berlin, 1977.
2. Roitt I, Brostoff J, Male D (eds.) *Immunology*, 3rd edn., Mosby, London, 1993.
3. Johnston RB Jr, Zucker-Franklin D. In: Zucker-Franklin D, Greaves MF, Grossi CE, Marmont AM (eds.), *Atlas of Blood Cells. Function and Pathology*, 2nd edn., Fischer, Stuttgart, pp 321–357, 1988.
4. Bennett JM, Catovsky D, Daniel MT et al. *Br J Haematol* 33: 451–458, 1976.
5. Bennett JM, Catovsky D, Daniel MT et al. *Br J Haematol* 51: 189–192, 1982.
6. Lange B, Valtieri M, Santoli D et al. *Blood* 70: 192–199, 1987.
7. Chen PM, Chiu CF, Chiou TJ et al. *Gann* 75: 660–664, 1984.
8. Sugita K, Sakakibara H, Eguchi M et al. *Int J Hematol* 55: 45–51, 1992.
9. Gattei V, Bernabei PA, Pinto A et al. *J Cell Biol* 116: 437–447, 1992.
10. Yamamoto K, Seto M, Iida S et al. *Blood* 83: 2912–2921, 1994.
11. Nagamura F, Tojo A, Tani K et al. *Blood* 86: 150a, 1995.
12. Andersson BS, Bergerheim USR, Collins VP et al. *Exp Hematol* 20: 361–367, 1992.
13. Beran M, Pisa P, O'Brien S et al. *Cancer Res* 53: 3603–3610, 1993.
14. Adachi N, Takaki K, Nakamura R et al. *Pediatr Res* 34: 258–264, 1993.
15. Ikushima S, Yoshihara T, Fujiwara F et al. *Acta Haematol Jpn* 53: 678–687, 1990.
16. Yanagisawa K, Horiuchi T, Fujita S. *Blood* 78: 451–457, 1991.
17. Yanagisawa K, Sato M, Horiuchi T et al. *Blood* 84: 84–93, 1994.
18. Herrmann R, Han T, Barcos MP et al. *Cancer* 46: 1383–1388, 1980.
19. Matsuo Y, MacLeod RAF, Uphoff CC et al. *Leukemia* 11: 1469–1477, 1997.
20. Ziegler-Heitbrock HWL, Thiel E, Fütterer A et al. *Int J Cancer* 41: 456,461, 1988.
21. Hu ZB, Ma W, Zaborski M et al. *Leukemia* 10: 1025–1040, 1996.
22. Kato Y, Ogura M, Okumura M et al. *Acta Haematol Jpn* 49: 277, 1986.
23. Wang C, Curtis JE, Minden MD et al. *Leukemia* 3: 264–269, 1989.
24. Drexler HG, Dirks W, MacLeod RAF et al. (eds.) *DSMZ Catalogue of Human and Animal Cell Cultures*, 7th edn., DSMZ, Braunschweig, 1999.
25. Pirruccello SJ, Jackson JD, Lang MS et al. *Blood* 80: 1026–1032, 1992.
26. Pirruccello SJ, Jackson JD, Sharp G. *Leukemia Lymphoma* 13: 169–178, 1994.
27. Hirose M, Minato K, Tobinai K et al. *Gann* 73: 735–741, 1982.
28. Nagai M, Seki S, Kitahara T et al. *Gann* 75: 1100–1107, 1984.
29. Tucker KA, Lilly MB, Heck L Jr et al. *Blood* 70: 372–378, 1987.
30. Bradley TR, Pilkington G, Garson M et al. *Br J Haematol* 51: 595–604, 1982.
31. Lasky SR, Bell W, Huhn RD et al. *Cancer Res* 50: 3087–3094, 1990.
32. Wiemann M, Hollmann A, Posner M et al. *Clin Res* 33: 460A, 1985.
33. Kimura Y, Toki H, Okabe K et al. *Jpn J Cancer Res* 77: 862–865, 1986.
34. Nakagawa T, Matozaki S, Murayama T et al. *Br J Haematol* 85: 469–476, 1993.
35. Tsuchiya S, Yamabe M, Yamaguchi et al. *Int J Cancer* 26: 171–176, 1980.
36. Ohno H, Doi S, Fukuhara S et al. *Int J Cancer* 37: 761–767, 1986.
37. Nosaka T, Ohno H, Doi S et al. *J Clin Invest* 81: 1824–1828, 1988.
38. Sundström C, Nilsson K. *Int J Cancer* 17: 565–577, 1976.
39. Ikeda T, Saski K, Ikeda K et al. *Blood* 91: 4543–4553, 1998.
40. Ohno H, Fukuhara S, Doi S et al. *Cancer Res* 46: 6400–6405, 1986.

41. Santoli D, Yang YC, Clark SC et al. *J Immunol* 139: 3348–3354, 1987.
42. Taetle R, Payne C, Dos Santos B et al. *Cancer Res* 53: 3386–3393, 1993.
43. Lyons AB, Ashman LK. *Leukemia Res* 11: 797–805, 1987.
44. Drexler HG, Zaborksi M, Quentmeier H. *Leukemia* 11: 701–708, 1997.
45. Schumann RR, Nakarai T, Gruss HJ et al. *Blood* 87: 2419–2427, 1996.
46. Blick MB, Andersson BS, Gutterman JU et al. *Leukemia Res* 10: 1401–1409, 1986.
47. Liu P, Claxton DF, Marlton P et al. *Blood* 82: 716–721, 1993.
48. Tanabe S, Zeleznik-Le NJ, Kobayashi H et al. *Genes Chromosomes Cancer* 15: 206–216, 1996.
49. Nakamaki T, Kawamata N, Schwaller J et al. *Br J Haematol* 91: 139–149, 1995.
50. Ohyashiki K, Ohyashiki JH, Sandberg AA. *Cancer Res* 46: 3642–3647, 1986.
51. Ohyashiki JH, Ohyashiki K, Toyama K et al. *Cancer Genet Cytogenet* 37: 193–200, 1989.
52. MacLeod RAF, Voges M, Drexler HG. *Blood* 82: 3221–3222, 1993.
53. MacLeod RAF, Hu ZB, Kaufmann M et al. *Genes Chromosomes Cancer* 16: 144–148, 1996.
54. Drexler HG, MacLeod RAF, Borkhardt A et al. *Leukemia* 9: 480–500, 1995.
55. Dreyling MH, Martinez-Climent JA, Zheng M et al. *Proc Natl Acad Sci USA* 93: 4804–4809, 1996.
56. Yanagisawa K, Hasegawa H, Fujita S. *Br J Haematol* 80: 293–297, 1992.
57. Takeda K, Minowada J, Bloch A. *Cancer Res* 42: 5152–5253, 1982.
58. Shimamoto T, Ohyashiki K, Ohyashiki JH et al. *Blood* 86: 3173–3180, 1995.
59. Abrink M, Gobl AE, Huang R et al. *Leukemia* 8: 1579–1584, 1994.
60. Hamaguchi M, Morishita Y, Takahashi I et al. *Blood* 77: 94–100, 1991.
61. Kawaguchi R, Hosokawa Y, Komine A et al. *Leukemia* 6: 1296–1301, 1992.
62. Tsuchiya S, Kobayashi Y, Goto Y et al. *Cancer Res* 42: 1530–1536, 1982.
63. Ikeda T, Ikeda K, Sasaki K et al. *Biochem Biophys Res Commun* 253: 265–272, 1998.
64. Karpas A, Khalid G, Burns GF et al. *Br J Cancer* 37: 308–315, 1978.
65. Dialynas DP, Tan PC, Huhn GD et al. *Cell Immunol* 177: 182–193, 1997.
66. Giller RH, Mori M, Hayward AR. *J Clin Immunol* 4: 429–438, 1984.
67. Kávai M, Ádány R, Pásti G et al. *Cell Immunol* 139: 531–540, 1992.
68. Townsend M, MacPherson J, Krilis S et al. *Br J Haematol* 85: 452–461, 1993.
69. Salahuddin SZ, Markham PD, McCredie KB et al. *Leukemia Res* 6: 729–741, 1982.
70. Wu KF, Rao Q, Zheng GG et al. *Leukemia Res* 18: 843–849, 1994.
71. Osgood EE, Brooke JH. *Blood* 10: 1010–1022, 1955.
72. Drexler HG, Häne B, Hu ZB et al. *Leukemia* 7: 2077–2078, 1993.
73. Nelson-Rees WA, Flandermeyer RR. *Science* 191: 96–97, 1976.
74. Ohta M, Furukawa Y, Ide C et al. *Cancer Res* 46: 3067–3074, 1986.
75. Dialynas DP, Lee MJ, Shao LE et al. *Blood* 88: 158a, 1996.
76. Tamura S, Sugawara M, Tanaka H et al. *Blood* 76: 501–507, 1990.
77. Iida S, Saito M, Okazaki T et al. *Leukemia Res* 16: 1155–1163, 1992.
78. Kojika S, Sugita K, Inukai T et al. *Leukemia* 10: 994–999, 1996.
79. Kimura Y, Miura I, Hashimoto K et al. *Blood* 76: 289a, 1990.
80. Tweeddale M, Jamal N, Nguyen A et al. *Blood* 74: 572–578, 1989.
81. Wang C, Koistinen P, Yang GS et al. *Leukemia* 5: 493–499, 1991.
82. Ono Y, Iwasaki H, Otsuka T et al. *Blood* 84: 574a, 1994.
83. Hamamoto Y, Nagai K, Muto M et al. *Arch Dermatol Res* 288: 225–229, 1996.
84. Lübbert M, Koeffler HP. *Blood Rev* 2: 121–133, 1988.
85. Lübbert M, Herrmann F, Koeffler HP. *Blood* 77: 909–924, 1991.

86. Drexler HG, Thiel E, Ludwig WD. *Leukemia* 7: 489–498, 1993.
87. Drexler HG. *Leukemia* 12: 845–859, 1998.
88. Hayhoe FGJ, Flemans RJ (eds.) *A Colour Atlas of Haematological Cytology*, 3rd edn., Wolfe, London, 1992.
89. Uphoff CC, Gignac SM, Metge K et al. *Leukemia* 7: 58–62, 1993.
90. Uphoff CC, Hu ZB, Gignac SM et al. *Leukemia* 8: 1510–1526, 1994.
91. Gignac SM, Hu ZB, Denkmann SA et al. *Leukemia Lymphoma* 22: 143–151, 1996.
92. Koeffler HP. *Blood* 62: 709–721, 1983.
93. Koeffler HP. *Semin Hematol* 23: 223–236, 1986.
94. Harris P, Ralph P. *J Leuk Biol* 37: 407–422, 1985.

# Chapter 8

# Erythroid – Megakaryocytic Cell Lines

Hans G. Drexler
*DSMZ-German Collection of Microorganisms and Cell Cultures, Department of Human and Animal Cell Cultures, Braunschweig, Germany; Tel: +49-531-2616-160; Fax: +49-531-2616-150; E-mail: hdr@dsmz.de*

## 1. INTRODUCTION

For reasons outlined below, leukemia cell lines derived from erythroid or megakaryocytic cells are combined in this chapter. These cell lines represent *in vitro* models of immature erythroid and megakaryocytic cells and their distinct or common precursors.

Morphologically, the erythrocytic series consists of a succession of cells which begins with a pronormoblast and ends with the erythrocyte (or red blood cell) [1]. These circulating red cells and their precursors may be considered as a functional unit which has been designated as the erythron. The cells of this unit are not restricted to those recognizable morphologically in the bone marrow and in the peripheral blood, but also include elements which are not morphologically identifiable, namely the committed precursors of the erythroid line, the existence of which has been demonstrated by functional assays [2].

The thrombocytic series is a succession of cells which starts with the basophilic megakaryoblast in the bone marrow and ends with the circulating thrombocyte or platelet [1]. In normal human bone marrow, the megakaryocyte is the largest cell, measuring 20–150 $\mu$m in diameter. The cells are polyploid, but not multinucleated, in contrast to the other giant bone marrow-derived cells, the osteoclasts. Normal megakaryopoiesis is maintained by morphologically unidentifiable precursors which are capable of differentiating into morphologically recognizable megakaryoblasts or of reproducing themselves [3].

As both erythrocytes and thrombocytes are terminally differentiated, anucleated end-stage cells, immature progenitor and precursor cells are considered to be the targets in the leukemogenic process. The acute myeloid leukemias originating in these cell lineages have been termed AML M6

259

(erythroid) and AML M7 (megakaryocytic) [4,5]. Erythroid and/or mega-karyocytic cells can also represent a subpopulation of the main leukemic population in the myeloid blast crisis of CML [6].

Over the last two decades, a large panel of erythroid-megakaryocytic leukemia cell lines has been established from patients with AML, CML in blast crisis or rare hematological disorders [7–9]. The features displayed by a cell line assigned to this category are not single lineage-specific, but in most instances extend to both lineages. Indeed, data from both normal and malignant cells support this notion. Multiple lines of evidence underscore the concept of a close relationship between erythroid and megakaryocytic lineages [10]. The two share a number of transcription factors (including NF-E2, GATA-1, GATA-2, SCL) [11–13]. Furthermore, erythroid and mega-karyocytic cell surface markers (including GlyA, CD41, CD42, CD61) are found on both types of leukemia cell and this dual expression is found on the same cell [14–17]. The cytokines EPO and TPO, originally considered to be cell lineage-specific, also have stimulatory effects on the megakaryocytic cell system and the erythron, respectively [18–20]. The receptor for EPO, which is the principal growth factor regulating the production of erythrocytes, has also been detected on megakaryocytes [21]. A bipotent normal erythroid-megakaryocytic progenitor could be isolated from human bone marrow [22]. Finally, most of the "erythroid" cell lines available display, or can be induced to display, features of megakaryocytic differentiation. The converse is true for cell lines initially thought to be exclusively megakaryocytic.

While a given cell line may show a preponderance of erythroid or megaka-ryocytic features, the notion of a close lineage relationship, extended here to the assignment of such cell lines to a common category, is borne out by the extensive published data, summarized here in Tables 1–6. Most erythroid-megakaryocytic cell lines were established in the 1980s (49%) and 1990s (47%), with K-562 being the oldest cell line [45].

## 2. CLINICAL CHARACTERIZATION

Forty-nine cell lines with erythroid and megakaryocytic characteristics are listed in Table 1. These cell lines were derived mainly from patients with CML in blast crisis (41%), *de novo* or secondary AML M6 (14%) or AML M7 (33%). It is of note that 4/7 AML M6 and 8/16 AML M7 cases were either secondary to or accompanied by other hematological disorders or (pre)malignancies. The ages of the patients ranged from 0.5 to 73 years with 3 infants (<1 year) and 10 other children. Specimens were obtained at diagnosis (n = 12), at relapse (n = 11) or in blast crisis (n = 20). Taking CML blast crisis as an indicator of relapse, then 63% of these cell lines were

*Table 1.* Erythroid-megakaryocytic cell lines: clinical characterization

| Cell line[a] | Patient[b] age/sex | Diagnosis[c] | Treatment status[d] | Specimen site[e] | Authentication[f] | Year est. | Culture medium[g] | Availability[h] | Primary Ref. |
|---|---|---|---|---|---|---|---|---|---|
| AP-217 | 42 M | CML-BC | BC | PB | yes | 1992 | RPMI 1640 + FBS | Author | 23 |
| AS-E2 | 62 M | AML M6 | R | BM | yes | 1993 | IMDM + FBS + EPO | Author | 24 |
| B1647 | 14 M | AML M2 | | BM | no | | IMDM + FBS | | 25 |
| CHRF-288-11[i] | 2 M | AML M7 (+ myelofibrosis) | T | Tu | yes | 1988 | Fischer + HS | Author | 26,27 |
| CMK[j] | 1 M | AML M7 (+ Down's syndrome) | R | PB | yes | 1985 | RPMI 1640 + FBS | DSMZ, IFO | 28–30 |
| CMS | 2 F | granulosarcoma → AML M7 | R | PB | yes | 1991 | RPMI 1640 + FBS | | 31 |
| CMY | 2 M | AML M7 (+ Down's syndrome) | T | BM | no | 1991 | RPMI 1640 + FBS | | 32 |
| ELF-153 | 48 M | Myelofibrosis → AML M7 | R | BM | no | 1988 | RPMI 1640 + FBS + GM-CSF | Author | 33,34 |
| F-36P[k] | 68 M | MDS (RAEB) → AML M6 | D | PE | no | 1989 | RPMI 1640 + FBS + GM-CSF or IL-3 | RIKEN | 35 |
| GRW | 4 F | AML M7 (+ Down's syndrome) | R | PB | yes | 1992 | α-MEM + FBS + SCF | Author | 36 |
| HEL | 30 M | Hodgkin → AML M6 (post-BMT) | R | PB | yes | 1980 | RPMI 1640 + FBS | ATCC, DSMZ, JCRB | 37 |
| HIMeg-1[l] | 25 F | CML | CP | PB | no | | IMDM/RPMI 1640 + FBS | Author | 38 |
| HML | 2 M | AML M7 (+ Down's syndrome) | D | PB | no | | α-MEM + FBS + GM-CSF | | 39 |
| HU-3[m] | 69 F | AML M7 | D | BM | no | 1991 | RPMI 1640 + human serum + GM-CSF | Author | 40–42 |
| JK-1 | 62 M | CML-ery BC | BC | Tu | no | 1987 | RPMI 1640 + FBS or human serum | DSMZ | 43 |
| JURL-MK1[n] | 73 M | CML-BC | BC | PB | no | 1993 | DMEM + FBS | Author | 44 |
| K-562 | 53 F | CML-BC | BC | PE | yes | 1970 | RPMI 1640 + FBS | ATCC, DSMZ, JCRB, RIKEN | 45 |
| KH184 | 25 M | Megakaryocytic sarcoma | R | PB | no | 1984 | RPMI 1640 + FBS | Author | 46 |
| KH88[o] | 70 M | CML-ery BC | BC | PB | yes | 1988 | IMDM + FBS | Author | 47 |
| KMOE-2[p] | 2 F | Acute erythremia (erythroblastosis) | D | PB | no | 1978 | Ham's F12 + FBS | DSMZ | 48 |
| LAMA-84[q] | 29 F | CML-my/meg BC | BC | PB | yes | 1984 | RPMI 1640 + FBS | DSMZ | 49,50 |

*Continued on next page*

Table 1. (continued)

| Cell line[a] | Patient[b] age/sex | Diagnosis[c] | Treatment status[d] | Specimen site[e] | Authentication[f] | Year est. | Culture medium[g] | Availability[h] | Primary Ref. |
|---|---|---|---|---|---|---|---|---|---|
| M-07e[r] | 0.5 F | AML M7 | D | PB | no | 1987 | IMDM + FBS + IL-3 | DSMZ | 42,51,52 |
| M-MOK[s] | 1 F | AML M7 | R | BM | no | 1989 | RPMI 1640 + FBS + GM-CSF | Author | 42,53 |
| MB-02 | 70 M | Myelofibrosis + myeloid metaplasia → AML M7 | D | PB | yes | 1988 | RPMI 1640 + human serum + GM-CSF | Author | 54 |
| MC3 | 51 F | CML-meg BC | BC | PB | yes | 1986 | RPMI 1640 + FBS | Author | 55 |
| MEG-01[t] | 55 M | CML-meg BC | BC | BM | no | 1983 | RPMI 1640 + FBS | ATCC, DSMZ, IFO | 56,57 |
| MEG-A2 | 24 M | CML-meg BC | BC | PB | yes | 1991 | IMDM + FBS | Author | 58 |
| MG-S | 3 M | AML M0 (+ Down's syndrome) | R | PB | no | 1991 | RPMI 1640 + FBS | | 59 |
| MHH 225 | 60 M | AML M7 | D | BM | yes | 1993 | RPMI 1640 + FBS or serum-free | Author | 60,61 |
| MKPL-1 | 66 M | AML M7 | D | BM | yes | 1989 | RPMI 1640 + FBS | Author | 62 |
| MOLM-1 | 41 M | CML-BC | BC | BM | no | 1988 | RPMI 1640 + FBS | Author | 63 |
| MOLM-7[u] | 44 M | CML-my BC | BC | PB | no | 1992 | RPMI 1640 + FBS | Author | 64 |
| M-TAT | 3 M | MDS (RAEBT) | R | PB | no | 1992 | RPMI 1640 + FBS + GM-CSF | | 65 |
| NS-Meg | 44 F | CML-meg BC | BC | PB | yes | 1992 | RPMI 1640 + FBS | Author | 66 |
| OCIM1 | 62 M | CLL → AML M6 | | | no | 1984 | IMDM + FBS | Author | 67 |
| OCIM2 | 56 M | MDS → AML M6 | | | no | 1984 | IMDM + FBS | Author | 67 |
| RM10 | 44 F | CML-BC | BC | BM | no | 1982 | IMDM or RPMI 1640 + FBS | Author | 68 |
| RS-1 | 2 F | AML M7 | R | PB | no | 1985 | RPMI 1640 + FBS | Author | 69 |
| SAM-1 | 22 M | CML-BC | BC | PB | yes | 1987 | RPMI 1640 + FBS | Author | 70 |
| SET-2 | 71 F | essential thrombocythemia (leukemic conversion) | D | PB | yes | 1995 | DMEM + FBS | Author | 71 |
| SKH1 | 62 F | CML-meg BC | BC | PB | no | 1985 | RPMI 1640 + FBS | Author | 72 |
| T-33 | 62 F | CML-meg BC | BC | PB | no | 1987 | RPMI 1640 + FBS or serum-free | Author | 73 |
| TF-1[v] | 35 M | AML M6 | D | BM | no | 1987 | RPMI 1640 + FBS + GM-CSF | ATCC, DSMZ | 42,74,75 |
| TS9;22 | 56 M | CML-BC | BC | PB | yes | 1992 | RPMI 1640 + FBS | Author | 76 |

*Continued on next page*

*Table 1.* (continued)

| Cell line[a] | Patient[b] age/sex | Diagnosis[c] | Treatment status[d] | Specimen site[e] | Authentication[f] | Year est. | Culture medium[g] | Availability[h] | Primary Ref. |
|---|---|---|---|---|---|---|---|---|---|
| UoC-M1 | 68 M | AML M1 | D | BM | yes | 1992 | McCoy's 5A + FBS | | 77 |
| UT-7[w] | 64 M | AML M7 | D | BM | no | 1988 | IMDM + FBS + GM-CSF or IL-3 or EPO | DSMZ | 78–81 |
| Y-1K | 57 F | CML-ery BC | BC | PB | no | 1990 | IMDM + FBS | | 82 |
| YN-1 | 35 M | CML-ery BC | BC | PB | no | 1980 | IMDM + FBS | | 82 |
| YS9;22 | 23 F | CML-BC | BC | PB | yes | 1992 | RPMI 1640 + FBS | Author | 76 |

[a] Names of cell lines are indicated as given in the original literature; subclones (variant cell lines derived from a parental cell line) and sister cell lines (derived independently from the same patient from different specimens or at different time points) are indicated for each cell line. [b] Age at the time of establishment of cell line. [c] Diagnoses are indicated as given in the original literature; → : disease progressed from a pre-malignancy/first malignancy into the final malignancy. [d] BC – at blast crisis; CP – in the chronic phase; D – at diagnosis (prior to therapy); R – at relapse; T – during therapy. [e] BM – bone marrow; PB – peripheral blood; PE – pleural effusion; Tu – tumor. [f] Evidence (e.g. cytogenetic marker chromosomes, immunoprofile, others) that this cell line was derived from the patient indicated. [g] Culture media as indicated in the original literature; cell line might also grow with other medium and/or supplements. [h] Availability from cell banks (ATCC; DSMZ; IFO; JCRB; RIKEN) or from the original investigator (author) [i] Cell line was established in 1985 from a solid tumor line (CHRF-288) that was passaged by heterotransplantation into nude mice. [j] CMK6 (poorly differentiated) and CMK11-5 (well differentiated) are subclones established from CMK. [k] A subclone termed F-36E is EPO-dependent/responsive and is grown with 20 U/ml EPO or 1 ng/ml GM-CSF or 1 ng/ml IL-3. [l] This subclone termed HIMeg-1 was established from the parental line HIMeg. [m] A TPO-dependent subclone termed HU-3/TPO was established. [n] A subclone termed JURL-MK2 was established. [o] Two subclones KH88 B4D6 and KH88 C2F8 with slightly different immunophenotypical and cytogenetic features were established. [p] Sister cell lines KMOE-1 and KMOE-3N from the same patient were established from the bone marrow in suspension culture and by heterotransplantation of KMOE-1 cells into an athymic nude mouse, respectively. [q] Subclones LAMA-87 (with an erythroid-eosinophilic phenotype) and LAMA-88 (with an eosinophilic-monocytic phenotype) were established by heterotransplantation of LAMA-84 into athymic mice. [r] This subclone termed M-07e was established from the parental growth factor-independent line M-07 at passage 15; TPO-dependent subclone M-07e/TPO was established. [s] A TPO-dependent subclone M-MOK/TPO was established. [t] A slightly different subclone named MEG-01s was established. [u] Seven cell lines from a single sample from the same patient were established: MOLM-7 and MOLM-11 represent the megakaryocytic lineage; MOLM-6, –8, –9, –10, –12 represent the myelocytic lineage. [v] Originally termed MFD-1; a TPO-dependent subclone TF-1/TPO was established. [w] EPO- and TPO-dependent subclones UT-7/Epo and UT-7/TPO were established.

*Table 2.*   Erythroid-megakaryocytic cell lines: immunophenotypic characterization

| Cell line | T-/NK-cell marker[a] | B-cell marker | Myelomonocytic marker | Erythroid-megakaryocytic marker | Progenitor/activation marker | Adhesion marker | Ref. |
|---|---|---|---|---|---|---|---|
| AP-217 | CD2− CD3− | CD19− CD22− | CD15+ CD33+ | CD36+ CD41a+ GlyA(+) | CD34− HLA-DR+ | CD11a+ CD11b+ | 23 |
| AS-E2 | CD2− CD3− | CD10− CD19− | CD13− CD14− CD33− | CD36+ CD41− GlyA+ | CD34− CD38+ CD71+ HLA-DR− | CD11b(+) | 24 |
| B1647 | CD3− CD4− CD8− CD56− CD57− | CD19− CD20− CD23− | CD14− CD16− CD33+ | CD41+ CD42b− GlyA+ vWF+ | CD34− CD38+ HLA-DR+ | CD11c− CD54+ | 25 |
| CHRF-288-11 | CD1− CD2− CD3− CD4− CD5− CD7− CD8− CD56− | CD10− CD19− CD20− | CD13+ CD14− CD33+ | CD36+ CD41a+ GlyA− PPO+ vWF+ | HLA-DR+ | | 27 |
| CMK[b] | CD1− CD2− CD3− CD4(+) CD7(+) CD8− CD56− | CD10− CD19− CD20− CD21− CD22− | CD13+ CD14(+) CD15− CD33+ | CD36+ CD41a+ CD42b(+) GlyA+ PPO+ vWF+ | CD34+ CD38− CD71+ HLA-DR+ | | 28,30 |
| CMS | CD1− CD2− CD3− CD4− CD5− CD7− CD8− | CD19− | CD13− CD33+ | CD41+ CD42b(+) GlyA− PPO− | CD34+ CD38− HLA-DR+ | | 31 |
| CMY | CD3− CD4+ CD7+ CD8− | CD10− CD19− CD20− | CD13+ CD14− CD33+ | CD41a+ GlyA− PPO+ | CD34+ | | 32 |
| ELF-153 | CD2− CD3− CD4+ | CD10− CD19− CD20− | CD13+ CD14− CD15− CD33+ | CD9− CD31+ CD36− CD41+ CD42a− CD42b− CD61+ CD62− CD63+ CD107a+ CD107b+ GlyA− PPO− vWF(+) | CD34+ CD38− HLA-DR+ | CD11b− CD11c− CD51+ | 33 |
| F-36P | CD4− CD5− CD8− | CD10− CD19− | CD13+ CD14− CD33+ | CD41a+ CD42b− GlyA+ PPO(+) | CD34+ | | 35 |
| GRW | CD3− CD4− CD7+ | CD10− CD19− sIg− | CD14− CD33+ | CD41a+ CD42b(+) CD61+ CD62P− GlyA− | CD34+ CD38+ CD71+ HLA-DR+ | | 36 |
| HEL | CD1− CD2− CD3− CD4+ CD5− CD7− CD8− CD28− CD56− TCRαβ− | CD10− CD19− CD20− CD21(+) CD22− Ig− | CD13+ CD14− CD15+ CD33+ | CD9+ CD36+ CD41a+ CD42a− CD42b(+) CD61+ CD62− GlyA+ PPO− vWF(+) | CD34− CD38− CD71+ HLA-DR(+) | CD11b+ CD44+ | 83,84 |
| HIMeg-1 | | | CD14− CD16+ CD33+ | CD41a(+) CD41b− GlyA+ PPO+ | CD34− | CD11c+ | 38 |
| HML | CD2− CD3− CD4+ CD8− | CD19− | CD13+ CD14+ CD15− CD33+ | CD36+ CD41b+ CD42b− GlyA+ vWF− PPO(+) | CD34− CD38− CD71+ HLA-DR+ | CD11b(+) | 39 |
| HU-3 | CD1a− CD2− CD3− CD4+ CD7− CD8− | CD10− CD19− CD21+ CD23− | CD13+ CD14+ CD15+ CD16− CD32+ CD33+ CD35+ | CD31+ CD36+ CD41a+ CD41b− CD42a− CD42b− CD61+ GlyA+ | CD34+ CD38+ CD71+ HLA-DR+ | CD11a− CD11b+ CD11c− CD18− CD44+ CD49b− CD49d+ CD49e+ CD49f+ CD51− CD54+ CD58+ CD11b− | 40,41 |
| JK-1 | CD2− CD3− CD4+ CD8− | CD10− CD19− CD20− CD21− | CD13+ CD14− | CD41+ CD42+ GlyA+ | HLA-DR− | CD11b− | 43,83 |
| JURL-MK1 | CD2− CD7− CD56− | CD19− CD24− CD40− | CD13+ CD33+ | CD9− CD36+ CD41a+ CD42b− CD61+ CD62P− CD63+ GlyA+ | CD30− CD34− CD69+ HLA-DR+ | CD11a− CD11b− CD11c− CD43+ CD44− CD54− CD62L− | 44 |

*Continued on next page*

*Table 2.* (continued)

| Cell line | T-/NK-cell marker[a] | B-cell marker | Myelomonocytic marker | Erythroid-megakaryocytic marker | Progenitor/activation marker | Adhesion marker | Ref. |
|---|---|---|---|---|---|---|---|
| K-562 | CD1– CD2– CD3– CD4– CD5– CD7– CD8– CD28– CD56– CD57– TCRαβ/δ– | CD10– CD19– CD20– CD21– CD22– CD80+ CD86– | CD13+ CD14– CD15+ CD33– | CD9+ CD36– CD41(+) CD42b– CD61(+) GlyA+ PPO– vWF– | CD34– CD38– CD71+ HLA-DR– TdT– | CD11b(+) CD44+ CD54+ | 28,83 |
| KH184 | CD2– CD4– CD5– CD7– | CD10– CD19– CD20– | CD13(+) CD14– CD33+ | CD36(+) CD41a+ CD42b+ GlyA– PPO– | CD34– HLA-DR(+) | | 46 |
| KH88 | CD1– CD2– CD3– CD4– CD8– CD56– | CD10– CD19– CD20– CD21– CD22– | CD13– CD14– CD15– CD33– | CD36+ CD41b– CD42– CD61– GlyA– PPO+ | CD34– CD38– CD71+ HLA-DR– | CD11b+ | 28,47 |
| KMOE-2 | CD3– | CD19– cy/sIg– | CD13+ | GlyA+ | TdT– | | 48,83 |
| LAMA-84[c] | CD2– CD3– CD4– CD5– CD7+ CD8– CD56– | CD10– CD19– CD20+ CD21– CD22– CD23– CD24+ CD37+ CD40+ | CD13+ CD14– CD15– CD16– CD17+ CD32+ CD33+ CD35+ CD65– CD66a– CD66b– CD68+ | CD9– CD31+ CD41a+ CD41b+ CD42b– CD61+ CD63+ GlyA– PPO+ vWF– | CD34– CD69+ CD71+ HLA-DR+ TdT– | CD11a+ CD11b+ CD11c– CD43+ CD44+ CD54– | 49,50,83,85 |
| M-07e[d] | CD3– | CD19– | CD13+ CD14– CD15(+) CD33+ | CD41a+ CD42b+ GlyA– PPO(+) | CD34(+) CD71+ HLA-DR– | CD44+ | 83,86 |
| M-MOK | CD1– CD3– CD4– CD7– CD8– | CD10– CD19– CD20– | CD13+ CD14– CD33+ | CD41b+ CD42b+ GlyA– PPO– | CD34+ HLA-DR– | CD11a+ CD11b– CD11c– CD18+ CD54+ | 53 |
| MB-02 | CD1a– CD2– CD3– CD4– CD7– CD8– | CD10– CD19– CD20– CD21+ CD23– | CD13+ CD14– CD15– CD16– CD32+ CD33+ CD35+ | CD31+ CD36+ CD41a– CD41b– CD42a– CD42b– CD61+ GlyA+ | CD34+ CD38+ CD45+ CD71+ HLA-DR+ | CD11a+ CD11b+ CD11c– CD18+ CD44+ CD49b– CD49d+ CD49e+ CD49f– CD51– CD54+ CD58+ | 54 |
| MC3 | CD3– CD7– | CD10– CD19+ CD20– cyIgM– | CD13+ CD14– CD33+ | CD41a+ CD42b+ | CD34+ CD38+ HLA-DR– | CD54+ | 55,87 |
| MEG-01 | CD1– CD2– CD3– CD4+ CD5– CD7– CD8– CD28– CD56– CD57(+) | CD10– CD19– CD20– CD21+ CD22– | CD13+ CD14– CD15+ CD16– CD33+ CD68– | CD9+ CD36+ CD41a+ CD42a– CD42b(+) CD61+ CD62P– GlyA– PPO+ vWF+ | CD34– CD38– HLA-DR+ TdT– | CD11b+ CD44+ | 28,57,84 |
| MEG-A2 | CD1– CD2– CD3– CD4+ CD5– CD7+ CD8– | CD10– CD19– CD20– | CD13+ CD14– CD33+ | CD41a+ CD42b– GlyA– PPO– vWF– | CD34+ HLA-DR+ | | 58 |
| MG-S | CD1a– CD2– CD3– CD4+ CD5– CD7+ CD8– CD56– CD57– | CD10– CD19– CD20– CD21– cyCD22+ CD24+ | CD13+ CD14– CD15(+) CD16– CD33+ | CD36+ CD41a+ CD42b– GlyA– PPO+ | CD34+ CD38+ HLA-DR+ TdT+ | CD11b+ CD11c– | 59 |
| MHH 225 | CD2– CD3– CD4– CD5– CD7– CD8– | CD10– CD19– CD20– CD21– | CD13+ CD15+ CD33+ CD65+ | CD41+ CD42b– CD62+ GlyA+ PPO+ | CD34+ CD38– HLA-DR+ TdT– | CD11a– CD11b– CD11c– CD54– | 60,61 |
| MKPL-1 | CD3– CD4– CD5– CD8– | CD10– CD19– CD20– | CD13+ CD33+ | CD36+ CD41a+ CD42b– CD61+ GlyA– PPO– vWF– | CD30– | | 62 |
| MOLM-1 | CD1– CD2– CD3– CD4(+) CD5– CD7+ CD8– CD57– | CD10– CD19– CD20– sIg– | CD13+ CD14– CD15– CD33+ | CD9+ CD41(+) CD42a– CD42b– CD61+ CD62(+) vWF– | CD34+ CD71+ HLA-DR+ TdT– | CD44+ | 63 |

*Continued on next page*

*Table 2.* (continued)

| Cell line | T-/NK-cell marker[a] | B-cell marker | Myelomonocytic marker | Erythroid-megakaryocytic marker | Progenitor/activation marker | Adhesion marker | Ref. |
|---|---|---|---|---|---|---|---|
| MOLM-7 | CD3− CD4+ CD7+ CD8− | CD10− CD19− CD20− CD24− | CD13+ CD14− CD15− CD33+ | CD9+ CD41a+ CD42a− CD42b+ CD61+ CD62− vWF+ GlyA+ | CD34+ CD71+ HLA-DR+ | CD11a(+) CD44+ | 64,84 |
| M-TAT | | | | | | | 65 |
| NS-Meg | CD3− CD4+ CD7+ CD56− | CD10+ CD19− | CD13+ CD14− CD33+ | CD36+ CD41a+ CD42a− CD61+ CD62− GlyA+ PPO+ GlyA+ | CD34+ HLA-DR+ | CD11b+ CD18+ | 66 |
| OCIM1 | CD1− CD2− CD3− CD4− CD5− CD7− CD8− | CD10− CD19− sIg− | CD13+ CD15− CD33+ | CD36+ CD41(+) CD42b− GlyA+ vWF− | CD34+ CD38− HLA-DR+ TdT− | | 67 |
| OCIM2 | CD1− CD2− CD3− CD4− CD5− CD7− CD8− | CD10− CD19− sIg− | CD13+ CD15− CD33+ | CD36+ CD41+ CD42b− GlyA+ vWF+ | CD34+ CD38− HLA-DR− TdT− | | 67 |
| RM10 | CD1a− CD2− CD3− CD4− CD8− CD57− | CD10+ CD19− sIg− | CD13+ CD14+ CD16− CD33+ | CD41− CD42− GlyA+ vWF− | HLA-DR+ | CD11b− | 68 |
| RS-1 | CD2− | CD20− | CD13+ CD33+ | CD41+ CD61+ | CD34+ | CD11b− | 69 |
| SAM-1 | CD2− CD3− CD4− CD5− CD8− CD57− | CD19− CD20− | CD13− CD33− | CD41+ CD42b+ GlyA+ PPO+ | CD34− CD38− HLA-DR− | CD11b− | 70 |
| SET-2 | CD2− CD3− CD4+ CD5− CD7+ | CD10− CD19− CD20− | CD13+ CD14− CD33+ | CD9− CD36+ CD41+ CD42b− CD61+ CD62P− GlyA− | CD34+ CD38+ CD71+ CD95− HLA-DR+ | | 71 |
| T-33 | CD3− CD4− CD5− CD8− CD57− | CD10− CD19− | CD13+ CD14+ CD15− CD33+ | CD41a+ CD42b+ CD61+ GlyA− PPO− vWF+ | HLA-DR+ | | 73 |
| TF-1 | CD3− CD4− CD5− CD7− CD8− CD57− | CD10− CD19− | CD13+ CD14− CD15− CD33+ | CD41+ CD42a(+) CD42b+ CD61+ CD62− GlyA(+) PPO− | CD34+ CD38+ HLA-DR+ | CD11a− CD11b+ CD11c+ CD18+ CD44+ CD54+ CD56− CD62L− CD106− | 75,83 |
| TS9;22 | CD1a− CD2− CD3− CD4− CD5− CD7− CD8− CD28− CD56− CD57− TCRαβ/δ− | CD10− CD19− CD20− CD21− CD22− CD24+ cy/sIg− | CD13− CD14− CD15− CD16− CD33+ | CD9+ CD36+ CD41a+ CD42b− CD61(+) GlyA− | CD34+ CD38+ CD71+ HLA-DR+ TdT− | CD11b− | 76 |
| UoC-M1 | CD1a− CD2− CD3− CD4− CD5− CD7+ CD8− CD56− | CD10− CD19− CD20− CD24+ sIg− | CD13+ CD14− CD15− CD33+ | CD41− CD42− CD61+ GlyA− | CD34+ CD38+ CD45+ HLA-DR+ | | 77 |
| UT-7 | CD2− CD3− | CD10− CD19− CD20− | CD13+ CD14− CD15+ CD16+ CD33+ | CD41a+ CD41b+ CD42b+ CD61+ GlyA+ PPO+ | CD34+ HLA-DR+ | CD11b+ CD44(+) | 78 |
| Y-1K | CD3− CD4+ CD8− | CD19− CD20− | CD13+ CD33+ | CD41+ GlyA+ | CD34+ HLA-DR+ | | 82 |
| YN-1 | CD3− CD4− CD8− | CD19− CD20− | CD13+ CD33+ | CD41− GlyA+ | CD34− HLA-DR− | | 82 |
| YS9;22 | CD1a− CD2− CD3− CD4− CD5− CD7− CD8− CD22− CD24− CD56+ CD57− | CD10− CD19− CD20− CD21− CD22− CD24− sIg− | CD13+ CD14− CD15− CD16− CD33+ | CD9+ CD36+ CD41a+ CD42b− CD61+ GlyA− | CD34+ CD38+ CD71+ HLA-DR− TdT− | CD11b− | 76 |

[a] +, strong, definite protein expression (mostly more than 10% cells positive); (+), weak protein expression, qualitatively and quantitatively (<10% cells positive); −, no protein expression. [b] Part of the immunophenotype is that of CMK11-5 subclone. [c] Extensive immunophenotype in ref. [85]. [d] Extensive immunophenotype on parental cell line M-07 in ref. [51].

*Table 3.* Erythroid-megakaryocytic cell lines: cytokine-related characterization

| Cell line | Cytokine receptor expression[a] | Cytokine production[a] | Proliferation response to cytokines[b] | Differentiation response to cytokines[b] | Dependency on cytokines[c] | Ref. |
|---|---|---|---|---|---|---|
| AP-217 | RT-PCR: EPO-R+; protein: MPL+ | | | | | 23 |
| AS-E2 | EPO-R+, IL-2Rα− | | EPO+ | | EPO | 24 |
| B1647 | EPO-R+, IL-2Rα−, IL-2Rβ−, KIT(+), MPL+ | | TPO+ | | | 25 |
| CHRF-288-11 | RT-PCR: FGF-R1+, FGF-R2+, GM-CSFRα+, IFNαR+, IFNβR+, IL-1R+, IL-6R+, Kit+, M-CSFR+, TNFαR+; protein: IL-2Rα−, MPL+ | RT-PCR: bFGF+, GM-CSF+, IFNα+, IL-1β+, IL-3+, IL-7+, IL-8+, IL-11+, SCF+, TGFβ+, TNFα+; ELISA: bFGF+ GM-CSF+ GM-CSF+ IFNα+, SCF+, TGFβ+, TNFα+ | | | | 88 |
| CMK | IL-2Rα+, IL-2Rβ−, MPL(+) | GM-CSF+, IL-1α+, IL-1β+, IL-6+, TGFβ+, TNFα+ | GM-CSF+, IL-3+, TPO+ | | | 30 |
| CMY | | | GM-CSF(+), IL-3(+), IL-6(+), TPO(+) | | | 32 |
| ELF-153 | EPO-R−, GM-CSFRα+, gp130+, IL-6Rα+, Kit+, MPL+ | | GM-CSF+, IL-3+, IL-4+, IL-6(+), PIXY-321+, SCF+ | | GM-CSF | 33,34 |
| F-36P | GM-CSFRα+, IL-2Rα− | | GM-CSF+, IFNγ+, IL-3+, IL-5+, PIXY-321+ | EPO+ | GM-CSF or IL-3 | 35,42 |
| GRW | RT-PCR: G-CSFR+, IL-1R+, IL-2Rα+, IL-3Rα+, IL-4Rα+, IL-6Rα+, IL-7Rα+, IL-9R+, IL-11R+, Kit+, TNFαR+ | PDGF+ | SCF+ | EPO+, GM-CSF+, IL-3+ | SCF | 36 |
| HEL | RT-PCR: IL-2Rα(+), IL-2Rβ+, IL-2Rα+, IL-4Rα(+), IL-9Rα+; protein: EPO-R+, IL-2Rα−, IL-2Rβ+, IL-2Rγ+, Kit+, MPL+ | | | | | 16,86 |
| HIMeg-1 | MPL+ | | IL-6+ | | | 38 |
| HML | gp130+, IL-6Rα−, Kit+ | | GM-CSF+, SCF+; inhibition: TGFβ+ | SCF + IL-5 or SCF + EPO | GM-CSF | 39 |
| HU-3 | IL-2Rα− | RT-PCR: GM-CSF+, IL-1β+, IL-6+, IL-7+, IL-10+, IL-13+, SCF+, TGFβ+, TNFα+ | EPO+, GM-CSF+, IFNβ+, IFNγ+, IL-1α+, IL-3+, IL-4+, IL-5+, IL-6+, LIF+, NGF+, OSM+, PIXY-321+, SCF+, TNFα+, TNFβ+, TPO+ | EPO+ | GM-CSF, IL-3 or TPO | 41, 42, 89 |
| JK-1 | EPO-R+, Kit+ | | EPO+ | | | 43 |
| JURL-MK1 | IL-2Rα+, IL-2Rβ−, Kit+ | | SCF(+) | | | 44 |

*Continued on next page*

*Table 3.* (continued)

| Cell line | Cytokine receptor expression[a] | Cytokine production[a] | Proliferation response to cytokines[b] | Differentiation response to cytokines[b] | Dependency on cytokines[c] | Ref. |
|---|---|---|---|---|---|---|
| K-562 | RT-PCR: IL-2Rα(+), IL-2Rβ(+), IL-2Rγ+, IL-4Rα(+), IL-9Rα+, IL-11Rα+ protein: EPO-R+, IL-2Rα−, IL-2Rβ(+), IL-2Rγ+, Kit(+) | PDGF+, TGFβ+ | | | | 16, 86 |
| M-07e | RT-PCR: IL-2Rα+, IL-2Rβ+, IL-2Rγ+, IL-11Rα+: protein: IL-2Rα−, IL-2Rβ+, IL-2Rγ+, MPL+ | | GM-CSF+, IFNα+, IFNβ+, IFNγ+, IL-2+, IL-3+, IL-4+, IL-6+, IL-9+, IL-15+, NGF+, PIXY-321+, SCF+, TNFα+, TPO+; inhibitory: TGFβ1+ | | GM-CSF, IL-3 or TPO | 42, 52, 89 |
| M-MOK | Kit+, MPL+ | | GM-CSF+, IFNβ+, IFNγ+, IL-3+, IL-4+, IL-9+, IL-13+, IL-15+, PIXY-321+, SCF+, TNFα+, TNFβ+, TPO+; inhibitory: TGFβ+ | | GM-CSF or TPO | 42, 53, 89, 90 |
| MB-02 | IL-2Rα− | RT-PCR: GM-CSF+, IL-1β+, IL-7+, IL-13+, TGFβ+ | EPO+, GM-CSF+, IL-1β+, IL-3+, PIXY-321+, SCF+, TNFα+, TNFβ+; inhibitory: IFNα+, IFNβ+, TGFβ+, | EPO+ | GM-CSF | 42, 54 |
| MC3 | RT-PCR: EPO-R+, MPL+ | | IL-1β+, IL-3+; inhibitory: TGFβ+ | | | 55 |
| MEG-01 | IL-2Rα−, IL-2Rβ− | RT-PCR: PDGF+; ELISA: β-TG+, PDGF+, PF-4+, TGFβ+ | | | | 57 |
| MEG-A2 | | RT-PCR: G-CSF+, GM-CSF+, IL-1α+, IL-1β+, IL-3+, IL-4+, IL-6+, M-CSF+, TPO+ | EPO+, GM-CSF+, IL-3 | EPO(+), GM-CSF(+), IL-3(+), IL-6(+) | | 58 |
| MG-S | IL-2Rα−, Kit+ | | GM-CSF+ | | | 59 |
| MHH 225 | IL-2Rα− | | SCF(+); inhibitory: IFNα(+), TNFα(+) | IL-4+ | | 60, 61 |
| M-TAT | EPO-R+ | | EPO+, GM-CSF+, IL-3+, SCF+ | | EPO, GM-CSF or SCF | 65 |
| NS-Meg | Northern: EPO-R+ | | EPO+, GM-CSF+, IL-3+ | EPO+ | | 66 |
| OCIM1 | EPO-R+, Kit+ | PDGF+ | | | | 16, 67 |
| OCIM2 | Kit+ | PDGF+ | | | | 16, 67 |
| RM10 | EPO-R+, IL-2Rα−, Kit− | | | | | 68 |
| RS-1 | | | IL-3+ | | | 69 |
| SAM-1 | EPO-R−, Kit− | | IL-6+ | | | 70 |

*Continued on next page*

*Table 3.* (continued)

| Cell line | Cytokine receptor expression[a] | Cytokine production[a] | Proliferation response to cytokines[b] | Differentiation response to cytokines[b] | Dependency on cytokines[c] | Ref. |
|---|---|---|---|---|---|---|
| SET-2 | gp130+, IL-2Rα−, IL-6Rα+, Kit+, MPL+ | IL-6(+) | GM-CSF(+), IL-3(+), IL-6(+) | | | 71 |
| T-33 | | | EPO+ | | | 73 |
| TF-1 | RT-PCR: GM-CSFRβ+, IL-1R II+, IL-4Rα+, IL-11Rα+, TGFβR II+, EPO-R+, GM-CSFRα+, IL-3Rα+, IL-5Rα+, MPL(+) protein: GM-CSFRα+, IL-3Rα+, IL-5α+, Kit+ | | CNTF+, EPO+, GM-CSF+, IFNγ+, IL-1+, IL-3+, IL-4+, IL-5+, IL-6+, IL-13+, LIF+, NGF+, OSM+, PIXY-321+, SCF+, TNFα+, TNFβ+, TPO+ inhibitory: IFNα+, IFNβ+, TGFβ+ | | GM-CSF, IL-3 or TPO | 42, 74, 75, 89 |
| UT-7 | EPO-R+, GM-CSFRα+ | ELISA: β-TG(+), PF-4(+) | EPO+, G-CSF+, GM-CSF+, IFNβ+, IFNγ+, IL-3+, IL-4+, IL-5+, IL-6+, NGF+, PIXY-321+, SCF+; inhibitory: TGFβ+ | | GM-CSF, IL-3, EPO or TPO | 42, 78-81, 89 |

[a] Receptor expression or cytokine production at the protein level (ELISA, McAb, RIA), unless otherwise indicated, e.g. at the mRNA level (by RT-PCR, Northern); +, strong, definite expression; (+), weak expression; −, no expression.

[b] Effects of cytokine exposure on proliferation or differentiation (−, no effect; +, positive effect).

[c] Following growth factor withdrawal, these cell lines will die by apoptosis.

established from patients with a refractory disease. Apart from tumors and pleural effusions, peripheral blood (n = 28) and bone marrow (n = 14) were the preferred sites from which specimens for cell culture were obtained.

While no authentication was provided for 28 cell lines, evidence for a derivation from the assumed patient was given in 21 instances, mostly in the form of a comparison of the patient's and cell line's cytogenetic profile, but also by comparing immunoprofiles and genotypic profiles (e.g. gene banding patterns in Southern blots). The media used were RPMI 1640 (n = 30), IMDM (n = 12), ($\alpha$-MEM (n = 2), Dulbecco's MEM (n = 2), Fisher's (n = 1), McCoy's 5A (n = 1), and Ham's F12 (n = 1). Usually, the cultures require addition of 10–20% FBS. Cell lines said to require horse serum or human serum (CHRF-288-11, HU-3, JK-1, MB-02) could also be adapted to grow with FBS alone. Only two cell lines could be cultured under serum-free conditions (MHH 225, T-33) (Appendix 2, p. 286). Several cell lines are constitutively growth factor-dependent, undergoing apoptosis in the absence of the respective cytokines: AS-E2, ELF-153, F-36P, GRW, HU-3, M-07e, M-MOK, MB-02, M-TAT, TF-1, and UT-7 [89] (Table 3; Appendix 3). Most of these cell lines can be obtained from the original investigators or cell banks (CMK, F-36P, HEL, JK-1, K-562, KMOE-2, LAMA-84, M-07e, MEG-01, TF-1, UT-7) (Appendix 1, p. 284).

Subclones with features significantly different from those of their parental lines have been derived from several cell lines. The differences concern mainly immunological and cytogenetic characteristics (parental lines: JURL-MK1, KH88, LAMA-84, MEG-01), stage of differentiation and ability to differentiate along a certain maturation pathway (parental lines: CMK, LAMA-84), and responsiveness/dependence on various growth factors (parental lines: F-36P, HU-3, M-07e, M-MOK, TF-1, UT-7). Of particular interest are cell lines derived from different patient sites or at different stages of the disease (parental lines: KMOE-2, MOLM-7).

## 3. IMMUNOPHENOTYPE

The erythroid-megakaryocytic cell lines display characteristic immunoprofiles (Table 2). Apart from the occasional positivity for CD4 and CD7, which are also expressed by myeloid precursor cells [110], the cells are usually negative for classical T-/NK-cell markers (including CD1, CD2, CD3, CD4, CD5, CD7, CD8, CD56, CD57, TCR) and B-cell markers (including CD10, CD19, CD20, CD21, Ig). Most cell lines express the two pan-myeloid surface antigens CD13 and CD33 and are generally negative for the granulocytic and monocytic markers CD15 and CD14, respectively.

*Table 4.* Erythroid-megakaryocytic cell lines: genetic characterization

| Cell line | Karyotype | Unique translocations (→ fusion genes) | Unique gene alterations, receptor gene rearrangements* | Ref. |
|---|---|---|---|---|
| AP-217 | 87, XXYY, −7, −9, der(5)t(5:?)(q31 or q32:?), t(9;22)(q34:q11), del(17)(p11), der(19)t(11;19)(q13;q13), der(21)t(21:?)(p13:?) | Ph+ t(9;22)(q34;q11) → *BCR-ABL* (b3-a2) fusion gene | | 23 |
| AS-E2 | 67–82, XXYY, −2, −2, −4, −5, −10, −10, −13, −13, +14, −15, −15, −16, −16, −17, −18, −19, −19, −22, −22, +mar, del(1)(p32p36), del(3)(p21), del(3)(q21), der(4)(1:4)(q21:p16)x2, add(8)(p23), add(8)(q24)x2, add(9)(q34), +del(9)(q11q22), add(10)(p14), add(11)(q23), i(11)(q10), add(15)(p12) | | | 24 |
| B1647 | 53, XY, +2, +5, +8, +13, −14, +19, +21, t(10:11)(p11:q21), +der(14)t(14:?)(p11:?) | | | 25 |
| CHRF-288-11 | 50, XY, 1p+, 6p−, +6q−, +8, −10, 12p+, −15, +17, +19p+ | | | 27 |
| CMK | 85-90<4n>XY, −X, −Y, −2, −3, +5, −6, −6, −8, +11, −15, −15, +16, −17, −19, +21, +22, +7-11mar, add(1)(p36), add(1)(q31), add(3)(q11), del(3)(p14x2−3, add(5)(q11), add(5)(q13), dup(8)(q11q21), add(8)(q13−21), del(8)(q11), del(9)(p21)x2, add(9)(q11)x2, del(10)(q22q24), der(11:17)(q10:q10). der(11)dup(11)(p13p15)t(5:11)(q11:p15)x1−2, del(11)(q23), add(12)(p13)x2, add(17)(p1?), add(18)(q23)x2−3, add(19)(p13), der(20)t(1:20)(q2?5:q1?2)x2, add(22)(q13) | | *P53* PM | 30, 83 |
| CMS | 46, X, del(X)(q23), der(2)t(1:2)(q22−24:q35), add(5)(q15), der(7)t(7:8)(p22:q21), add(17)(p13), add(19)(q13) | | *P53* PM | 31 |
| CMY | del(17p) | | | 32 |
| ELF-153 | 43, XY, −7, −14, −17, del(5)(q13q31), t(12;14)(p11.2;q11.2) | | | 33 |
| F-36P | 43, Y, Xp+, −5, −7, −13, −16, −17, −19, −21, 2q−, 9p+, 10q+ | | *P16INK4A* DD: *P53* PM | 35, 91 |

*Continued on next page*

*Table 4.* (continued)

| Cell line | Karyotype | Unique translocations (→ fusion genes) | Unique gene alterations, receptor gene rearrangements* | Ref. |
|---|---|---|---|---|
| GRW | 44, XX, −2, −4, −5, −7, −18, −19, +2mar, add(4)(p14), add(7)(q36). +der(15)t(15;21)(q10;q10)x2 | | | 36 |
| HEL | 63(60–64)<3n>XXYY, −2, −9, −10, −10, −11, −14, −16, −16, −17, −19, +20, +21, +r, der(2)t(2;7)(q14;q32), add(3)(p14/21.3), der(4)t(4;17)(q13.3;q12), add(5)(q12), der(6)t(1;6)(p31.3;p23), der(6)t(3;6)(p21;q21), der(7)t(2;7)(q14;q32)t(7;?18)(p14;q21), der(9)qdp(9)(p11p24)t(9;20)(p24;?q11)t(11;20)(q13;?q13), add(17)(p12), add(20)(q11), r(20)(p11q11), dup(21)(qter;q11), add(22)(p11); masked 5q− and 20q− | | P15INK4B DD, P16INK4A DD | 83, 91–94 |
| HML | 47, XY, +21, del(3)(q24q26.1) | | | 39 |
| HU-3 | 73(70–76)XXX, +7, +7, +8, +8, −17, −17, −22 | | | 41 |
| JK-1 | 48(45–48)<2n>XY, +8, +22, t(1;7)(q10;q10), der(9)t(9;22)(q34;q11), der(22)t(9;22)(q34;q11)x2; carries two Ph | Ph+ t(9;22)(q34;q11) → BCR-ABL (b2-a2) fusion gene | P16INK4A DG | 43, 83, 92, 94 |
| JURL-MK1 | 39(33–43)<2n>XY, −4, −5, −9, −11, −12, −18, −19, +mar, der(9)t(9;22)(q34;q11), i(17), der(22)t(9;22)(q34;q11) | Ph+ t(9;22)(q34;q11) → BCR-ABL (b3-a2) fusion gene | | 44 |
| K-562 | 61-68<3n>XX, −X, −3, +7, −13, −18, +3mar, del(9)(p11/13), der(14)t(14;?)(p11;?), der(17)t(17;?)(p11/13;?), der(?;18)t(15;?18)(q21;?q12), del(X)(p22), 2 markers appear from FISH to have arisen from Ph | Ph+ t(9;22)(q34;q11) → BCR-ABL (b3-a2) fusion gene | P15INK4B DD, P16INK4A DD; ABL amplified | 83, 91–94 |
| KH184 | 44, X, −Y, −1, −2, −2, −6, −16, −17, −18, +19, −20, −20, −20, −22, −22, +der(?Y)t(?Y;22)(p11;q11), t(1;8)(q11;q11), +der(2)t(2;?)(p23;?), +der(6)t(6;?)(p23;?), del(9)(q13q22), t(10;17)(p12;q11), t(11;19)(q23;q13.3), +der(17)t(17;?), +der(20)t(3;20)(q13;p13), +der(22)t(22;?)(p13;?) | | | 46 |

*Continued on next page*

*Table 4.* (continued)

| Cell line | Karyotype | Unique translocations (→ fusion genes) | Unique gene alterations, receptor gene rearrangements* | Ref. |
|---|---|---|---|---|
| KH88 | 71, XY, +Y, +3, +5, +6, +8, −9, +10, −12, −14, −14, −15, −15, −17, −18, −20, +21, −22, +3mar, del(X)(q22q28), del(2)(p11.1), add(7)(p11.2), del(9)(p13), t(9;22)(q34;q11), i(11)(q10), der(17)t(2;17)(p13;p13)x2, add(18)(p11.2), +der(19)t(7;19)(q10;q10), +add(19)(p13) | Ph+ t(9;22)(q34;q11) → *BCR-ABL* (b3-a2) fusion gene | | 46 |
| KMOE-2 | 82(80–88)<4n>XX, −X, −X, −10, −13, −14, −16, −18, −20, −22, +3mar, der(1)t(1;?)(q21:?)x2, del(7)(p15)x2, del(9)(p12), i(17q) | | | 83 |
| LAMA-84 | 73/74(69–77)<3n>XX-X, +1, −2, +5, +6, +8, +13, −14, +17, +17, −18, +22, +mar, del(7)(p15), der(9)t(9;22)(q34;q11)x2, i(11q), add(13)(q33), del(17)(p12), der(22)t(9;22)(q34;q11)x4; carries Ph (4 copies) | Ph+ t(9;22)(q34;q11) → *BCR-ABL* (b3-a2) fusion gene | *P15INK4B* DD, *P16INK4A* DD; *P53* PM | 49, 83, 93 |
| M-07e | 46(45–46)<2n>XX, t(11;21)(p11;p13), add(13)(p13). add(22)(p13) | | *BCL2* G, *BCR* G, *IGL* G, | 51, 83 |
| M-MOK | 46, XX, 2q+, 14p+ | | *IGH* G, *TCRB* G | 53 |
| MB-02 | 43–50, XY, −1, −3, −4, −10, −11, −12, −14, −14, +15, +16, +6mar, +der(1)t(1;2)(q42;q23), +del(3)(q21). +der(3)t(3;?)(p25;?) | | | 54 |
| MC3 | 55, XX, +6, +8, +11, +15, +19, +2mar, t(1;5)(q25;q15). t(9;22)(q34;q11) | Ph+ t(9;22)(q34;q11) → *BCR-ABL* (b3-a2) fusion gene | *IGH* G, *IGK* G; *P53* R | 55 |
| MEG-01 | 54(53–56)<2n>XY, +6, +19, +19, +21, +3–4mar, t(1;15)(p13;p13), ?inv(3)(p25q26), i(4q), add(5)(p15), der(9)t(9;22)(q34;q11)x2, add(10)(p14), dup(13)(q13q33–34), add(14)(p11), der(22)t(9;22)(q34;q11) | Ph+ t(9;22)(q34;q11) → *BCR-ABL* (b2-a2) fusion gene | *P15INK4B* DD, *P16INK4A* DD; *P53* PM | 56, 83, 92, 94 |
| MEG-A2 | 55(51–56)XY, +1, +6, +8, +11, +18, −20, +21, +21, t(1;6)(p13;q22), t(9;22)(p34;q11), del(17)(p11), +dic(19)t(8;19)(p22;q13), +dic(20)(8;20)(p22;q13), +der(22)t(9;22)(p34q11) | Ph+ t(9;22)(q34;q11) → *BCR-ABL* (b3-a2) fusion gene | | 58 |

*Continued on next page*

Table 4. (continued)

| Cell line | Karyotype | Unique translocations (→ fusion genes) | Unique gene alterations, receptor gene rearrangements* | Ref. |
|---|---|---|---|---|
| MG-S | 46, XY, −4, −12, −17, −17, +21, +2mar, +der(17)t(4;17)(4qter → 4q21::17p11 → 17qter), 7p+, 7q+, 13p+ | | | 59 |
| MHH 225 | del(7)(p13), t(9;21)(q10;q10), 11q+, 9p−, r(21), t(9;11)(p?;q23) | t(9;11)(p?;q23) → MLL-AF9 fusion gene? | | 61 |
| MKPL-1 | 92, +3mar | | | 62 |
| MOLM-1 | 72, X, +1, +2, +2, +2, +4, +4, +6, +8, +8, +10, +11, +12, +13, +13, +14, +15, +16, +19, +19, +20, +21, +21, +22q−, inv(3)(q21q26), +inv(9)t(9;22)(q34;q10), +der(17)t(17;?)(p1?1;?) | Ph+ t(9;22)(q34;q11) → BCR-ABL (b2-a2) fusion gene; inv(3)(q21q26) → EVI1 overexpression | | 63, 91, 95 |
| NS-Meg | 53, XY, +6, +7, +8, +9, +11, +14, +19, +21, t(9;22)(q34;q11), der(17)t(1;17)(p11;p13), +der(22)t(9;22)(q34;q11) | Ph+ t(9;22)(q34;q11) → BCR-ABL (b3-a2) fusion gene | | 66 |
| RM10 | 44–212(63) XX, +4, −8, +9p−, −11, +12, +13, +14q+, +15, +16, +18q+, +18q+, +19, +20, +21, +22q−, +3mar, +i(11q). Ph+ | Ph+ t(9;22)(q34;q11) → BCR-ABL (M-bcr) fusion gene | ABL amplified | 68 |
| RS-1 | 46, XX, −16, +mar | | | 69 |
| SAM-1 | t(9;22)(q34;11) | Ph+ t(9;22)(q34;q11) → BCR-ABL (b3-a2) fusion gene | | 70 |
| SET-2 | 47–48, X, −5, −9, −14, −21, −22, +1-3r, +5-7mar, add(X)(p21), del(X)(p21), add(1)(p36), add(4)(p1?2), add(7)(q32), del(10)(q24), add(12)(p13), add(13)(q3?2), add(16)(q22), add(17)(q25), add(19)(p13) | Ph+ t(9;22)(q34;q11) → BCR-ABL (b3-a2) fusion gene | | 71 |
| SKH1 | t(3;21)(q26;q22), t(9;22)(q34;q11) | Ph+ t(9;22)(q34;q11) → BCR-ABL (b3-a2) fusion gene; t(3;21)(q26;q22) → EVI1-AML1 fusion gene | | 72, 91 |
| T-33 | 51, XX, 1p+, +4, +9, +9, +19, +Ph, t(9;22)(q34;q11) | Ph+ t(9;22)(q34;q11) → BCR-ABL (b3-a2) fusion gene | | 73 |

Continued on next page

*Table 4.* (continued)

| Cell line | Karyotype | Unique translocations (→ fusion genes) | Unique gene alterations, receptor gene rearrangements* | Ref. |
|---|---|---|---|---|
| TF-1 | 52–57<2n>XY/XXY, +3, +5, +6, +8, +12, +15, +19, +19, +20, +20, +3mar, der(1)?dup(1)(p21p31)t(1;8)(p36;q11), t(2;12)(q32;q14), t(3;12)(p13–14;p12–13), add(3)(q21), add(5)(q11–13), der(8)t(1:8)(p36;q11), der(12)t(3;12)(p13–14;p12–13)t(1;12)(q31–32;q24), add(14)(p12), iso(17)(q10), add(17)(q21), add(19)(q13), trp(19)(q12;q13.3), der(21)t(19;21)(q13.1;q22)dup(19)(q13.1q13.3)t(11;19)(q13;q13.3), der(22)t(19;22)(q11;p11) | | | 75, 83 |
| TS9;22 | 43, XY, –7, –9, –10, –13, –14, –15, add(1)(p22), add(4)(p15), t(9;22)(q34;q11), add(17)(p12), +3mar | Ph+ t(9;22)(q34;q11) → BCR-ABL (b3-a2) fusion gene | | 76 |
| UoC-M1 | 42, X, –Y, –7, –9, –16, –19, +21, der(9)t(Y:9)(q1?2:p22)t(9;19)(q1?2;p12 or q12), del(5)(q1?2q3?4), der(5)t(5;9)(p1?5;q1?3), dic(11)t(9;11;19), del(14;21)(q10;q10)del(14)(q1?3), der(17)t(7;17)(p14;p12), del(19)(p1?1q1?2), der(21)t(11;21)(q22;q22)dup(21)(q11q22), +der(21)t(16;21)(p11;p11), der(22)t(19;22)(p12 or q12;p1?1) | *MLL* amplification | | 77 |
| UT-7 | 92(92–96)XXYY, –2, –2, –3, +6, –11, –11, –13, –13, –13, –14, –17, –17, –18, –18, –19, –20, –21, –21, +18mar, +der(2)t(2;5)(p11;q11)x2 | | *P16INK4A* DD; *P53* PM | 78, 91 |
| Y-1K | 66(64–67)XXX, t(9;22) | Ph+ t(9;22)(q34;q11) → BCR-ABL (b3-a2) fusion gene | | 82 |
| YN-1 | 53, XY, +8, +8, +14, +14, +19, +21, +22q–, t(9;22)(q34;q11) | Ph+ t(9;22)(q34;q11) → BCR-ABL (b3-a2) fusion gene | | 82 |
| YS9;22 | 46, X, –X, +19, der(1)t(1;3)(p32;p12), dic(3)(q26;p12), t(9;22)(q34;q11), add(16)(p13) | Ph+ t(9;22)(q34;q11) → BCR-ABL (b3-a2) fusion gene | | 76 |

* Receptor gene arrangements: D – deleted; G – germline; PM – point mutation; R – rearranged.

*Table 5.* Erythroid-megakaryocytic cell lines: functional characterization

| Cell line | Doubling time | EBV status | Cytochemistry | Inducibility of differentiation | Heterotransplantability into mice | Special functional features | Ref. |
|---|---|---|---|---|---|---|---|
| AP-217 | 24–30 h | | | TPA → meg differentiation; hemin, retinoic acid → ery differentiation | | $\beta$-, $\gamma$-globin mRNA+; cloning/colony formation in methylcellulose; PO: c-myc mRNA+; TF: GATA-1 mRNA+ | 23, 96 |
| AS-E2 | 49 h | EBV– | ANBE–, Benzidine+, CAE–, MPO–, PAS(+) | | | $\gamma$-globin mRNA+; colony formation in methylcellulose; TF: GATA-1 mRNA+, GATA-2 mRNA+ | 24 |
| B1647 | | | | hemin, hydroxyurea, TPA → ery differentiation | | $\alpha$-, $\beta$-globin mRNA+; $\gamma$-globin+; $\alpha$-tubulin+; colony formation in agar | 25 |
| CHRF-288-11 | | | MPO– | TPA → meg differentiation | | $\alpha$-granules+ | 27 |
| CMK | 40–50 h | EBV– HTLV-I– HIV– | ACP+, ALP–, ANAE+, MPO–, PAS+ | TPA → meg differentiation | into nude mice | $\alpha$-granules+; demarcation system+; colony formation in methylcellulose; PO: c-sis mRNA+; TF: GATA-1 mRNA+, GATA-2 mRNA+ | 28–30,83 |
| CMS | 42 h | | ACP+, ALP–, ANAE+, MPO–, PAS+ | TPA → multinucleated cells | | colony formation | 31 |
| CMY | 46 h | | ACP+, ALP–, ANBE+ (NaF resistant), MPO–, PAS+ | cytokines → meg differentiation | | colony formation in methylcellulose; $\alpha$-granules+ | 32 |
| ELF-153 | 36 h | | Benzidine– | | into nude mice | $\epsilon$-, $\gamma$-globin mRNA–; demarcation system+; cloning/colony formation in methylcellulose; TF: GATA-1 mRNA+, GATA-2 mRNA+ | 33, 34 |
| F-36P | | | | EPO → ery differentiation | | | 35 |
| GRW | 91 (40–120) h | | | EPO, GM-CSF, IL-3, SCF → ery, meg or baso differentiation | | | 36 |
| HEL | 24–36 h | EBV– HTLV-I– HIV– | ACP+, ANAE+, ANBE+, Benzidine–, CAE–, MPO–, PAS+, SBB– | hemin → ery differentiation; TPA → meg differentiation | | $\alpha$-, $\beta$-, $\gamma$-, $\epsilon$-, $\zeta$-globin+; $\delta$-aminolevulin synthase mRNA+, Hb Bart's+; clonable in methylcellulose; TF: GATA-1 mRNA+, GATA-2 mRNA+, SCL mRNA+ | 68, 82, 83, 97 |
| HIMeg-1 | | | | TPA → mono differentiation; leukocyte conditioned medium, retinoic acid, vit. D3 → meg differentiation | | | 38 |
| HML | | | ACP+, ALP–, ANBE+, CAE–, MPO–, PAS+, SBB– | TPA → meg differentiation; SCF + IL-5 + retinoic acid → eosino differentiation; SCF + EPO → ery differentiation | | eosino POX–, major basic protein(+), Biebrich scarlet–, Luxol fast blue–, toluidine blue–, PF4–, $\beta$-TG– | 39 |

*Continued on next page*

*Table 5.* (continued)

| Cell line | Doubling time | EBV status | Cytochemistry | Inducibility of differentiation | Heterotransplant-ability into mice | Special functional features | Ref. |
|---|---|---|---|---|---|---|---|
| HU-3 | | | | TPA → meg differentiation; EPO → ery differentiation | | clonable; TF: NF-E2+ | 41 |
| JK-1 | 48 h | EBV− HTLV-I− HIV− | ACP+, ANBE−, Benzidine+, CAE−, MPO−, PAS+ | spontaneous ery differentiation; δ-ALA → ery differentiation; (ARA-C, DMSO, hemin, Na butyrate no effect) | | β-, δ-globin mRNA+; HbF+; (red) colony formation in methylcellulose | 43, 68, 83 |
| JURL-MK1 | 48 h | | ALP−, ANAE(+), MPO−, PAS+ | TPA → meg differentiation; (ARA-C, ATRA, DMSO, hemin no effect) | | | 44 |
| K-562 | 24–30 h | EBV− HTLV-I− HIV− | ALP−, ANBE−, Benzidine+, MPO−, PAS+ | TPA → meg differentiation; hemin → ery differentiation | into nude or SCID mice | α-, γ-, ε-, ζ-globin mRNA+; δ-aminolevulin synthase mRNA+; Hb Bart's+, Hb Portland+, Hb Gower I+, Hb Gower II+, HbF+; colony formation in agar; no α-granules, no demarcation system; target for NK assay; no α-granules, no demarcation system; TF: GATA-1 mRNA+, GATA-2 mRNA+, SCL mRNA+ | 45, 68, 82, 83, 97 |
| KH184 | 36–60 h | | ACP+, ANAE−, CAE−, MPO−, PAS− | (TPA no effect) | | globin mRNA−; no α-granules, no demarcation system | 46 |
| KH88 | 20–26 h | | ACP+, ANAE+, Ferritin−, Lactoferrin−, MPO− | TPA → meg differentiation; hemin → ery differentiation; (DMSO, retinoic acid no effect) | | HbA+, HbF+ | 47 |
| KMOE-2 | 24 h | EBV− HTLV-I− HIV− | ANAE−, Benzidine−, MPO−, PAS+, SBB− | Na butyrate → ery differentiation | into nude mice | α-, β-, γ-, δ-globin mRNA+; HbA+; colony formation in agar | 48, 68, 83, 99 |
| LAMA-84 | 30 h | EBV− HTLV-I− HIV− | ACP+, CAE+, Lactoferrin−, Luxol−, MPO−, PAS+, SBB(+), Vimentin+ | hemin → ery differentiation; DMSO, Na butyrate, TPA → meg differentiation; (retinoic acid, vit. D3 no effect) | into estrone-treated nude mice | α-, β-, γ-globin mRNA+; colony formation in agar; no α-granules, no demarcation system; no NBT reduction, no phagocytosis, no lysozyme | 49, 50, 83, 85 |
| M-07e | 32–50 h | EBV− HTLV-I− HIV− | ACP+, ANAE+, ANBE−, CAE−, MPO−, PAS− | TPA → meg differentiation; | into SCID mice | cloning/colony formation in methylcellulose; spontaneous production of platelet-like particles; demarcation system+ | 51, 52, 83, 98 |
| M-MOK | | EBV− | ANBE−, CAE−, MPO−, PAS−, SBB− | TPA → meg differentiation; (ATRA not effective) | | TF: GATA-1 mRNA−, GATA-2 mRNA+, GATA-3 mRNA+ | 53 |

*Continued on next page*

*Table 5.* (continued)

| Cell line | Doubling time | EBV status | Cytochemistry | Inducibility of differentiation | Heterotransplant-ability into mice | Special functional features | Ref. |
|---|---|---|---|---|---|---|---|
| MB-02 | | EBNA– | ACP+, ANAE–, Benzidine–, CAE–, MPO–, PAS+ | TPA → meg differentiation; DMSO, EPO → ery differentiation | | β-, γ-, ε-globin mRNA+, ζ-globin mRNA–; HbF+; not clonable; colony formation in methylcellulose or agar; no phagocytosis; TF: NF-E2+ | 54 |
| MC3 | | | ACP+, ANBE–, CAE–, MPO–, PAS–, SBB– | TPA → meg differentiation | | TF: GATA-2 mRNA+ | 87 |
| MEG-01 | 36–48 h | EBV– HTLV-I– HIV– | ACP+, ALP–, ANAE+, ANBE–, CAE–, MPO–, PAS(+) | TPA → meg differentiation | | no α-granules, no demarcation system; no lysozyme, no phagocytosis | 28, 56, 57, 83, 84 |
| MEG-A2 | 26–30 h | EBV– | ACP+, ALP–, ANAE–, ANBE+, CAE–, MPO–, PAS+ | cytokines, TPA → meg differentiation; (DMSO, hemin, retinoic acid no effect) | | no α-granules, demarcation system+ | 58 |
| MG-S | 36 h | | ACP+, ANBE–, CAE–, MPO–, PAS+ | TPA → meg differentiation | | no α-granules, no demarcation system | 59 |
| MHH 225 | 38–48 h | | ACP(+), ANAE–, ANBE–, PAS+, SBB– | IL-4 → myeloid differentiation | | SCF protects against apoptosis | 60, 61 |
| MKPL-1 | 30 h | EBNA– | ACP+, ALP–, ANAE+, ANBE+, CAE–, MPO–, PAS+ | | into nude mice | α-granules+, demarcation system+ | 62 |
| MOLM-1 | 43 h | | MPO– | TPA → meg differentiation | | no NBT reduction, no phagocytosis | 63 |
| MOLM-7 | 24–60 h | | MPO– | TPA → meg differentiation | | | 64, 84 |
| M-TAT | | | Benzidine+ | EPO → ery differentiation | | γ-globin mRNA+, δ-aminolevulin synthase mRNA+; TF: GATA-1+, GATA-2+ | 65 |
| NS-Meg | | | ACP+, ANAE+, ANBE–, Benzidine–, MPO–, PAS+, SBB– | cytokines, TPA → meg differentiation; EPO → ery differentiation; (δ-ALA, DMSO, hemin, Na butyrate no effect) | | α-, γ-globin mRNA+; induction of HbF; α-granules+, demarcation system+; TF: GATA-1 mRNA+; spontaneous production of platelet-like particles | 66 |
| OCIM1 | | | ANAE+, Benzidine+, CAE–, MPO–, PAS+ | δ-ALA → ery differentiation; TPA → macro/meg differentiation | into SCID mice | α-, γ-, ε-globin+; induction of Hb Bart's, HbF; cloning/(red) colony formation in methylcellulose | 67, 68 |
| OCIM2 | | | ANAE+, Benzidine+, CAE–, MPO–, PAS+ | δ-ALA → ery differentiation; TPA → macro/meg differentiation | | α-, γ-, δ-, ε-, ζ-globin+; induction of Hb Bart's, HbF, Hb Portland; cloning/(red) colony formation in methylcellulose | 67, 68, 98 |

*Continued on next page*

*Table 5.* (continued)

| Cell line | Doubling time | EBV status | Cytochemistry | Inducibility of differentiation | Heterotransplant-ability into mice | Special functional features | Ref. |
|---|---|---|---|---|---|---|---|
| RM10 | 18–20 h | EBNA– HTLV-I– | ACP–, ANBE–, Benzidine+, CAE–, MPO–, PAS+, SBB– | AraC, hemin → ery differentiation; TPA → mono differentiation; (δ-ALA, DMSO, Na butyrate, retinoic acid, vitamin D3 no effect) | | α-, γ-, ε-, ζ-globin mRNA+; Hb Bart's+, Hb Portland+, Hb Gower1+; no phagocytosis | 68 |
| RS-1 | 72 h | | ANAE–, CAE–, MPO– | retinoic acid, TPA inhibit proliferation | | colony formation in methylcellulose (with IL-3); emeripolesis+ | 69 |
| SAM-1 | | | | TPA → differentiation | | γ-globin mRNA+; TF: GATA-1 mRNA+, GATA-2 mRNA+, GATA-3 mRNA– | 70 |
| SET-2 | 36 h | EBV– | ANBE+, CAE–, MPO–, PAS(+) | TPA → meg differentiation | | spontaneous production of platelet-like particles; α-granules+; β-TG+, PF-4+ | 71 |
| T-33 | 24–36 h | | ACP+, ALP+, ANAE+, ANBE(+), CAE–, GLC+, MPO–, PAS+, SBB+ | TPA → meg differentiation | | α-granules+, demarcation system+; HbA– | 73 |
| TF-1 | 70 h | EBV– HTLV-I– HIV– | ACP–, ANAE–, CAE–, Fe–, MPO–, PAS+ | δ-ALA, hemin → ery differentiation; TPA → macro/meg differentiation; (DMSO, retinoic acid, vit. D3 no effect) | into SCID mice | α-, β-, γ-globin mRNA+; induction of HbF, HbA; cloning/colony formation in agar/methylcellulose | 74, 75, 83 |
| TS9.22 | | | ACP+, ANAE+ (NaF inhibitable), ANBE–, MPO– | | | PO: EVI1 mRNA–; TF: GATA-1 mRNA+, GATA-2 mRNA+, SCL mRNA+ | 76, 97 |
| UoC-M1 | | | | | | no demarcation system | 77 |
| UT-7 | 36–48 h | EBV– | ACP+, ALP–, ANAE+, ANBE+, CAE–, MPO–, PAS(+), SBB– | TPA → meg differentiation | | α-granules+, no demarcation system; clonable; PO: c-myb mRNA+; TF: GATA-1 mRNA+, GATA-2 mRNA+ | 78 |
| Y-1K | 62 h | | ALP+, ANBE+, Benzidine–, CAE–, MPO–, PAS+ | (hemin, EPO no effect) | | γ-globin mRNA+; δ-aminolevulin synthase mRNA+; TF: GATA-1 mRNA+ | 82 |
| YN-1 | 20–24 h | | ALP–, ANBE–, Benzidine+, CAE–, MPO–, PAS+ | hemin, Ara C → ery differentiation (EPO no effect) | | γ-globin mRNA+; δ-aminolevulin synthase mRNA+; Hb Portland+, HbF+, HbA(+); TF: GATA-1 mRNA+ | 82 |
| YS9.22 | | | | | | PO: EVI1 mRNA+; TF: GATA-1 mRNA+, GATA-2 mRNA+, SCL mRNA+ | 76, 97 |

PO – (proto)-oncogenes; TF – transcription factors.

*Table 6.* Erythroid-megakaryocytic cell lines: unconfirmed cell lines (not immortalized, not characterized, not verified, other)

| Cell line | Patient | Features | Ref. | Remarks* |
|---|---|---|---|---|
| AML-HJ | from BM of AML M7 | CD9(+) CD33+ CD34+ CD36(+) CD41(+) CD42(+) CD61(+) GlyA(+) HLA-DR+ | 100 | insufficiently characterized |
| B-403 | from PB of AML | CD41a+ CD42+ CD71+ HLA-DR+; ANAE+ EBNA− MPO− PAS+; produces PDGF; polyploid | 101 | probably EBV+ B-LCL |
| DAMI | from PB of 57 M with AML M7 in 1986 | extensive immunological and cytogenetic characterization (e.g. CD41a+ CD42b+ GlyA+) | 102 | cross-contaminated: in reality cell line HEL [103] |
| EST-IU | from BM of secondary AML | | 104, 105 | not immortalized cell line |
| KG-91 | from PB of AML | CD34+ CD71+ GlyA+; HbF+, Benzidine+; $\gamma$-, $\alpha$-, $\zeta$-globin+; AraC, hemin, Na butyrate → ery differentiation | 106 | insufficiently characterized |
| KOPMK 53 | from CML-my BC | ANAE− MPO−; CD13+ CD33+ CD41+ | 107 | insufficiently characterized |
| MEG-J | from PB of 24 M with CML-BC | CD41+ CD42+ HLA-DR− PPO+; IL-11 mRNA+; responsive to IL-11 | 108 | insufficiently characterized |
| P-320 | from PB of AML M7 | CD41a+ CD42+ CD71+ HLA-DR+; ANAE+ EBNA− MPO− PAS+; produces PDGF; polyploid | 101 | probably EBV+ B-LCL |
| S-1214 | from PB of ALL | CD41a+ CD42+ CD71+ HLA-DR+; ANAE+ EBNA− MPO− PAS+; produces PDGF; polyploid | 101 | probably EBV+ B-LCL |
| TK91 | from CML-BC | CD13+ CD33+ CD34+ CD36− CD41(+) GlyA− HLA-DR− | 28 | insufficiently characterized |
| TW14-8 | from PB of 25 M with AML M6 (with Rothmund-Thomson syndrome) | CD13+ CD33+ CD34+ HLA-DR+; responsive to IL-5 | 109 | insufficiently characterized |

* For most cell lines, the insufficient characterization concerns the description of essential features, including clinical data, authentication, immunoprofile and/or cytogenetics.

A unique erythroid lineage-specific marker is GlyA which was detected on many cell lines (25 positive, 19 negative cell lines). The panel of megakaryocytic surface proteins is considerably larger and includes CD31, CD36, CD41, CD42 and CD61. These latter antigens are different types of the so-called platelet glycoproteins. As expected, all cell lines express at least one of these markers. The "von Willebrand Factor" antigen is also associated with cells committed to megakaryocytic differentiation, but was detected on only a few cell lines (on 9/18 tested, 50%). Another informative megakaryocyte-specific marker is the platelet peroxidase (PPO) which requires electron microscopic detection with specific antibodies: 18 (75%) cell lines are positive, 6 cell lines are negative. In light of the close relationship of immature erythroid and megakaryocytic cells with presumably a common immunologically and functionally definable precursor, it is not surprising that 8/25 (32%) cell lines tested for both markers are GlyA+ PPO+.

Expression of the progenitor cell marker CD34 (27/41 [65%] cell lines are positive) confirms the immaturity of the cell lines. This notion is underlined by the detection of HLA-DR on 30/41 (73%) cell lines. Finally, most cell lines examined express one or more surface adhesion molecules.

A typical immunoprofile of an erythroid-megakaryocytic cell line is: T-antigen negative, B-antigen negative, CD13+, CD14–, CD15–, CD33+, CD34+, CD41+, CD42+, CD61+, GlyA+, HLA-DR+, PPO+.

# 4. CYTOKINE-RELATED CHARACTERIZATION

Two cytokines uniquely associated with the erythroid and megakaryocytic cell lineages promote survival, proliferation and differentiation at various stages of maturation: EPO and TPO [18,20,111,112]. Appropriately, several cell lines express the receptors for EPO and TPO, the latter also known as MPL [20,113] (Table 3). With regard to cytokine production, PDGF mRNA expression and protein production is of particular interest and seen in several cell lines [16]. As most cell lines grow autonomously in culture, without the addition of external growth factors apart from those present in the FBS, it is no surprise that the majority of the cell lines do not respond to cytokine exposure, either with cellular proliferation or differentiation.

An exception to the above rule are the so-called growth factor-dependent leukemia cell lines (Appendix 3, p. 287). These cell lines will die by apoptosis within 1–3 days in the absence of specific growth factors, although FBS can delay this factor withdrawal-induced cell death. Twenty-two percent of the erythroid-megakaryocytic cell lines are constitutively growth factor-dependent, mainly on GM-CSF or IL-3. The cytokine dependency of some of these cell lines could be switched to TPO [42,80,89]. In addition, these factor-dependent cell lines are proliferatively responsive to a large panel of other cytokines [89]. The differentiation-inducing effects of cytokines on erythroid-megakaryocytic cell lines are rather limited, reflecting their primary action in the enhancement of survival and proliferation.

## 5. GENETIC CHARACTERIZATION

Detailed cytogenetic karyotypes are available for 44 of the 49 erythroid-megakaryocytic cell lines (Table 4). As already seen in other leukemia-lymphoma cell lines, there are complex numerical and structural chromosomal aberrations. Among the many abnormalities, the occurrence of the so-called Philadelphia chromosome stands out. The majority of the erythroid-megakaryocytic cell lines have been established from patients with CML, a disease which is highly associated (>90%) with the unique translocation (9;22). This translocation and the resulting *BCR-ABL* fusion gene are seen in many cell lines [114–116] (Appendix 4). The t(3;21) and the inv(3) alterations causing the *EVI1-AML1* fusion gene and *EVI1* overexpression have been described for cell lines MOLM-1 and SKH1, respectively. Amplifications of the *MLL* and *ABL* genes in the cell lines UoC-M1 and K-562, respectively, are of further note.

Homozygous or heterozygous deletions of the tumor suppressor genes *P15INK4B* and *P16INK4A*, both mapped to chromosome region 9p21 [117,118], have been found in several cell lines (including F-36P, HEL, JK-1, K-562, LAMA-84, MEG-01, and UT-7). Their functional homologue genes *P18INK4C* and *P19INK4D* were not deleted or otherwise altered in any of the cell lines. Point mutations or other alterations of the *P53* tumor suppressor gene [119] were reported for cell lines CMK, CMY, F-36P, LAMA-84, MC3, MEG-01, and UT-7.

## 6. FUNCTIONAL CHARACTERIZATION

The doubling times of the cell lines listed in Table 5 are mostly in the range of 1–2 days. None of the cell lines tested is positive for EBV. In cytochemical staining, these cell lines showed the following results: 22/24 (91%) ACP+, 2/13 (15%) ALP+, 16/24 (66%) ANAE+, 9/23 (39%) ANBE+, 8/14 (57%) benzidine+, 1/24 (4%) CAE+, 0/37 MPO+, 29/33 (87%) PAS+, and 2/12 (16%) SBB+. Thus, the typical cytochemical profile of an erythroid-megakaryocyte cell line is ACP+ PAS+ ANAE± ANBE± benzidine± ALP– CAE– MPO– SBB–.

The majority of the cell lines have been subjected to induction of differentiation using pharmacological (AraC, DMSO, sodium butyrate, TPA) or physiological reagents ($\delta$-ALA, hemin, retinoic acid, vitamin D3). Depending on the inducer used and the stage of differentiation at which the cells are arrested, the cells can be triggered to differentiate along the megakaryocytic, erythroid, or monocytic/macrophage cell axes. AraC, $\delta$-ALA and hemin induce erythroid differentiation, whereas DMSO, sodium butyrate and retinoic acid promote either erythroid or megakaryocytic differentiation, presumably depending on the stringency of commitment to one cell lineage or the other. Besides vitamin D3, TPA is the most effective inducer of megakaryocytic differentiation. However, some TPA-treated cell lines can acquire monocytic/macrophage-associated features [120]. Bipotency with regard to induced differentiation along the erythroid or megakaryocytic lineages was reported for 16 cell lines (Table 5).

Nine cell lines have been reported to be heterotransplantable in nude or SCID mice (Appendix 5, p. 289). Specific functional features of erythroid cells are the expression of hemoglobin (detectable at the protein level by benzidine staining or isoelectric focusing) and of the various globin chains (at the mRNA or protein level). Of the various globin chains, the embryonic chains predominate, including Hb Gower I/II, Hb Bart's, Hb Portland, and fetal hemoglobin HbF. Rare cell lines express adult HbA. The various types of hemoglobin are composed of heme and different globin chains, i.e. $\alpha$-, $\beta$-, $\gamma$-, $\delta$-, $\epsilon$- and $\zeta$-globin. Specific morphological features of megakaryocytes are $\alpha$-granules and demarcation membranes. Of the cell lines that were examined for these two specific parameters, 11/17 (64%) lines showed $\alpha$-granules and/or demarcation membranes and 19/21 (90%) lines expressed hemoglobin or globin (Table 5). Finally, several cell lines express the GATA-1, GATA-2 and SCL transcription factors that are associated with both cell lineages.

## 7. UNCONFIRMED CELL LINES

Several cell lines with erythroid-megakaryocytic characteristics not listed in Table 1 have been reported in the literature (Table 6). However, these cell lines are either not sufficiently characterized, not immortalized (EST-IU), not of leukemic origin (B-403, P-320, S-1214), or are cross-contaminated (DAMI). For the insufficiently characterized cell lines (AML-HJ, KG-91, KOPMK 53, MEG-J, TK-91, TW14-8), essential descriptive parameters such as clinical data, authentication, immunoprofile, and karyotype have yet to be published.

# 8. APPENDICES

*Appendix 1.* Myeloid cell lines available from major cell banks

| Cell line | Cell type | Cell bank[1] | Catalogue no. |
|---|---|---|---|
| AML-193 | monocytic | ATCC | CRL 9589 |
| CMK | erythroid-megakaryocytic | DSMZ | ACC 392 |
| | | IFO | IFO 50428/30 |
| CTV-1 | monocytic | DSMZ | ACC 40 |
| EM-2 | myelocytic | DSMZ | ACC 135 |
| EM-3 | myelocytic | DSMZ | ACC 134 |
| EOL-1 | myelocytic (eosinophil) | DSMZ | ACC 386 |
| | | ECACC | ECACC 94042252 |
| | | RIKEN | RCB0641 |
| F-36E | erythroid-megakaryocytic | RIKEN | RCB 0776 |
| F-36P | erythroid-megakaryocytic | RIKEN | RCB 0775 |
| GDM-1 | myelocytic | DSMZ | ACC 87 |
| HEL | erythroid-megakaryocytic | ATCC | TIB 180 |
| | | DSMZ | ACC 11 |
| | | ECACC | ECACC 92111706 |
| | | JCRB | JCRB 0062 |
| HL-60 | myelocytic | ATCC | CCL 240 |
| | | DSMZ | ACC 3 |
| | | ECACC | ECACC 88120805 |
| | | IFO | IFO 50022 |
| | | JCRB | JCRB 0085 |
| | | RIKEN | RCB 0041 |
| HNT-34 | myelocytic | RIKEN | RCB 1296 |
| JK-1 | erythroid-megakaryocytic | DSMZ | ACC 347 |
| JOSK-I | monocytic | DSMZ | ACC 155 |
| JOSK-M | monocytic | DSMZ | ACC 30 |
| K-562 | erythroid-megakaryocytic | ATCC | CCL 243 |
| | | DSMZ | ACC 10 |
| | | JCRB | JCRB 0019 |
| | | RIKEN | RCB 0027 |
| Kasumi-1 | myelocytic | DSMZ | ACC 220 |
| KG-1 | myelocytic | ATCC | CRL 8031/CCL 246 |
| | | DSMZ | ACC 14 |
| | | JCRB | JCRB 0611 |
| | | RIKEN | RCB 1166 |
| KG-la | myelocytic | ATCC | CCL 246.1 |
| | | DSMZ | ACC 421 |

*Continued on next page*

*Appendix 1.* (continued)

| Cell line | Cell type | Cell bank[1] | Catalogue no. |
|---|---|---|---|
| KMOE-2 | erythroid-megakaryocytic | DSMZ | ACC 37 |
| KMT-2 | monocytic | RIKEN | RCB 0712 |
| KU-812 | myelocytic (basophil) | DSMZ | ACC 378 |
| | | ECACC | ECACC 90071807 |
| | | IFO | IFO 50363 |
| | | JCRB | JCRB 0104 |
| | | RIKEN | RCB 0495 |
| KU-812E | myelocytic (basophil) | ECACC | ECACC 90071803 |
| | | JCRB | JCRB 0104.1 |
| | | RIKEN | RCB 0496 |
| KU-812F | myelocytic (basophil) | ECACC | ECACC 90071804 |
| | | JCRB | ICRB 0104.2 |
| | | RIKEN | RCB 0497 |
| KY821 | myelocytic | JCRB | JCRB 0105 |
| LAMA-84 | erythroid-megakaryocytic | DSMZ | ACC 168 |
| M-07e | erythroid-megakaryocytic | DSMZ | ACC 104 |
| MEG-01 | erythroid-megakaryocytic | ATCC | CRL 2021 |
| | | | ACC 364 |
| IFO | | DSMZ | IFO 50151 |
| ML-2 | monocytic | DSMZ | ACC 15 |
| Mono Mac 1 | monocytic | DSMZ | ACC 252 |
| Mono Mac 6 | monocytic | DSMZ | ACC 124 |
| MUTZ-2 | myelocytic | DSMZ | ACC 271 |
| MUTZ-3 | monocytic | DSMZ | ACC 295 |
| MV4-11 | monocytic | ATCC | ATCC 9591 |
| | | DSMZ | ACC 102 |
| NB4 | myelocytic | DSMZ | ACC 207 |
| OCI/AML-2 | monocytic | DSMZ | ACC 99 |
| OCI/AML-5 | myelocytic | DSMZ | ACC 247 |
| P31/Fujioka | monocytic | JCRB | JCRB 0091 |
| P39/Tsugane | monocytic | JCRB | JCRB 0092 |
| PLB-985 | monocytic | DSMZ | ACC 139 |
| RC-2A | monocytic | DSMZ | ACC 6 |
| SCC-3 | monocytic | JCRB | JCRB 0115 |
| SKM-1 | monocytic | JCRB | JCRB 0118 |
| SPI-801 | erythroid-megakaryocytic | DSMZ | ACC 86 |
| SPI-802 | erythroid-megakaryocytic | DSMZ | ACC 92 |

*Continued on next page*

*Appendix 1.* (continued)

| Cell line | Cell type | Cell bank[1] | Catalogue no. |
|---|---|---|---|
| TF-1 | erythroid-megakaryocytic | ATCC | CRL 2003 |
| | | DSMZ | ACC 334 |
| THP-1 | monocytic | ATCC | TIB 202 |
| | | DSMZ | ACC 16 |
| | | JCRB | JCRB 0112 |
| | | RIKEN | RCB 1189 |
| TMM | EBV+ B-LCL | DSMZ | ACC 95 |
| U-937 | monocytic | ATCC | CRL 1593 |
| | | DSMZ | ACC 5 |
| | | IFO | IFO 50038 |
| | | JCRB | JCRB 9021 |
| | | RIKEN | RCB 0435 |
| UT-7 | erythroid-megakaryoctic | DSMZ | ACC 137 |
| YNH-1 | myelocytic | RIKEN | RCB 1291 |

[1] ATCC – American Type Culture Collection; DSMZ – German Collection of Microorganisms and Cell Cultures; ECACC – European Collection of Animal Cell Cultures; IFO – Institute for Fermentation Osaka (now = HSRRB-Health Science Research Resources Bank; JCRB – Japanese Collection of Research Bioresources; RIKEN – Cell Bank.

*Appendix 2.* Myeloid cell lines growing in serum-free media

| Cell line | Cell type | Culture medium |
|---|---|---|
| AML-193 | monocytic | serum-free: IMDM + 2 ng/ml GM-CSF or 3 U/ml IL-3 |
| GF-D8 | myelocytic | serum-free: RPMI 1640 + 50 ng/ml GM-CSF or + 20% FBS] |
| MHH 225 | erythroid-megakaryocytic | serum-free: RPMI 1640 [or + 10% FBS] |
| KU-812-F | myelocytic | serum-free: RPMI 1640 [or + 10% FBS] |
| MV4-11 | monocytic | serum-free: IMDM + 5 ng/ml GM-CSF [or + 10% FBS without GM-CSF] |
| NKM-1 | myelocytic | serum-free: RPMI 1640 [or + 10% FBS] |
| NOMO-1 | monocytic | serum-free: RPMI 1640 [or + 10% FBS] |
| PLB-985 | monocytic | serum-free: RPMI 1640 [or + 10% FBS] |
| T-33 | erythroid-megakaryocytic | serum-free: RPMI 1640 [or + 10% FBS] |

*Appendix 3.* Growth factor-dependent myeloid cell lines

| Cell line | Cell type | Absolute dependency on cytokines |
|---|---|---|
| AML-193 | monocytic | GM-CSF or IL-3 |
| AS-E2 | erythroid-megakaryocytic | EPO |
| ELF-153 | erythroid-megakaryocytic | GM-CSF |
| F-36P | erythroid-megakaryocytic | EPO, GM-CSF or IL-3 |
| FKH-1 | myelocytic | G-CSF or GM-CSF |
| GF-D8 | myelocytic | GM-CSF or IL-3 |
| GM/SO | myelocytic | GM-CSF |
| GRW | erythroid-megakaryocytic | SCF |
| HML | erythroid-megakaryocytic | GM-CSF |
| HU-3 | erythroid-megakaryocytic | GM-CSF, IL-3 or TPO |
| IRTA17 | myelocytic | G-CSF, GM-CSF or SCF |
| M-07e | erythroid-megakaryocytic | GM-CSF, IL-3 or TPO |
| M-MOK | erythroid-megakaryocytic | GM-CSF or TPO |
| MB-02 | erythroid-megakaryocytic | GM-CSF |
| MDS92 | myelocytic | IL-3 |
| M-TAT | erythroid-megakaryocytic | EPO, GM-CSF or SCF |
| MUTZ-2 | myelocytic | 5637 CM or SCF |
| MUTZ-3 | monocytic | 5637 CM, GM-CSF, IL-3 or PIXY-321 |
| OCI/AML-1 | myelocytic | 5637 CM or G-CSF |
| OCI/AML-4 | myelocytic | grows better with GM-CSF, IL-3 or SCF |
| OCI/AML-5 | myelocytic | 5637 CM or GM-CSF or IL-3 |
| OHN-GM | myelocytic | GM-CSF |
| OIH-1 | myelocytic | G-CSF, GM-CSF |
| SKNO-1 | myelocytic | GM-CSF |
| TF-1 | erythroid-megakaryocytic | GM-CSF, IL-3 or TPO |
| UCSD/AML1 | myelocytic | GM-CSF |
| UG3 | monocytic | GM-CSF or IL-3 |
| UT-7 | erythroid-megakaryocytic | EPO, GM-CSF, IL-3 or TPO |
| YNH-1 | myelocytic | G-CSF, GM-CSF or IL-3 |

*Appendix 4.* Myeloid cell lines with translocations and fusion genes

| Chromosomal abnormality | Fusion genes | Cell line | Cell type | Remarks |
|---|---|---|---|---|
| t(1;12)(q25;p13) | *ETV6-ARG* | HT93 | myelocytic | |
| t(3;3)(q21;q26) | *EVI1* | HNT-34 | myelocytic | EVI1 overexpression |
| | | UCSD/AML1 | myelocytic | EVI1 overexpression |
| inv(3)(q21q26) | *EVI1* | Kasumi-4 | myelocytic | EVI1 overexpression |
| | | MOLM-1 | erythroid-megakaryocytic | EVI1 overexpression |
| | | MUTZ-3 | monocytic | |
| t(3;7)(q27;q22) | *EVI1* | Kasumi-3 | myelocytic | EVI1 overexpression |
| t(3;21)(q26;q22) | *EVI1-AML1* | SKH1 | erythroid-megakaryocytic | |
| t(4;11)(q21;q23) | *MLL-AF4* | KOCL-48 | monocytic | |
| | | MV4-11 | monocytic | |
| t(6;9)(p23;q34) | *DEK-CAN* | FKH-1 | myelocytic | |
| t(6;11)(q27;q23) | *MLL-AF6* | ML-2 | monocytic | |
| | | CTS | myelocytic | |
| t(8;21)(q22;q22) | *ETO-AML1* | Kasumi-1 | myelocytic | |
| | | SKNO-1 | myelocytic | |
| t(9;11)(p22;q23) | *MLL-AF9* | IMS-M1 | monocytic | |
| | | Mono Mac 6 | monocytic | |
| | | THP-1 | monocytic | |
| | | UG3 | monocytic | |
| ins(11;9)(q23;p22p23) | *MLL-AF9* | MOLM-13 | monocytic | |
| t(9;22)(q34;q11)Philadelpha | *BCR-ABL* | AP-217 | erythroid-megakaryocytic | b3-a2 fusion |
| | | AR230 | myelocytic | e19/e18-a2 fusion |
| | | CML-C-1 | myelocytic | |
| | | EM-2 | myelocytic | b3-a2 fusion |
| | | GM/SO | myelocytic | b3-a2 fusion |
| | | HNT-34 | myelocytic | M-bcr/m-bcr fusion |
| | | JK-1 | erythroid-megakaryocytic | b2-a2 fusion |
| | | JURL-MK1 | erythroid-megakaryocytic | b3-a2 fusion |
| | | K-562 | erythroid-megakaryocytic | b3-a2 fusion |
| | | Kasumi-4 | myclocytic | b2-a2 fusion |
| | | KBM-5 | monocytic | b3-a2 fusion |
| | | KBM-7 | myelocytic | b2-a2 fusion |
| | | KCL-22 | myelocytic | b2-a2 fusion |
| | | KH88 | erythroid-megakaryocytic | b3-a2 fusion |
| | | KOPM-28 | myelocytic | b3-a2 fusion |
| | | KOPM 30 | myelocytic | el-a2 fusion |
| | | KOPM 55 | myelocytic | |
| | | KT-1 | myelocytic | b2-a2 fusion |
| | | KU-812 | myelocytic | b3-a2 fusion |
| | | KYO-1 | myelocytic | b2-a2 fusion |
| | | LAMA-84 | erythroid-megakaryocytic | b3-a2 fusion |
| | | MC3 | erythroid-megakaryocytic | b3-a2 fusion |
| | | MEG-01 | erythroid-megakaryocytic | b2-a2 fusion |
| | | MEG-A2 | erythroid-megakaryocytic | b3-a2 fusion |
| | | MOLM-1 | erythroid-megakaryocytic | b2-a2 fusion |
| | | RM10 | erythroid-megakaryocytic | M-bcr fusion |
| | | RWLeu-4 | monocytic | b2-a2 fusion |
| | | SAM-1 | erythroid-megakaryocytic | b3-a2 fusion |
| | | SKH1 | erythroid-megakaryocytic | b3-a2 fusion |
| | | T-33 | erythroid-megakaryocytic | b3-a2 fusion |
| | | TS9;22 | erythroid-megakaryocytic | b3-a2 fusion |
| | | Y-1K | erythroid-megakaryocytic | b3-a2 fusion |
| | | YN-1 | erythroid-megskaryocytic | b3-a2 fusion |
| | | YOS-M | myelocytic | b2-a2 fusion |
| | | YS9;22 | erythroid-megakaryocytic | b3-a2 fusion |
| t(10;11)(p14;q23) | *MLL-CALM* | U-937 | monocytic | not *MLL-AF10* |
| t(12;22)(p13;q12) | *ETV6/TEL-MN1* | MUTZ-3? | monocytic | |
| | | UCSD/AML1 | myelocytic | |
| t(15;17)(q22;q21) | PML-RARA | Ei501 | myelocytic | |
| | | HT93 | myelocytic | bcr3 fusion |
| | | NB4 | myelocytic | bcrl-2 fusion |
| | | UF-1 | myclocytic | |
| inv(16)(p13q22) | *CBFB-MYHII* | ME-1 | monocytic | |
| t(i6;21)(p11;q22) | *TLS/FUS-ERG* | IRTA17 | myelocytic | |
| | | YNH-1 | myelocytic | |

*Appendix 5.* Heterotransplantable myeloid cell lines

| Cell line | Cell type | Heterotransplantation |
|---|---|---|
| AML-CL[1] | myelocytic | into SCID mice |
| AML-PS[1] | myelocytic | into SCID mice |
| CHRF-288-11 | erythroid-megakaryocytic | into nude mice |
| CML-C-1 | myelocytic | into nude mice |
| ELF-153 | erythroid-megakaryocytic | into nude mice |
| EM-2 | myelocytic | into SCID mice |
| HL-60 | myelocytic | into nude mice or SCID mice |
| HMC-1 | mast cell | into nude mice |
| K-562 | erythroid-megakaryocytic | into nude mice or SCID mice |
| KBM-3 | monocytic | into nude mice or SCID mice |
| KBM-5 | monocytic | into SCID mice |
| KBM-7 | myelocytic | into nude mice |
| KCL-2 | myelocytic | into newborn hamsters |
| KG-1 | myelocytic | into nude or SCID mice |
| KMOE-2 | erythroid-megakaryocytic | into nude mice |
| LAMA-84 | erythroid-megakaryocytic | into estrone-treated nude mice |
| M-07e | erythroid-megakaryocytic | into SCID mice |
| MKPL-1 | erythroid-megakaryocytic | into nude mice |
| OCIM2 | erythroid-megakaryocytic | into SCID mice |
| OMA-AML-1 | monocytic | into nude or SCID mice |
| P39/Tsugane | monocytic | into nude mice |
| PLB-985 | monocytic | into nude mice |
| RWLeu-4 | monocytic | into nude mice |

[1] Cell lines can be maintained and serially transplanted, but do not grow long-term *in vitro*

*Appendix 6.* Cross-contaminated cell lines

| Purported cell line | Purported cell type | Subclone of cell line | Actual cell type |
|---|---|---|---|
| DAMI | erythroid-megakaryocytic | HEL | erythroid-megakaryocytic |
| J-111 | monocytic | HELA | cervix carcinoma |
| JOSK-I | monocytic | U-937 | monocytic |
| JOSK-K | monocytic | U-937 | monocytic |
| JOSK-M | monocytic | U-937 | monocytic |
| JOSK-S | monocytic | U-937 | monocytic |
| LR10.6 | precursor B-cell | NALM-6 | precursor B-cell |
| MOBS-1 | monocytic | U-937 | monocytic |
| MOLT-15 | T-cell | CTV-1 | monocytic |
| PBEI | precursor B-cell | NALM-6 | precursor B-cell |
| SPI-801 | T-cell | K-562 | erythroid-megakaryocytic |
| SPI-802 | T-cell | K-562 | erythroid-megakaryocytic |

# Abbreviations

ACP – acid phosphatase
ADCC – antigen-dependent cell-mediated cytotoxicity
ALL – acute lymphoblastic leukemia
ALP – alkaline phosphatase
AML – acute myeloid leukemia
ANAE – $\alpha$-naphthylacetate esterase
ANBE – $\alpha$-naphthylbutyrate esterase
AraC – cytosine arabinoside
ATCC – American Type Culture Collection
ATRA – all-trans retinoic acid
AUL – acute undifferentiated leukemia
$\beta$-TG – $\beta$-thromboglobulin
baso – basophil
BC – blast crisis (of CML)
BCGF – B-cell growth factor
BCP – B-cell precursor
bFGF – basic fibroblast growth factor
BM – bone marrow
BMT – bone marrow transplantation
CAE – naphthol AS-D chloroacetate esterase
cALL – common acute lymphoblastic leukemia
CLL – chronic lymphocytic leukemia
CM – conditioned medium
CML – chronic myeloid leukemia
CMML – chronic myelomonocytic leukemia
CP – chronic phase (of CML)
CSF – colony-stimulating factor or cerebrospinal fluid
cyIg – cytoplasmic immunoglobulin
D – at diagnosis of disease
$\delta$-ALA – $\delta$-aminolevulinic acid
DMEM – Dulbecco's modified essential medium
DMSO – dimethylsulfoxide
DSMZ – Deutsche Sammlung von Mikroorganismen und Zellkulturen
dt – doubling time
EBNA – Epstein-Barr virus nuclear antigen
EBV – Epstein-Barr virus
ECACC – European Collection of Animal Cell Cultures
EDF – erythroid differentiation factor
EGF – epidermal growth factor
ELISA – enzyme-linked immunoassay
eosino – eosinophil
EPO – erythropoietin
ery – erythroid
F – female
FAB – French–American–British morphological AML/ALL/MDS classifications
FBS – fetal bovine serum
FL – FLT3 ligand
FN – fibronectin
G-CSF – granulocyte CSF
GLC – $\beta$-glucuronidase
GlyA – glycophorin A
GM-CSF – granulocyte-macrophage CSF

HS – horse serum
hu – human
IFN – interferon
Ig – immunoglobulin
IGF – insulin-like growth factor
IGH/K/L – immunoglobulin heavy/kappa light/lambda light chain gene
IL – interleukin
IMDM – Iscove's modified Dulbecco's medium
jCML – juvenile chronic myeloid leukemia
JCRB – Japanese Cancer Research Resources Bank
LCL – lymphoblastoid cell line
LGL – large granular lymphocytes
LIF – leukemia inhibitory factor
lym – lymphoid
M – male
M1 – (immature) myeloblastic AML
M2 – myeloblastic AML
M3 – promyelocytic AML
M4 – myelomonocytic AML
M5 – monocytic AML
M6 – erythroid AML
M7 – megakaryocytic AML
macro – macrophage
M-CSF – macrophage CSF
MDR – multiple drug resistance
MDS – myelodysplastic syndromes
meg – megakaryocytic
MEM – minimum essential medium
MIP – macrophage inflammatory protein
MLR – mixed lymphocyte reaction
mono – monocytic
MPO – myeloperoxidase
MPD – myeloproliferative disorder
MSE – monocyte-specific esterase
my – myeloid
NBT – nitroblue tetrazolium
NCS – newborn calf serum
neutro – neutrophil
NGF – nerve growth factor
NHL – Non-Hodgkin's lymphoma
NK – natural killer
OSM – oncostatin M
PAS – periodic acid Schiff
PB – peripheral blood
PDGF – platelet-derived growth factor
PE – pleural effusion
PF-4 – platelet factor-4
Ph – Philadelphia chromosome
PPO – platelet peroxidase
R – at relapse of disease
RAEB – refractory anemia with excess of blasts
RAEBT – refractory anemia with excess of blasts in transformation
RARS – refractory anemia with ring sideroblasts
RIKEN – Riken (Japanese cell bank)

RIA – radioimmunoassay
RPMI – RPMI 1640 medium
RT-PCR – reverse transcriptase-polymerase chain reaction
SBB – Sudan black B
SCF – stem cell factor
sIg – surface immunoglobulin
T – during therapy of disease
t-AML/MDS – therapy-related AML/MDS
TCRA/B/G/D - T-cell receptor $\alpha$, $\beta$, $\gamma$, $\delta$-chain
TGF – transforming growth factor
TNF – tumor necrosis factor
TPA – phorbol ester 12-0-tetradecanoyl phorbol-13 acetate
TPO – thrombopoietin
TRAP – tartrate-resistant acid phosphatase
Tu – tumor
Vit. D3 – vitamin D3
vWF – von Willebrand factor

# References

1. Bessis M (ed.) *Blood Smears Reinterpreted*, Springer, Berlin, 1977.
2. Castoldi GL, Beutler E. In: Zucker-Franklin D, Greaves MF, Grossi CE, Marmont AM (eds.), *Atlas of Blood Cells. Function and Pathology*, 2nd edn., Fischer, Stuttgart, pp 47–156, 1988.
3. Zucker-Franklin D. In: Zucker-Franklin D, Greaves MF, Grossi CE, Marmont AM (eds.), *Atlas of Blood Cells. Function and Pathology*, 2nd edn., Fischer, Stuttgart, pp 619–693, 1988.
4. Bennett JM, Catovsky D, Daniel MT et al. *Br J Haematol* 33: 451–458, 1976.
5. Bennett JM, Catovsky D, Daniel MT et al. *Ann Int Med* 103: 460–462, 1985.
6. Bennett JM, Catovsky D, Daniel MT et al. *Br J Haematol* 87: 746–754, 1994.
7. Hassan HT, Freund M. *Leukemia Res* 19: 589–594, 1995.
8. Hassan HT, Drexler HG. *Leukemia Lymphoma* 20: 1–15, 1995.
9. Hoffman R. *Blood* 74: 1196–1212, 1989.
10. McDonald TP, Sullivan PS. *Exp Hematol* 21: 1316–1320, 1993.
11. Romeo PH, Prandini MH, Joulin V et al. *Nature* 344: 447–449, 1990.
12. Ohyashiki K, Ohyashiki JH, Shimamoto T et al. *Leukemia Lymphoma* 23: 431–436, 1996.
13. Ito E, Kasai M, Toki T et al. *Leukemia Lymphoma* 23: 545–550, 1996.
14. Long MW, Heffner CH, Williams JL et al. *J Clin Invest* 85: 1072–1084, 1990.
15. Nakahata T, Okumura N. *Leukemia Lymphoma* 13: 401–409, 1994.
16. Papayannopoulou T, Raines E, Collins S et al. *J Clin Invest* 79: 859–866, 1987.
17. Tani T, Ylänne J, Virtanen I et al. *Exp Hematol* 24: 158–168, 1996.
18. Kaushansky K. *Blood* 86: 419–431, 1995.
19. Kaushansky K, Broudy VC, Grossman A et al. *J Clin Invest* 96: 1683–1687, 1995.
20. Drexler HG, Quentmeier H. *Leukemia* 10: 1405–1421, 1996.
21. Fraser JK, Tan AS, Lin FK et al. *Exp Hematol* 17: 10–16, 1989.
22. Debili N, Coulombel L, Croisille L et al. *Blood* 88: 1284–1296, 1996.
23. Mossuz P, Prandini MH, Leroux D et al. *Leukemia Res* 21: 529–537, 1997.
24. Miyazaki Y, Kuriyama K, Higuchi M et al. *Leukemia* 11: 1941–1949, 1997.
25. Bonsi L, Grossi A, Strippoli P et al. *Br J Haematol* 98: 549–559, 1997.

26. Witte DP, Harris RE, Jenski LJ et al. *Cancer* 58: 238–244, 1986.
27. Fugman DA, Witte DP, Jones CLA et al. *Blood* 75: 1252–1261, 1990.
28. Toba K, Kishi K, Koike T et al. *Exp Hematol* 24: 894–901, 1996.
29. Komatsu N, Suda T, Moroi M et al. *Blood* 74: 42–48, 1989.
30. Sato T, Fuse A, Eguchi M et al. *Br J Haematol* 72: 184–190, 1989.
31. Sato S, Ohnuma N, Tanabe M et al. *Leukemia Lymphoma* 36: 397–404, 2000.
32. Miura N, Sato T, Fuse A et al. *Int J Mol Med* 1: 559–563, 1998.
33. Mouthon MA, Freund M, Titeux M et al. *Blood* 84: 1085–1097, 1994.
34. Hassan HT, Hanauske AR, Lux E et al. *Int J Exp Pathol* 76: 361–367, 1995.
35. Chiba S, Takaku F, Tange T et al. *Blood* 78: 2261–2268, 1991.
36. Zipursky A, Vanek W, Grunberger T et al. *Blood* 82: 122a, 1993.
37. Martin P, Papayannopoulou T. *Science* 216: 1233–1235, 1982.
38. Cheng T, Erickson-Miller CL, Li C et al. *Leukemia* 9: 1257–1263, 1995.
39. Ma F, Koike K, Higuchi T et al. *Br J Haematol* 100: 427–435, 1998.
40. Morgan D. *Blood* 82: 373a, 1993.
41. Morgan D, Class R, Soslau G et al. *Exp Hematol* 25: 1378–1385, 1997.
42. Drexler HG, Zaborski M, Quentmeier H. *Leukemia* 11: 541–551, 1997.
43. Okuno Y, Suzuki A, Ichiba S et al. *Cancer* 66: 1544–1551, 1990.
44. Di Noto R, Luciano L, Lo Pardo C et al. *Leukemia* 11: 1554–1564, 1997.
45. Lozzio CB, Lozzio BB. *Blood* 45: 321–334, 1975.
46. Yasanuga M, Ryo R, Sawada YKS et al. *Ann Hematol* 68: 145–151, 1994.
47. Furukawa T, Koike T, Ying W et al. *Leukemia* 8: 171–180, 1994.
48. Okano H, Okamura J, Yagawa K et al. *J Cancer Res Clin Oncol* 102: 49–55, 1981.
49. Seigneurin D, Champelovier P, Mouchiroud G et al. *Exp Hematol* 15: 822–832, 1987.
50. Champelovier P, Valiron O, Michèle J et al. *Leukemia Res* 18: 903–918, 1994.
51. Avanzi GC, Lista P, Giovinazzo B et al. *Br J Haematol* 69: 359–366, 1988.
52. Avanzi GC, Brizzi MF, Giannotti J et al. *J Cell Physiol* 145: 458–464, 1990.
53. Itano M, Tsuchiya S, Minegishi N et al. *Exp Hematol* 23: 1301–1309, 1995.
54. Morgan DA, Gumucio DL, Brodsky I. *Blood* 78: 2860–2871, 1991.
55. Okabe M, Kunieda Y, Nakane S et al. *Leukemia Lymphoma* 16: 493–503, 1995.
56. Ogura M, Morishima Y, Ohno R et al. *Blood* 66: 1384–1392, 1985.
57. Ogura M, Morishima Y, Okumura M et al. *Blood* 72: 49–60, 1988.
58. Abe A, Emi N, Kato H et al. *Leukemia* 9: 341–349, 1995.
59. Higashitani A, Yokoyama S, Shimizu Y et al. *J Jpn Pediatr Soc* 99: 920–928, 1995.
60. Hassan HT, Grell S, Borrmann-Danso U et al. *Hematol Oncol* 12: 61–66, 1994.
61. Hassan HT, Petershofen E, Lux E et al. *Ann Hematol* 71: 111–117, 1995.
62. Takeuchi S, Sugito S, Uemura Y et al. *Leukemia* 6: 588–594, 1992.
63. Matsuo Y, Adachi T, Tsubota T et al. *Human Cell* 4: 261–264, 1991.
64. Tsuji-Takayama K, Kamiya T, Nakamura S et al. *Human Cell* 7: 167–171, 1994.
65. Minegishi N, Minegishi M, Tsuchiya S et al. *J Biol Chem* 269: 27700–27704, 1994.
66. Tsuyuoka R, Takahashi T, Suzuki A et al. *Stem Cells* 13: 54–64, 1995.
67. Papayannopoulou T, Nakamoto B, Kurachi S et al. *Blood* 72: 1029–1038, 1988.
68. Hirata J, Sato H, Takahira H et al. *Leukemia* 4: 365–372, 1990.
69. Skinnider LF, Pabello P, Zheng CY. *Acta Haematol* 98: 26–31, 1997.
70. Kamesaki H, Michaud GY, Irving SG et al. *Blood* 87: 999–1005, 1996.
71. Uozumi K, Otsuka M, Ohno N et al. *Leukemia* 14: 142–152, 2000.
72. Mitani K, Ogawa S, Tanaka T et al. *EMBO J* 13: 504–510, 1994.
73. Tange T, Nakahara K, Mitani K et al. *Cancer Res* 48: 6137–6144, 1988.
74. Kitamura T, Tojo A, Kuwaki T et al. *Blood* 73: 375–380, 1989.

75. Kitamura T, Tange T, Terasawa T et al. *J Cell Physiol* 140: 323–334, 1989.
76. Ohyashiki K, Ohyashiki JH, Fujieda H et al. *Leukemia* 8: 2169–2173, 1994.
77. Allen RJ, Smith SD, Moldwin RL et al. *Leukemia* 12: 1119–1127, 1998.
78. Komatsu N, Nakauchi H, Miwa A et al. *Cancer Res* 51: 341–348, 1991.
79. Komatsu N, Yamamoto M, Fujita H et al. *Blood* 82: 456–464, 1993.
80. Komatsu N, Kunitama M, Yamada M et al. *Blood* 87: 4552–4560, 1996.
81. Komatsu N, Kirito K, Shimizu R et al. *Blood* 89: 4021–4033, 1997.
82. Endo K, Harigae H, Nagai T et al. *Br J Haematol* 85: 653–662, 1993.
83. Drexler HG, Dirks W, MacLeod RAF et al. (eds.) *DSMZ Catalogue of Human and Animal Cell Cultures*, 7th edn., DSMZ, Braunschweig, 1999.
84. Micallef M, Matsuo Y, Takayama K et al. *Hematol Oncol* 12: 163–174, 1994.
85. Blom T, Nilsson G, Sundström C et al. *Scand J Immunol* 44: 54–61, 1996.
86. Schumann RR, Nakarai T, Gruss HJ et al. *Blood* 87: 2419–2427, 1996.
87. Okabe M, Kunieda Y, Shoji M et al. *Leukemia Res* 19: 933–943, 1995.
88. Sandrock B, Hudson KM, Williams DE et al. *In Vitro Cell Dev Biol* 32A: 225–233, 1996.
89. Drexler HG, Zaborski M, Quentmeier H. *Leukemia* 11: 701–708, 1997.
90. Quentmeier H, Zaborski M, Graf G et al. *Leukemia* 10: 297–310, 1996.
91. Ogawa S, Hirano N, Sato N et al. *Blood* 84: 2431–2435, 1994.
92. Nakamaki T, Kawamata N, Schwaller J et al. *Br J Haematol* 91: 139–149, 1995.
93. Aguiar RCT, Sill H, Goldman JM et al. *Leukemia* 11: 233–238, 1997.
94. Shiohara M, Spirin K, Said JW et al. *Leukemia* 10: 1897–1900, 1996.
95. Ogawa S, Kurokawa M, Tanaka T et al. *Oncogene* 13: 183–191, 1996.
96. Berthier R, Mossuz P, Valiron O et al. *Blood* 84: 154a, 1994.
97. Shimamoto T, Ohyashiki K, Ohyashiki JH et al. *Blood* 86: 3173–3180, 1995.
98. Kamel-Reid S, Letarte M, Sirard C et al. *Science* 246: 1597–1600, 1989.
99. Kaku M, Yagawa K, Nakamura K et al. *Blood* 64: 314–317, 1984.
100. Scheid C, Fuchs M, Winter S et al. *Blood* 86: 778a, 1995.
101. Morgan DA, Brodsky I. *J Cell Biol* 100: 565–573, 1985.
102. Greenberg SM, Rosenthal DS, Greeley TA et al. *Blood* 72: 1968–1977, 1988.
103. MacLeod RAF, Dirks WG, Reid YA et al. *Leukemia* 11: 2032–2038, 1997.
104. Roth BJ, Sledge GW Jr, Straneva JE et al. *Blood* 72: 202–207, 1988.
105. Sledge GW Jr, Glant M, Jansen J et al. *Cancer Res* 46: 2155–2159, 1986.
106. Kollia P, Politou M, Tassiopoulou A et al. *Blood* 90: 156b, 1997.
107. Kojika S, Sugita K, Inukai T et al. *Leukemia* 10: 994–999, 1996.
108. Kobayashi S, Teramura M, Sugawara I et al. *Blood* 81: 889–893, 1993.
109. Winkler U, Rieder H, März W et al. *Blood* 90: 236b, 1997.
110. Drexler HG. *Leukemia* 1: 697–705, 1987.
111. Barber DL, D'Andrea AD. *Semin Hematol* 29: 293–304, 1992.
112. Youssoufian H, Longmore G, Neumann D et al. *Blood* 81: 2223–2236, 1993.
113. Graf G, Dehmel U, Drexler HG. *Leukemia Res* 20: 831–838, 1996.
114. Drexler HG, MacLeod RAF, Borkhardt A et al. *Leukemia* 9: 480–500, 1995.
115. Drexler HG, Borkhardt A, Janssen JWG. *Leukemia Lymphoma* 19: 359–380, 1995.
116. Drexler HG, MacLeod RAF, Uphoff CC. *Leukemia Res* 23: 207–215, 1999.
117. Siebert R, Willers CP, Opalka B. *Leukemia Lymphoma* 23: 505–520, 1996.
118. Drexler HG. *Leukemia* 12: 845–858, 1998.
119. Drexler HG, Fombonne S, Matsuo Y et al. *Leukemia* 14: 198–206, 2000.
120. Koeffler HP. *Blood* 62: 709–721, 1983.

# Chapter 9

# Non-Hodgkin's B-Lymphoma Cell Lines

Ramzi M. Mohammad, Frances W.J. Beck and Ayad M. Al-Katib
*Department of Hematology and Oncology, Wayne State University and Karmanos Cancer Institute, Lande Research Building Room 317, 550E Canfield, Detroit, MI 48201, USA. Tel: +1-313-577-7919; Fax: +1-313-577-7925; E-mail: mohammad@karmanos.org*

## 1. INTRODUCTION

The non-Hodgkin's lymphomas (NHL) comprise a group of B- or T-lymphocytic malignancies that most commonly originate in lymph nodes. These lymphomas are thought to represent lymphocyte populations that are arrested at a certain stage of the differentiation pathway [3,72,77].

There are several different classification schemes for B-NHL [9,18,30, 37,52,59,63]. Until the 1960s, the classification of the NHL included the terms reticulum cell sarcoma, lymphosarcoma, and giant follicular lymphoma [10]. Later, Rappaport classified lymphomas based on the cell growth and morphology [9,10,47,89]. Other classification systems have been introduced to include types of NHL which did not fit the Rappaport classification. The most widely used classification in the United States is that of Lukes-Collins [60], which was based on the immunologic profile of the lymphoma cell. A number of other classification schemes were reported in Europe [18,31,52]. Recently, two additional classifications, the International Working Formulation (WF) and the Revised European-American Lymphoma (REAL) have been developed [11,35,36,94]. This chapter will use the most recent NHL classification, the REAL, but will include, where applicable, the older classifications, including Rappaport, Lukes-Collins and the International Working Formulation.

In Table 1 we present the classification of non-Hodgkin's B-lymphoma according to the Revised European American Lymphoma (REAL) system and compare the previous classification of the International Working Formulations and the Rappaport classification to this present and newest classification. In Table 2, we list the chromosomal abnormalities frequently detected in NHL. Data listed in Tables 3–5 detail the general characteristics of individual B-NHL cell lines based on REAL classification.

*J.R.W. Masters and B.O. Palsson (eds.), Human Cell Culture Vol. III, 295–320.*
© 2000 *Kluwer Academic Publishers. Printed in the Netherlands.*

*Table 1.* Classifications of non-Hodgkin's B-lymphoma according to the Revised European–American Lymphoma (REAL), Working Formulation (WF) and Rappaport

| Revised European–American lymphoma (REAL) | Working formulation (WF) | Rappaport |
|---|---|---|
| (1) Precursor B-lymphoblastic lymphoma/leukemia (B-LBL) | Lymphoblastic | Lymphoblastic |
| (2) B-cell chronic lymphocytic leukemia/prolymphocytic leukemia/Small lymphocytic lymphoma | Small lymphocytic consistent with CLL<br>Small lymphocytic, plasmacytoid | Well-differentiated lymphocytic diffuse |
| (3) Lymphoplasmacytoid lymphoma | Small lymphocytic, plasmacytoid<br>Diffuse, mixed small and large cell | Well-differentiated lymphocytic, plasmacytoid<br>Diffuse mixed lymphocytic and histiocytic |
| (4) Mantle cell lymphoma | Small lymphocytic<br>Diffuse, small cleaved cell<br>Follicular, small cleaved cell<br>Diffuse, mixed small and large cell<br>Diffuse, large cleaved cell | Intermediately or poorly differentiated<br>lymphocytic, diffuse or nodular |
| (5) Follicular center lymphoma, follicular | | |
| – Grade I | Follicular, predominantly small cleaved cell | Nodular, poorly differentiated, lymphocytic |
| – Grade II | Follicular, mixed small and large cell | Nodular, mixed lymphocytic & histiocytic |
| – Grade III | Follicular, predominantly large cell | Nodular histiocytic |
| (6) Follicular center lymphoma, diffuse, small cell (provisional) | Diffuse, small cleaved cell<br>Diffuse, mixed small and large cell | Diffuse lymphocytic<br>Poorly differentiated |
| (7) Extranodal marginal zone B-cell lymphoma (low grade B-cell lymphoma of MALT type) | Small lymphocytic<br>Diffuse, small cleaved cell<br>Diffuse, mixed small and large cell | Not specifically listed, diffuse mixed<br>lymphocytic and histiocytic |
| Nodal marginal zone B-cell lymphoma (provisional) | Small lymphocytic<br>Diffuse, small cleaved cell<br>Diffuse, mixed small and large cell<br>Unclassified | Nodular or diffuse |

*Continued on next page*

*Table 1.*  (continued)

| Revised European–American lymphoma (REAL) | Working formulation (WF) | Rappaport |
|---|---|---|
| (8) | Splenic marginal zone B-cell lymphoma (provisional) | Small lymphocytic | Not specifically listed, well-differentiated lymphocytic (WDL) or WDL-plasmacytoid |
| | | Diffuse, small cleaved cell | |
| (9) | Diffuse large B-cell lymphoma | Diffuse, large cell | Diffuse histiocytic |
| | | Large cell immunoblastic | |
| (10) | Burkitt's lymphoma | Small non-cleaved cell, Burkitt's | Undifferentiated lymphoma, Burkitt's type |
| (11) | High-grade B-cell lymphoma, Burkitt-like (provisional) | Small non-cleaved cell, non-Burkitt's | Undifferentiated; non-Burkitt's; kind not listed |
| | | Diffuse, large cell | |
| | | Large cell immunoblastic | |

*Table 2.* Chromosomal abnormalities frequently detected in NHL

| Chromosome No. | Type of abnormality | Histological subtypes |
|---|---|---|
| 1 | q-Armbreak | Diffuse large cell |
| 2 | q-Armbreak | Diffuse large cell |
| 3 | Trisomy | Follicular mixed |
|   | t(3;14) | Diffuse large cell lymphoma |
| 6 | q-Armbreak | Diffuse large cell, immunoblastic |
|   |   | Follicular lymphomas |
| 7 | Trisomy | Follicular large cell |
| 8 | Trisomy, | Follicular mixed, small non-cleaved |
|   | t(2;8) |   |
|   | t(8;14) |   |
|   | t(8;22) |   |
| 11 | t(11;14) | Small lymphocytic, mantle cell |
| 12 | Trisomy | Small lymphocytic |
| 14 | q-Armbreak | Most subtypes, follicular lymphomas |
|   |   | Diffuse large cell |
| 17 | I(17q) | Follicular lymphomas, diffuse large cell |
| 18 | t(14;18) | Follicular lymphomas, diffuse large cell |
|   | Trisomy | Diffuse large cell, follicular lymphoma |
| 21 | Trisomy | Diffuse large cell |

The establishment of human EBV-negative (non-transformed) B-NHL cell lines has been difficult. Fifty-nine well-characterized immortal B-NHL cell lines will be described in this chapter. The oldest continuous B-NHL cell line is the Burkitt's lymphoma cell line Raji, reported in 1964 by Pulvertaft [88], but is not listed here because it is EBV+. The oldest established EBV-negative cell line is SKW-4, established from a diffuse histiocytic lymphoma in 1966 [81].

## 2. IMMUNOPHENOTYPE

Approximately 80% of all lymphoid tumors are of B-cell origin, while the remaining 20% are of T-cell origin [27]. The use of a large number of monoclonal antibodies directed against specific antigens known to be associated with certain stages of B-NHL development has greatly contributed to the identification and classification of NHL [6,41,86,97].

*Table 3.* EBV-negative non-Hodgkin's B-lymphoma cell lines: clinical characterization

| Cell line[a] | Donor age[b]/sex | Diagnosis[c] | Specimen site est. | Year | Culture medium[d] | Source[e] |
|---|---|---|---|---|---|---|
| (1) Precursor B-lymphoblastic lymphoma | | | | | | |
| Karpas 1106 | 23 F | Mediastinal lymphoblastic B-NHL | Pleural and ascitic fluid | 1983 | RPMI 1640 with 10% FBS | 76 |
| (2) Chronic lymphocytic leukemia (CLL) | | | | | | |
| WSU-CLL | 66 M | CLL | PB mononuclear cells | 1995 | RPMI 1640 with 10% FBS | 74 |
| TANOUE | 11 M | B cell leukemia | PB mononuclear cells | 1995 | RPMI 1640 with 10% FBS | 24 |
| (3) lymphoplasmacytoid lymphoma | | | | | | |
| FM | 43 M | DML | ND | 1988 | RPMI 1640 with 20% FBS | 28 |
| NU-DHL-1 | ND | DHL | Lymph node | 1983 | RPMI 1640 with 10% FBS | 100 |
| SK-DHL-2 | 39 M | DHL | Peritoneal fluid (ascitic) | 1983 | RPMI 1640 with 20% FBS and 40% autologous peritoneal fluid | 82 |
| SK-DHL-2A | 39 M | DHL | Peritoneal fluid (ascitic) | 1983 | RPMI 1640 with 20% FBS and 40% autologous peritoneal fluid | 82 |
| SK-DHL-2B | 39 M | DHL | Peritoneal fluid (ascitic) | 1983 | RPMI 1640 with 20% FBS and 40% autologous peritoneal fluid | 82 |
| SKW-4 | ND, M | DHL | Pleural fluid | 1966 | RPMI 1640 with 10% FBS | 81 |
| SU-DHL-1 | 10 M | DHL | Pleural fluid | 1977 | RPMI 1640 with 20% FBS and 10% human serum | 25 |
| SU-DHL-2 | 73 F | DHL | Pleural fluid | 1977 | RPMI 1640 with 20% FBS and 10% human serum | 25 |
| SU-DHL-3 | 35 M | DHL | Peritoneal fluid | 1977 | RPMI 1640 with 20% FBS and 10% human serum | 25 |

*Continued on next page*

*Table 3.* (continued)

| Cell line[a] | Donor age[b]/sex | Diagnosis[c] | Specimen site est. | Year | Culture medium[d] | Source[e] |
|---|---|---|---|---|---|---|
| (3) lymphoplasmacytoid lymphoma (continued) | | | | | | |
| SU-DHL-4 | 38 M | DHL | Peritoneal fluid | 1977 | RPMI 1640 with 20% FBS and 10% human serum | 25 |
| SU-DHL-5 | 17 F | DHL | Lymph node | 1977 | RPMI 1640 with 20% FBS and 10% human serum | 25 |
| SU-DHL-6 | 43 M | DHL | Peritoneal fluid | 1977 | RPMI 1640 with 20% FBS and 10% human serum | 25 |
| SU-DHL-7 | 47 F | DHL | Pleural fluid | 1977 | RPMI 1640 with 20% FBS and 10% human serum | 25 |
| SU-DHL-10 | 25 M | DHL | Pleural fluid | 1977 | RPMI 1640 with 20% FBS and 10% human serum | 25 |
| (4) Mantle cell lymphoma (ML) | | | | | | |
| HF-4a | ND | ML | PB mononuclear cell | 1993 | DMEM with 10% FBS | 48 |
| JeKo-1 | 78 F | Retroperitoneal mass | PB mononuclear cell | 1998 | RPMI 1640, Ham's F-12, HBSS (2:1:1) | 46 |
| SP-53 mantle cell | 58 F | Intermediate lymphocytic lymphoma | PB mononuclear cell | 1988 | RPMI 1640 with 20% FBS | 14 |
| (5) Follicular follicular-Grade I | | | | | | |
| CJ | 54 F | Nodular small cleaved lymphoma | Lymph node | 1989 | RPMI 1640 with 20% FBS | 28 |
| Follicular follicular-Grade II | | | | | | |
| WSU-FSCCL | 30 M | FSCCL | PB mononuclear cell | 1993 | RPMI 1640 with 20% FBS | 73 |
| FL-18 | 68 M | FSCCL | Lymph node | 1982 | RPMI 1640 with 10% FBS | 17 |
| ONHL-1 | 58 M | FSCCL | Bone marrow mononuclear cells | 1980 | RPMI 1640 with 10% FBS | 64 |

*Continued on next page*

Table 3. (continued)

| Cell line[a] | Donor age[b]/sex | Diagnosis[c] | Specimen site est. | Year | Culture medium[d] | Source[e] |
|---|---|---|---|---|---|---|
| (5) Follicular follicular-Grade III | | | | | | |
| U-937 | 37 M | Histiocytic lymphoma | Pleural effusion | 1976 | RPMI 1640 with 10% FBS | DSMZ-93 |
| JOSK-M | 37 M | Histiocytic lymphoma | PB mononuclear cell | 1984 | RPMI 1640 with 10% FBS | DSMZ-84 |
| WSU-NHL | 46 F | NHL | Pleural fluid | 1988 | RPMI 1640 with 20% FBS | DSMZ-71 |
| (6) Follicular center lymphoma | | | | | | |
| BALM-3 | 63 F | DPDL | Pleural fluid (effusion) | 1979 | RPMI 1640 10% FBS and antibiotics | 58 |
| BALM-4 | 63 F | DPDL | Pleural fluid (effusion) | 1979 | RPMI 1640 10% FBS and antibiotics | 58 |
| BALM-5 | 63 F | DPDL | Pleural fluid (effusion) | 1979 | RPMI 1640 10% FBS and antibiotics | 58 |
| (7) Extranodal marginal zone | | | | | | |
| No cell lines | | | | | | |
| (8) Splenic marginal zone | | | | | | |
| No classified cell lines | | | | | | |
| (9) Diffuse large cell lymphoma (DLCL) | | | | | | |
| DS | 70 F | Large immunoblastic B-cell lymphoma | Bone marrow | 1993 | Serum free culture medium | 50 |
| HOB1 | 24 M | Immunoblastic B-lymphoma | Gingival lesions | 1989 | RPMI 1640 with 10% FBS | 38 |
| a. HOB1/VCR1.0 | | | | 1993 | RPMI 1640 with 10% FBS | 51 |
| b. HOB1/ADR | | | | 1995 | RPMI 1640 with 10% FBS | 12 |
| WSU-DLCL | 32 F | Diffuse histiocytic lymphoma | Pleural fluid | 1991 | RPMI 1640 with 10–20% FBS | 75 |
| WSU-DLCL2 | 40 M | Diffuse large cell lymphoma | Pleural fluid | 1996 | RPMI 1640 with 10% FBS | 2 |
| HBL-1 | 65 M | Diffuse large cell lymphoma | Pleural effusion | 1984 | RPMI 1640 with 10% FBS | 1 |
| HBL-2 | 65 M | Diffuse large cell lymphoma | Pleural effusion | 1984 | RPMI 1640 with 10% FBS | 1 |
| HF-1 | | Diffuse large non-cleaved cell lymphoma | Lymph node | 1993 | DMEM with 10% FBS | 48 |
| KAL-1 | 37 M | Diffuse large cell lymphoma | Pleural effusion | 1988 | RPMI 1640 with 10% FBS | 43 |
| LNPL | 18 M | Diffuse large cell lymphoma | | 1980 | RPMI 1640 with 10% FBS | 16 |
| OZ | 32 M | Diffuse large cell lymphoma | Bone marrow mononuclear cells | 1997 | RPMI 1640 with 20% FBS | 79 |

*Continued on next page*

*Table 3.* (continued)

| Cell line[a] | Donor age[b]/sex | Diagnosis[c] | Specimen site est. | Year | Culture medium[d] | Source[e] |
|---|---|---|---|---|---|---|
| (10) Burkitt's lymphoma (BL) | | | | | | |
| BJAB | 5 F | Burkitt's lymphoma | Tumor tissue | 1979 | RPMI 1640 with 10% FBS | DSMZ-68 |
| BL-41 | 8 M | Burkitt's lymphoma | Tumor tissue | 1987 | RPMI 1640 with 5–10% FBS | DSMZ |
| BL-70 | 16 M | Burkitt's lymphoma | Tumor tissue | 1985 | RPMI 1640 with 10% FBS | DSMZ |
| CA-46 | young, M | Burkitt's lymphoma | Ascites fluid | 1980 | RPMI 1640 with 20% FBS | ATCC-DSMZ-61 |
| CW 678 | ND | Burkitt's lymphoma | ND | | RPMI 1640 with 10% FBS | 7 |
| DG-75 | young, ND | Burkitt's lymphoma | ND | 1981 | RPMI 1640 with 10% FBS | 69 |
| MANCA | young, ND | Burkitt's lymphoma | ND | 1985 | RPMI 1640 with 10% FBS | 33 |
| Ramos (RA-1) | 3 M | Burkitt's lymphoma | | 1973 | RPMI 1640 with 10% FBS | ATCC |
| Sc-1 | 67 M | Burkitt's lymphoma | Ascitic fluid | 1977 | RPMI 1640 with 10% FBS | 95 |
| ST486 | | Burkitt's lymphoma | Ascitic fluid | 1977 | RPMI 1640 with 20% FBS | ATCC |
| WSU-BL | 41 M | Burkitt's lymphoma | Peritoneal fluid (ascites) | 1989 | RPMI 1640 with 10% FBS | |
| (11) High-grade B-cell lymphoma | | | | | | |
| MC 116 | ND, M | Undifferentiated lymphoma | Pleural effusion | 1980 | RPMI 1640 with 20% FBS | ATCC |
| Others | | | | | | |
| U-698-M | 7 M | B cell lymphoma | Involved tonsil | 1974 | RPMI 1640 with 10% FBS | DSMZ 81 |
| Karpas 422 | 72 F | Non-Hodgkin's lymphoma | Pleural fluid | 1989 | RPMI 1640 with 20% FBS | 22 |
| HF-1 | | Nodular and diffuse Centroblastic lymphoma | Lymph node | 1993 | DMEM with 10% FBS | 48 |

*Table 3.* (continued)

| Cell line[a] | Donor age[b]/sex | Diagnosis[c] | Specimen site est. | Year | Culture medium[d] | Source[e] |
|---|---|---|---|---|---|---|
| Others (continued) | | | | | | |
| MHH-PREB-1 | 5 M | Non-Hodgkin's lymphoma | Lymph node | 1994 | RPMI 1640 with 7.5–10% FBS | DSMZ |
| MN-60 | 20 M | B cell leukemia | PB | 1982 | Ham's F10 with 10% FBS | DSMZ-90 |

[a] Cell line names are given as listed in the original literature.
[b] Age of patient at the time of establishment.
[c] Diagnosis is indicated as given in the original reference.
[d] Cell line might also grow in other culture media.
[e] Availability from cell bank as indicated (ATCC, DSMZ, author, or not listed).

**Abbreviations:** BM – bone marrow, PB – peripheral blood, CLL – chronic lymphocytic leukemia, DML – diffuse mixed lymphoma, DHL – diffuse histiocytic lymphocytic lymphoma, LIB – large cell immunoblastic lymphoma, ML – mantle cell lymphoma, FSCCL – follicular small cleaved cell, NHL – nodular histiocytic (follicular) lymphoma, DPDL – diffuse poorly differentiated lymphocytic lymphoma, DLCL – Diffuse large cell lymphoma, BL – 'Burkitt's lymphoma', NPDL – nodular poorly differentiated lymphocytic lymphoma, ATCC – American Tissue Culture Collection, DSMZ – Deutsche Sammlung von Mikroorganismen und Zellkulturen GmbH, ND – not determined.

*Table 4.* EBV-negative non-Hodgkin's B-lymphoma cell lines: genetic characterization

| Cell line | Primary ref. | Doubling time | EBV status | Karyotype |
|---|---|---|---|---|
| **(1) Precursor B-lymphoblastic lymphoma** | | | | |
| Karpas 1106 | Blood 84:3422; 1994 | 20–24 hr | Negative | 49, X, del(2)(p11.2,p13.3), der(3)t(2:3)(p13.3;p25.1)+t(9p),ins(12;?)(q13.1;q13.3), del(14)(q11.2;q13.1), del(15)(q11.2;q15.3), der(18)t(x:13:18)(q28;q21.3;q12.1) −20 del(20)(q13.1;q13.3)X2, der(X)t(X;13:18)(q28;q21.3;q12.1)+ix(p) |
| **(2) Chronic lymphocytic leukemia (CLL)** | | | | |
| WSU-CLL | Leukemia 10:130; 1996 | 18–24 hr | Negative | 45, X, del(3)(p14;p24), t(4;12;12)(q31;q22;p13), t(5;12)(q31;p13), add(16)(q24) X2, t(18;21)(q12;p12) |
| TANOUE | Leukemia Res 19:249; 1995 | 40 hr | Negative | human hyperdiploid karyotype with 12% polyploidy; 47/48X, −Y/XYqh+, +7, +14, dupt(1)(q21.1/21.2;q23.1/23.2), t(2;4)(q2?2;q2?5), del(6)(q27), t(8;14)(q24;q32) |
| **(3) Lymphoplasmacytoid lymphoma** | | | | |
| FM | Blood 75:1311; 1989 | ND | ND | t(14;18), +3, +9, +12, del 7 |
| NU-DHL-1 | Blood 63:140; 1984 | ND | Negative | ND |
| SK-DHL-2 | Cancer Genet Cytogenet 12:39; 1984 | 15 hr | Negative | t(8;14) |
| SK-DHL-2A | Cancer Genet Cytogenet 12:39; 1984 | 18 hr | Negative | diploid 46, XY t(8q;14q) |
| SK-DHL-2B | Cancer Genet Cytogenet 12:39; 1984 | 18 hr | Negative | tetraploid 82-92, XXYY t(8q;14q), t(8q;14q) |
| SKW-4 | Hematol Oncol 1:277; 1983 | ND | Negative | t(6p;1q), t(1p;2p), t(?;14q), t(?;7q), 7p+ t(q;11q) |
| SU-DHL-1 | Cancer 42:2379; 1978 | ND | Negative | 83,XXYY, −1, −2, −3, −4, −5, −7, −8, −9, −10, −11, +12, −13, −13, −14, −15, −16, −17, −21, +8mar, 6q−, t(y;?), t(Y;?) |
| SU-DHL-2 | Cancer 42:2379; 1978 | ND | Negative | 51, XX, +4, +9, +11, +13, −22, +2mar,6q−, 18q− |
| SU-DHL-3 | Cancer 42:2379; 1978 | ND | Negative | 47, XY, −4, +12, −14, +2mar, 2p−, t(11;?), t(13;?), 18q− |
| SU-DHL-4 | Cancer 42:2379; 1978 | ND | Negative | 47, XY, +8, t(3;?), t(14;18)(q32;q21) |
| SU-DHL-5 | Cancer 42:2379; 1978 | ND | Negative | 47, XX, +12, 2q−, 6q− |
| SU-DHL-6 | Cancer 42:2379; 1978 | ND | Negative | 47, X, −4, +6, +7, −8, −17, −22, +i(17q), +3mar, 6p−, 6q−, 7q−, 9p−, 18q−, t(11;?)(q25;?), t(14;18)(q32;q21) |
| SU-DHL-7 | Cancer 42:2379; 1978 | ND | Negative | 44, XX, −10, −12, −13, −18, −21, −22, +4mar, 2q−, 6q−, t(1;X)(p11;q28), t(14;?)(q32;?) |
| SU-DHL-10 | Cancer 42:2379; 1978 | ND | Negative | 96, XXYY, −14, −14, +6mar, t(7;?), t(11;Y)(q23;q11), t(11;Y)(q23;q11), 18q−, 18q− |
| **(4) Mantle cell lymphoma (ML)** | | | | |
| HF-4a | Eur J Haematol 52: 65, 1994 | ND | ND | 47-52, XX, +X, +1, t(1;8)(p21;q24), −2, del(3)(q21), +add(7)(q36), t(14;18) t(1;6), t(1;8), t(3;19) |
| JeKo-1 | Br J Haematology 102:1323; 1998 | 33 hr | Negative | 41, XO (85%):+add(1)(p13), add(1)(q12), −2, add(3)(q27), add(5)(p13), −6, add(7)(q22), −8, −9, add(9)(q34), add(10)(p15), add(11)(p11), −12, −13, −14, −14, −16, −16, −20, add(21)(p13), −22, +mar1, +mar2, +mar3, +mar4, +mar5, +mar6, devoid of t(11;14) |
| **(5) Follicular-Grade I** | | | | |
| SP-53 mantle cell | Cancer 64:1248; 1989 | ND | Negative | 46, XX (64%) t(11;14)(q13;q32) |
| CJ | Blood 75:1311; 1990 | ND | Negative | 51, XX, +X, −4, +12, +del(2)(p21), +del(2)(q32), +del(3)(q21), +del(3)(q21), der(4)t(1,4)(q23;q35), t(8;22)(q24;q11), del(10)(q22;q24), t(14;18)(q32;q21) |

*Continued on next page*

*Table 4.* (continued)

| Cell line | Primary ref. | Doubling time | EBV status | Karyotype |
|---|---|---|---|---|
| (5) Follicular-Grade II | | | | |
| WSU-FSCCL | *Cancer Genet Cytogenet* 70:62; 1993 | 26 hr | Negative | 46–47, t(14;18)(q32;q21), t(8;11)(q24;q21) |
| FL-18 | *Blood* 70:1619, 1987 | 26 hr | Negative | 49, XY, +7, +12, −17, +del(X)(p11), del(6)(p11), dir ins (13;8) q(14;q22;q24) t(8;22)(q24;q13), t(14;18)(q32;q21)+der(17), t(17;?)(q23;?) |
| ONHL-1 | *Int J Cancer* 46:1107; 1990 | 25 hr | Negative | 46, X, −Y, −3, −8, −9, −10, −11, −13, −14, −15, −16, −18, −20, t(2;12)(p11.2;q24.1), +der(3)t(3;9;10)(p21.1;p22;q22), del(5)(q22;q31.1), +del(6)(q15;q22.2), +der(8)t(8;?)(p23.1;?), +der(9)t(3;9;10)(p21.1;p22;q22), +der(11)t(11;?)(q25;?), +der(14)t(14;?)(q32.3;?), t(15;16)(q21.1;q22), +der(15)t(15;?)(p13;?), +der(16)t(16;?)(p13.1;?), t(17;17)(p13;q21.1), +der(18)t(18;?)(q21;?), +der(20)t(20;?)(q13.3;?), +der(21)t(8;21)(q13;p13), +mar1, +masr2 |
| (5) Follicular-Grade III | | | | |
| U-937 | *Int J Cancer* 17:565; 1976 | 30–40 hr | Negative | human flat-moded hypotriploid karyotype: 63(58-69)XXY, −2, −4, −6, +7, −9, −20, −21, +3mar, t(1;12)(q21;p13), der(5)t(1;5)(p22;q35), add(9)(q22), t(10;11)(p14;q23), i(11q), t(12p), add(16)(q22), add(19)(q13); carries t(10;11)(seen in AML M5); t(1;5) resembles variant of t(2;5)/(histiocytic lymphoma) |
| JOSK-M | *Cancer Res* 46: 3067, 1986 | ND | Negative | hyperdiploid with 4% polyploidy; 55 (51–75)<2n>XXY, +X, +3, +7, +t(1;5;7)(p21-22;q35;p11), der(3), t(1;3)(q21;q26-27), der(6)t(6;?), der(10)t(10;?)(p13;?), del(11)(q22-23), der(12)(12;?)(p13;?) |
| WSU-NHL | *Leukemia Res* 12:833; 1988 | 57 hr | Negative | 45, XX, t(14;18)(q32;q21) |
| (6) Follicular center lymphoma | | | | |
| BALM-3 | *Int J Cancer* 24:572; 1979 | 36–48 hr | Negative | 52, XX, +7, +8, +8, t(14;18)(q32;q21), +19, +M1, +M2 |
| BALM-4 | *Int J Cancer* 24:572; 1979 | 60–72 hr | Negative | 52, XX, +7, +8, +8, t(14;18)(q32;q21), +19, +M1, +M2 |
| BALM-5 | *Int J Cancer* 24:572; 1979 | 60–72 hr | Negative | 52, XX, +7, +8, +8, t(14;18)(q32;q21), +19, +M1, +M2 |
| (9) Diffuse large cell lymphoma (DLCL) | | | | |
| DS | *Leukemia* 8:1164; 1994 | 24 hr | Negative | 48, XX, t(1;6)(q12;q26), add(2)(q37), +del(7)(q21;q32), t(8;14)(q24;q32)+ inv(12)(p12.3;q24), der(13)(q12;p12), t(14;18)(q32;q21), del(20)(p11;p13), +ace |
| HOB1 | *Br J Cancer* 61:655; 1990 | 22 hr | Negative | hypodiploid 45 (22–73) with multiple abnormalities: t(2;4)(p21>cen>qter;q26), t(3;4;18)(p25;q21;q21), del(2)(p12;p25), t(8,14), +13, +20, +17, +21 |
| (a) HOB1/VCR1.0 | *Cancer Lett* 73:105; 1993 | | | |
| (b) HOB1/ADR | *FEBS Lett* 373:285; 1995 | | | |
| WSU-DLCL | *Cancer* 69:1468; 1992 | 20 hr | Negative | 70–85, XXXXXX, 14q+, t(q8;q10), structural abnormalities of 3;4 |
| WSU-DLCL2 | *Clinical Cancer Res* 4:1305; 1998 | 18 hr | Negative | 48, XY, t(1;2)(p36.1;q37), der(3)t(3;7)(q13;p15), t(4;14)(q27;132), +t(7p), der(7)t(3;7)(q21;q11.2), +8, t(14;18)(q32;q21), der(15)(q26.1), del(16)(q22), del(17)(q25) |
| HBL-1 | *Cancer* 61:483; 1988 | 18–24 hr | Negative | 44, X, t(6; 14; 16), t(12;7), t(14;17), t(16;17) |
| HBL-2 | *Cancer* 61:483; 1988 | 18–24 hr | Negative | 46, X, t(6; 9; 11), t(11;14)(q13;q32), t(11;?), t(14;15), t(18;9), t(9;22) |
| HF-1 | *Eur J Haematol* 52:65; 1994 | ND | ND | ND |
| KAL-1 | *Cancer Res* 51: 5392; 1991 | 20 hr | Negative | 46, XY, dup (1)(q21;q32), t(8;22)(q24;q11) |
| LNPL | *Cancer Res* 42:1368; 1982 | 25 hr | Negative | t(7;2) and t(8;14) |
| OZ | *Hematological Oncology* 15:109; 1997 | 36 hr | Negative | 47, XY, del(2)(q35), der(8)t(8;12)(p21;q13.3), add(10)(p11.12), +add(12)(q24), t(14;18)(q32;q21), add(17)(p11) |

*Continued on next page*

*Table 4.* (continued)

| Cell line | Primary ref. | Doubling time | EBV status | Karyotype |
|---|---|---|---|---|
| **(10) Burkitt's lymphoma (BL)** | | | | |
| BJAB | *Biomedicine* 22:276; 1979 | ND | Negative | human pseudodiploid karyotype with 1.5% polyploidy; 46(44–46)XX, +7, −9, der(7)t(7:?)(p21:?), der(8)t(2:8)(p12:q24) |
| BL-41 | *IARC Scientific Publ* 59:309; 1985 | 30 hr | Negative | human hyperdiploid karyotype with 7% polyploidy; 48(42-49)XY, +7, −13, +2mar, add(8)(q24), t(8;14)(q24:q32), der(15)t(13:15)(q13:p11),add(17)(q12), subclonal rearrangements at 1q23, 7p22, 11q13 |
| BL-70 | *IARC Scientific Publ* 60:309; 1985 | 44 hr | Negative | human hyperdiploid karyotype with 4% polyploidy; 47(42-49)XY, +7, inv(1)(p21:q21), del(2)(q33), t(8;14)(q24:q32), t(12:22)(q21.1:q13.2) |
| CA-46 | *J Natl Cancer Inst* 64:477; 1980 | 16 hr | Negative | human near-diploid karyotype with about 3% triploidy; 46(45-48)<2n>X/XY, dup(1)(q21:q32), dup(7)(q12:q22), t(8;14)(q24:q32); additional rearrangements were present in a sideline, viz. t(6:13)(p21:q32) and del(11)(p11); a supernumerary dmin was present in about 25% metaphases |
| CW 678 | *J Immunol* 129: 1336, 1982 | ND | Negative | t(8;14) |
| DG-75 | *Haematol Blood Transf* 26:322; 1981 | ND | Negative | t(8;14) |
| MANCA | *Nature* 314:366; 1985 <br> *JNCI* 37:547; 1966 | 22 hr | Negative | t(8;14) |
| Ramos (RA-1) | *Intervirology* 5:319; 1975 | 18–24 hr | Negative | human hypodiploid 45 |
| Sc-1 | *NJC* 39:89; 1987 | | | 49, XY, t(14:17)(q32:q21), +3, +7, +8 |
| ST486 | *Magrath JNCI* 64:465; 1980 | 19 hr | Negative | human 48, XX, +7, +t(7:17)(7qter-7q22::17qter-17pter); t(8;14)(8pter-8q23::18q11-18qter) |
| WSU-BL | *J Immunol* 129:1336; 1982 | 19hr | | |
| | *Cancer* 64:1041; 1989 | | Negative | 53, XY, t(8;14)(q24:q32) with 1q+, 2q+, +7, +13, +14q+, +18, +19q+, +21 |
| **(11) High-Grade B-cell lymphoma** | | | | |
| MC 116 | *Magrath* 64: 1980 | 26 hr | Negative | human 45XO/46XY, dupl (q21:32), t(8:14)(8pter-8q23: 14q32-14qter; 14 pter-14q32:8q23-8qter), del10 |
| **Others** | | | | |
| U-698-M | *Int J Cancer* 13:808; 1974 | 48 hr | Negative | human hyperdiploid karyotype with 5.5% polyploidy; 49(44-50)XY, +3, +7, −14, +mar, dupl(1)(q43q21.2), der(2)t(2:3)(p16:p11), add(3)(p11), del(6)(q15:q22), del(9)(p22), dupl(11)(q23:q13), add(13)(p12), add(16)(q24); carries large submetacentric dupl(1) marker |
| Karpas 422 | *Blood* 75:709; 1990 | | Neg./Pos. | 46, XX, t(2:10)(p23:q22.1), t(4:11)(q21.3:q23.1), t(4:16)(q21.3:p13.1), t(14:18)(q32.1:q21.3) |
| HF-1 | *Eur J Haematol* 52:65; 1994 | | | 52, XX, +X, t(2:8)(p12:q24), +?i(5)(p10), +?i(5)(p10), +?i(5)(p10), del(7)(q35), +add(7)(q32), add(9)(q34), +12, t(14:18)(q32:q21), +21[20] |
| MHH-PREB-1 | not published | 20–40 hr | Negative | human hyperdiploid karyotype with 12% polyploidy; 48<2n>XYY, +21, ?ins(2:2)(q21:p15:p22), t(8:14)(q24:q32) |
| MN-60 | *Leukemia Res* 6:685; 1982 | | Negative | human near-diploid karyotype; 46(45-47)<2n>XY, dupl(1)(q21:q41), del(6)(q21), t(8;14)(q24-q32), i(13q) |

ND — not determined.

*Table 5.* EBV-negative non-Hodgkin's B-lymphoma cell lines: immunophenotypic characterization

| Cell line | Immunophenotype | Additional comments |
|---|---|---|
| **(1) Precursor B-lymphoblastic Lymphoma** | | |
| Karpas 1106 | CD3–, CD5–, CD10–, CD19+, CD20+, CD23–, CD37+, FMC7+, IgM–, IgD–, IgG+, IgA–, Ig lambda+, Ig kappa– | Bcl-2– |
| **(2) Chronic lymphocytic leukemia (CLL)** | | |
| WSU-CLL | CD3–, CD5–, CD10+, CD19+, CD20+, CD22+, CD37+, CD45R+, HLA-DR+, sm/cyIgG+, sm/cylamda+ | Bcl-2+, P53–, C-MYC+ |
| TANOUE | CD3–, CD10+, CD13–, CD19+, CD20+, CD34–, CD37+, HLA-DR+, sm/cyIgG–, sm/cyIgM+, sm/cykappa(+), sm/cylambda+ | |
| **(3) Lymphoplasmacytoid lymphoma** | | |
| FM | CD3–, CD19+, CD20+, HLA-DR+, sm/cyIgM+, sm/cykappa+ | |
| NU-DHL-1 | CD3+, CD20+, CD19–, CD10–, sm/cylambda+, HLA-DR+, sm/cykappa– | |
| SK-DHL-2 | CD3–, CD20+, CD21–, CD19+, CD10+, HLA-DR+, sm/cyIgM+, sm/cylambda+ | |
| SK-DHL-2A | CD3–, CD20+, CD21–, CD19+, CD10+, HLA-DR+, sm/cyIgM+, sm/cylambda+ | |
| SK-DHL-2B | CD3–, CD20+, CD21–, CD19+, CD10+, HLA-DR+, sm/cyIgM+, sm/cylambda+ | |
| SKW-4 | CD3–, CD20+, CD19+, CD10–, sm/cyIgM+, HLA-DR+, sm/cykappa+ | |
| SU-DHL-1 | CD3–, CD19–, CD10–, negative T-cell markers, smIgM–, smIgE–, smIgA–, smkappa–, smlambda– | |
| SU-DHL-2 | CD3–, CD19–, CD10–, negative T-cell markers, smIgM–, smIgE–, smIgA–, smkappa–, smlambda– | |
| SU-DHL-3 | CD3–, CD19–, CD10–, negative T-cell markers, smIgM–, smIgE–, smIgA+, sm/cykappa+, smlambda– | |

*Continued on next page*

*Table 5.* (continued)

| Cell line | Immunophenotype | Additional comments |
|---|---|---|
| **(3) Lymphoplasmacytoid lymphoma (continued)** | | |
| SU-DHL-4 | CD3–, CD19–, CD10–, negative T-cell markers, smIgM–, smIgE+, smIgA–, sm/cykappa+, smlambda– | |
| SU-DHL-5 | CD3–, CD19–, CD10–, negative T-cell markers, smIgM+, smIgE–, smIgA–, smkappa–, sm/cylambda+ | |
| SU-DHL-6 | CD3–, CD19–, CD10–, negative T-cell markers, smIgM+, smIgE–, smIgA–, smkappa–, sm/cylambda+ | |
| SU-DHL-7 | CD3–, CD19–, CD10–, negative T-cell markers, smIgM–, smIgE+, smIgA+, smkappa–, smlambda+ | |
| SU-DHL-10 | CD3–, CD19–, CD10–, negative T-cell markers, smIgM–, smIgE+, smIgA–, smkappa–, smlambda+ | |
| **(4) Mantle cell lymphoma (ML)** | | |
| HF-4a | CD19+, CD20+, CD22+, CD39+, CD45+, smIgG+, HLA-DR+, sm/cykappa+, sm/cyIgM+, sm/cyIgD+ | Bcl-2+ |
| JeKo-1 | CD3–, CD5+, CD10–, CD19+, CD20+, CD23–, sm/cyIgM+ | |
| SP-53 mantle cell | CD2–, CD3–, CD4–, CD5+, CD10–/+, smIgM+, sm/cylamda+, sm/cykappa–, sm/cymu+, Leu-12+, OKIa1+ | |
| **5) Follicular-Grade I** | | |
| CJ | CD2–, CD3–, CD19+, CD20+, CD10+, sm/cyIg+, HLA-DR+, Tac(IL-2R)– | |
| **(5) Follicular-Grade II** | | |
| WSU-FSCCL | CD2–, CD3–, CD4–, CD5–, HLA-DR+, CD10+, CD19+, CD20+, leu-10+, CD22+, CD37+, CD38+, CD11b–, Cdw13–, CD11c–/+, CDw14–, MY9, sm/cyIgM+, sm/cykappa+, sm/cylambda–, sm/cyIgD– | Bcl-2+ |

*Continued on next page*

*Table 5.* (continued)

| Cell line | Immunophenotype | Additional comments |
|---|---|---|
| **(5) Follicular-Grade II (continued)** | | |
| FL-18 | CD2−, CD3, HLA-DR+, CD10+, CD19+, CD20+, CD22+, CD38+, OKT9+, sm/cykappa+ and Mu+ | |
| ONHL-1 | CD2−, HLA-DR+/−, CD20+, CD19−, CD24+, CD10−, PCA-1−, CD38−, sm/cykappa−/+, sm/cylamda− | |
| **(5) Follicular-Grade III** | | |
| U-937 | CD3−, CD13+, CD14−, CD15+, CD19−, CD33+, CD34−, CD68+ | |
| JOSK-M | CD3−, CD13+, CD14−, CD15+, CD19−, CD33+, CD34− | |
| WSU-NHL | CD2−, CD3−, CD19+, CD20+, CD10+, CD21−,Leu-10+, Leu12+, Leu14+, Leu16+, BL1+, BL4+, BL7+, HLA-DR+, sm/cyIgG+, sm/cylambda+ | |
| **(6) Follicular center lymphoma** | | |
| BALM-3 | sm/cyIg+, Ia antigen weakly positive, CALLA-negative | |
| BALM-4 | sm/cyIg+, Ia antigen weakly positive, CALLA-negative | |
| BALM-5 | sm/cyIg+, Ia antigen weakly positive, CALLA-negative | |
| **(9) Diffuse large cell lymphoma (DLCL)** | | |
| DS | CD2−,CD3−, CD4−, CD5−, CD13−,CD15−, CD19+, CD10+, CD22−, CD33−, CD37+, CD45+, HLA-DR+,TdT−, lambda rearranged | Bcl-2+, C-MYC+ |
| HOB1 | CD2−, CD3−, CD4−, CD8−, CD10−, HLA-DR+, CD21−, CD20+, CD19+, Leu 14, TdT−, negative Igs | |
| (a) HOB1/VCR1.0 | CD2−, CD3−, CD4−, CD8−, CD10−, HLA-DR+, CD21−, CD20+, CD19+, Leu 14, TdT−, negative Igs | |

*Continued on next page*

*Table 5.* (continued)

| Cell Line | Immunophenotype | Additional comments |
|---|---|---|
| (9) Diffuse large cell lymphoma (DLCL) (continued) | | |
| (b) HOB1/ADR | CD2−, CD3−, CD4−, CD8−, CD10−, HLA-DR+, CD21−, CD20+, CD19+, Leu 14, TdT−, negative Igs | |
| WSU-DLCL | CD2−, CD3−, CD4−, CD5−, CD8−, CD19+, CD20+, CD21−, CD22+, BL4+, BL7+, HLA-DR+, Leu-10+, CD22+/−, CD37+, CD38−, CD10−/+, sm/cyIgM+, sm/cykappa+, sm/cylambda−, sm/cyIgG− | |
| WSU-DLCL2 | CD2−, CD3−, CD4−, CD5−, CD8−, CD10+, CD19+, CD20+, CD21−, CD22+,CD37+, CD45R−/+, HLA-DR+, Leu-10+, sm/cyIgG+, sm/cylambda+, sm/cykappa− | Bcl-2+, P53+, C-MYC+ |
| HBL-1 | CD3−, CD20+, CD24+, HLA-DR+, sm/cyIgM+, sm/cykappa+ | |
| HBL-2 | CD3−, CD20+, CD24+, HLA-DR+, sm/cyIgM(D)+, sm/cylambda+ | |
| HF-1 | CD19+, CD20+, CD22+, CD39+, CD45+, smIgG+, HLA-DR+, sm/cykappa+, sm/cyIgM−, sm/cyIgD− | Bcl-2+ |
| KAL-1 | CD3−, CD20+, CD19+, CD10+, HLA-DR+, sm/cyIgM+, sm/cylambda+ | |
| LNPL | CD10+, HLA-DR+, sm/cykappa+, mu+ | |
| OZ | CD2−, CD3−, CD4−, CD5−, CD10+, CD13−, CD19+, CD20−, CD24+, CD33−, CD38+, HLA-DR+, smIg-mix− | Bcl-2+, p26+ |
| (10) Burkitt's lymphoma (BL) | | |
| BJAB | CD3−, CD10+, CD13−, CD19+, CD37+, HLA-DR+, sm/cyIgG−, sm/cyIgM+, sm/cykappa+, sm/cylambda− | |
| BL-41 | CD3−, CD10+, CD13−, CD19+, CD20+, CD37+, HLA-DR+, sm/cyIgM+, sm/cyIgG−, sm/cykappa+, sm/cylambda− | |

*Continued on next page*

Table 5.   (continued)

| Cell Line | Immunophenotype | Additional comments |
|---|---|---|
| (10) Burkitt's lymphoma (BL) (continued) | | |
| BL-70 | CD3−, CD10−, CD13−, CD19+, CD20+, CD37+, HLA-DR+, sm/cyIgM+, sm/cyIgG−, sm/cykappa+, sm/cylambda− | |
| CA-46 | CD3−, CD10+, CD13−, CD19+, CD37+, HLA-DR+, sm/cyIgM+, sm/cyIgG−, sm/cykappa+ | |
| DG-75 | CD3−, CD19+, CD20+, SmIg+, cyIg+, HLA-DR+, CD10+, TdT− | |
| MANCA | CD19+, CD20+, CD22, CD10+, Leu10+, Leu12+, Leu14+, Leu16+, BL1+, BL4+, BL7+, HLA-DR+, sm/cyIgM+, sm/cylambda+ | |
| Ramos (RA-1) | CD3−, CD19+, CD20+, HLA-DR+, sm/cykappa+, cymu+ | |
| Sc-1 | HLA-DR+, CD19+, CALLA+ (CD10+), CD20+, FMC7−, OKT10+, sm/cylamda+ | |
| ST486 | CD3−, CD19+, CD20+, HLA-DR+, sm/cyIgM+, sm/cykappa+ | |
| WSU-BL | HLA-DR+, B1+, B4+, CALLA+, sm/cyIgM+, sm/cylamda+, BL3+, BL4+ | Bcl-2+ |
| (11) High-Grade B-cell lymphoma | | |
| MC 116 | CD3−, CD19+, CD20+, HLA-DR+, smIgM+, sm/cylamda+, C mu+ | |
| Others | | |
| U-698-M | CD3−, CD13−, CD19+, CD37+, HLA-DR+ | |
| Karpas 422 | CD19+, CD37+, sm/cyIgM+, smIgG+, smIgD+, CDw52+ | |
| HF-1 | CD19+, CD20+, CD22+, CD39+, CD45+, smIgG+, HLA-DR+, sm/cykappa+, | Bcl-2+ |
| MHH-PREB-1 | CD3−, CD10+, CD13+, CD19+, CD20+, CD37+, HLA-DR+, sm/cyIgG−, sm/cyIgM+, sm/cykappa−, sm/cylambda+ | |
| MN-60 | CD3−, CD10+, CD13−, CD19+, CD37+, HLA-DR+, sm/cyIgG−, sm/cyIgM+ | |

(+), strong, definite protein expression (more than 50% positive) moderate expression (20–50% positive); (−/+), weak expression (less than 20% positive).

The cell surface antigens expressed during B-cell differentiation are represented in the B-NHL cell lines. These are CD21 (B2) [96], BL4 [40] and BL7 [13]. CD11c and CD22 are two markers that were first reported in 1982 [92]. While CD11c is expressed on monocytes and CD22 on B-lymphocytes, the co-expression of the two antibodies was initially thought to be specific for hairy cell leukemia (HCL). Since then, however, a new subset of non-Hodgkin's lymphoma, the monocytoid B-cell lymphoma (MBCL), has been described that also co-expresses CD11c and CD22 [23,78]. Both HCL and MBCL also express acid phosphatases (AcP). However, such expression can be inhibited by tartrate (tartrate sensitive) in MBCL but not in HCL [45].

The most common forms of NHL are follicular small cleaved cell lymphoma (REAL; Mantle cell lymphoma, 40%) and follicular mixed small cleaved and large cell lymphomas (both now classified as follicular center lymphomas, 20-40%). These lymphomas are high grade, express the B-cell antigens CD19, CD20 and CD22, and are CD5-negative. The expression of other antigens varies among cell lines representing B-NHL. More than 80% of splenic lymphoma with villous lymphocytes (SLVL) are CD24+ and FMC7+ and express membrane CD22 [65]. Mantle cell NHL is almost always CD5+ and CD43+ [98] and cell lines isolated from these patients and some SLVL exhibit overexpression of cyclin D1, unlike other B-cell NHLs [15]. B-NHL cell lines established from patients with various grades of B-cell malignancy are presented in Tables 3–5.

## 3. GENETIC CHARACTERIZATION

The first detection of chromosomal abnormality in NHL was the demonstration of the reciprocal translocation of genetic material between chromosomes 8 and 14. This translocation is frequent in patients with diffuse small non-cleaved cell lymphoma of the Burkitt's type [56,62]. Translocation (8;14)(q24;q32) is associated with small non-cleaved cell lymphoma of Burkitt's type; t(14;18)(q32;q21) with follicular B-NHL and trisomy 12 and t(11;14)(q11;q32) with small lymphocytic lymphoma (Table 2). There are correlations between the cytogenetic findings and the prognosis of patients with NHL [54].

## 4. CLINICAL CHARACTERISTICS

The histopathological diversity of the NHL is also reflected in its clinical characteristics. The natural history of NHL, clinical management and treatment philosophy varies among different types of NHL. One advantage of the

Working Formulation is the grouping of NHL into three categories based on their clinical behavior.

## 5. CLASSIFICATION OF EBV-NEGATIVE B-NHL CELL LINES ACCORDING TO THE REVISED EUROPEAN AND AMERICAN LYMPHOMA (REAL) SYSTEM

Fifty-nine well-characterized (EBV-) B-NHL cell lines are listed in Tables 3–5. The tables show stage, cell line name, donor age, sex, specimen site, year established, culture medium, source, reference, doubling time, EBV status, karyotype, and immunophenotype. For most cell lines, where there is insufficient information, this concerns descriptions of the essential features and data collected on immunophenotypic, genetic and clinical characteristics. As a result, we are limited in allocating such cell lines to the proper NHL stage or category.

To represent the immunophenotypes, the notations and abbreviations are as follows: +, over 90% of the cases positive; +/−, over 50% of the cases positive; −/+, less than 50% of the cases positive; −, less than 10% of the cases positive; IgH-R and IgL-R, Ig heavy/light chain genes rearranged; smIg, surface Ig; cyIg, cytoplasmic Ig; CD, cluster of differentiation.

### 5.1. Precursor B-Lymphoblastic Lymphoma/Leukemia (B-LBL)

Example: Karpas 1106. Rappaport: lymphoblastic and formerly diffuse poorly differentiated lymphocytic [PDL]). Lukes-Collins: not defined. Working Formulation: lymphoblastic.

Children are more commonly affected than adults by this disease. B-LBL accounts for less than 20% of lymphoblastic lymphoma. The Karpas 1106 cell line was established from a 23-year-old patient with mediastinal lymphoblastic B-NHL [76].

The tumor cells are characteristically CD19+ CD79a+ CD22+ CD20 −/+ CD10+/− HLA-DR+ smIg- cyMu−/+ CD34+/−, and may express CD13 and/or CD33 [20,44,87]. Ig heavy chain genes are usually rearranged: light chain genes may be rearranged [49]. Karpas 1106 is a B-NHL lymphoblastic cell line with the immunophenotype CD5−, CD10−, CD19+, IgG+, and lambda+ (Table 5).

Cytogenetic abnormalities in B-LBL are variable [53,000]. Karpas 1106 has complex chromosomal abnormalities.

## 5.2. B-Cell Chronic Lymphocytic Leukemia (B-CLL)/Prolymphocytic Leukemia (B-PLL)/Small Lymphocytic Lymphoma (B-SLL)

Examples: TANOUE and WSU-CLL. Rappaport: well-differentiated lymphocytic, diffuse. Lukes-Collins: small lymphocyte B, B-CLL. Working Formulation: small lymphocytic, consistent with CLL.

The majority of the cases occur in adults, although TANOUE was established from an 11 year old. The B-CLL comprises >90% of CLL in United States and Europe. Most patients have bone marrow (BM) and peripheral blood (PB) involvement and tumors can invade nodes, spleen, and liver. Extra-nodal infiltrates may also be evident [94].

WSU-CLL and TANOUE lines were established from peripheral blood (PB) and are relatively slow-growing cells with doubling time up to 40 hours (Table 4). There is a shortage of B-PLL and B-SLL cell lines.

Typical tumor cells of B-CLL have faint smIgM, are smIgD+/−, (cyIg −/+), B-cell-associated antigen CD19+ CD20+ CD79a+ CD5+ CD23+ CD11c−/+ and CD10− [34,91,103]. WSU-CLL and TANOUE cell lines are CD5- and CD10+ (Table 5). Cases of B-PLL may be CD5-, have strong smIg and more often express CD22 [8]. Ig heavy and light chain genes are rearranged.

## 5.3. Lymphoplasmacytoid Lymphoma/Immunocytoma

Examples: Nu-DHL-1, SK-DHL-2, SKW-3, SU-DHL, FM. Rappaport: well-differentiated lymphocytic, plasmacytoid, diffuse mixed lymphocytic and histiocytic. Lukes-Collins: plasmacytic-lymphocytic. Working Formulation: small lymphocytic, plasmacytoid, diffuse mixed small and large cells.

These tumors involve BM, lymph nodes and spleen, and less frequently PB or extranodal sites. Most of the patients have a serum spike of monoclonal IgM. 17 cell lines are listed, most of which were established from patients with diffuse histiocytic lymphocytic (DHL) lymphoma, aged between 17 and 73 years. Generally, the disease is well represented by the cell lines.

The cells have surface and in some types, cytoplasmic Ig, usually of IgM type. Cells usually lack IgD, but are B-cell-associated antigens+ (CD19, 20, 22, 79a). CD5− CD10− CD43+/− and CD11c may be faintly positive in some cases [34,91,103]. A lack of CD5 and the presence of strong cyIg are useful to distinguish them from B-CLL. The Ig heavy and light chain genes are rearranged, although no specific abnormality is known.

The main cytogenetic abnormalities involve chromosomes 8 and 14.

## 5.4. Mantle Cell Lymphoma

Examples: JeKo-1, SP-53, HF-4. Rappaport: intermediate or poorly differen-tiated lymphocytic, diffuse or nodular (ILL/IDL/PDL). Lukes-Collins: small cleaved follicular center cell (FCC). Working Formulation: small cleaved cell, diffuse or nodular; rarely diffuse mixed or large cleaved cell.

The tumor occurs in older adults with a high male to female ratio. Sites involved include lymph nodes, spleen, BM, PB (from which JeKo-1 and SP-53 were derived) and extra nodal sites such as the gastrointestinal tract [83]. The growth pattern of mantle cell lymphoma is usually diffuse or vaguely nodular. The well-defined follicles characteristic of follicular lymphomas are rarely seen.

The tumor cells are smIgM+, usually IgD+, $\lambda > \kappa$, B-cell−antigen+; CD5+ CD10−/+ CD23− CD43+ CD11c−. The absence of CD23 is use-ful in distinguishing mantle cell lymphoma from B-CLL. CD5 is useful in distinguishing mantle cell lymphoma from follicle center and marginal zone lymphomas. Two of the three cell lines are CD5+ and have a low level of CD10 expression. The JeKo-1 cell line is CD23− and smIgM+.

t(11;14) involves the Ig heavy chain locus at bcl-1 locus on the long arm of chromosome 11, and this translocation is present in the JeKo-1 and SP-53 cell lines.

## 5.5. Follicle Center Lymphoma, Follicular

Examples: CJ; (grade I); WSU-FSCCL, FL-18, ONHL-1 (grade II); WSU-NHL, JC (grade III). Rappaport: nodular PDL mixed lymphocytic-histiocytic, or histiocytic. Lukes-Collins: small cleaved, large cleaved or large non-cleaved FCC, follicular. Working Formulation: follicular, small cleaved, mixed, or large cell.

The terms follicular lymphoma, grades I, II and III are analogous to the terms used to classify other tumor types. The pattern of growth can be follicular or follicular and diffuse, and is of prognostic significance.

The cell lines are CD19−, CD34−, CD13+ and CD33+ for grade I and CD10+ and CD19+ for grade-III.

Cytogenetic abnormalities mostly involve chromosomes 10, 11 and 14.

## 5.6. Follicle Center Lymphoma, Diffuse

Examples: BALM-3, BALM-4, BALM-5. Rappaport: diffuse poorly differ-entiated lymphocytic. Lukes-Collins: diffuse small cleaved FCC. Working Formulation: diffuse small cleaved cell.

Follicle center lymphoma, diffuse, affects predominantly adults, with equal male/female incidence [94]. This disease accounts for 40% of adult NHL in the United States [53].

BALM-3, 4 and 5 are slow-growing cell lines with doubling time from 36 to 72 hours.

The tumor cells are usually smIg+ (smIgM+/− IgD>IgG>IgA), B-cell-associated antigen +, CD10+/− CD5− CD23−/+ CD11c−. BALM-3, 4 and 5 cell lines show sm/cyIg+ and CD10−. A lack of CD5 and CD43 is useful in distinguishing follicle center lymphoma from mantle cell lymphoma, and the presence of CD10 can be useful in distinguishing it from marginal zone cell lymphomas.

All the cell lines have t(14;18). This translocation, involving a rearrangement of Bcl2, is present in 70 to 95% of the cases [39,66]:

## 5.7. Extranodal and Nodal Marginal Zone B-Cell Lymphoma

Example: none. Rappaport: (not specifically listed) well-differentiated lymphocytic (WDL) or WDL-plasmacytoid, IDL, ILL, PDL, mixed lymphocytic-histiocytic (nodular or diffuse). Lukes-Collins: small lymphocyte B, lymphocytic-plasmacytic, small lymphocyte B, monocytoid. Working Formulation: (not specifically listed) SLL (some CLL, some plasmacytoid), small cleaved or mixed small and large cell (follicular or diffuse).

There are two major clinical presentations. (1) Extranodal marginal zone lymphomas (low grade marginal zone or MALT type) are tumors of adults. Many of the patients have a history of autoimmune disease such as Sjogren's syndrome or Hashimoto's Thyroiditis and (2) Nodal marginal zone lymphomas. There are no representative cell lines for this type of B-NHL.

Typically, the tumor cells express smIg (M>G or A), lack IgD, and about 40% are cyIg+. B-cell-associated-antigens are expressed and the tumors are CD5− CD10− CD23- CD43−/+ CD11c+/−.

No rearrangements of bcl-2 or bcl-1 are seen [85]. Trisomy 3 chromosome or t(11;18) have been reported [4,26].

## 5.8. Splenic Marginal Zone B-Cell Lymphoma

Example: none. Rappaport: (not specifically listed) well-differentiated lymphocytic (WDL) or WDL-plasmacytoid. Lukes-Collins: small lymphocyte B, lymphocytic-plasmacytic, small lymphocyte B, monocytoid. Working Formulation: (not specifically listed) SLL.

Patients with this B-NHL have BM and PB involvement. The course of the disease is indolent, and splenoctomy may be followed by prolonged remission. There are no cell lines derived from this disease.

Tumor cell antigen expression is similar to that of extra-nodal and nodal marginal zone B-cell lymphomas.

No molecular genetic changes have been identified, as the tumors are not well studied.

## 5.9.  Diffuse Large B-Cell Lymphoma

Examples: HBL-1, KAL-1, LNPL, WSU-DLCL, WSU-DLCL2. Rappaport: diffuse histiocytic, occasionally diffuse mixed lymphocytic-histiocytic. Lukes-Collins: large cleaved or large non-cleaved FCC, B-immunoblastic. Working Formulation: diffuse large cell cleaved, non-cleaved or immunoblastic; occasionally diffuse mixed small and large cell.

Large B-cell lymphomas constitute 30–40% of adult NHLs. Patients typically present with a rapidly enlarging, often symptomatic, mass at a single nodal or extranodal site. LCL are aggressive but potentially curable [94].

The tumor cells are smIg+/−, cyIg−/+, B-cell-associated-antigens+ CD45+/− CD5−/+ CD10−/+ [21]. Most of the cell lines listed under this stage are CD5− and CD10+/− with B-cell-associated antigens+.

The *bcl-2* gene is rearranged in about 30% of these tumors [57,99]. C-*myc* is reported to be rearranged in some cases [102]. The cell lines carry t(8;14) and t(14;18).

## 5.10.  Burkitt's Lymphoma

Examples: BJAB, BL-41, CA-46, WSU-BL. Rappaport: undifferentiated lymphoma, Burkitt's type. Lukes-Collins: small non-cleaved FCC. Working Formulation: small non-cleaved cell, Burkitt's type.

Burkitt's lymphoma is most common in children, and 7 of the 11 cell lines were derived from tumors in children. Adult cases are often associated with immune deficiency. The tumor is very aggressive, but potentially curable.

The established cell lines provide a good representation of Burkitt's lymphoma. The cell doubling time is short, less than 24h in 6 of the 12 cell lines in which it was measured.

The tumor cells are smIgM+, B-cell-associated-antigens+ CD10+ CD5− CD23− [29]. The cell lines are smIg+/− (may have cyIg), B-cell-associated-antigens+ (11/11 of cell lines tested).

Most cases have a translocation of c-*myc* from chromosome 8 to the Ig heavy chain region on chromosome 14 (t(8;14), which is present in 7 of 11 cell lines. Less commonly, c-*myc* is translocated to light chain loci on 2 [t(2,8) or 22 t(8,22)]. Epstein-Barr virus genomes can be demonstrated in the tumor cells in most Burkitt's lymphoma arising in Africans and in 25% to 40% of cases associated with acquired immune deficiency syndrome [5,32,67].

### 5.11. High Grade B-Cell Lymphoma, Burkitt's-like

Example: MC116. Rappaport: undifferentiated; non-Burkitt's. Lukes-Collins: small non-cleaved FCC. Working Formulation: small non-cleaved cell, non-Burkitt's.

Tumors in this category are relatively uncommon and occur mostly in adults, sometimes with a history of immunosuppression. Cases in children appear to behave similarly to classic Burkitt's tumor [42], whereas in adults they appear to be highly aggressive.

Tumor cells are smIg +/− (may have cyIg), B-cell associated antigens+ CD5−, and usually CD10+ [29].

C-*myc* is rearranged in approximately 30% of cases [101].

### ACKNOWLEDGEMENTS

This work was supported by Grant CA79837 from the US National Cancer Institute and from the Leukemia Society of America grant No. 6323-99. The authors would like to thank Dr. Ishtiaq Ahmad for his technical help.

### References

1. Abe M, Nozawa Y, Wakasa H et al. *Cancer* 61: 483, 1988.
2. Al-Katib A, Smith MR, Kamanda WS et al. *Clinical Cancer Res* 4: 1305, 1998.
3. Al-Katib A, Mohammad RM. In: Valeriote, FA, Nakeff A, Valdivieso M (eds.), *Basic and Clinical Applications of Flow Cytometry*, Kluwer Academic Publishers, Boston, 1992.
4. Auer IA, Gascoyne RD, Connors JM et al. *Annals of Oncology* 8: 979, 1997.
5. Ballerini P, Gaidano G, Gong J et al. *Blood* 81:166, 1993.
6. Battifora H, Trowbridge IS. *Cancer* 51: 816, 1983.
7. Benjamin D, Magrath IT, Maguire R et al. *J Immunol* 129: 1336, 1982.
8. Bennett J, Catovsky D, Daniel M-T et al. *J Clin Pathol* 42: 567, 1989.
9. Brown TC , Peters MV, Bergsagel DE et al. *Br J Cancer* 31 (Suppl II): 174, 1975.
10. Byme Jr GE. *Cancer Treat Rep* 61: 935, 1977.
11. Chan JK, Banks PM, Cleary ML et al. *Am J Clin Pathol* 103: 543, 1995.
12. Chao CC. *Cancer Lett* 73: 105, 1993.
13. Cossman J, Neckers LM, Arnold A et al. *N Engl J Med* 307: 1251, 1982.
14. Daibata M, Kubonishi I, Eguchi T et al. *Cancer* 64: 1248, 1989.
15. De Boer CJ, Schuuring E, Dreef E et al. *Blood* 86: 2715, 1995.
16. Dillman RO, Handley HH, Royston I. *Cancer Res* 42: 1368, 1982.
17. Doi S, Ohno H, Talsumi E et al. *Blood* 70: 1619, 1987.
18. Dorfman RF. In: *The Reticuloendothelial System*, International Academy of Pathology, No. 16, Baltimore, Williams & Wilkins, p262, 1975.
19. Dorfman RF. *Lancet* 1: 1295, 1974.

20. Drexler H, Thiel E, Ludwig W-D. *Leukemia* 5: 637, 1991.
21. Doggett R, Wood G, Horning S et al. *Am J Pathol* 115: 245, 1984.
22. Dyer M, Fischer P, Nacheva E et al. *Blood* 75: 709, 1990.
23. Efremidis AP, Haubenstock H, Holland JF et al. *Blood* 66: 953, 1985.
24. El-Sonbaty et al. *Leukemia Res* 19: 249, 1995.
25. Epstein AL, Leu YR, Kim H et al. *Cancer* 42: 2379, 1978.
26. Fiun T, Isaacson P, Wotherspoon A. *J Pathol* 170: 335, 1995.
27. Foon KA, Todd RF. *Blood* 68: 1, 1986.
28. Ford R, Goodacre A, Ramirez I et al. *Blood* 75: 1311, 1990.
29. Garcia C, Weiss L, Warnke R. *Hum Pathol* 17: 454, 1986.
30. Gaul EA, Mallony TB. *Am J Pathol* 18: 381, 1942.
31. Gerard-Marchant R et al. *Lancet* 2: 406, 1974.
32. Hamilton-Dutoit S, Pallesen G, Franzmann M et al. *Am J Pathol* 138: 149, 1991.
33. Hann SR, Thompson CB, Eisenman RN. *Nature* 314: 366, 1985.
34. Harris N, Bhan A. *Hum Pathol* 16: 829, 1985.
35. Harris NL, Jaffe ES, Stein H et al. *Blood* 84: 1361, 1994.
36. Harris NL, Jaffe ES, Stein H et al. *Blood* 85: 857, 1995.
37. Henry K, Bennett MH, Farrer-Brown G. In: Mathe C, Seligmann M, Tubiana M (eds.), *Recent Results in Cancer Research*, New York, Springer-Verlag, pp 38–56, 1978.
38. Ho Y-S, Sheu L-F, Ng J-A et al. *Br J Cancer* 61: 655, 1990.
39. Hockenbery D, Zutter M, Hickey W et al. *PNAS USA* 88: 6961, 1991.
40. Hokland P, Ritz J, Schlossman SF et al. *J Immunol* 135: 1746, 1985.
41. Hsu S-M. *Am J Clin Pathol* 80: 429, 1983.
42. Hutchison R, Murphy S, Fairclough D et al. *Cancer* 64: 23, 1989.
43. Ichinose I, Nakano S, Ohshima KI et al. *Cancer Res* 51: 5392, 1991.
44. Janossy G, Bollum F, Bradstock K et al. *Blood* 56: 430, 1980.
45. Janossy G, Bollum FJ, Bradstock KF et al. *J Immunol* 123: 1525, 1979.
46. Jeon HJ, Kim CW, Yoshino T, et al. *Br J Haematol* 102: 1323, 1998.
47. Jones SE, Fuks Z, Bull M et al. *Cancer* 31: 806, 1973.
48. Knuutila S, Klefstrom J, Szymanska J et al. *Eur J Hematol* 52: 65, 1994.
49. Korsmeyer S, Hieter P, Pavetch J et al. *Proc Natl Acad Sci USA* 78: 7096, 1981.
50. Lambrechts AC, Dorssers LCJ, Hupkes PE et al. *Leukemia* 8: 1164, 1994.
51. Lee WP. *FEBS Lett* 373: 285, 1995.
52. Lennert K , Mohri N, Stein H et al. *Br J Haematol* 31 (Suppl): 193, 1975.
53. Lennert K , Feller A. *Histopathology of Non-Hodgkin's Lymphomas*, 2nd edn., Springer-Verlag, New York, 1992.
54. Lenoir GM, Preud'homme JL, Bernheim A et al. *Nature* 298: 474, 1982.
55. Lenoir et al. *IARC Scientific Publ* 60: 309, 1985.
56. Levine EG, Arthur DC, Frizzer G et al. *Blood* 66: 1414, 1985.
57. Lipford E, Wright J, Urba W et al. *Blood* 70: 1816, 1987.
58. Lok MS, Koshiba H, Han T et al. *Int J Cancer* 24: 572, 1979.
59. Lukes RJ , Butler JJ. *Cancer Res* 26: 1063, 1966.
60. Lukes RJ, Collins RD. *Cancer* 34: 1488, 1974.
61. Magrath IT, Freeman CB, Pizzo P et al. *J Natl Cancer Inst* 64: 477, 1980.
62. Manolova Y, Manolova G, Kieler J et al. *Hereditas* 90: 5, 1979.
63. Mathe G et al. In: *WHO International Histological Classification of Tumors*, No. 59, 813, 1981.
64. Matsumara I, Tomaki T, Katagiri S et al. *Int J Cancer* 46: 1107, 1990.
65. Matutes E, Morilla R, Owusu-Ankomah K et al. *Blood* 83: 1558, 1994.

66.  McDonnell T, Deane N, Platt F et al. *Cell* 57: 79, 1989.
67.  Meeker T, Shiramizu B, Kaplan L et al. *Blood* 81: 166, 1993.
68.  Menezes J, Bourkas AE. *Biomedicine* 31: 2, 1979.
69.  Minowada J. *Haematol Blood Trans* 26: 322, 1981.
70.  Mohamed AN, Mohammad RM, Koop BF et al. *Cancer* 64: 1041, 1989.
71.  Mohamed AN, Al-Katib A., *Leukemia Res* 12: 833, 1988.
72.  Mohammad RM, Al-Katib A. *Leukemia Res* 17: 1, 1993.
73.  Mohammad RM, Mohamed AN, Smith MR et al. *Cancer Genet Cytogenet* 70: 62, 1993.
74.  Mohammad RM, Mohamed AN, Hamdan MY et al. *Leukemia* 10: 130, 1996.
75.  Mohammad RM, Mohamed AN, KuKuruga M et al. *Cancer* 69: 1468, 1992.
76.  Nacheva E, Dyer MJS, Metivier C et al. *Blood* 84: 3422, 1994.
77.  Nadler LM. In: Wilson JD, Braunwald E, et al (eds.), *Harrison's Principles of Internal Medicine* 12th edn., Vol. 2, McGraw-Hill, New York, 1991.
78.  Nadler LM, Anderson KC, Marti G et al. *J Immunol* 131: 244, 1983.
79.  Nagai M, Fujtta M, Ohmori M et al. *Hematol Oncol* 15: 109, 1997.
80.  Nilsson K, Sundstrom C. *Int J Cancer* 13: 808, 1974.
81.  Nilsson K, Klareskog L, Ralph P et al. *Hematol Oncol* 1: 277, 1983.
82.  Nishikori M, Hansen H, Jhanwar S et al. *Cancer Genet Cytogenet* 12: 39, 1984.
83.  O'Briain D, Kennedy M, Daly P et al. *Am J Surg Pathol* 13: 691, 1989.
84.  Ohta M, Furukawa Y, Ide C et al. *Cancer Res* 46: 3067, 1986.
85.  Pan L, Diss T, Cunningham D et al. *Am J Pathol* 135: 7, 1989.
86.  Preud'homme JL, Seligmann M. *Blood* 40: 777, 1972.
87.  Pesando J, Ritz J, Lazarus H et al. *Blood* 54:1240, 1979.
88.  Pulvertaft *Lancet* 1: 238, 1964.
89.  Rappaport H. In: *Atlas of Tumor Pathology*, sec 3, fasc 8, Washington DC, US Armed Forces Institute of Pathology, 1966.
90.  Roos G, Adams A, Giovanella B, et al. *Leukemia Res* 6: 685, 1982.
91.  Stein H, Lennert K, Feller A et al. *Adv Cancer Res* 42: 67, 1984.
92.  Sugawara I. *Cellular Immunol* 72: 88, 1982.
93.  Sundstrom C, Nilsson K. *Int J Cancer* 17: 565, 1976.
94.  *The Non-Hodgkin's Lymphoma Classification Project*: National Cancer Institute. *Cancer* 49: 2112, 1982.
95.  Th'ng, KH. *Int J Cancer* 39: 89, 1987.
96.  Wang CY, Al-Katib A, Lane Cl et al. *J Exp Med* 158: 1757, 1983.
97.  Warnke RA, Gatter KC, Falini B et al. *N Engl J Med* 309: 1281, 1983.
98.  Weisenburger DD, Armitage JO. *Blood* 87: 4483, 1996.
99.  Weiss L, Warnke R, Sklar J et al. *N Engl J Med* 317: 1185, 1987.
100. Winter JN, Variakojis D, Epstein AL. *Blood* 63: 140, 1984.
101. Yano T, van Krieken J, Magrath I et al. *Blood* 79: 1282, 1992.
102. Yunis J, Mayer M, Amesen M. *N Engl J Med* 320: 1047, 1989.
103. Zukerberg L, Medeiros L, Ferry J et al. *Am J Clin Pathol* 100: 373, 1993.

# Chapter 10

# Mature T-Cell Malignancies

Martin J.S. Dyer
*Academic Hematology and Cytogenetics, Haddow Laboratories, Institute of Cancer Research, Sutton, Surrey SM2 5NG, UK. Tel: +44-208-722-4254; Fax: +44-208-770-1208; E-mail: mdyer@icr.ac.uk*

## 1. INTRODUCTION

Malignancies of mature, post-thymic T-cells are rare in comparison with their B-cell counterparts and are highly heterogenous. They continue to pose major clinical problems both in terms of diagnosis and management. This is in part due to their rarity, but also to the fact that diagnosis requires detailed immunophenotypic and genotypic analyses to demonstrate lineage, clonality and stage of differentiation. In many cases, these data are not available. More extensive study of the pathogenesis of the various types of malignancy remains hampered by the lack of suitable cell lines.

The purpose of this chapter is to review some of the functions of mature T-cells and advances in our understanding of the different forms of mature T-cell malignancy, to describe some of the derived cell lines and to place these in the current scheme of classification.

## 2. FUNCTIONS OF MATURE T-CELLS

T-lymphocytes derive from hematopoietic precursor cells within the bone marrow, which initially migrate to the thymus. Here, the T-cell receptor for antigen (TCR) proteins are first expressed following *TCR* gene rearrangement. A complex process of both positive and negative selection of antigen- and self-reactive T-cells occurs through interaction of the T-cell precursors with thymic stromal and antigen presenting cells. Two different T-cell lineages can be identified on the basis of their expression of TCR proteins composed of either TCR$\alpha/\beta$ or TCR$\gamma/\delta$ heterodimers. These lineages have different functions and tissue distributions. In clinical specimens, affiliation

321

*J.R.W. Masters and B.O. Palsson (eds.), Human Cell Culture Vol. III, 321–337.*
© 2000 *Kluwer Academic Publishers. Printed in the Netherlands.*

to either lineage can be ascertained either by the use of monoclonal anti-bodies (MAB) specific for constant epitopes within the TCR proteins or by using DNA methods to detect clonal rearrangements within the *TCR* genes. The *TCRδ* gene segments are located entirely within the *TCRα* complex and rearrangement of *TCRα* results in complete deletion of the *TCRδ* sequences. Mature T-cells of the TCRγ/δ lineage comprise about 5% of total peripheral blood T-cells and malignancies of this lineage are uncommon and have some distinct properties, as discussed below.

On emerging from the thymus, mature T-cell subpopulations express a panoply of surface membrane proteins that reflect their functions. The functions of many of these proteins have now been identified. Some of those that have been used clinically are shown in Table 1 (reviewed in Barclay et al. 1998). These molecules can be used to differentiate the malignancies of T-cell precursors (T-cell lymphoblastic leukemias and lymphomas) from the various malignancies of post-thymic T-cells and from malignancies of other related lineages, notably malignancies of natural killer (NK) cells. Some forms of T-cell malignancy may co-express both T-cell and NK lineage markers. Clinically, the most widely utilized of these proteins are the CD4 and CD8 molecules, which broadly divide mature T-cells into those which mediate B-cell "help" and those which mediate T-cell cytolysis respectively. In contrast to thymic malignancies that are often CD4/CD8 double positive, mature post-thymic T-cells usually express only one or other of these molecules, although in some instances, notably in T-cell prolymphocytic leukemia, co-expression of CD4 and CD8 may be observed. Assessment of the expression of the nuclear enzyme terminal deoxynucleotidyl transferase (TdT) may be necessary to distinguish malignancies of T-cell precursors from those of mature T-cells.

Mature T-cells migrate to a number of peripheral lymphoid sites, including spleen and lymph nodes, but also to more "specialized" sites such as the skin and intestinal epithelia; T-cells in these sites may differ from those elsewhere. They are competent to perform a number of different effector functions including mediation of:

- B-cell "help" to produce specific antibodies (predominantly a function of CD4+ subpopulation).

- Cytolysis of virally infected/bacterially infected cells as well as al-logeneic and malignant cells (predominantly a function of CD8+ subpopulation).

- Stimulation of monocytes/macrophages in the inflammatory response.

These subjects are discussed in detail in Paul [37].

*Table 1.* Clinically utilized T-cell and NK differentiation antigens

| CD No. | Cellular distribution | Functions |
|---|---|---|
| 1 | Cortical thymocytes | Antigen presentation |
| 2 | Thymocytes and mature T-cells | Adhesion molecule – binds CD58 |
| 3[a] | Thymocytes and mature T-cells | Component of the antigen receptor |
| 4 | Thymocytes/mature T-cell subsets | Component of TCR/coreceptor for MHC class II |
| 5 | Thymocytes, T-cells, some B-cells | T-cell activation – binds CD72 |
| 7 | Thymocytes and mature T-cells | Unknown |
| 8 | Thymocytes/mature T-cell subsets | Component of TCR/Coreceptor for MHC class I |
| 25 | Immature thymocytes/activated T-cells | Component of IL-2R |
| 28 | Thymocytes and mature T-cells | T-cell activation – binds B7.1/2 |
| 52 | All lymphocytes | Unknown |
| 56 | NK cells | Adhesion molecule – NCAM isoform |
| 57 | NK cells, some mature T-cells | Unknown |
| TdT | Lymphoid precursors | Adds in non-templated nucleotides during recombination |
| HLA-DR | Mature/activated T-cells | Antigen presentation |

[a] Surface CD3 expression seen in most mature T-cell malignancies *in vivo*, whereas expression is limited to the cytoplasm in malignancies of T-cell precursors.

## 3. CLASSIFICATION OF MATURE T-CELL MALIGNANCIES

Malignancies of mature T-cells can be objectively diagnosed with MAB specific for T-cell differentiation antigens and PCR or DNA blot methods for the detection of clonal *TCR* gene rearrangements. Using these techniques, a number of distinct clinical entities have been recognised. These can be divided into those that present with a primarily leukemic picture and those that present with primarily lymph nodal or extra-nodal infiltration (Table 2). The interested reader is referred to a monograph [28] and to several papers [4,20,39].

Given the distinctiveness of the individual diseases, the cell lines are described within the most appropriate clinical entity. Only those diseases from which cell lines have been derived are mentioned in detail below. T-cell prolymphocytic leukemia is also discussed as this disease, despite its highly aggressive nature, has repeatedly failed to yield cell lines, and the lack of such lines remains a major deficiency.

*Table 2.*   Classification of mature T-cell malignancies

*(A) Predominantly leukemic*

| | |
|---|---|
| Adult T-cell leukemia/lymphoma | (ATLL) |
| T-cell prolymphocytic leukemia | (T-PLL) |
| T-cell large granular lymphocytic leukemia | (T-LGL) |
| NK/T-cell leukemia/lymphoma | |

*(B) Predominantly nodal*

Nodal peripheral T/NK cell lymphoma unspecified
Anaplastic large cell lymphoma
Angioimmunoblastic T-cell lymphoma

*(C) Predominantly extranodal*

| | |
|---|---|
| Mycosis fungoides/Sézary syndrome | (MF/SS) |
| Enteropathy-associated intestinal T-cell lymphoma | |
| Primary cutaneous CD30+ve lymphoproliferative disorders | |
| Subcutaneous panniculitis-like T-cell lymphoma | |
| NK/T-cell lymphoma – nasal type | |
| Hepatosplenic T-cell lymphoma | |
| Extranodal peripheral T-cell lymphoma unspecified | |

## 3.1.  Mature T-Cell Leukemias

A comparison of the typical immunophenotypes of the mature T-cell leukemias is shown in Table 3. Distinction from malignancies of T-cell precursors is usually made from the cytological appearances in conjunction with expression of surface membrane CD3 and HLA-DR and absence of CD1 and TdT.

*Table 3.*   Comparison of the common immunophenotypes of mature, post-thymic T-cell leukemias

| Disease | sCD3 | TCRα/β | CD5 | CD7 | CD4+/CD8− | CD4+/CD8+ | CD4−/CD8+ | CD25 |
|---|---|---|---|---|---|---|---|---|
| T-PLL | ++ | ++ | ++ | ++ | ++ | + | − | − |
| T-LGL | ++ | ++ | ++ | ++ | − | − | ++ | − |
| ATLL | +/− | +/− | ++ | +/− | ++ | − | − | ++ |
| MF/SS | ++ | ++ | ++ | +/− | ++ | − | − | − |

### 3.1.1. Adult T-Cell Leukemia/Lymphoma (ATLL)

This disease is intimately associated with the human retrovirus, HTLV-1, although the role of the virus in the etiology of the disease, like EBV in the pathogenesis of Burkitt lymphoma, remains obscure. ATLL is clustered within regions in which HTLV-1 is endemic, including south-western Japan and the Caribbean basin. All cases show clonal HTLV-1 proviral integration and have serological evidence for HTLV-1 infection. About 60% of patients present with a leukemic form of the disease, whilst the remainder present with lymphomatous disease. Hypercalcemia is common at presentation. A characteristic feature is high expression of CD25, a component of the IL-2 receptor.

Many cell lines used in the study of HTLV-1 infection and its role in neoplastic transformation of T-cells were produced by the co-culture of un-infected lymphocytes with virus-producing cells (for example, Miyoshi et al. [31]). These cell lines, which include MT-2, are not discussed further. There is also in the literature a large number of cell lines derived directly from patients with leukemia and lymphoma, predominantly from Japanese patients (see for example Nakao et al. [34] and references therein; Morita et al. [33]; Sagawa et al. [41]). Most of these are poorly characterized. In contrast, the HUT-102 cell line has been extensively studied. This cell line was derived from the primary culture of lymph node cells from a patient with ATLL, initially in the presence of IL-2. The origin of this cell line and the HUT-78 cell line and their uses in the isolation and characterization of the HIV and HTLV-1 have been reviewed recently (Bunn and Foss [7], both cell lines originally described in Gazdar et al. [16]).

The ATLL cell line MU is of some interest [24]. Despite being derived from the peripheral blood of a patient with ATLL in leukemic phase, the cell line fails to express any of the anticipated range of T-cell differentiation antigens *in vitro*. The only evidence that this cell line was derived from the leukemic cells is the presence of identical *TCRβ* rearrangement in both the primary cells and the derived cell line. There is also another interesting cell line, KHM-3S, derived from a Japanese patient with small cell lung cancer, which nevertheless expresses CD25 and CD56, but is otherwise negative for all other hematopoietic markers such as CD45 [26]. This cell line shows monoclonal proviral integration, but most surprisingly, clonal *TCR* rearrangements. The tropism of HTLV-1 is not limited to T-cells alone and the cell of origin of both cell lines is not obvious. The presence of the clonal *TCRβ* rearrangements would suggest a T-cell origin but the complete lack of expression of all T-cell differentiation antigens is unusual.

In contrast, an rIL2-dependent cell line from a patient with apparent ATLL (WHN2) has been described which, despite retaining the phenotype, karyotype and clonal *TCR* gene rearrangements of the primary tumor, failed to demonstrate any clonal HTLV-1 proviral integration [21].

### 3.1.2.  T-cell Prolymphocytic Leukemia (T-PLL)

T-PLL is a very aggressive disease which typically presents in the sixth or seventh decade with extremely high white cell counts (up to $10^{12}$/litre), organomegaly, usually splenomegaly and lymphadenopathy [21,28]. The disease is often completely resistant to intensive chemotherapy but frequently remits with the humanized Mab, CAMPATH-1H [38]. A disease with identical cytology, immunophenotype, karyotype and molecular features has been seen in adult patients with ataxia-telangiectasia [45].

Cytogenetically, T-PLL is characteristically associated with rearrangements of chromosome 14 involving the $TCR\alpha/\delta$ locus at 14q11.2 and the *TCL1* locus at chromosome 14q32.1, centromeric of the *IGH* locus, which is located at the telomere at 14q32.3. These rearrangements take the form of either t(14;14)(q11.2;q32.1) or inv(14)(q11.2;q32.1). They are not, however, specific for T-PLL, and identical translocations and inversions have been reported in T-cell precursor ALL. Cases which lack rearrangements of the *TCL1* locus may exhibit rearrangement of the *MTCP1* gene, a protein homolog of *TCL1*, on chromosome Xq28 as either t(X;14)(q28;q11.2) or involving the $TCR\beta$ locus on 7q35 as t(X;7)(q28;q35). Acquired abnormalities within the ataxia telangiectasia or *ATM* gene are common in T-PLL, and homozygous mutations and structural rearrangements within the gene may be seen in most cases of sporadic T-PLL [46,47].

Despite the aggressive nature of the disease, there are no cell lines available from T-PLL. SKW-3 was derived from a 68-year-old Japanese man with some form of mature T-cell leukemia, but whether or not this truly represents T-PLL cannot be determined. There is no reference available that describes the establishment of this line. Initially, SKW-3 co-expressed CD4 and CD8, as is seen in some cases of T-PLL, but with further *in vitro* growth, expression of these molecules has been lost (Larsen et al. [25] and references therein). This cell line is of some interest as it has a t(8;14)(q24.1;q11.2) chromosomal translocation involving the MYC oncogene on chromosome 8 with the $TCR\delta/\alpha$ locus on chromosome 14 [15,29,44]. This translocation is therefore a variant of the *MYC* translocations involved in Burkitt and other forms of mature B-cell malignancy that involve *MYC* with the various immunoglobulin loci. Chromosome abnormalities of chromosome 8, but not usually translocations involving *MYC*, are seen commonly in T-PLL.

A related cell line may be the Kit-225 cell line, which was also derived from an elderly Japanese man with T-cell lymphocytosis [19]. This cell line is rIL2 dependent and has a complex karyotype with, like SKW-3, a breakpoint on chromosome 3q27. Both Kit-225 and SKW-3 are HTLV-1 negative.

We have examined some of the derived T-cell precursor ALL/lymphoblastic lymphoma cell lines with either inv(14) or t(14;14) to determine if any of these might have been derived from patients with T-PLL. SUP-T11 (Chapter 4) for example was derived from a 78-year-old man in whom the diagnosis of T-PLL might be anticipated. However, this cell line has the phenotype of TCP-ALL. HT-1 is another TCP-ALL cell line with inv(14) [1]. It has been claimed that it is possible to grow T-PLL cells in immunodeficient mice although, despite repeated attempts, we have been unable to grow xenografts. The development of a T-cell lymphoproliferative disorder similar to T-PLL in mice transgenic for the *TCL1* oncogene may allow further advances in our understanding of this disease.

### 3.1.3. T-Cell Large Granular Lymphocytic Leukemia (T-Cell LGL)

This disease is characterized by the cytological appearances of the malignant cells within the blood and marrow. There are T-cell (sCD3+ve) and NK (sCD3-ve) forms of the disease. Patients present with only mild lymphocytosis and often with cytopenias. Lymphadenopathy is uncommon. Clonality studies are therefore essential to demonstrate this leukemia. The disease usually follows an indolent course, although the NK form may be more aggressive. There has been no consistent cytogenetic abnormality detected.

No typical human T-cell LGL cell lines have been reported, although cell lines with an NK-LGL phenotype exist (Chapter 4). Interestingly, 50% of the Fisher strain of laboratory rats develop T-cell LGL and cell lines have been derived. The nature of the genetic abnormalities underlying this remains unknown. One IL-2 dependent cell line (EBT-8) has been derived from a patient with Epstein–Barr viral-associated disease. Although described as LGL and although the phenotype of the cells is consistent with this diagnosis, the presence of EBV in the cells is not characteristic.

### 3.1.4. NK/T-Cell Leukemia/Lymphoma

These malignancies co-express differentiation antigens of both T-cell and NK lineages. Some of these are EBV positive and are mentioned below. Others are EBV negative and often pursue an aggressive and leukemic pattern. They can be distinguished from malignancies of the NK lineage by the presence of clonal *TCR* gene rearrangements. A cell line (MTA) from one such patient has been described and may be useful in determining the pathogenesis of this disease [13]. This cell line co-expresses CD2, sCD3, CD4 and CD56 and has clonal *TCR* rearrangements.

### 3.1.5. Sézary-Cell-Like Leukemia

This disease is very rare and related to T-PLL. No cell lines are known to exist, although the HUT-78 cell line, with its highly complex karyotype, may have been derived from such a patient.

## 3.2. Nodal T-Cell Lymphomas

### 3.2.1. Nodal Peripheral T-NHL – Unspecified

This is something of a "waste-basket" for T-cell lymphomas that do not readily fit into a more specific category and, unfortunately, is the largest diagnostic group. There are no consistent immunophenotypic, cytogenetic or molecular features. CD4 expression is more common than CD8, but expression of this and other T-cell differentiation antigens can change during the course of the disease; loss or lack of expression of CD7 is common. Most of the described T-cell lines fall into this category, although it should be noted that the description of the cell lines is in nearly all cases inadequate in one or more aspects and limited to a single report.

### 3.2.2. Anaplastic Large Cell Lymphoma (ALCL)

ALCL are characterized by chromosomal translocations involving the *ALK* gene that encodes a tyrosine kinase and is located on chromosome 2p23. The most common translocation is the t(2;5)(p23;q35), which involves *ALK* with the nucleophosmin or *NPM* gene on chromosome 5. Unlike most other translocations that are specific for one hematopoietic lineage and often for one specific disease, *ALK* translocations are seen in lymphomas of both B and T-cell lineages. There are several *ALK*+ve T-cell lines available: these are discussed in Chapter 11. It is noteworthy that the cell line HPB-MLp-W, derived from a patient with a poorly characterized T-NHL, expresses CD30. This cell line does not appear to show the karyotypic changes typical of ALCL, but it has not been studied with modern cytogenetic methods. Whether it expresses ALK has not been established.

## 3.3. Extranodal T-Cell Lymphomas

### 3.3.1. Mycosis Fungoides/Sézary Syndrome

This is primarily a cutaneous disorder. Mycosis fungoides often presents with localized skin lesions and pruritus. Sézary syndrome may be considered as the leukemic manifestation of Mycosis fungoides, although there may also be lymph nodal involvement. Transformation to large-cell T-NHL may occur in about 10% of patients. Most, if not all cases are CD4+ve and lack expression of CD25. Association of MF/SS with retroviruses has been claimed, but has not been confirmed in large studies.

The HUT-78 cell line was derived from a patient with typical Sézary syndrome and exhibits a number of interesting features. It is the only malignant lymphoid cell line to express large amounts of the CAMPTH-1 (CD52) antigen. This molecule, whose functions are unknown, is expressed at very high levels ($> 5 \times 10^5$ molecules per cell comprising about 2% of the cell surface) on nearly all malignant lymphocytes of both T-cell and B-cell lineages and has been used as a target for antibody therapy. However, for reasons which remain obscure, the protein is lost during *in vitro* culture in most cell lines with the exception of HUT-78, which retains levels of expression comparable to those seen *in vivo*. HUT-78 is therefore useful in modelling therapy with CD52 antibodies.

HUT-78 exhibits an extermely complex karyotype which has not yet been adequately studied [9]. Nevertheless, molecular studies have shown a number of interesting abnormalities, although none are characteristic of "regular" Sézary syndrome. Firstly, HUT-78 exhibits a potent synergistic combination of deregulated *MYC* expression and p53 mutation [10] that may explain the rapid growth of the cells. The *MYC* expression derives from a chromosomal translocation t(2;8)(q34;q24) involving the *MYC* oncogene with a novel locus, TCL4 on chromosome 2q34 [14]. Both the *MYC* translocation and the p53 mutation (analogous to that seen in Burkitt lymphoma) appear to be unusual events in the pathogenesis of MF/SS and indeed of all T-cell malignancies. The possible involvement of the *TCL4* gene in other T-cell malignancies remains to be determined.

HUT-78 also exhibits carboxy-terminal deletion of two molecules involved in signal transduction. The *NFκB2/Lyt-10* gene, which maps to chromosome 10q24, was cloned through its direct involvement in the t(10;14)(q24;q32) seen in a subset of aggressive B-cell lymphomas (reviewed in Neri et al. [35]). HUT-78 exhibits a carboxy-terminal truncation of this molecule [48]. *c-CBL*, a ring-finger gene that maps to human chromosome 11q23.3 has also been shown to undergo deletion through rearrangement with unknown sequences [6]. Neither genetic rearrangement appears to be common. Further molecular genetic dissection of the chromosomal abnormalities of HUT-78 is warranted.

### 3.3.2. Enteropathy-Associated Intestinal T-Cell NHL

This disease is strongly associated with celiac disease (gluten-sensitive enteropathy) and appears to arise from T-cells present within the mucosa which are predominantly CD8+ve/CD103+ve. Disease is usually localized to the small bowel and patients may present with perforation or obstruction. At least one cell line, (OCI-Ly 17) has been derived from a patient with this disease, but it is poorly characterized and the clonal relationship with the primary tumor was not demonstrated.

### 3.3.3. Hepatosplenic T-NHL

Historically, this disease was described as a disease of T-cells of the TCR$\gamma/\delta$ lineage although cases of the $\alpha/\beta$ lineage have now been reported [11]. This disease occurs mostly in young males, with marked hepatosplenomegaly occurring in the absence of significant lymphadenopathy. There is a characteristic immunophenotype: CD2+, CD3+, CD7+, CD56+ with expression of TCR$\gamma/\delta$ proteins. Isochromosome 7q(iso(7q)) has been suggested to be a recurrent event [42,43]. One cell line (HPB-MLp-W) with this cytogenetic abnormality, on the background of a highly complex karyotype, has been reported. However, from the clinical data presented it is very unlikely that the cell line was derived from a case of hepatosplenic T-NHL.

## 3.4. EBV-Related T-Cell Lymphomas

Although in lymphocytes EBV has been classically associated with infection of B-cells, it is now clear that EBV can infect T-cells through low level expression of CD21 on these cells. In some cases this can give rise to EBV-associated lymphomas. Syndromes include:

- Post-transplant T-cell lymphoproliferative disorders, although these remain rare in contrast with their B-cell counterparts [17].
- Nasal T-cell/NK lymphomas are strongly associated with EBV. These occur primarily in the Orient. There is one recently described cell line (HANK-1), although from the presence of the germline *TCR* genes it is likely that the cell line belongs to the NK rather than the T-cell lineages [22].
- Severe chronic active EBV infection leading to EBV-positive T-NHL [22].

EBV positive normal T-cell lines have been described. The EBV-associated LGL cell line EBT-8 has been mentioned above. Details of the EBV-positive DEGLIS cell line which co-expresses T-cell and B-cell lineage antigens and exhibits a most unusual combination of *TCR* and *IG* gene rearrangements are given in the tables. The cell of origin of this apparently unique cell line is not known, but the authors suggest that it arose in a "cell broadly committed to the lymphoid lineage".

## 4. CELL LINES DERIVED FROM PATIENTS WITH MATURE T-CELL MALIGNANCIES

There are at least two problems in discussing cell lines derived from patients with mature T-cell malignancies. Firstly, the classification presented above is recent, and therefore the precise categorization of cell lines established

*Table 4.*  Mature T-cell malignant cell lines: clinical characterization

| Cell line | Patient age/sex | Diagnosis | Treatment status | Specimen Site | Authentication | Year est. | Culture medium | Availability | Primary ref. |
|---|---|---|---|---|---|---|---|---|---|
| HUT 78 | 53 M | Sézary syndrome | NK | PB | NO | 1977 | RPMI 1640 10% FCS | ATCC | 16 |
| HUT 102 | 26 M | ATLL leukemic phase | NK | LN | NO | 1978 | RPMI 1640 10% FCS | ATCC | 16 |
| KARPAS 384 | 48 M | T-NHL NOS— subcutaneous → leukemic transformation | PROGR | PB | YES | 1987 | RPMI 1640 10% FCS | Author | 12 |
| SKW-3 | 61 M | Mature T-cell leukemia | NK | PB | NO | NK | RPMI 1640 10% FCS | DSMZ | Unknown |
| SMZ-1 | 46 M | T-NHL NOS/stage IV/ systemic lupus erythematosis (SLE) | DIAGN | ASCITES | NO | 1992 | RPMI 1640 15% FCS | Author | 30 |
| ST-4 | 12 M | T-NHL NOS | DIAGN | LN | NO | 1984 | RPMI 1640 10% FCS | Author | 2 |
| PFI-285 | 13 M | ?T-NHL NOS/stage IV | 2ND REL | PB | NO | Unknown | RPMI 1640 10% FCS | Author | |
| OC-LY 17 | 72 M | T-NHL NOS/celiac disease | DIAGN | PB | NO | 1987 | IMDM 20% human plasma | Author | 8 |
| MU | 37 M | ATLL leukemic phase | NK | PB | YES | 1986 | RPMI 1640 20% FCS | Author | 24 |
| HPB-MLp-W | 50 F | T-NHL NOS/stage IV/CD30+ve | DIAGN | LN | NO | 1986 | RPMI 1640 10% FCS | Author | 32 |
| EBT-8 | 36 M | T-cell LGL(?) EBV+ve | NK | PB | NO | 1994 | RPMI 1640 10% FCS +40U/mL rIL2 | Author | 3 |
| T-34 | 73 M | T-NHL NOS | 1ST REL | LN | YES | 1986 | RPMI 1640 20% FCS | Author | 36 |
| HANK-1 | 46 F | Nasal type T/NK NHL retroperitoneal mass | DIAGN | LN | YES | 1995 | Cosmedium 20% human plasma+ 100U/ml rIL2 | Author | 22 |
| DEGLIS | 68 M | Mediastinal mass | DIAGN | LN | YES | 1990 | Iscoves's 10% FCS | Author | 40 |
| KIT-225 | 62 M | Mature T-cell leukemia | DIAGN | PB | YES | 1985 | RPMI 1640 10% FCS + 10U/mL rIL2 | Author | 19 |

Table 5. Mature T-cell malignant cell lines: immunophenotypic characterization

| Cell line | T-cell marker | B-cell marker | Myelomonocytic marker | Non-lineage/activation markers |
|---|---|---|---|---|
| HUT 78 | CD3+, CD4+, CD8− | NT | NT | CD25− HLA-DR+ |
| HUT 102 | CD4+, CD8− | NT | NT | CD25+, HLA-DR+ |
| KARPAS 384 | CD1a−, CD2−, sCD3+, CD4−, CD5−, sCD7−, cCD7+, CD8−, TCRα/β−, TCRγ/δ− | CD10−, CD19−, CD20− | NT | CD25−, CD38−, CD45+, CD45RO−, CD52−, HLA-DR+ |
| SKW-3 | CD2+, CD3−, CD4−, CD5+, CD6+, CD7+, CD8−, TCRα/β− | NT | CD13− | |
| SMZ-1 | CD1b−, CD2+, sCD3+, CD4+, CD5−, CD7+/−, CD8−, CD28+/− | CD10−, CD19−, CD20− | CD13−, CD14−, CD15−, CD16− | CD25−, CD71+, HLA-DR+ |
| ST-4 | CD1+, CD2−, sCD3−, CD4−, CD7+, CD8− | CD10− | NT | CD25−, HLA-DR− |
| PFI-285 | CD1+/−, CD2+, sCD3−, CD4+/−, CD5+, CD7+, CD8+/− | CD10−, CD19− | CD15−, CD16−, CD36− | CD25−, HLA-DR− |
| OC-LY 17 | CD2+, sCD3−, CD4+, CD5−, CD7−, CD8− | CD10−, CD19−, CD20−, sIg− | NT | |
| MU | CD1−, CD2−, CD3−, CD7−, CD8− | CD10−, CD19−, CD20−, sIg− | CD11−, CD13+, CD14−, CD15−, CD16 | CD25+/−, CD34−, HLA-DR+/− |
| HPB-MLp-W | CD1a−, CD2+, sCD3−, cCD3+, CD4+, CD5−, CD7−, CD8+, TCRα/β− TCRγ/δ− | CD9−, CD10−, sIg− | CD13−, CD14−, CD15−, CD16− | CD25+, CD30+, HLA-DR+ |
| EBT-8 | CD2+, sCD3+, CD4−, CD8+, TCRα/β+, TCRγ/δ− NK markers CD56, CD57− | CD19−, CD20−, CD21−, CD23− | NT | CD25+, HLA-DR+ |
| T-34 | CD1a−, CD2+/−, sCD3+, CD4+/−, CD8− | CD10−, CD19−, sIg− | NT | CD25+/− |

*Continued on next page*

*Table 5.* (continued)

| Cell line | T-cell marker | B-cell marker | Myelomonocytic marker | Non-lineage/ activation markers |
|---|---|---|---|---|
| HANK-1 | CD1−, CD2+, CD3− CD4−, CD5−, CD7+, CD8−, cCD3ε+ NK markers CD56+, CD57−, | CD10− | NT | CD25−, HLA- DR+ |
| DEGLIS | CD1−, CD2+, cCD3+/−, CD4+, CD5−, CD7+, CD8− TCRα/β− | CD10−, CD19+, CD20+, CD21−, CD22−, CD23+, CD37+ | CD11−, CD14−, CD15−, CD16− CD68+ | CD25−, CD30+, CD70+, CD45+, CD45RO− TdT+ |
| KIT-225 | SCD3+, CD4+, CD6+, CD8− | NT | NT | CD25+, HLA-DR+ |

*Table 6.* Mature T-cell malignant cell lines: genetic characterization

| Cell line | Cytogenetic karyotype | Chromosomal rearrangements of potential interest | Antigen receptor gene rearrangements | Ref. |
|---|---|---|---|---|
| HUT 78 | *Extremely complex* – see Chen [9] | *MYC/TCL4* rearrangement Truncation of *NFκB2* and *c-CBL* | | |
| HUT 102 | Poorly characterized | Unknown | | |
| KARPAS 384 | 47, XO, +20, t(1:2)(q11:q35), t(2:1:14)(q35:q11- q32.1:q22.1), t(7:14)(p13:q11.2), inv(7)(p13:q22.1), int deletion(12)(q24.1q24.3)+ marker | 14q11.2 = location of the *TCRα/δ* locus. t(7:14)(p13:q11.2) does not involve the *TCRγ* locus | Belongs to the TCRγ/δ lineage (a) 3 TCR Jδ3 rearrangements presumably marking involvement in 14q11.2 translocation (b) *TCRγ9/JγP/Cγ1* rearrangement | 12 |
| SKW-3 | 43-48 XY, +8, −14, t(3:3)(q11:q27), der(8)t(8:14)(q24:q11)x2, t(8:11)(p21:p12), der(12)t(12:?)(q24:?), der(14)t(8:14)(q24:q11) | *MYC* translocated to the *TCRα/δ* locus | | 15 |
| SMZ-1 | 47 XY, t(6:14)(p21.1:q24), +8, del(9)(p13q22), der(9)t(1:9)(p12:p13), −10, der(17)(?::p11- q23::?), +21 | Unknown | | |
| ST-4 | 47 X - Y, +X, t(2:6)(q21:q23)+der(2)t(2:6)(q21:q23) | | | |
| PFI-285 | Normal karyotype in 29/44 metaphases – remainder = tetraploid only with no structural changes | | | |
| OC-LY 17 | Not determined | | | |
| MU | *Clonal evolution* 54 XY, −1, +7, +8, −14, +19, +20, +21, + der(1)t(1;?)(q36:?), + del(6)(q13q25), +der (14)t(14:?)(q32:?)+ two markers | Del(6q) and abnormalities of 14q32 commonly seen in T-cell NHL | Identical *TCRCβ* rearrangements seen in patient material and cell line | |
| HPB-MLp-W | 88, XX, +3q, +3q, −8p. iso(7q), +8 markers | Unknown | | |
| EBT-8 | 48 XY, +2, der(3)t(3:3)(p25:q21), del(11)(q23), +17, der(22)t(1:22)(q11:p11) Del(11)(q23) similar to that seen in T-PLL and may involve the *ATM* locus | Unknown EBV monoclonal integration | | |

*Continued on next page*

*Table 6.* (continued)

| Cell line | Cytogenetic karyotype | Chromosomal rearrangements of potential interest | Antigen receptor gene rearrangements | Ref. |
|---|---|---|---|---|
| T-34 | *Extremely complex*<br>Multiple abnormalities of chromosomes # 1, 2 , 3, 8, 11, 12, 13, 14, 15, 16, and 19 Del(11)(q21q23) similar to that seen in T- PLL and may involve the *ATM* locus | 8- to 16-fold *MYC* amplification | | |
| HANK-1 | 48 XX, +2, del(16q13), +21 | Unknown | | |
| DEGLIS | 46 XY, −2, +der(2)(?::2p23-2q37::?), −4, + der(4)t(4:?)(q34:?), iso(6p), +del(6p22), −8, +der(8)t(8:?)(q24:?), t(13;15)(p11:p11), t(21:22)(p11:p11) | Possible breaks within *ALK* locus at 2p23 and *MYC* locus at 8q24 | *IGJH* and *TCRCβ* rearrangements | |
| KIT-225 | *Complex*<br>47 XY, −5, −6, −14, +19, ins inv(1)(pter-p36.3::p34.1->p31.2::p36.3->p34.1::p31.2->p22.1::q12-> p22.1::q12-qter) inv(3)(p26q27), +der(5)t(5;7)(p15.3:q22), +der(6)t(6:?)(q21:?), +der(14)t(6;14)(q21:p12) | Unknown – abnormalities of chromosome 1p22 described in Sézary syndrome | | |

several years ago may not be possible. Thus, there may be some confusion between cell lines derived from patients with T-cell precursor ALL, T-cell lymphoblastic lymphoma and post-thymic T-cell leukemia. Although immunophenotypic studies may allow clarification, changes associated with *in vitro* culture may obscure the precise derivation. Some T-cell malignancies lose surface antigen expression with *in vitro* culture (see for example Ohno et al. [36]). As with B-cell malignancies, loss of surface antigen receptor expression due to ongoing rearrangements or mutations at the *TCR* loci may result in loss of surface CD3 expression and may cloud the issue. Secondly, many of the cell lines listed have been inadequately studied both in terms of expression of T-cell differentiation antigens and *TCR* gene rearrangements. The latter demands the use of a series of probes to both the constant and joining gene segments; a single probe is insufficient. Thus, in many cases, it is not possible to determine the precise stage of T-cell differentiation from which the cells are derived. In the case of PFI-285 and ST-4 it seems likely from the immunophenotype (CD1 positivity and lack of HLA-DR expression) that the cell lines were derived from a lymphoblastic lymphoma/ T-cell precursor ALL rather than a mature T-cell malignancy.

Finally, most of the cell lines described have been derived from Oriental patients where the pattern of T-cell disease differs substantially from that in the Occident, are the subject of only a single report and their availability in some instances is not known. There remains a requirement for cell lines from many of the recently recognised diseases of mature T-cells.

## References

1.  Abe M et al. *Cancer* 69: 1235–1240, 1992.
2.  Arione R et al. *Cancer Res* 48: 1312–1318, 1988.
3.  Asada H et al. *Leukemia* 8: 1415–1423, 1994.
4.  Ascari S et al. *Annals Oncol* 8: 583–592, 1997.
5.  Barclay AN et al. *The Leucocyte Antigen Factsbook*, 2nd edn., Academic Press, London, 1997.
6.  Blake TJ and Langdon WY. *Oncogene* 7: 757–762, 1992.
7.  Bunn PA and Foss FM. *J Cell Biochem Supplement* 24: 12–23, 1996.
8.  Chang H et al. *Leukemia Lymphoma* 8: 97–107, 1992.
9.  Chen TR. *J Natl Cancer Inst* 84: 1922–1926, 1992.
10. Cheng J and Haas M. *Mol Cell Biol* 10: 5502–5509, 1990.
11. Cooke CB et al. *Blood* 88: 4265–4274, 1996.
12. Dyer MJS et al. *Leukemia* 7: 1047–1053, 1993.
13. Emi N et al. *Int J Hematol* 69: 180–185, 1999.
14. Finger L et al. *Proc Natl Acad Sci USA* 85: 9158–9162, 1988.
15. Finver SN et al. *Proc Natl Acad Sci USA* 85: 3052–3056, 1988.
16. Gazdar AF et al. *Blood* 55: 409–417, 1980.
17. Hanson MN et al. *Blood* 88: 3626–3633, 1996

18. Helgestad J et al. *Eur J Haematol* 44: 9–17, 1990.
19. Hori T et al. *Blood* 70: 1069–1072, 1987.
20. Jaffe ES, Krenacs L and Raffeld M. *Annals Oncol* 8 (Suppl. 2): S17–S24, 1997.
21. Kagami Y et al. *Jpn J Cancer Res* 84: 371–378, 1993.
22. Kagami Y et al. *Br J Haematol* 103: 669–677, 1998.
23. Kanegane H et al. *Blood* 91: 2085–2091, 1998.
24. Koizumi S et al. *J Natl Cancer Inst* 84: 690–693, 1992.
25. Larsen CJ et al. *Leukemia* 2: 247–248, 1988.
26. Matsuzaki H et al. *Jpn J Cancer Res* 83: 450–457, 1992.
27. Matutes E et al. *Blood* 78: 3269–3274, 1991.
28. Matutes, E. *T-cell Lymphoproliferative Disorders: Classification, Clinical and Laboratory Aspects*, Harwood Academic Publishers, 1999.
29. McKeithan TW et al. *Proc Natl Acad Sci USA* 83: 6636–6640, 1986.
30. Miyanishi S and Ohno H. *Cancer Genet Cytogenet* 59: 199–205, 1992.
31. Miyoshi I et al. *Nature* 294: 770–771, 1981.
32. Morikawa S et al. *Leukemia Res* 15: 381–389, 1991.
33. Morita M et al. *Blood* 77: 1766–1775, 1991.
34. Nakao Y et al. *J Natl Cancer Inst* 78: 1079–1086, 1987.
35. Neri A et al. *Leukemia Lymphoma* 23: 43–48, 1996.
36. Ohno H et al. *Cancer Res* 48: 4959–4963, 1988.
37. Paul WE (ed.) *Fundamental Immunology*, 4th edn., Lippincott-Raven, Philadelphia/New York, 1998.
38. Pawson R et al. *J Clin Oncol* 15: 2667–2672, 1997.
39. Pileri SA et al. *Annals Oncol* 9: 797–801, 1998.
40. Al Saati T et al. *Blood* 80: 209–216, 1992.
41. Sagawa K et al. *Kurume Med J* 42: 149–160, 1995.
42. Salhany KE et al. *Blood* 89: 3490–3491, 1997.
43. Salhany KE et al. *Hum Pathol* 28: 674–685, 1997.
44. Shima EA et al. *Proc Natl Acad Sci USA* 83: 3439–3443, 1986.
45. Taylor AMR et al. *Blood* 87: 423–438, 1996.
46. Vorechovsky I et al. *Nat Genet* 17: 96–99, 1997.
47. Yuille MAR et al. *Oncogene* 16: 789–796, 1998.
48. Zhang J et al. *Oncogene* 9: 1931–1937, 1994.

# Chapter 11

# Hodgkin's Disease

Andrea Staratschek-Jox, Jürgen Wolf and Volker Diehl
*Department of Internal Medicine I, HS16, University of Cologne, D-50924 Cologne,*
*Germany. Tel: +49-221-478-4400; Fax: +49-221-478-5455*

## 1. INTRODUCTION

The microscopic appearance of Hodgkin's disease (HD) tissue is a small number of lymphoma cells, the so-called Hodgkin and Reed-Sternberg (H-RS) cells, surrounded by a non-neoplastic cellular environment consisting mostly of T-lymphocytes [47].

The first cell line (L428) was derived from a pleural effusion obtained from a 37-year-old woman with relapse of nodular sclerosing Hodgkin's disease [52]. The L428 cell line was considered to be of H-RS cell origin for the following reasons: the clonal cell population expressed the H-RS cell-associated clusters of differentiation CD15 and CD30, cytogenetic analysis revealed a grossly aberrant karyotype, the cell line did not harbor EBV and tumor development was observed in nude mice after intracranial inoculation. These four criteria are used to judge whether the continuous cell line was established from the HD cells. The establishment of such lines is a rare event and only 17 cell lines have been described. One of these, known as Co or Cole, is cross-contaminated and is in fact the T-ALL derived cell line CCRF-CEM [15].

A cell line that grows out from a culture of HD affected tissue or effusion does not, as a rule, represent an H-RS cell population, since other cells present can give rise to a continuous cell line. For only one cell line, L1236 [69] was derivation from H-RS cells unequivocally demonstrated by amplification of identical Ig gene rearrangements from the cell line and from single H-RS cells microdissected from a section of a bone marrow biopsy from the patient [31]. For all of the other cell lines, there is no such authentication.

From one cell line (SBH-1), the histology of the lymphoma tissue was not available [6], and thus there is no proof that the patient suffered from HD. Another cell line, (HKB-1), was derived from a recurrence of HD in a patient who initially presented with a large cell anaplastic lymphoma at the same

339

 *J.R.W. Masters and B.O. Palsson (eds.), Human Cell Culture Vol. III, 339–353.*
© 2000 *Kluwer Academic Publishers. Printed in the Netherlands.*

site [66]. Consequently, the cell line may have been derived from the ALCL, rather than the HD.

The clinical characterization, immunophenotype, cytokine expression, chromosomal aberrations and growth characteristics of the 16 cell lines which may represent HD cells are summarized in Tables 1–5 and discussed below.

## 2. CLINICAL CHARACTERIZATION

Two further cell lines (L591, L540) were established from HD by the same group that developed L428 [8]. The three lines were derived from pleural effusions or bone marrow aspirate obtained from young women with progressive HD of nodular sclerosis subtype. A further 13 HD derived cell lines have been reported, 11 of which were derived from young patients suffering from nodular sclerosis HD, and in the other 2 cases, the line was derived from HD of mixed cellularity. This reflects the incidence of the histological HD subtypes among young adults, with most having nodular sclerosis HD [2]. Like the first three cell lines, most of the subsequent lines grew from HD-affected material obtained from pretreated patients during relapse or progressive disease. Eleven cell lines were established from either pleural effusion, pericardial effusion or peripheral blood (see Table 1).

## 3. IMMUNOPHENOTYPE

The vast majority of H-RS cells and the HD cell lines express CD30. This antigen is also expressed on activated or transformed (with human T-lymphotrophic virus-1 or Epstein–Barr virus) T and B lymphocytes [55], activated [45] and differentiated macrophages [1], and on the tumor cells of anaplastic large cell lymphoma (ALCL), which can also be called Ki-1-lymphoma [43]. This reaction pattern makes CD30 antibodies a valuable diagnostic tool. The CD30 antigen is a 120 kDa, membrane-bound, phosphorylated glycoprotein with a non-phosphorylated, 84 kDa, intracellular apoprotein and a 90 kDa degradation residue released into the supernatant [21]. Additionally, an independently synthesized 57 kDa intracellular molecule has the same antigenicity. The gene coding for CD30 has been cloned and identified as a member of the TNF-receptor superfamily [16]. The CD30 ligand has also been cloned [57]. The interaction of CD30 with its ligand is thought to be involved in the regulation of apoptosis and proliferation of activated lymphatic cells.

*Continued on next page*

*Table 1.* Hodgkin's disease derived cell lines: clinical characterization

| Cell line | Patient age/sex | Diagnosis/staging | Treatment status | Specimen site | Culture medium | Cell of origin IG−, TCR rearrangement | Primary ref. |
|---|---|---|---|---|---|---|---|
| HuT$_{11}$ | 6/F | HD, mixed cellularity; IIA, first diagnosis | no treatment | lymph node | McCoy's medium 5A, 10%FCS | B-cell Cγ, Cκ expression | 50 |
| L428 | 37/F | HD, nodular sclerosis, IVB, relapse | mustine, oncovin, procarbazine, prednisone | pleural effusion | RPMI 1640, 10%FCS, 4 mmol/L L-glutamine | B-cell JH R/D, Cκ G/G, Cλ G/G [59] | 52 |
| L540 | 20/F | HD, nodular sclerosis, IVB | | bone marrow | RPMI 1640, 10%FCS, 4 mmol/L L-glutamine | T-cell JH G/G, Cκ G/G, Cλ G/G,TcRα R, TCRβ R, TCRγ R/R [59] | 8 |
| L591 | 31/F | HD, nodular sclerosis/IVB | | pleural effusion | RPMI 1640, 10%FCS, 4 mmol/L L-glutamine | B-cell JH R/D, [59] | 8 |
| SU/RH-HD1 | 12/M | HD, nodular sclerosis | | spleen | RPMI-1640, 15%FCS, 0.3% L-glutamine | | 44 |
| DEV | 51/M | HD, nodular sclerosis, IV, relapse | radiation | pleural effusion | RPMI 1640, 10%FCS | B-cell expression of IGA2 | 46 |
| KM-H2 | 32/M | HD, mixed cellularity, IV, relapse | prednosine, vinblastine, cyclophosphamide | pleural effusion | RPMI 1640, 10%FCS, | B-cell JH R/D | 29 |
| HDLM-2 | 74/M | HD, nodular sclerosis | | pleural effusion | RPMI 1640, 10%FCS | T-cell JH G/G, Cκ G/G, TCRβ R/R, TCRγ R/R [14] | 12 |
| Sup-HD1 | 34/M | HD, nodular sclerosis, IV, relapse | procarbazine, melphalan, velban/adriamycin, bleomycin, velban, dacarbazine, aggressive chemotherapy and irradiation in relapse | pleural effusion | McCoy 5A, 10% FCS, 10ng/ml insulin like growth factor | B-cell JH R/D, Cκ R/R, Cλ G, TCRβ R | 38 |

*Table 1.* (continued)

| Cell line | Patient age/sex | Diagnosis/staging | Treatment status | Specimen site | Culture medium | Cell of origin IG−, TCR rearrangement | Primary ref. |
|---|---|---|---|---|---|---|---|
| ZO | 26/F | HD, nodular sclerosis, IV, progressive disease | mustine, oncovin, procarbazine, prednisone | pericardial effusion | RPMI 1640, 20% FCS, add lymphocult | B-cell JH R/R, Cκ R/R, Cλ G | 49 |
| HO | | HD, nodular sclerosis | | lymph node | RPMI 1640, 10%FCS | T-cell JH G, TCRγ R, TCRCβ1 R/R, TCRCβ2G/G | 26 |
| HD-70 | 69/M | HD, nodular sclerosis, IV, first diagnosis | cyclophosphamide, vincristine, procarbazine, prednisolone/doxorubicin, bleomycin, vincristine, dexamethasone | peripheral blood | RPMI 1640, 20%FCS, 10% human cord blood serum | B-cell JH R/R, Cκ R/G, TcRα G, TCRβ G, TCRγ G | 30 |
| HD-Myz | 29 | HD, nodular sclerosis, IVB, relapse | combined chemo- and radiotherapy | pleural effusion | RPMI 1640, 20% FCS, 2 mM L-glutamine | non-B- non-T-cell JH G, TCRβ G/G | 3 |
| SBH-1 | 78/F | HD, no histology available, pleural effusion contained H-RS cells | no treatment | pleural effusion | RPMI 1640, 10%FCS | B-cell JH R/R, Cκ R/D, Cλ R/R | 6 |
| L1236 | 34/M | HD, mixed cellularity, IV, relapse | radiation, 3 cycles COP/ABVD. High dose chemotherapy autologous bone marrow transplantation | peripheral blood | RPMI 1640, 10%FCS, 4 mmol/L L-glutamine | B-cell JH R/R, Cκ R/G, Cλ G/G | 69 |
| HKB-1 | 14/F | initial diagnosis: large cell anaplastic lymphoma developing on the basis of a preexisting HD, in relapse: relapse HD, nodular sclerosis, IV | combined chemo- and radiotherapy | intrapulmonary tumor biopsy during relapse | RPMI 1640, 10% FCS, glutamine | B-cell JH R, TCR G | 66 |

R – rearranged, G – germline, D – deleted. Cell lines are listed in order of publication.

*Table 2.* Hodgkin's disease derived cell lines: immunophenotypical characterization

| Cell line | T/NK cell marker | B-cell marker | Myelomonocytic marker | Other markers | CD30/CD15 | Ref. |
|---|---|---|---|---|---|---|
| HuT$_{11}$ | | Cγ+, Cα−, Cμ−, Cδ−, Cε−, Cκ+, Cλ− | | | | 58 |
| L428 | CD1−, CD2−, CD3−, CD4−, CD5−, CD6−, CD7−, CD8−, CD16−, Leu-7−, Leu-19− | IgM−, IgA−, IgG− Cκ−, Cλ−, CD19−, CD20−, CD21−, CD22−, CD24− | CD11a−, CD11b−, CD11c−, CD13−, CD14−, CD33− | CD9−, CD10−, CD34−, CD71+, Ki-67+, HLA-A, B, C−, HLA-DP+, HLA-DQ+, HLA-DR+, | CD30+, CD15+ | 8,14,17 |
| L591 | CD2+, CD3−, CD4−, CD8− | CD19+, CD20+, IgM−, IgA−, IgG− Cκ−, Cλ−, | | HLA-DR+ | CD30+, CD15+ | 8,53 |
| L540 | CD2+, CD3−, CD4+, CD5−/+, CD7−, CD8− | IgM−, IgD−, IgG−, IgA− Cκ−, Cλ−, CD22−, CD20−, CD19− | CD11b+, CD11c−, CD14+ | Ki-M1 (+), CD10−, HLA-DR+ | CD30+, CD15+ | 17,49 |
| SU/RH-HD1 | CD5−, CD8−, CD4− | Leu-10−, IgG+, IgM−, | | Leu-7−, HLA-I+, HLA-DR+ | CD15−, CD30− | 10,44 |
| DEV | Leu7−, OKT1−, CD3−, CD4−, OKT5−, CD1a−, CD8−, OKT11− | Cκ−, Cλ−, Cα+, CD20+, CD24−, B2− | OKM1−, OKM2− | CD10−, HLA-DR−, CD71+, | CD30+, CD15+ | 46,49 |
| KM-H2 | CD1−, CD2−, CD3−, CD4−, CD5−, CD6−, CD7−, CD8−, CD16−, Leu-7−, Leu-19− | CD19−, CD20−, CD21−/+, CD22−, CD24− | CD11b−, CD11c−, CD13−, CD14−, CD33− | CD9+, CD10−, CD11a−, CD10−, CD34−, CD71+, HLA class A,B,C+, HLA-DQ+, HLA-DP+, HLA-DR+, Ki-67+ | CD30+, CD15+ | 14,29,49 |
| HDLM-2 | CD1−, CD2+, CD3−, CD4−, CD5−, CD6−, CD7−, CD8−, CD16−, Leu-7−, Leu-19- | CD19−, CD20−, CD21−, CD22−, CD24− | CD11b−, CD13−, CD14−, CD33− | CD9−, CD10−, CD11a−, CD34−, CD71+, HLA-A,B,C+, HLA-DP+, HLA-DQ+, HLA-DR+, Ki-67+ | CD30+, CD15+ | 14 |

*Continued on next page*

*Staratschek-Jox et al.*

*Table 2.*  (continued)

| Cell line | T/NK cell marker | B-cell marker | Myelomonocytic marker | Other markers | CD30/CD15 | Ref. |
|---|---|---|---|---|---|---|
| Sup-HD1 | CD2–, CD3–, CD4–, CD5–, CD7–, CD8–, CD16–, CD56– | CD19–, CD20–, CD21–, CD22–, CD37– | CD11c (+), CD14–, CD68– | | CD30–/CD15+ | 38 |
| Zo | CD2–, CD3–, CD4+, CD7–, CD8– | CD19–, CD20–, CD21– | CD11c– | HLA class I+, HLA class II+, CD45– | CD30+, CD15+ | 49 |
| Ho | CD1–, CD2–, CD3+, CD4+, CD7+, CD8–, CD38 (+) | CD19–, CD22–, CD23–, Cκ–, Cλ–, | | | CD30+, CD15– | 26 |
| HD-70 | CD1a–, CD2–, CD3–, CD4–, CD5–, CD8–, CD45RO– | CpCκ+, CpCα+, CpCλ–, CpCδ–, CpCγ–, CpCμ–, CD19–, CD20–, CD21–, CD24–, CD38–, | CD13–, CD14–, CD33–, CD36–, Leu-M2–, Leu M3– | | CD30+, CD15+ | 30 |
| HD-Myz | CD1a–, CD2–, CD3–, CD3–, CD4–, CD5–, CD7–, CD8– | CD19–, CD20–, CD21–, CD22–, CD23–, CD37–, CD38–, CD39–, IgG–, IgD–, IgM– | CD11b–, CD13+, CD14–, CD68+ | CD10+, CD29+, CD33–, CD34–, CD40–, CD56–, CD71+, CD76–, CD77–, HLA-DR+, MHC I+, CD95– | CD15–, CD30– | 3 |
| SBH-1 | CD3–, CD5–, CD7– | CD19+, CD20+, CD22+ | CD13–, CD33–, CD14– | CD10–, CD45+, CD71+, EMA–, HLA-DR– | CD30+, CD15+ | 6 |
| L1236 | CD3–, CD5–, CD8–, CD45–, CD45Ro–, TCRγ delta–, CD16– | CD23+, CD19–, CD20–, CD38–, s-Igκ–, s-Igλ– | CD33–, CD14– | CD34–, CD54+, CD58+, CD71+, CD80+, CD86+, HLA-DP+, HLA-DR+, CD10– | CD30+, CD15+ | 69 |
| HKB-1 | CD3–, CD4–, CD8–, CD16–, CD56– | CD19+, CD20+, CD21–, CD23+ | CD14– | | CD30+, CD15– | 66 |

Table 3. Hodgkin's disease derived cell lines: cytokine related characterization

| Cell line | Cytokine receptor expression | Cytokine production | Ref. |
| --- | --- | --- | --- |
| L428 | IL-2R+, IL-6R+ | TGFβ+, TNF+, IL-1+, IL-4+, IL-6+, TARC+, IL-13+, IL-5+ | 7,17,19,28,34,39,40,42,64 |
| L591 | IL-2R+, IL-6R+ | IL-1+, IL-6+, TNF+, TARC+ | 10,19,28,40,53,64 |
| L540 | IL-2R+, IL-6R− | IL-1+, IL-6+/−, TNF+, TARC+ | 7,17,19,28,40,64 |
| SU/RH-HD-1 | IL-1+ | | 10,44 |
| DEV | IL-2R+ | | 53 |
| KM-H2 | IL-2R+, IL-6R− | TNFα+, TNFβ+, M-CFS+, IL-6+, IL-13+, IL-5+ | 19,22,23,28,34,49 |
| HDLM-2 | IL-2R+, IL-6R+ | IL-6+, IL-13+ | 13,28,34 |
| Sup-HD1 | IL-2R+ | INFγ+, IL-2− | 38 |
| Zo | IL-2R+ | | 49 |
| Ho | IL-2R− | IL-6−, TNF+ | 19,26 |
| HD-70 | IL-2R− | | 30 |
| HD-Myz | IL-1R+, IL-2R−, IL-3R(α-chain)−, IL-4R−, IL-6R+, IL-7R−, IL-8R+ | IL-1α+, IL-1β+, IL-2−, IL-3−, IL-4−, IL-5+, IL-6+, IL-7+, IL-8+, IL-10+, GM-CSF−, TGFβ−, TNFα−, TNFβ− | 3 |
| SBH-1 | IL-1R−, IL-2R+, IL-3R−, IL-4R+, IL-6R+, IL-7R+,TNFR+, G-CSFR− | TGFβ+, TGFα+, IL-1α−, IL-1β−, IL-2−, IL-3−, IL-4−, IL-5−, IL-6−, IL-7−, IL-9−, G-CSF−, SCF−, γIFN− | 6 |
| L1236 | IL-2R− | IL-6+, IL-8+, IL-10+, INFγ+, TNFα+, TGFβ+, GM-CFS+, IL-2−, IL-4−, IL-7−, TARC+ | 64,69 |
| HKB-1 | IL-2R+ | IL-1β−, IL-2−, IL-6+, IL-10−, IL-12−, INFγ− | 66 |

*Table 4.* Hodgkin's disease derived cell lines: genetic characterization

| Cell line | Representative karyotype | Ref. |
|---|---|---|
| L428 | 48, −9, −13, +12, dup(1)(p22p32), t(2p25:?), +del(2)(q33), +t(6q23:?), dup(7)(q22q36orq11, 2q32), t(9p24:?), del(11)(q21), +del(12)(q15), t(13p12:?), t(9p12:?), del(11)(q21), del(12)(q15), t(13p12:?), t(14q32:?), del(21)(q21) | 18 |
| L540 | 66, +2, +3, +4, +7, +9, +11, +12, +13, +17, +18, +19, +20, +del(1)(p22), +t(2;10?)(q33;q11.2), +del(5)(q15), t(8q24.1orq23.2:?), +t(11q23:?), +del(11)(q21), +t(12q22:?), +(15;1?)(p12:p31.2or22.3?), +t(21p12:?) | 18 |
| L591 | 46, dup(7)(q32q36), t(14q32.1:?) | 18 |
| DEV | 48, XXY, −2, +12, +mar, t(3;14)(3;22), −del 3, t(3;7) | 46 |
| KM-H2 | 50th passage: 44, X, −Y, −5, −7, −11, −15, −15, −16, −16, −17, −17, +20, −22, 2q+, 4q+, 5p+, 6p+, 7q+, 10p+, 14p+, +8mar | 29 |
| HDLM-2 | 36 chromosomes, non-random chromosome aberrations: 1p32, 3q13, 3q27, 6q23, 7p14, 9p11, 9p12, 11p11, 12q24, 19p13, 22q11 | 11 |
| Sup-HD1 | 44, X, Y−, −1, −2, −4, −7, −9, −13, dup(1)(p13q32), +der(1)t(1;1;6)(q44->q25::p34->q32::p25), del(2)(p23p25), +der(2)t(2;7)(p25;p15), del(5)(p13p15.3), xder(7), dic(4;7)(q31;p15), del(8)(p21p23), t(8;22)q22;q13), del(11)q23q25), t(11:?; 11:?)(p15:?;q23:?), t(14:?)(p11.2-q11.2:?), del(21)q21q22.3), +2mar | 38 |
| Zo | 53, XX, iso 1q, t(1;13), iso 2p, −3, 4q−, 6p−, t(7;17), −15, 16p+, 16q−, +17, +20, +5 markers | 49 |
| HD-70 | 73, Y−, +1, +1, +2, +3, +5, +7, +9, +10, −13, −13, +16, +19, +20, +21, +ins(6;?)(p21;?), +der(7)t(7;?)(q22;?), +der(8)t(8;14)(q24?:q32, +der(11)t(11;1)(p13;q13), +der(11)t(11;1)(p13;q13), +der(11)t(11:?), +der(12)t(12;?)(p13:?)(qcen->qzer), +der(17)t(17:?)(p13:?), +2mar1, +mar2, +mar3, +mar4, +mar5 | 30 |
| SBH-1 | common chromosomal aberrations: del(3)(p11:p25), del(4)(p12p15), del (4)(q21q28), del(6)(q21q28), +7, +dup(8)(q13q22), dup(9)(p13q22), del(11)(q23), +7, +dup(8)(q13q22), dup(9)(p13q22), del(11)(q23), add(12)(p13), i(15)(q10), t(14;18)(q32;q21) | 6 |
| L1236 | 65, +5, +der(1)t(1;14)(p34:?), +der(1)t(1;8)(p22:?), +der(1)dup(1)(q21q44)add(1)(p31-32), +dup(2)(p15p23), +dup(2)(p15p23)+der(3)t(3;16)(p25:?), +der(4)t(4;8)(q31:?), +der(6)dup(6)(p11p25), +der(6)t(1;6)(p34:q15), del(7)(q11), der(7)t(7;17)(p22:?), +der(7)t(7:?)(p22;q22), +der(7), der(8), del(10)(q11), add(11)(p13), +del(11)(q13), +der(12), +del(12)(q22), +del(12)(q15), der(14)t(1;14)(p34-35;q22), del(14)(q22-24), +der+(15)T(15;29)(q22;q11), +der(16), del(17)(q23), del(17)(q23), +der(19)t(7;19)(?;p13), add(20)t(15;20)(q22;q11), +add(20)t(15;20)(q22;q11), +mar1, +mar2 | 69 |
| HKB-1 | 46, X, der(X)add(X)(p11)add(q21), t(1;13)(q2?5:q21), add(2)(p23), t(6;14)(p21;q22-23), t(8;14)(q24:q32) | 66 |

*Table 5.* Hodgkin's disease derived cell lines: functional characterization

| Cell line | Doubling time | EBV status | Cytochemistry | Heterotransplantation into mice | Ref. |
|---|---|---|---|---|---|
| HuT$_{11}$ | 12 hr | EBV– | acid phosphatase+, alkaline phosphatase–, $\alpha$-naphthol esterase+, leucine aminopeptidase+, peroxidase– | hamster: s.c: tumor development in 100% of hamsters | 50 |
| L428 | 42–46 hr | EBV– | PAS–, peroxidase+/–, $\alpha$-naphthylesterase+, acid phosphatase+, naphthol-chloracetate esterase– | nude mice: i.c: tumor development in 2/2 mice, s. c.: no tumor development, Scid mice: s.c.: tumor development in 3/5 mice, i.v.: tumor development in 1/10 mice | 8, 32, 51, 52, 65 |
| L540 | | EBV– | naphthol-chlorate esterase–, peroxidase–, acid $\alpha$-naphthyl acetate esterase+, alkaline phosphatase–, acid phosphatase+ | nude mice: i.c.: tumor development in 24/25 mice, s.c.: tumor development in 23/25 mice, Scid mice: i.v.: tumor development in 4/10 mice | 8, 32, 51 |
| L591 | | EBV+ | naphthol-chlorate esterase–, peroxidase, acid $\alpha$-naphthyl acetate esterase+, alkaline phosphatase–, acid phosphatase+ | nude mice: s.c.: no tumor development, i.c.: tumor development in 100% of mice, Scid mice: s.c.: tumor development in 1/5 mice, i.p.: no tumor development | 8, 65 |
| SU/RH-HD1 | | EBV– | peroxidase–, non-specific esterase+ | nude mice: i.c: tumor development, s.c.: tumor development in 1/14 mice | 10, 44 |
| DEV | 73 hr | EBV– | acid phosphatase+, alkaline phosphatase–, 5-nucleotidase–, aminopeptidase–, peroxidase–, $\alpha$-naphthylesterase– | Scid mice: s.c.: tumor development in 3/5 mice, i.v.: no tumor development | 46, 65 |

*Continued on next page*

*Table 5.* (continued)

| Cell line | Doubling time | EBV status | Cytochemistry | Heterotransplantation into mice | Ref. |
|---|---|---|---|---|---|
| KM-H2 | 60 hr | EBV– | peroxidase–, acid phosphatase+, alkaline phosphatase–, α-NAE+, CAE– | nude mice: s.c.: no tumor development, Scid mice: s.c.: tumor development in 3/5 mice, i.v.: tumor development in 1/10 mice | 29, 32, 65 |
| HDLM-2 | 72–129 hr | EBV– | terminal deoxynucleotide transferase– | Scid mice s.c.: tumor development in 1/5 mice, i.v. no tumor development | 13,65 |
| Sup-HD1 | 72–96 hr | EBV– | adenosine deaminase+, nucleoside phosphorylase+, terminal deoxynucleotidyl transferase– | nude mice: s.c.: no tumor development, i.p.: no tumor development | 38 |
| Zo | growth: slowly | EBV– | | nude mice: s.c. tumor development | 49 |
| Ho | | EBV– | | | 26 |
| HD-70 | | EBV– | acid phosphatase+, α-naphthyl butyrate esterase+, periodic acid-Schiff+, peroxidase–, sudan black–, AS-D chloroacetate esterase–, alkaline phosphatase– | newborn hamster: i.p.: tumor development in 5/5 hamsters | 30 |
| HD-Myz | | EBV– | | Scid mice: i.v.: tumor development, s.c.: tumor development | 3 |
| SBH-1 | 48–60 hr | | peroxidase–, sudan black–, acid phosphatase+, α-naphthyl acetate+ | Scid mice: i.v.: tumor development, i.p.: tumor development | 6 |
| L1236 | | EBV– | | Scid mice: i.v.: tumor development in 2/3 mice, s.c.: tumor development in 3/3 mice | 69 |
| HKB-1 | | EBV– | | | 66 |

i.c. – intracerebral, s.c. – subcutaneous, i.p. – intraperitoneal, i.v. – intravenous.

In addition to CD30, most of the H-RS cell lines express CD15 and the transferrin receptor (CD71) (see Table 2). Besides these, no characteristic pattern of marker expression can be defined either for classical H-RS cells or for the cell lines. 11 of the HD derived cell lines appear to be of B-cell origin, while 3 of the lines show a T-cell phenotype, as defined by Southern blot analysis of rearranged Ig or T-cell receptor (TCR) genes or expression of the respective rearranged genes (see Table 1). One cell line, HDMyz, has neither an Ig gene rearrangement nor a TCR rearrangement. In only 5 of the 11 B-cell derived cell lines was expression of at least one B-cell marker (such as CD19, CD20 or CD21) and/or expression of Ig heavy and/or Ig light chain constant region genes detected. In addition, most of the T-cell lines do not express T-cell associated markers (such as CD3, CD4, CD5 or CD8). In conclusion, in most cases, immunophenotype analysis does not define the cell of origin either of H-RS cells or their cell lines.

# 4. CYTOKINE RELATED CHARACTERIZATION

HD shares many of the clinical and biological characteristics of an inflammatory process, including fluctuating fever, nightsweats and elevated serum levels of IL-2 receptor [20]. In affected lymph nodes the H-RS cells are surrounded mostly by T-lymphocytes [47]. This observation led to the hypothesis that T-cells are attracted by cytokines secreted from H-RS cells. In support of this observation, TARC expression was detected in 4 HD lines [64]. TARC is a chemokine that attracts TH2-cells and its expression in H-RS cells might account for the T-cell infiltration.

Cytokine expression in HD-derived cell lines is heterogenous, and most cytokines are undetectable (see Table 3). Most H-RS cells and HD cell lines do express IL-2R. Binding of IL-2 to the IL-2R leads to activation and cell proliferation. Expression of TNF$\alpha$ and TNF$\beta$ was observed in some of the cell lines, and might account for the symptoms. TGF$\beta$ is expressed in some cell lines, and in H-RS cells *in vivo* may result in suppression of T helper cell function.

# 5. GENETIC CHARACTERIZATION

Primary H-RS cells show numerical and structural chromosomal aberrations in most cases [67]. However, no H-RS cell specific chromosomal aberration has been detected [54,60,61], although cytogenetic analysis is difficult to perform due to the scarcity of the cells. Cytogenetic analysis of most of the HD derived cell lines has found grossly aberrant karyotypes (see Table 4).

In addition, many of the chromosomal partners involved in translocations could not be identified using standard banding techniques. Since H-RS cells are usually derived from germinal center B cells, the frequent occurrence of chromosomal breaks within the chromosomal region 14q32 in both H-RS cells [4,47] as well as in B-cell HD derived lines may be important. The detection of chromosomal breaks affecting this region carrying the Ig heavy chain locus is also a common feature of other lymphomas (for example Burkitt's lymphoma [58] and follicular lymphoma [63]. The changes lead to oncogene deregulation; for example, c-myc in Burkitt's lymphoma [5] and bcl-2 in follicular lymphoma [62]. The characterization of translocations involving 14q32 in H-RS cells may provide new insights into the mechanism of transformation of H-RS cells.

## 6. FUNCTIONAL CHARACTERIZATION

In thymus aplastic T-cell deficient nude mice, HD-derived cell lines only grow after intracranial inoculation [9] (see Table 5). In contrast, most of the HD-derived cell lines grow in SCID (severe combined immunodeficient) mice after subcutaneous inoculation [65]. SCID mice, due to a genetic recombinase defect, lack functional T- and B-cells. The HD-derived cell lines L540 [32], HD-MyZ [3] and L1236 [69] disseminate intralymphatically after inoculation into SCID-mice. This experimental model for the *in vivo* growth of H-RS cells has been used for the preclinical testing of new immuno-therapeutic modalities, such as immunotoxins [68]. The histology of the xenografts in SCID mice resembles that of anaplastic large cell lymphoma. The typical features of HD, a few tumor cells surrounded by a large excess of reactive cells, are not present. This difference is probably due to the absence of T-cells in SCID mice, and is a major limitation of the model for HD. The transplantation of biopsy tissue into SCID mice was unsuccessful, since outgrowth of H-RS cells was never observed [33].

All HD derived cell lines but one (L591) are negative for the Epstein–Barr virus (EBV) (see Table 5). This is in contrast to the detection of EBV in H-RS cells in about half the cases of HD in industrialized countries [24].

## 7. PERSPECTIVES AND CONCLUSIONS

Microdissection of single H-RS cells from frozen lymph node sections and the subsequent analysis of these single cells using polymerase chain reaction (PCR) is allowing H-RS cells to be characterized genetically [36]. Analysis of rearranged immunoglubulin (Ig) genes shows that H-RS cells are clonally derived from germinal center B-cells in most cases [35]. Also, somatic mutations have been detected, rendering potentially functional $V_H$ gene rearrangements non-functional, thus preventing Ig gene expression in H-RS cells [35]. Since B-cells that do not express an antibody undergo apoptosis within the germinal center, it was speculated that H-RS cells are "crippled" germinal center B-cells that escape programmed cell death by a yet unidentified mechanism.

The Ig gene rearrangement has been used as a clonal marker to detect H-RS cells belonging to the same clone in other tissues obtained from the same patient during the course of the disease. Using this approach, it was demonstrated that H-RS cells clonally expand, leading to disseminated disease and to relapse of the lymphoma after clinical remission. This approach was also used to show that L1236, which grew from the peripheral blood of a patient suffering from relapse of mixed cellularity HD [69], is unequivocally derived from the H-RS cells in that patient. Sequence analysis of the rearranged Ig genes showed that Ig gene expression in L1236 cells as well as in the H-RS biopsy is prevented by a somatic mutation within the promoter region of the potentially functional $V_H$ gene rearrangement [27].

H-RS cells, in most cases, represent clonal B-cells. Therefore, the origin of HD derived cell lines with a T-cell genotype remains to be elucidated. Some T-cell derived HD have been described [37,56], although the finding of rearranged T-cell receptor genes in H-RS cells is rare, even in those HD tissues in which the H-RS cells express several T-cell associated marker genes [37,56]. Thus, the T-cell HD cell lines could also be derived from T-cells surrounding H-RS cells.

The derivation of the cell line HD-Myz, which shows a non-T, non-B cell phenotype [3], from H-RS cells is unlikely, since a myeloid derivation of H-RS cells has never been proven.

The B-cell genotype by itself does not provide unequivocal evidence for the derivation of HD derived B-cell cell lines from B-cell H-RS cells. It is essential to show clonal derivation of each cell line from the H-RS cells of the HD tissue of the patient from whom the cell line was derived.

Enrichment of H-RS cells has been achieved, providing viable purified H-RS cell populations for analysis [25], and reducing the need for cell lines. The broader application of molecular analysis of single microdissected H-RS cells will provide new insights into the pathogenesis of HD and avoid the possible disadvantages of analyzing cell populations that have been selected during *in vitro* culture.

# References

1. Andreesen R, Brugger W, Löhr GW, Bross KJ. *Am J Path* 134: 187–192, 1989.
2. Alexander FE. In: Jarret RF (ed.) *Etiology of Hodgkin's Disease*, NATO ASI Series, Series A: Life Sciences, Vol. 280, Plenum Press New York, 1995.
3. Bargou RC, Mapara MY, Zugck C et al. *J Exp Med* 177:1257–1268, 1993.
4. Cabanillas F, Pathak S, Trujillo J et al. *Blood* 71: 1615–1617, 1988.
5. Dalla-Favera R, Bregni M, Erikson J et al. *Proc Natl Acad Sci USA* 79: 7824–7827, 1982.
6. DeCoteau JF, Reis MD, Griesser H et al. *Blood* 85: 2829–2838, 1995.
7. Diehl V, Burrichter H, Schaadt M et al. *Hematol Oncol* 1: 139–147, 1983.
8. Diehl V, Kirchner HH, Burrichter H et al. *Cancer Treat Rep* 66: 615–632, 1982.
9. Diehl V, Kirchner HH, Schaadt M et al. *J Cancer Res Clin Oncol* 101: 111–124, 1981.
10. Diehl V, Pfreundschuh M, Fonatsch C et al. *Cancer Surv* 4: 399–419, 1985.
11. Drexler HG. *Leukemia Lymphoma* 9: 1–25, 1993.
12. Drexler HG, Gaedicke G, Lok MS et al. *Leukemia Res* 10: 487–500, 1986.
13. Drexler HG, Gignac SM, Hoffbrand AV et al. *Recent Results Cancer Res* 117: 75–82, 1989.
14. Drexler HG, Leber BF, Norton J et al. *Leukemia* 2: 371–376, 1988.
15. Drexler HG, Dirks WG, MacLeod RAF. *Leukemia* 13: 1601–1607, 1999.
16. Dürkop H, Latza U, Hummel M et al. *Cell* 68: 421–427, 1992.
17. Falk MH, Tesch H, Stein H et al. *Int J Cancer* 40: 262–269, 1987.
18. Fonatsch C, Diehl V, Schaadt M et al. *Cancer Genet Cytogenet* 20: 39–52, 1986.
19. Foss HD, Herbst H, Oelmann E et al. *Br J Haematol* 84: 627–635, 1993.
20. Gause A, Roschansky V, Tschiersch A et al. *Ann Oncol* 2 (Suppl. 2): 43–47, 1991.
21. Hansen H, Lemke H, Bredfeldt G et al. *Biol Chem Hoppe Seyler* 370: 409–416, 1989.
22. Hsu PL, Hsu SM. *Am J Pathol* 135: 735–745, 1989.
23. Hsu PL, Lin YC, Hsu SM. *Int J Hematol* 54: 315–326, 1991.
24. Hummel M, Anagnostopoulos I, Dallenbach F et al. *Br J Haematol* 82: 689–694, 1992.
25. Irsch J, Nitsch S, Hansmann ML et al. *Proc Natl Acad Sci USA* 95: 10117–1022, 1998.
26. Jones DB, Furley AJ, Gerdes J et al. In: Diehl V, Pfreundschuh M, Loeffler M (eds.), *Recent Results Cancer Res*, Springer-Verlag, Berlin 117: 62–66, 1989.
27. Jox A, Zander T, Küppers R et al. *Blood* 93: 3964–3972, 1999.
28. Jucker M, Abts H, Li W et al. *Blood* 77: 2413–2418, 1991.
29. Kamesaki H, Fukuhara S, Tatsumi E et al. *Blood* 68: 285–292, 1986.
30. Kanzaki T, Kubonishi I, Eguchi T et al. *Cancer* 69: 1034–1041, 1992.
31. Kanzler H, Hansmann ML, Kapp U et al. *Blood* 87: 3429–3436, 1996.
32. Kapp U, Dux A, Schell-Frederick E et al. *Ann Oncol* 5 (Suppl. 1):121–126, 1994.
33. Kapp U, Wolf J, Hummel M et al. *Blood* 82: 1247–1256, 1993.
34. Kapp U, Yeh WC, Patterson B et al. *J Exp Med* 189: 1939–1946, 1999.
35. Küppers R, Rajewsky K. *Annu Rev Immunol* 16: 471–493, 1998.
36. Küppers R, Rajewsky K, Zhao M et al. *Proc Natl Acad Sci USA* 91: 10962–10966, 1994.
37. Müschen M, Rajewsky K, Bräuniger A et al. *J Exp Med* 1999.
38. Naumovski L, Utz PJ, Bergstrom SK et al. *Blood* 74: 2733–2742, 1989.
39. Newcom SR, Ansari AA, Gu L. *Blood* 79: 191–197, 1992.
40. Newcom SR, Gu L. *J Clin Pathol* 48: 160–163, 1995.
41. Newcom SR, Kadin ME, Ansari AA, Diehl V. *Clin Invest* 82: 1915–1921, 1998.
42. Newcom SR, Kadin ME, Phillips C. *Int J Cell Cloning* 6: 417–431, 1988.
43. O'Connor NT, Stein H, Gatter KC et al. *Histopathology* 11: 733–740, 1987.

44. Olsson L, Behnke O, Pleibel N et al. *J Natl Cancer Inst* 73: 809–830, 1984.
45. Pfreundschuh M, Mommertz E, Meissner M et al. *Anticancer Res* 8: 217–224, 1988.
46. Poppema S, de Jong B, Atmosoerodjo J et al. *Cancer* 55: 683–690, 1985.
47. Poppema S, Kaleta J, Hepperle B. *J Natl Cancer Inst* 84: 1789–1793, 1992.
48. Poppema S, Kaleta J, Hepperle B, Visser L. *Ann Oncol* 3 (Suppl. 4): 5–8, 1992.
49. Poppema S, Visser L, de Jong B et al. In: Diehl V, Pfreundschuh M, Loeffler M (eds.), *Recent Results Cancer Res*, Springer-Verlag Berlin 117: 67–74, 1989.
50. Roberts AN, Smith KL, Dowell BL, Hubbard AK. *Cancer Res* 38: 3033–3043, 1978.
51. Schaadt M, Burrichter H, Pfreundschuh M et al. In: Diehl V, Pfreundschuh M, Loeffler M (eds.), *Recent Results Cancer Res*, Springer-Verlag Berlin 117: 53–66, 1989.
52. Schaadt M, Fonatsch C, Kirchner H, Diehl V. *Blut* 38: 185–190, 1979.
53. Schaadt M, v Kalle C, Tesch H et al. *Cancer Rev* 10: 108–122, 1988.
54. Schouten HC, Sanger WG, Duggan M et al. *Blood* 8: 2149–2154, 1989.
55. Stein H et al. *Blood* 66: 848–858, 1985.
56. Stein H, Seitz V, Hummel M et al. *Ann Oncol* 10 (Suppl. 3): 49, 1999.
57. Smith CA, Gruss HJ, Davis T et al. *Cell* 73(7): 1349–1360, 1993.
58. Taub R, Kirsch I, Morton C et al. *Proc Natl Acad Sci USA* 79: 7837–7841, 1982.
59. Tesch H, Jücker M, Falk MH et al. *Hematol Oncol* 6: 223–231, 1988.
60. Thanagavelu M, Le Beau MM. *Hemat Oncol Clin North Am* 3: 221–236, 1989.
61. Tilly H, Bastard C, Delastre T et al. *Blood* 6: 1298–1304, 1991.
62. Tsujimoto Y, Cossman J, Jaffe E, Croce CM. *Science* 228(4706): 1440–1443, 1985.
63. Tsujimoto Y, Finger LR, Yunis J, Nowell PC, Croce CM. *Science* 226(4678): 1097–1099, 1984.
64. Van den Berg A, Visser L, Poppema S. *Am J Pathol* 154: 1685–1691, 1999.
65. Von Kalle C, Wolf J, Becker A et al. *Int J Cancer* 52: 887–891, 1992.
66. Wagner HJ, Klintworth F, Jabs W et al. *Med Pediatr Oncol* 31: 138–143, 1988.
67. Weber-Matthiesen K, Deerberg J, Poetsch M et al. *Blood* 86: 1464–1468, 1995.
68. Winkler U, Gottstein C, Schön G et al. *Blood* 83: 466–475, 1994.
69. Wolf J, Kapp U, Bohlen H et al. *Blood* 87: 3418–3428, 1996.

Chapter 12

# CD30-Positive Anaplastic Large Cell Lymphoma Cell Lines

Hermann Herbst and Hans G. Drexler
*Gerhard-Domagk Institute of Pathology, University of Münster, Domagkstr 17, 48129
Münster, Germany, and DSMZ-German Collection of Microorganisms & Cell Cultures,
Department of Human and Animal Cell Cultures, Braunschweig, Germany. Tel:
+49-251-835-6752; Fax: +49-251-835-5460; E-mail: herbsth@uni-muenster.de.*

## 1. INTRODUCTION

In the late 1970s, the application of the monoclonal antibody Ki-1 (CD30) to anaplastic malignancies led to the recognition of a new entity of malignant lymphomas, CD30 (Ki-1 antigen)-positive anaplastic large cell lymphoma ("Ki-1 lymphoma", ALCL) [52,53]. On the basis of conventional histology alone, ALCL had previously been diagnosed as Hodgkin's sarcoma, malignant histiocytosis, malignant fibrous histiocytoma, or even as non-hematopoietic malignancies such as undifferentiated sarcoma, undifferentiated carcinoma, or amelanotic melanoma. These lymphomas were recognized as a distinct entity of high grade B- and T-cell lymphomas in the updated Kiel classification of non-Hodgkin lymphomas (NHL) [51] and are included in the Revised European-American Lymphoma (REAL) classification as a type of high grade T-/null-cell lymphoma and as a variant of diffuse large cell lymphomas of B-cell type [23]. As there is some overlap with Hodgkin's disease (HD), distinguishing ALCL with features of HD from lymphocyte depleted HD is a matter of ongoing debate.

## 2. CHARACTERISTICS OF ALC LYMPHOMAS

Clinically, ALCL presents most frequently in lymph nodes, often with subsequent infiltration of extranodal tissue. Among primary extranodal ALCL, cutaneous lesions are the most prevalent. ALCL may arise secondary to HD, lymphomatoid papulosis, mycosis fungoides, pleomorphic T-cell lymphoma

355

*J.R.W. Masters and B.O. Palsson (eds.), Human Cell Culture Vol. III, 355–370.*
© 2000 *Kluwer Academic Publishers. Printed in the Netherlands.*

or T-cell lymphoma of angioimmunoblastic (AILD) type. The age distribution of primary ALCL revealed a bimodal pattern similar to HD, whereas ALCL arising simultaneously with or subsequent to other lymphomas showed a single peak in the fifth decade [52,53].

Histologically, the tumors are characterized by a preferential perifollicular involvement of lymph nodes by tumor cells often growing in coherent sheets with initial sparing of germinal centers, sinusoidal dissemination, and occasional foci of necrosis. The tumor cell morphology comprises a spectrum ranging from large pleomorphic cells with abundant, often basophilic, cytoplasm and irregularly shaped nuclei containing multiple small nucleoli or a single prominent, often rod-shaped nucleolus, to cells with more regular, rounded nuclei frequently containing a single nucleolus. The cells have a high mitotic rate and in many ALCL cases multinucleated tumor cells, often resembling Reed-Sternberg cells, may be found [52,53].

Unlike HD, where CD30 staining is a useful but not absolutely necessary diagnostic adjunct, the diagnosis of ALCL by definition requires the expression of the CD30 antigen in all of the tumor cells. Other activation antigens such as the low-affinity interleukin-2 receptor (CD25), class II histocompatibility antigens (HLA-DR), CD70 and proliferation-associated antigens such as the transferrin receptor (CD71) are usually found on ALCL tumor cells. Because activation antigens are not lineage specific markers, interest has centered on the study of cell type characteristic molecules. Similar to antigen- or mitogen-activated peripheral blood lymphocytes, which cease to express CD45 molecules, the three forms of the leukocyte common antigen (CD45, CD45RA, and CD45RO) are variably expressed on ALCL cells. Early lymphoid antigens present on precursor B- and T-cells, such as CD10 or terminal nucleotidyl transferase (TdT), or macrophage antigens, are usually not expressed by CD30+ malignancies. ALCL of T-cell type occurs more frequently than ALCL of B-cell type, and few cases do not express B- and T-lymphoid marker molecules if a sufficient panel of antibodies is applied. Primary cutaneous ALCL are generally of T-cell type [27]. T- and null-cell ALCL have a phenotype of cytotoxic cells with expression of granzyme B, T-cell-restricted intracellular antigen (TIA)-1 and granule membrane (GMP-17) proteins as well as perforin transcripts [19]. The clinical relevance of these phenotypic details has been challenged by the finding of a generally better outcome in primary ALCL of either T- or B-cell type, i.e., independent of the immunophenotype, in adults as compared to other diffuse large cell lymphomas [56].

In contrast to HD, which appears to be a neoplasm of constitutively cytokine-secreting cells, a limited body of data is available on the expression of cytokines in ALCL. Although CD25 (IL-2 receptor)-positive, these lymphomas do not express IL-2 [43], thus excluding the possibility of an autocrine regulatory loop involving that interleukin. Similar to HD, IL-6 expression is a common finding in ALCL [36]. IL-9 expression has been found in a large proportion of HD cases and ALCL in the tumor cells, but not in other lymphoid malignancies [37]. IL-9 expression has been considered a characteristic feature of CD30-positive lymphomas with possible implications for autocrine growth control.

T-cell receptor gene rearrangements are the most frequent finding in ALCL. A minority of cases show rearranged immunoglobulin genes, and a few cases show rearrangements within both T-cell receptor and immunoglobulin loci. However, in approximately 30% of the cases, antigen receptor genes were found in germline configuration, although the cases had a tumor cell content sufficient for reliable analysis of DNA extracts, suggesting that some ALCL may represent genotypically immature lymphomas despite their display of an activated lymphoid phenotype [27].

It is not surprising that a number of ALCL cases, usually of B- or T-cell type respectively, are associated with the lymphotropic viruses, Epstein–Barr virus (EBV) and HTLV (human T-lymphotropic virus)-I, because both viruses are strong inducers of CD30 in lymphoid cells [3,25,28]. Moreover, the predominantly perifollicular distribution of EBV-infected cells in infectious mononucleosis tonsils mirrors the distribution of ALCL cells in early lesions. The expression of the latent protein (LMP)-1, an EBV gene product with transforming potential, and even of the nuclear antigen, EBNA-2, in occasional cases of B-ALCL, suggested a potential etiological role for the virus in a proportion of ALCL [25,33].

Cytogenetic analyses revealed involvement of a site at 2p23, often in a reciprocal translocation involving chromosome 5 [t(2;5)(p23;q35)], unique to ALCL associated with a T- or null-cell phenotype and genotype. The t(2;5) fuses the *ALK* (anaplastic lymphoma kinase) and the *NPM* (nucleophosmin) genes, leading to the formation of a chimaeric NPM-ALK-protein (p80) consisting of the terminal portion of NPM linked to the cytoplasmic domain of the neural receptor tyrosine kinase ALK [29,40]. It was subsequently shown that retrovirus-mediated gene transfer of *NPM-ALK* causes lymphoid malignancies in mice, indicating that the translocation indeed resulted in a dominant acting oncogene [32]. Moreover, the cloning of this novel breakpoint provided the basis for detecting the t(2;5) by reverse transcription of transcripts and subsequent polymerase chain reaction (RT-PCR) as well as of truncated *ALK* transcripts and protein by *in situ* hybridization with *ALK*-specific probes and immunohistology using polyclonal and monoclonal antibodies, respectively. The occurrence in HD of the t(2;5) and transcripts derived from the fused genes are a matter of scientific debate. On balance, if it occurs in this context, the t(2;5) is a very rare finding in HD. The same

seems to be true for lymphomatoid papulosis and primary cutaneous ALCL [26].

Recently, variants of the t(2;5) were described which involve genes other than NPM as partners for ALK, namely a t(1;2)(q25;p23), t(1;2)(q21;p23), and t(2;3)(p23,q21), all of which produce specific transcripts containing a truncated form of ALK [39,47]. The cryptic inv(2)(p23q35) abnormality was found to define another subtype of ALK-positive ALCL [59]. In t(1;2)(q25)(p23), the fusion partner was characterized as tropomyosin-3 (TPM3), which was previously found in papillary thyroid carcinomas where it forms a fusion with the TRK kinase gene [34]. Moreover, using ALK-specific antibodies, a novel subtype of diffuse large-cell lymphoma was defined which displays expression of the full-length ALK kinase, but lacks both the t(2;5) and CD30 expression [14].

## 3. DEFINITION OF ALCL CELL LINES

The phenotypic heterogeneity of ALCL calls for a restrictive approach to assigning a cell line to the group of *bona fide* ALCL cell lines. First, such lines should display strong expression of CD30 protein. However, this feature is shared with numerous leukemia and lymphoma cell lines, mitogen activated lymphocytes as well as EBV−, HTLV-I/II−, and HHV-8-positive cell lines, and is therefore not a sufficient criterion. Second, these lines should display the t(2;5) or an equivalent variant translocation resulting in a constitutive activation of the ALK gene. Because of the specificity of this abnormality for ALCL, a cell line may be considered an ALCL cell line even if biopsy material is not available for review. This could apply to a number of cell lines derived from tumors diagnosed in the past as malignant histiocytosis according to now obsolete classification schemes.

ALK-positive ALCL is a subset within the spectrum of these lymphomas. Therefore, cell lines corresponding to ALK-negative ALCL without genetic changes involving 2p23 presumably exist. However, in these cases the histological and phenotypic details should unequivocally meet all of the criteria for the histological diagnosis of ALCL, and the cell elements *in vitro* should closely mirror the phenotypic, genotypic and karyotypic characteristics of the tumor cells *in vivo*.

## 4. ALCL CELL LINES

Similar to the HD-derived cell lines, the ALCL cell lines listed in Tables 1–5 were obtained mainly from effusions rather than from cultivation of disaggregated solid biopsy material. AMS3 was derived from a solid tissue biopsy that was heterotransplanted into nude mice and passaged several times prior to *in vitro* cultivation [50]. Many cell lines lead to tumor formation when heterotransplanted into nude or SCID mice. For many of the cell lines the currently published body of information is fragmentary. For example, UCONN-L2 and L82 were briefly mentioned in reports focusing on cloning of the t(2;5) breakpoint and on its expression in HD, respectively [40,41]. All of these cell lines share the characteristic translocation t(2;5)(p23;q35) as evidenced by karyotypic analysis and/or, in cases of a cryptic t(2;5), by analysis of NPM-ALK chimaeric gene transcripts. Moreover, expression of ALK gene products was also verified by *in situ* hybridization with *ALK*-specific RNA probes, and immunocytochemistry using ALK-specific antibodies [24,46]. The cell lines lack all of those translocations frequently found in germinal center cell, mantle cell, diffuse large B-cell, and Burkitt's lymphomas, such as t(14;18), t(11;14), t(3;14), t(3;22), t(8;2), t(8;14) and t(8;22).

## 5. IMMUNOPHENOTYPE

ALCL cell lines, by definition, express the CD30 antigen ("Ki-1"-antigen), a member of the TNF receptor superfamily and receptor to the CD30 ligand, CD30L. Paralleling the phenotype of Hodgkin- and Reed-Sternberg (HRS) cells and HD-derived cell lines, other markers characteristic for mitogen-activated lymphocytes ("activation markers") are regularly found on ALCL cell lines such as the IL-2 receptor (CD25, Tac antigen), the CD70 antigen ("Ki-24"-antigen, a member of the TNF family and ligand to CD27), and HLA class II-antigens (HLA-DR) (Table 2). The transferrin receptor (CD71), a widely expressed proliferation-associated marker, is also found on virtually all ALCL cell lines. CD40, another member of the TNF receptor superfamily, is expressed by Karpas 299, but may not be common to all ALCL cell lines because, and in contrast to HRS cells, less than 40% of ALCL biopsies display CD40 protein [9]. CD80 (B7 antigen), a co-regulator of T-cell activation in concert with CD86, was not expressed on three cell lines studied, despite its apparently regular expression in ALCL as previously suggested by Delabie et al. [12].

Although fragmentary in some instances, the published phenotypes of the 11 cell lines listed in Tables 1–5 reflect heterogeneous expression of T-, B-, and/or myelomonocytic markers. CD3 and T-cell receptor (TcR) transcripts or proteins are found in only a few of the lines, in some instances requiring stimulation by phorbol esters, such as SR786 [54]. There are a number of discrepancies between published phenotypic details and those evaluated at the

*Table 1.* ALCL cell lines: clinical characterization

| Cell | Patient age/sex | Diagnosis | Previous treatment | Specimen site | Culture medium | Ig-, TcR rearrangement status* | Primary ref. |
|---|---|---|---|---|---|---|---|
| SU-DHL-1 | 10 M | diffuse large cell lymphoma, non-cleaved | none (at presentation) | pleural effusion | RPMI 1640 + 10% FCS | TcRβ1 G/D, TcRβ2 R/G,IgH G/G | 16,17,39 |
| Karpas 299 | 25 M | Ki-1+ high grade T-cell lymphoma | extensive chemotherapy | pleural effusion | RPMI 1640 + 15% FCS | TcRβ R | 18 |
| SR768 | 11 M | Ki-1+ large T-cell lymphoma | | pleural effusion | RPMI 1640 + 15% FCS | TcRβ R, IgH G | 5,54 |
| SUP-M2 | 5 F | malignant histiocytosis | chemotherapy (CHOP) | cerebrospinal fluid | | TcRβ1 R/R, TcRβ2 G/G, IgH G/G | 39 |
| JB6 | 12/M | ALCL, advanced | | peripheral blood | RPMI 1640 + 15% FCS | TcRβ R, IgH G | 30 |
| DEL | 51/M | malignant histiocytosis | none | | RPMI 1640 + 15% FCS | TcRβ G/G, IgH R/G, IgLCκ G/G | 4,20 |
| KI-JK | Child | ALCL | none | pleural effusion | RPMI 1640 + 10% FCS | TcRβ G, IgH G | 49 |
| UCONN-L2 | | relapsed ALCL | | | | | 40 |
| AMS3 | 23 M | ALCL | | "5 × 5 cm tumor in the back" | McCoy's 3A + 15% FCS grows only in SCID mice | | 50 |
| WSU-ALCL | 18 | primary T-ALCL | | lymph node | RPMI 1640 + 10% FCS | | 2 |
| L82 | 24 F | relapsed ALCL | | pleural effusion | | TcRβ R, IgH G | 41 |

* R – rearranged, G – germline, D – deleted.

Table 2. Phenotypic characteristics of ALCL cell

| Cell line | p80 (IH) | ALK (ISH) | NPM-ALK | Activation markers | | | | | | Lymphoid markers | T-cell markers | | | | | | | |
|---|---|---|---|---|---|---|---|---|---|---|---|---|---|---|---|---|---|---|
| | | | | CD30 | CD25 | CD70 | CD71 | CD80 | HLA-DR | CD45 | CD2 | CD3 | CD4 | CD5 | CD7 | CD8 | CD43 | CD45R0 |
| SU-DHL-1 | + | + | + | +++ | +++ | ++ | +++ | - | +++ | | - | - | - | +++ | - | - | | |
| Karpas 299 | + | + | + | +++ | +++ | ++ | +++ | - | +++ | + | - | - | + | +++ | - | - | ++ | |
| SR786 | + | | + | +++ | +++ | | +++ | | +++ | + | - | - | + | - | - | - | | + |
| SUP-M2 | + | | + | +++ | +++ | | +++ | | +++ | | + | - | + | - | - | - | | |
| JB6 | + | + | + | +++ | +++ | ++ | +++ | - | +++ | | - | - | + | - | - | - | | |
| DEL | + | + | + | +++ | +++ | ++ | +++ | | +++ | + | - | - | - | - | - | - | | - |
| KI-JK | + | | + | +++ | +++ | | +++ | | +++ | | - | + | - | + | +++ | - | | - |
| UCONN-L2 | + | | + | +++ | +++ | | +++ | | +++ | | | - | - | - | + | | | |
| AMS3 | + | | + | +++ | +++ | | +++ | | +++ | + | + | + | - | | | - | | |
| WSU-ALCL | + | | + | +++ | - | | +++ | | +++ | | + | + | + | +++ | +++ | - | | |
| L82 | + | | + | +++ | +++ | | +++ | | +++ | | + | + | + | + | | - | | |

*Continued on next page*

Table 2. (continued)

| Cell line | B-cell markers | | | | | | Myelomonocytic markers | | | | | | | Progenitor markers | | Other markers |
|---|---|---|---|---|---|---|---|---|---|---|---|---|---|---|---|---|
| | CD19 | CD20 | CD21 | CD22 | CD23 | CD45RA | CD11b | CD11c | CD13 | CD14 | CD15 | CD33 | CD64 | CD10 | CD34 | EMA |
| SU-DHL-1 | – | – | – | – | | | – | – | – | – | – | – | | – | – | + |
| Karpas 299 | – | – | – | – | – | – | – | + | ++ | – | – | – | – | – | – | + |
| SR786 | +++ | + | – | – | | | +++ | | + | – | + | ++ | | – | – | |
| SUP-M2 | – | – | – | – | | | – | – | + | – | – | ++ | | – | – | + |
| JB6 | | +/– | | | | | – | | +++ | – | ++ | – | | – | – | |
| DEL | – | – | – | – | | | – | – | ++ | – | ++ | ++ | | – | – | |
| KI-JK | +++ | +++ | – | | – | | +++ | | +++ | – | + | +++ | | – | – | |
| UCONN-L2 | – | ++ | | | | | ++ | | – | – | – | – | | – | – | |
| AMS3 | – | – | | | | – | | | | | | | | | | |
| WSU-ALCL | – | – | | | | | – | | – | – | +++ | – | | ++ | | |
| L82 | | | | | | | | | | | | | | – | | |

| Cell line | References |
|---|---|
| SU-DHL-1 | [15,24,39]; H. Dürkop, personal communication, DSMZ database |
| Karpas 299 | 15,18,24]; H. Dürkop, personal communication, DSMZ database |
| SR786 | [5,15,54] |
| SUP-M2 | [35,39] |
| JB6 | [15,24]; H. Dürkop, personal communication; DSMZ database |
| DEL | [4,15]; DSMZ database |
| KI-JK | [15,49]; DSMZ database |
| UCONN-L2 | [15,41]; DSMZ database |
| AMS3 | [50] |
| WSU-ALCL | [2,15] |
| L82 | [41]; H. Merz, personal communication |

*Table 3.* ALCL cell lines: cytokine, cytokine receptor and c-onc gene products

| Cell line | Cytokine receptor gene expression | Cytokine gene expression | Expression of other c-onc genes | Ref. |
|---|---|---|---|---|
| SU-DHL-1 | IL-1-R, IL-2-Rα, IL-6-R, TNFα-R, c-fms | IL-1β, IL-7, IL-10, TNFα | bcl-6 | 1,8 |
| Karpas 299 | IL-2-Rα, c-met | HGF | bcl-6 | 8,45,57 |
| SR768 | IL-2-Rα, c-met | IL-2, TGFβ, HGF, CCR4, CCR8 | c-myc, H-ras | 45,54,55 |
| SUP-M2 | IL-1-R, IL-2-Rα, IL-6-R, TNFα-R, c-fms | | | 1 |
| JB6 | IL-1-R, IL-2-Rα, IL-6-R, TNFα-R | | | 1 |
| DEL | IL-2-Rα, c-fms | | bcl-6, c-fgr, c-myb, c-myc, c-pim, K-ras | 8,20 |
| KI-JK | IL-2-Rα | | | 1,49 |
| UCONN-L2 | c-met | HGF | | 45 |
| AMS3 | | | | |
| WSU-ALCL | | | c-myc, bcl-2 | 2 |
| L82 | | | | |

*Table 4.* ALCL cell lines: genetic characterization

| Cell line | Representative karyotype | Ref. |
|---|---|---|
| SU-DHL-1 | 74(67-75)<3n>XX, −Y, +1, +2, +3, +5, −7, +12, −16, −18, +19, −20, +21, +3mar, del(1)(p21), t(2;5)(p23;q35)x2, del(6)(q23)x1–2, add(8)(p12), add(9)(p21), del(10)(p14), add(12)(q24), add(14)(p12), add(16)(q24), dup(19)(q13.1qter); sdl with der(12)t(12;19(q13;q13); carries two copies of balanced t(2;5) | 39; DSMZ database |
| Karpas 299 | 44, XY, −10, −22, rcp t(1;17)(p21;p11.2); rcp t(2;5)(p23.2;q35.3); rcp t(3;6)(p23;p12.2); t(13;15)(p12;q13.1); t(14;22)(p12;q11.21); mar22:der(22)22p12-q11.1 | 18 |
| SR768 | 70-84<3n>XX, +1, +2, −4, +5, +6, +7, +8, −13, −13, +14, −18, +19, +22, +6-9mar, add(1)(q11), del(1)(p11)/der(?)t(1;?)(q11;?), der(2)t(2;5)(p23;q35)inv(2)(p23q14)x2, del(4)(q22), der(5)t(2;5)(p23;q35)x2, del(7)(q21), der(9)t(1;9)(q11;p24)x2, der(12)t(12;13)(q24.32;q11)x1–2, del(13)(q13q31), add(14)(p11)/der(?)t(14;?)(q11;?)x2, del(21)(q22) | DSMZ database |
| SUP-M2 | 47, XX, +X, −9, +der(1)t(1;?)(q44;?), del(1)(p34), t(2;5)(p23;q35) | 39 |
| JB6 | 49, XY, +1, der(2)t(2;5)(p23;q35)t(2;21)(q14;q11), der(5)t(2;5)(p23;q35), +8, +13, add(15)(q26), der(21)t(2;21)(q32;q11), i(22)q(10) | 42 |
| DEL | 74 (74-77) <3n>XXY, +1, +3, +5, +5, +6, +7, −10, +13, −18, +20, t(5:6)(q35;p21) x2, add (10)(q24) x1–2, der(13)t(1;13)(q21;p11)t(1;13)(q21;q24)x2; add(16)(q23), add (19)(p13) | DSMZ database |
| KI-JK | | |
| UCONN-L2 | Pseudotetraploid with modal no. of 82, cryptic t(2;5) | 49 |
| AMS3 | 46, XY, t(2;5)(p23;q35), del(6)(q13q23) | 50 |
| WSU-ALCL | 45, 10p− | 2 |
| L82 | | |

*Table 5.* ALCL cell lines: functional characterization

| Cell line | Doubling time | Viruses | Cytochemistry | Heterotransplantation into mice | Ref. |
|---|---|---|---|---|---|
| SU-DHL-1 | 15–18 hr | EBV– (EBER–), HBV–, HCV–, HHV-8–, HIV–, HTLV-I/II– | ACP +, ALP–, ANAE +, ANBE +, CAE–, GLC +, lysozyme–, MGP +, MPO +, PAS +, SDB–, TRAP– | tumor growth in nude mice | DSMZ database |
| Karpas 299 | 18 hr | EBV– (EBNA–, EBER–), HBV–, HCV–, HHV-8–, HTLV-I | ACP +, ANAE +, ANBE +, CAE–, MPO–, PAS +, SDB–, elastase–, muraminidase–, | | 18; DSMZ database |
| SR768 | 30 hr | EBV–, HBV–, HCV–, HHV-8–, HIV–, HTLV-I/II– | | | DSMZ database |
| SUP-M2 | 15–18 hr | | | | DSMZ database |
| JB6 | 18 hr | EBV– (EBER–) | | tumor growth in SCID mice | DSMZ database |
| DEL | 18–30 hr | EBV– (EBER–), HBV–, HCV–, HHV-8–, HIV–, HTLV-I/II– | ACP+, ALP–, ANAE(+), CAE–, MPO–, oil red O+, PAS (+) | tumor growth in nude mice | DSMZ database |
| KI-JK | 72 hr | EBV+ (LMP1, EBNA2, EBER) | ANBE +, lysozyme + | tumor growth in nude mice | 49 |
| UCONN-L2 | | | | | |
| AMS3 | | | | established from heterotransplanted tumor tissue | 50 |
| WSU-ALCL | | | | | |
| L82 | | | | | |

ACP – acid phosphatase; ALP – alkaline phosphatase; ANAE – alpha-naphthylacetate esterase; ANBE – alpha-naphthylbutyrate esterase; CAE – chloro-acetate esterase; MPO – myeloperoxidase; PAS – periodic acid-Schiff; GLC – β-glucuronidase; MGP – methyl green pyronine; MPO – myeloperoxidase; SDB – sudan black; TRAP – tartrate-resistant acid phosphatase.

DSMZ. In part, these discrepancies may be related to differences in culture conditions, such as differences in fetal bovine serum (FBS) supplementation, and may merely reflect inducibility of certain genes.

Rearrangements of TcR genes are observed in most cell lines and correlate with TcR gene expression in SR786 and Karpas 299. Expression of B-lineage antigens CD19 and CD20 is restricted to three of the lines, UCONN-L2, KI-JK, and SR786; in the latter two, on a background of CD3 expression. Immunoglobulin heavy or light chain expression was not observed, even in DEL which carries a monoallelic IgH gene rearrangement [20]. Many of the cell lines display a spectrum of myelomonocytic markers, most notably CD11b (integrin $\alpha$M, CR3), CD13 (aminopeptidase M), and CD33 (gp67 sialoadhesin). CD15 (Lewis X antigen, X hapten) expression, which is described as a distinguishing immunohistological feature of HRS cells, but not of typical ALCL, is surprisingly found in the majority of the ALCL cell lines. On balance, most lines can be assigned to a lymphoid, mainly T-cell lineage on the immunophenotypic and gene rearrangement data.

Among markers typically expressed in ALCL biopsy material is epithelial membrane antigen (EMA), an epithelial sialomucin encoded by the MUC1 gene on chromosome 1. It has been suggested that the t(2;5) might promote in lymphoid cells, by an unknown mechanism, expression of EMA and of CD30, also encoded on chromosome 1 [6]. EMA is expressed on SU-DHL-1, Karpas 299 and SUP-M2. Restin (Reed-Sternberg-cell intermediate filament-associated protein) is found in HRS cells and ALCL in most instances, and was also detected in Karpas 299 cells [13].

## 6. CYTOKINES, CYTOKINE RECEPTORS, C-ONC GENES

In contrast to HD-derived cell lines, few studies have described the expression of cytokines and cytokine receptors (with the exception of CD25, CD30, CD40, CD70, see above) by ALCL cell lines (Table 3). Expression of interleukin (IL)-2 was demonstrated for SR768 and may result in an autocrine stimulatory loop. The same may be true for HGF (hepatocyte growth factor, scatter factor) and its receptor, c-Met, in Karpas 299, SR768, and UCONN-L2 [45]. Although the frequent expression of IL-6, IL-9 and IL-10 in HRS cells and ALCL biopsy material might suggest that expression of these cytokines may be a regular finding in ALCL cell lines, published data are limited to a report of expression IL-10 transcripts in no more than 1% of SU-DHL-1 cells [7], and to a paper describing the presence of IL-9 transcripts in an ALCL cell line designated SKA (perhaps identical to SR768) [37]. Several ALCL cell lines, among them Karpas 299, JB6 and SUP-DHL-1, were tested for CD30 ligand (CD30L) expression [21]. These lines did not express CD30L. However, some of the cell lines listed as ALCL lines were probably EBV-positive lymphoblastoid cell lines obtained from ALCL tissue (H. Merz, personal communication). In contrast to HD-derived cell lines, it was found that CD30L produced antiproliferative effects in ALCL cell lines that could be blocked by addition of soluble CD30 [22].

Among receptors that may engage in paracrine loops is the proto-oncogene c-fms (CSF-1 receptor), displayed by SU-DHL-1 and DEL, and the chemokine receptors CCR4 and CCR8, present on Karpas 299 [57]. The cell line DEL has been studied more extensively for its expression of proto-oncogenes, and c-fgr, Ki-ras, c-myb, and c-myc have been found [20]. The significance of these findings is not clear. However, the expression of c-fgr has been claimed to point to a histiocytic origin for DEL. c-kit, encoding a membrane receptor tyrosine kinase, has been found in 11 of 16 ALCL biopsies [44] and seems to be exclusive to this lymphoma and HRS cells of HD. Corresponding data on ALCL cell lines are not yet available.

Interestingly, the zinc-finger protein Bcl-6, which has been identified in a proportion of diffuse large B-cell lymphomas by virtue of its involvement in translocations affecting chromosomal band 3q27, is often expressed by ALCL, and has been demonstrated in Karpas 299, SU-DHL-1, and DEL [8]. The biological significance of Bcl-6 expression, which is physiologically re-stricted to a small fraction of normal, resting CD4-positive T lymphocytes, is currently unknown. It is speculated that it may be related to the maintenance of the activated state [8]. Alterations of the PTEN tumor suppressor gene were not observed in Karpas 299, JB6, SR786, or KI-JK, indicating that abnormalities of the PTEN gene may not be relevant for ALCL [48].

# 7. VIRUSES

The association of ALCL with EBV has been documented by a number of studies. EBV gene expression in rare cases of T-ALCL, virtually never t(2;5)-positive, seems to be largely restricted to the EBV-encoded small nuclear RNA transcripts, EBER-1 and -2, whereas B-ALCL additionally express LMP-1. This is complemented by expression of EBNA-2 in a few cases, pro-ducing the phenotype of EBV-immortalized lymphoblastoid cell lines (type III latency). Unexpectedly, the cell line KI-JK, established from an ALCL of a child, displayed the full spectrum of latent EBV gene expression as well as NPM-ALK fusion transcripts and p80-specific immunostaining [15,49]. This finding underlines previous suggestions that EBV may be an etiologic factor relevant for at least a proportion of ALCL cases.

## 8. ALK-NEGATIVE ALCL CELL LINES

Few cell lines that have been considered to represent *in vitro* equivalents of ALCL have been reported, and none have been widely used. The cell lines McG-1 and McG-2 (also known as Mac-1 and Mac-2; [10]) were established from a patient with CD30-positive cutaneous T-cell lymphoma [31]. For the two cell lines, FE-PD and HKB-1, the derivation from either ALCL or HRS cells is not clear. FE-PD displays a t(1;8) and carries, with its immunoglobulin genes in germline configuration, a TcRβ rearrangement [11]. HKB-1 has been mentioned in the previous chapter on HD-derived cell lines [58]. These lines should be used with hesitation when drawing conclusions as to the distinguishing features of HD and ALCL. The cell line Michel, cited in several papers by Gruss and co-workers, is (though CD30-positive) not an ALCL cell line. Rather, it is an EBV-positive lymphoblastoid cell line (H. Merz, personal communication), and conclusions drawn from published data on the use of this cell line require particular caution.

## 9. CONCLUSION

ALCL cell lines have been instrumental for cloning of the genes involved in the t(2,5) and have served as controls for subsequent studies on biopsy material. Future studies may center on the definition of differences from HD-derived cell lines that may ultimately help to elucidate the molecular mechanisms leading to the diverse morphology and clinical presentation of the CD30-positive malignancies, HD and ALCL [27].

## ACKNOWLEDGEMENTS

The authors are indebted to Drs. Horst Dürkop, Berlin, Germany, and Hartmut Merz, Lübeck, Germany, for communicating unpublished phenotypic data on Karpas 299, SU-DHL-1, and JB6 (H.D.) as well as on the lines L82 and Michel (H.M.).

### References

1.  Al Hashmi I, Cohen A, Freedman MH, Osman A et al. *Blood* 82: 132a, 1993.
2.  Al-Katib A, Abubakr Y, Mohamed A, Maki A et al. *Blood* 84: 638a, 1994.
3.  Anagnostopoulos I, Hummel M, Kaudewitz P, Herbst H et al. *Am J Pathol* 137: 1317–1322, 1990.

4.  Barbey S, Gogusev J, Mouly H, Le Pelletier O et al. *Int J Cancer* 45: 546–553, 1990.
5.  Beckwith M, Longo DL, O'Connell CD, Moratz CM et al. *J Natl Cancer Inst* 82: 501–509, 1990.
6.  Benharroch D, Meguerian-Bedoyan Z, Lamant L, Amin C et al. *Blood* 91: 2076–2084, 1998.
7.  Boulland ML, Meignin V, Leroy-Viard K, Copie-Bergman C et al. *Am J Pathol* 153: 1229–1237, 1998.
8.  Carbone A, Gloghini A, Gaidano G, Dalla-Favera R et al. *Blood* 90: 2445–2450, 1997.
9.  Carbone A, Gloghini A, Gattei V, Aldinucci D et al. *Blood* 85: 780–789, 1995.
10. Davis TH, Morton CC, Miller-Cassman R, Balk SP et al. *N Engl J Med* 326: 1115–1122, 1992.
11. Del Mistro A, Leszl A, Bertorelle R, Calabro ML et al. *Leukemia* 8: 1214–1219, 1994.
12. Delabie J, Ceuppens JL, Vandenberghe P, de Boer M et al. *Blood* 82: 2845–2852, 1993.
13. Delabie J, Shipman R, Brüggen J, de Strooper B et al. *Blood* 80: 2891–2896, 1992.
14. Delsol G, Lamant L, Mariamé B, Pulford K et al. *Blood* 89: 1483–1490, 1997.
15. Dirks WG, Zaborski M, Jäger K, Challier C et al. *Leukemia* 10: 142–149, 1996.
16. Epstein AL, Kaplan HS. *Cancer* 34: 1851–1872, 1974.
17. Epstein AL, Levy R, Kim H, Henle W et al. *Cancer* 42: 2379–2391, 1978.
18. Fischer P, Nacheva E, Mason DY, Sherrington PD et al. *Blood* 72: 234–240, 1988.
19. Foss HD, Demel G, Anagnostopoulos I, Araujo I et al. *Pathobiology* 65: 83–90, 1997.
20. Gogusev J, Barbey S, Nezelof C. *Int J Cancer* 46: 106–112, 1990.
21. Gruss HJ, Boiani N, Williams DE, Armitage RJ et al. *Blood* 83:2045–2056, 1994.
22. Gruss HJ, DaSilva N, Hu ZB, Uphoff CC et al. *Leukemia* 8: 2083–2094, 1994.
23. Harris NL, Jaffe E, Stein H, Banks PM et al. *Blood* 84: 1361–1392, 1994.
24. Herbst H, Anagostopoulos I, Heinze B, Dürkop H et al. *Blood* 86: 1694–1700, 1995.
25. Herbst H, Dallenbach F, Hummel M, Niedobitek G et al. *Blood* 78: 2666–2673, 1991.
26. Herbst H, Sander C, Tronnier M, Kutzner H et al. *Br J Dermatol* 137: 680–686, 1997.
27. Herbst H, Stein H, Niedobitek G. *Crit Rev Oncogenesis* 4: 191–239, 1993.
28. Herbst H, Stein H. *Leukemia Lymphoma* 9: 321–328, 1993.
29. Iwahara T, Fujimoto J, Wen D, Cupples R et al. *Oncogene* 14: 439–449, 1997.
30. Kadin ME, Cavaille-Coll MW, Morton CC. *Blood* 76: 354a, 1990.
31. Kadin ME, Sako D, Morton C, Newcom SR et al. *Lab Invest* 58: 45, 1988.
32. Kuefer MU, Look AT, Pulford K, Behm FG et al. *Blood* 90: 2901–2910, 1997.
33. Kuze T, Nakamura N, Hashimoto Y, Abe M et al. *J Pathol* 180: 236–242, 1996.
34. Lamant L, Dastugue N, Pulford K, Delsol G et al. *Blood* 93: 3088–3095, 1999.
35. Mason DY, Bastard C, Rimokh R, Dastugue N et al. *Br J Haematol* 74: 161–168, 1990.
36. Merz H, Fliedner A, Orscheschek K, Binder T et al. *Am J Pathol* 139: 1173–1180, 1991.
37. Merz H, Houssiau FA, Orscheschek K, Renauld JC et al. *Blood* 78: 1311–1317, 1991.
38. Mitev L, Christova S, Hadjiev E, Guenova M et al. *Leukemia Lymphoma* 28: 613–616, 1998.
39. Morgan R, Smith SD, Hecht BK, Christy V et al. *Blood* 73: 2155–2164, 1989.
40. Morris SW, Kirstein MN, Valentine MB, Dittmer KG et al. *Science* 263: 1281–1284, 1994.
41. Orscheschek K, Merz H, Hell J, Binder T et al. *Lancet* 345: 87–90, 1995.
42. Ott G, Katzenberger T, Siebert R, DeCoteau JF et al. *Genes Chromosomes Cancer* 22: 114–121, 1998.
43. Peuchmaur M, Emilie D, Crevon MC, Solal-Celigny P et al. *Am J Pathol* 136: 383–390, 1990.
44. Pinto A, Gloghini A, Gattei V, Aldinucci D et al. *Blood* 83: 785–792, 1994.

45.  Pons E, Uphoff CC, Drexler HG. *Leukemia Res* 22: 797–804, 1998.
46.  Pulford K, Lamant L, Morris SW, Butler LH et al. *Blood* 89: 1394–1404, 1997.
47.  Rosenwald A, Ott G, Puford K, Katzenberger T et al. *Blood* 94: 362–364, 1999.
48.  Sakai A, Thieblemont C, Wellmann A, Jaffe ES et al. *Blood* 92: 3410–3415, 1998.
49.  Shimakage M, Dezawa T, Tamura S, Tabata T et al. *Intervirology* 36: 215–224, 1993.
50.  Shiota M, Fujimoto J, Semba T, Satoh H et al. *Oncogene* 9: 1567–1574, 1994.
51.  Stansfeld AG, Diebold J, Noel H, Kapanci Y et al. *Lancet* 1: 292–293, 1988.
52.  Stein H, Herbst H, Anagnostopoulos I, Niedobitek G et al. *Ann Oncol* 2 (Suppl. 2): 33–38, 1991.
53.  Stein H, Mason DY, Gerdes J, O'Connor N et al. *Blood* 66: 848–849, 1985.
54.  Su IJ, Balk SP, Kadin ME. *Am J Pathol* 132: 192–198, 1988.
55.  Su IJ, Kadin ME. *Am J Pathol* 135: 439–445, 1989.
56.  Tilly H, Gaulard P, Lepage C, Dumontet C et al. *Blood* 90: 3727–3734, 1997.
57.  van den Berg A, Visser L, Poppema S. *Am J Pathol* 154: 1685–1691, 1999.
58.  Wagner HJ, Klintworth F, Jabs W, Lange K et al. *Med Pediatr Oncol* 31: 138–143, 1998.
59.  Wlodarska I, de Wolf-Peeters C, Falini B, Verhoef G et al. *Blood* 92: 2688–2695, 1998.

# Chapter 13

# Authentication and Characterization

Roderick A.F. MacLeod and Hans G. Drexler
*DSMZ – German Collection of Microorganisms and Cell Cultures, Mascheroder Weg 1b,*
*38124 Braunschweig, Germany. Tel: +49-531-2616-160; Fax: +49-531-2616-150;*
*E-mail: rml@dsmz.de*

## 1. BACKGROUND

Despite their many and varied origins from different stages and pathways of differentiation, human hematopoietic tumor cell lines outwardly betray little of their individual characteristics. These characteristics must be elucidated by performing a battery of assays before the derivation can be confirmed and the potential value of the cell line assessed. In addition, given the unacceptably high incidence of cross-contamination occurring among new cell lines [22,57], originators must prove authenticity.

The most versatile and informative tests for hematopoietic cell lines are immunophenotyping and cytogenetic analysis. While some characteristics of tumors are almost invariably retained by derived cell lines (such as primary chromosome rearrangements), others may change during disease progression or undergo apparent modification *in vitro* (such as immunoprofiles, which may not coincide in every detail). Thus, rather than passively adopting the diagnosis provided for the donor patient, classification of hematopoietic cell lines must take into account the results of characterization, possibly resulting in subsequent reassignment to a different lineage.

The emergence of a correlation between the morphological French-American-British (FAB) system subdividing AML [3] and the results of cytogenetic investigation encouraged the subsequent scheme widely adopted for classifying lymphoid malignancies – Morphologic, Immunologic, Cytogenetic (MIC) – to incorporate chromosomal findings. Where a particular chromosome rearrangement exhibits unusual consistency and specificity, the existence of that rearrangement (whether diagnosed cytogenetically or by PCR) is the main diagnostic criterion, as for example the presence of t(11;14)(q13;q32) in mantle cell lymphoma, or t(2;5)(p23;q35) in anaplastic large cell lymphoma (ALCL).

*J.R.W. Masters and B.O. Palsson (eds.), Human Cell Culture Vol. III, 371–397.*
© 2000 *Kluwer Academic Publishers. Printed in the Netherlands.*

The presence of a particular chromosome change provides strong evidence that the original tumor and its derived cell line share common ancestral origins. By the same token, the failure of a cell line to retain a primary chromosome change previously detected in the donor patient makes a common origin highly unlikely, thus permitting reactive host and neoplastic cells to be distinguished.

It is seldom sufficiently appreciated that a significant number of cell lines have been misclassified or misidentified completely due to cross-contamination by another cell line. Provision of adequate evidence of authenticity is often crucial to understanding the significance of findings obtained using cell lines. For instance, the most frequently cited human endothelial cell line, ECV-304, is spurious, having been cross-contaminated by the bladder cancer cell line, T-24 [15].

Cell lines bearing rarer recurrent chromosome translocations constitute an irreplaceable resource for cloning the genes involved and mapping rearrangements at the molecular level. Analysis of cell lines described prior to the advent of molecular cytogenetics has identified several such lines.

This chapter will deal with the detection and identification of false cell lines using both DNA profiling and cytogenetic methods, and illustrate some uses of cytogenetics in characterizing human hematopoietic tumor cell lines.


## 2. AUTHENTICATION

A chronic problem in cell culture, which extent is seldom fully appreciated, is the cross-contamination of one cell line by another [58]. According to a recent survey, performed using the most sensitive identification methods currently available, approximately one in six new human tumor cell lines are impostors, having been cross-contaminated by other cell lines [57]. Although the problem among human hematopoietic cell lines is of comparable magnitude, the greater availability and informativeness of karyotypes among this group may facilitate the speedier detection and removal of false examples [22]. The vast majority of cross-contaminations in our survey involved intraspecies contamination of one human cell line by another. Another survey concluded that about a third of cell lines in circulation were false [39].

The number of human leukemia-lymphoma cell lines is estimated at more than one thousand [19], indicating the high level of sensitivity to be demanded of any system adopted for detecting and identifying cross-contamination by recipients of cell lines. Unfortunately, the number of cell lines which may be confidently identified from published references is limited. Rapid advances in molecular genetics have hitherto hampered the general adoption of standardized DNA fingerprinting methods for cell line authentication, and informative published DNA fingerprints are still conspicuously rare. Although detailed karyotypes are increasingly provided for many cell lines, most pre-date the advent of sensitive molecular methods such as fluorescence *in situ* hybridization (FISH) needed for interpreting all but the most straightforward structural chromosome rearrangements. As a result, karyotypes of a signific-ant number of the older cell lines may be incomplete or even misleadingly inaccurate.

The paucity of reports documenting the provenance or authenticity of cell lines suggests a lack of general awareness of the extent of cross-contamination among contributors and editors alike. Confirming this view, and despite the rapid expansion in the numbers and varieties of available cell lines, many of the same contaminants are involved as when the prob-lem was first encountered some twenty years ago. The vast majority of cross-contamination occurs intraspecies, with a distinct tendency for the contaminating cell lines to mimic the supposed attributes of their targets, While deliberate fraud cannot be excluded, our experience indicates that carelessness prevails over deceit as the main cause of the problem.

## 2.1. Distortions Attributable to Cross-Contaminated Cell Lines

U-937, a misidentified cell line masquerading as a representative *in vitro* model of macrophage-monocyte differentiation, was widely distributed be-fore cross-contamination by the chronic myeloid leukemia (CML) cell line K-562 was discovered [77]. Misidentified samples of U-937 and many other cell lines have been used in countless published studies, but in contrast to other types of scientific fraud, reports are rarely retracted or publicly qualified in the light of the use of the false cell line.

A second type of error occurs when misidentification involves mischar-acterization. This error is most problematic when the disease has few *in vitro* models, and so the reporting of inconsistent data is less apparent. For example, one study of Hodgkin's disease (HD) was in fact carried out using a T-cell acute lymphoblastic leukemia (ALL) cell line [30] and an-other contamination involved the well-known acute megakaryocyte leukemia (AMegL) cell line, DAMI [32], now known to have been cross-contaminated at initiation [56] by the classic erythroleukemia cell line, HEL [59]. Unfortu-nately, at the time, DAMI was one of few AMegL cell lines freely available, having been deposited by its originators with the ATCC, resulting in its rapid and widespread distribution. DAMI continues to be used despite its fictitious claim to represent AMegL, registering more than 100 retrievable citations on Evaluated Medline, which is an underestimate of the actual usage. Given their close developmental proximity, it is likely that many findings performed using erythrocytic cells are indeed applicable to megakaryocytes: hence, it is doubly unfortunate that, had its true origin been known at the time, the validity of many experiments inadvertently performed using the DAMI-subclone of HEL would have remained untainted.

Misclassification may also occur in the absence of cross-contamination. Perhaps the most widely used and best-known hematopoietic cell line is HL-60, described as originating from a case of acute promyelocytic leukemia (AML-M3) [10,29] – a misclassification which persists despite having been revised on morphological grounds to AML-M2 by the original investigators [11]. This reassignment gains crucial support from the results of cytogenetic investigations confirming the absence of t(15;17)(q22;q21), a rearrangement confined to AML-M3 and now known to occur in practically all cases with this disease subtype [5,49].

A further type of distortion occurs with the unwitting use of misidentified subclones. For example, the observation of induction of *RALDH2* expression by *TAL1* in T-cell leukemia was based on inducibility in "both" CCRF-CEM and MKB-1 cell lines [74] the latter being, unknown to these authors, a subclone of the former (Table 1). This type of distortion may lead to ill-founded conclusions. For example, a novel leukemia subtype was identified by the occurrence of molecularly identical chromosome translocations in three supposedly independent acute leukemia cell lines NALM-6, PBEI and LR10.6 [94], subsequently shown to be genetically identical by both karyotyping and DNA profiling [22,57]. Thus, both PBEI and LR10.6 are cross-contaminants of NALM-6, the first of the three to be established and, thereafter, widely distributed.

It is a cause for concern that false cell lines should remain undetected for so long despite yielding inappropriate data. Is is likely that negative results go unreported or, if reported, simply ignored. Those least likely to learn of doubts concerning particular cell lines, such as beginners or associates, are also those least empowered to act as "whistleblowers", and there appears to be a conspiracy of silence over this form of passive scientific fraud. Thus, it is often left to cell repositories like the ATCC or DSMZ, which perform multiparameter cell line authentication, to pronounce on matters concerning the true identity of cell lines with any degree of objectivity [54].

Restrictions on distribution are also a barrier to authenticating cell lines. By sidestepping checks by other users or cell repositories, restricted cell lines may evade authentication altogether. For example, the singular and restricted cell line AG-F [30], was purportedly established from an HD patient after previous treatment for high grade neuroblastoma. Uniquely among hematopoietic cell lines, it was reported to retain the MYC-N amplification seen in the tumor of origin, despite displaying a typical T-cell immunoprofile. The well-documented karyogram accompanying the original description of AG-F reveals a tetraploid karyotype which is an almost exact two-fold iteration

*Table 1.* False or misidentified cell lines

| False cell lines | | Actual cell lines | |
|---|---|---|---|
| Name | Classification | Name | Classification |
| (a) *Cross contaminations occurring in originators' laboratories*: | | | |
| 207* | BCP-ALL | REH | BCP-ALL |
| AG-F | Hodgkin's disease | CCRF-CEM | T-ALL |
| DAMI | AML-M7 | HEL | AML-M6 |
| ECV-304 | endothelial | T-24 | bladder carcinoma |
| F2-4B6, F2-4E5 | thymic epithelial | SK-HEP-1 | liver carcinoma |
| FQ, Rb, SpR | Hodgkin's disease | monkey | ? |
| HS-SULTAN | multiple myeloma | Jiyoye | Burkitt's lymphoma |
| J-111 | monocytic leukemia | HeLa | cervix carcinoma |
| JOSK-I | AML-M4 | U-937 | histiocytic lymphoma |
| JOSK-K | AML-M5 | U-937 | histiocytic lymphoma |
| JOSK-M | CML-BC | U-937 | histiocytic lymphoma |
| JOSK-S | AML-M5 | U-937 | histiocytic lymphoma |
| LR10.6 | BCP-ALL | NALM-6 | BCP-ALL |
| MKB-1 | T-ALL | CCRF-CEM | T-ALL |
| MOLT-15 | T-ALL | CTV-1 | AML-M5 |
| OCI-AML-12 | AML | OCI-AML-1 | AML-M4 |
| P1-4D6, P1-4E5 | thymic epithelial | SK-HEP-1 | liver carcinoma |
| SPI-801/802 | T-ALL | K-562 | CML-BC |
| (b) *Cross-contaminations occurring subsequently*: | | | |
| 207§ | BCP-ALL | CCRF-CEM | T-ALL |
| CO | Hodgkin's disease | CCRF-CEM | T-ALL |
| KE-37§ | T-ALL | CCRF-CEM | T-ALL |
| KM-3 | BCP-ALL | REH | BCP-ALL |
| MB-02§ | AML-M7 | HU-3 | AML-M7 |
| RC-2A | AML-M4 | CCRF-CEM | T-ALL |

The table lists some false cell lines of hematologic interest and the authentic prototypes from which they are derived. Part (a) lists spurious cell lines, i.e. cross-contaminations known to have occurred in originators' laboratories and except where indicated (*) may be taken to affect most or all stocks in circulation. Part (b) lists subsequent misidentifications *probably* occurring in recipient laboratories, although only for those labelled (§) can the existence of authentic stocks be confirmed. In most intraspecies contaminations, identities were confirmed by both DNA profiling and cytogenetics. Modified from Drexler et al. [22] and MacLeod et al. [57].

of the major diploid clone present in the "classic" T-cell line CCRF-CEM established some thirty years earlier and widely distributed [57]. CCRF-CEM and AG-F are the only two cell lines reported to display t(8;9)(p11;p24), pointing again to cross-contamination.

In the absence of a central register of cell line descriptors (DNA profiles and karyotypes), which might enable originators of cell lines and those reviewing manuscripts to check for untoward matches indicative of cross-contamination, false cell lines will continue to be described. There also seems to be an information barrier concerning the existence of false cell lines, perhaps because publication of cross-contamination incidents occurs haphazardly and in specialist journals [58]. For this reason, the DSMZ will, in future, list false cell lines on its website (www.dsmz.de). Secondly, even widespread publicity may be insufficient in itself. Despite repeated alerts to the risks of HeLa contamination and the availability of a variety of methods for its detection, our data show it remains the most prolific contaminant, at levels comparable to those reported 15 years ago. Thirdly, most reports describing new cell lines lack DNA profiling to document identity with the biopsy material, and detailed karyotypes are provided in few cases.

## 2.2. Prevention of Cross-Contamination

Some obvious precautions may be adopted to reduce the risks of receiving false cell lines. Cell lines are usually obtained from three different sources: friends and colleagues, cell repositories, and the orginators themselves. While the originators might be thought to offer the most impeccable provenance, this is the very group least likely to suspect cross-contamination. Furthermore, cell lines obtained from originators are in about a third of all cases contaminated with mycoplasma [90]. Cell lines obtained from friends or colleagues compound the problems caused by originators by adding to the risks of misidentification and cross-contamination at each remove. For cell repositories, the obvious consequences of late discovery means that the risks of cross- contamination among stocks distributed to investigators must be kept to a minimum. This awareness has led such institutions to undertake systematic identity testing programs among cell lines [35].

The risks of cross-contamination may be enhanced among unpublished cell lines. Those expecting to acquire for the first time new or untested cell lines from the originators can, and should, check if their genetic identity with biopsy material has been confirmed. Failing that, a well-documented karyotype, with accompanying karyogram, should be made available. Reports claiming unprecedented success in establishing cell lines where others have repeatedly failed should raise the suspicion of cross-contamination. For example, the reported establishment of a large series of thymic epithelial cell lines by "spontaneous immortalization" [26] have now all been shown to be identical cross-contaminants [57]. Reports describing the serial establishment of several cell lines from the same type of tumor should be treated with similar caution. In the same study, we found that multiple instances of cross-contamination appeared to be clustered within certain institutions, more than half arising within six laboratories. Untoward cytogenetic similarities among cell lines of shared provenance is a further ground for suspicion as, with few exceptions, notably t(8;14)(q24;q32) in cell lines derived from Burkitt's lymphoma and other B-cell neoplasms, or t(9;22) in CML cell lines [18,20,21], recurrent chromosome changes usually occur too rarely in cell lines to be a conspicuous feature. Reports describing consistent chromosome changes should be treated with scepticism unless cross-contamination has been excluded by DNA fingerprinting. The similar rearrangements described in the four monocytic leukemia cell lines, JOSK-I/K/M/S, [72] were due to multiple cross-contamination by U-937 rather than the operation of consistent primary chromosome change. Instructively, the only sizeable panel of cell lines whose establishment was verified by DNA profiling passing through our hands [82] was also the only example totally free from cross-contamination when retested by us.

## 2.3. Detection of Cross-Contamination

A further problem contributing to the difficulty of detecting intraspecies cross-contamination is that, until recently, the necessary methods for its accurate detection were unavailable. The first breakthrough came with the discovery of stably inherited polymorphic variant satellite DNA regions which led eventually to the development of forensic DNA profiling [42]. Specific probes for polymorphic single loci were used for cell line identification by Masters et al. [60]. DNA profiling has three advantages. Firstly, it may be used by originators of new cell lines to confirm the identity of a new cell line with the primary culture or the biopsy material. Secondly, it may be used to confirm that subsequent passages remain free from cross-contamination by the simple expedient of comparing their DNA profiles with those of standards prepared fom early passage material. Thirdly, it is now being used at the DSMZ to detect false cell lines by comparing DNA profiles of candidate cell lines during accession with those already held. The effectiveness of such positive vetting relies on the construction of a searchable databank compiled from DNA profiles stored on disk after digitization which is being made freely available to assist other investigators wishing to check the identity of their cell lines [16].

## 2.4. DNA Profiling

A number of routine DNA profiling methods available for cell line authentication rely on either Southern blotting or electrophoresed PCR products, which detect one or more polymorphic loci to reveal "signature" profiles. Traditional multilocus profiling by Southern analysis detecting restriction fragment length polymorphisms (RFLP) potentially offers the highest levels of discrimination but is relatively slow, requires more DNA, and is less reproducible and more resource-consuming than PCR fingerprinting. In addition, the multilocus profiles so generated are ill-suited to the automation essential for archival comparison, and important information conveyed by band intensity levels may be lost on digitization, and undue weight automatically attached to invariant high molecular weight bands. Single-locus profiling, on the other hand, is flexible, as the desired sensitivity level may be tailored to the needs of the user by judicious choice of allele number and, in addition, the band profiles are ideally suited to digitization.

Before choosing a particular profiling system, two problems peculiar to tumor cell culture must be considered. First, loss of heterozygosity (LOH), taken to indicate the presence of tumor suppressor genes in the region affected, is known to occur widely in many different types of tumor and derived cell lines and, if chromosomally generated (for example by deletion or sister chromatid exchange), is likely to affect neighboring loci. A polymorphic locus favored in identity testing, D17S5 (*YNZ22*), is located on the short-arm region of chromosome 17 and may be lost together with *TP53*. Enhanced LOH simultaneously reduces the numbers of alleles available for comparison, encouraging "false positive" type errors. In addition, some cell lines may be prone to LOH, as we and others have observed among different samples of U-937 cells [57,84]. Curiously, the LOH we observed in U-937 occurred in samples obtained directly from its originator and affected all four polymorphic loci tested: D1S80 at 1p35-36, *ApoB* at 2p23-24, D2S44 at 2q21, and D17S5 at 17p13.3, whereas indirectly sourced material (from Japan) had retained heterozygosity at both D1S80 and D2S44.

A second problem concerns the instability experienced by microsatellite repeat loci in some tumor cells. It has been observed recently that microsatellite instability and frameshift mutations in *BAX* and transforming growth factor-$\beta$ RII genes, though relatively uncommon in leukemias, may arise within hematopoietic cell lines at establishment or during subsequent culture [66]. We have observed significant variation in DNA profiles of $(GTG)_5$ microsatellite loci between subclones of some cell lines, confirming that instability may complicate identification based on microsatellite loci. In contrast to LOH, microsatellite instability tends to cause "false negative" errors by disguising similarities between subclones. In the cross-contamination involving subclones of NALM-6 referred to above, microsatellite instability is thought to have hidden the common identities of the cell lines involved (P. Marynen, personal communication).

Taking these and other difficulties into account, for routine DNA profiling of cell lines, we have adopted a system measuring RFLP affecting minisatellite variable number tandem repeat (VNTR) loci by multiplexed PCR amplification [16] based on the work of King et al. [45]. In this system the number of polymorphic loci chosen for testing is tailored to the required degree of sensitivity. The four VNTR loci used for routine DNA profiling at the DSMZ are described in Table 2. Positive matches can arise between candidate cell lines undergoing authentication and those previously validated, usually suggesting cross-contamination of the former by the latter. "False positives" are excluded by multilocus fingerprinting using the $(GTG)_5$ microsatellite probe and by comparison with the results of karyotyping which is carried out in parallel, principally for reasons of authentication.

In our experience, multiparameter authentication of human cell lines combining cytogenetics with DNA profiling is not only prudent but, because the two methods are complementary, may be the only way to detect and identify the culprit cell line [31]. The principal objection to using cytogenetic methods for authentication, its lower cost-effectivness when compared to DNA profiling, disappears if it is being used for characterization.

## 2.5. Cytogenetic Authentication

The human haploid karyotype has about 300 bands even in mediocre chromosome preparations. Allowing for an observer error of plus/minus one band, this resolves to about 100 microscopically distinct zones. Thus, a conservative estimate of the number of different two break chromosome rearrangements involving different chromosomes (i.e. most chromosome translocations) that can be detected is in the order of $100 \times 100$ (i.e. 10,000). Therefore, the chance of identical non-primary translocations arising by chance in different cell lines is in the order of 0.01% (1 in 10,000). In addition to being invaluable for identifying cell lines and cross-contamination, it is also possible to identify secondary translocations present in subclones, as seen with CCRF-CEM (Table 1). This permits the subclones to be distinguished cytogenetically and is an advantage over DNA fingerprinting and profiling.

The classic CML cell line, K-562, uniquely carries two marker chromosomes in which *BCR-ABL* fusion effected by t(9;22)(q34;q11) is amplified in tandem [18,21]. Their occurrence, therefore, in another pair of cell lines, SPI-801 and SPI-802 [22], supposedly derived from T-ALL, is spurious and actually due to cross-contamination by K-562 [31]. Unsurprisingly therefore,

*Table 2.*   VNTR loci used for authentication at DSMZ

| Probe [ref.] | VNTR type | Alias | Chromosome location | Repeat-length (bp) | Product-length (bp) | Heterozygosity (%) |
|---|---|---|---|---|---|---|
| Apo-B [6] | minisatellite | – | 2p23-p24 | 15 | 522–909 | 80 |
| D1S80 [75a] | minisatellite | *MCT118* | 1p35-p36 | 16 | 400–940 | 79 |
| D17S5 [95] | minisatellite | *YNZ22* | 17p13.3 | 70 | 170–1080 | 78 |
| D2S44 [70] | minisatellite | *YNH24* | 2q21 | 31 | 600–>5000 | 94 |
| (GTG)$_5$ [79a] | microsatellite | – | subtelomeric | 3 | 10–>15000 | >99.9 |

The table lists salient attributes of four single-locus minisatellite probes used in multiplexed PCR detection and microsatellite (GTG)$_5$ multilocus probe used to detect polymorphic variation. Table modified from Dirks et al. [16].

excluding autologous cell lines derived from the same patient and those with near-normal karyotypes, no two tumor cell lines in the DSMZ collection are karyotypically identical.

The main hindrances to using cytogenetics for identification are both technical and interpretational. Unlike DNA extraction and purification, for which standard methods apply irrespective of cell type, those intending to obtain usable chromosome preparations from cell lines must be prepared to try a variety of methods (for example, the use of different hypotonic treatments). Only after obtaining adequate chromosome preparations may the problems of analysis be addressed. Cell lines show many different types of alteration: primary changes affecting specific oncogene rearrangements, secondary changes whose molecular consequences may be unknown but which are believed to promote tumor progression, random changes (for example due to DNA instability), and those possibly associated with adaptation to growth in culture. The most complex karyotypes among lymphomas and leukemias are those derived from HD or AML-M7, respectively, in which full analysis may be impractical even when augmented by FISH.

Under the common notation used for transcribing pictorial karyograms into written format, the International System for Chromosome Nomenclature (ISCN), chromosome rearrangements (markers) unidentifiable by banding analysis are binned together as "mar", irrespective of size, shape or form, reducing their informativeness for subsequent detection of untoward karyotypic similarities (indicative of cross-contamination). Although the compositions of marker chromosomes are theoretically solvable with the help of FISH, in the absence of additional clues, trial and error may require an impractical number of repeated attempts before the correct informative combination of chromosome painting probes is arrived at.

However, the latest technical developments spawned by FISH obviate at a stroke the necessity for such laborious and frustrating procedures. These methods generally involve hybridizing chromosomes with sets of whole chromosome (painting) probes combinatorially labelled so that each homolog (or part thereof) is identifiable via a unique spectral signature. This requires the use of sensitive cameras and optical devices for distinguishing the labelling spectra as well as specialized computer software. Spectral karyotyping (SKY) utilizes interferometry and Fourier analysis of a spectroscopic image to assess the color composition of individual light pixels [80]. Multiplex FISH (M-FISH) involves merging images for each of five or more fluorochromes collected separately through the appropriate passband filter sets [83]. In both systems, material originating from the different homologs are pseudocolored to highlight their differences for subsequent manipulation and documentation. Among more complex karyotypes, weeks or even months of work and the materials needed for scores of experiments are therefore replaced by a single procedure. The relative advantages and disadvantages of both systems have been discussed elsewhere [52]. Neither system dispenses with the need for conventional banding analysis, however, as the chromosome painting probes used are not informative regarding the intrachromosomal locations of the rearrangements detected, and although a multicolor FISH system with band-specific probes (exploiting cross-hybridization with prosimian chromosome painting probes) has been recently developed [68,69], it is probably unsuitable for tumor cell karyotyping as breakpoints juxtaposed by translocation may share common syntenic origins and spectral signatures.

A second problem concerns so-called "chromosome instability". Although vanishingly few of the necessary longitudinal studies using cloned cell lines have been reported, the charge of instability is widely voiced regarding cell lines. The existence in a tumor cell line of a complex series of chromosome rearrangements, while often taken as evidence of "instability", is, of course, no proof that the rearrangements in question actually arose *in vitro*. In fact, comparison of three highly complex cell lines (HDLM-1/2/3) established independently from a patient with HD has revealed a high degree of similarity indicating relative stability *in vitro* [57a]. It is therefore probable that many instances of supposed chromosome instability may be due to the differential expansion of clones which had acquired their distinctive features *in vivo*. After comparing several subclones of the classic T-ALL cell line, CCRF-CEM [28] whose identities were all confirmed by DNA fingerprinting, we observed minor distinctions between the karyotypes in all examples except in one case (MKB-1) where the differences were extensive. MKB-1 is near-tetraploid and was reported to display a complex series of changes: 89, XX, −X, −X, −2, −4, t(5;6)(q13;q21–23), add(7)(q35), del(7)(q22), −9, inv del(9)(q1?3-p1?3::q1?3-qter)[sic], t(10;14)(q24;q11.2), del(12)(q21), del(13)(q12), dup(13)(q22qter), −16, −20, +21, der(21)t(21;21)(p11;q11), +22, del(22)(q11) by its originators (renotated from Matsuo et al. [61]). On the other hand, CCRF-CEM and its remaining subclones masquerading as 207, AG-F, KE-37, RC-2A, though distinct, display relatively simple, closely related karyotypes including a common change, t(8;9)(p11;p24) which is absent from MKB-1 (Table 1). This discrepancy prompted our examination of early-passage stocks of CCRF-CEM which revealed the presence of two distinct subclones, with and without the t(8;9), which had expanded differentially to give rise to the two families of subclones. Interestingly, Molenaar et al. [66] have shown that MKB-1, unlike the parent CCRF-CEM, exhibits microsatellite instability, suggesting a possible mechanism underlying its apparent chromosome instability. It remains to be ascertained whether this is a general mechanism leading to the acquisition of additional chromosome changes *in vitro*.

Chromosome instability may not be a general feature of hematopoietic cell lines, but some types of cell line, not necessarily those established the longest, do have both aneuploid and structurally rearranged karyotypes. Our experiences suggest that in the different lineages of hematopoietic tumor cell lines, control of numerical and structural rearrangement has been relinquished differentially. Of all types of cell line, those derived from mature B-cells retain the most stringent control of both ploidy and chromosome structure, followed by T-cell lines which, though often tetraploid at later passage, display relatively few structural rearrangements, while myeloid cell lines tend to lose control of both – the most extreme karyotypic deviants being those assigned to the erythoid-megakaryocytic pathway. Recent studies have shown that numerical chromosome change, as well as structural changes targeted at specific oncogenes, may be a key step in carcinogenesis [8,50,51].

The third main difficulty is that because of their complexity and unavoidable ambiguity in some cases, ISCN karyotypes resist digitization, preventing the karyotypic identification of false cell lines by computer. The problems inherent in the computerized "understanding" of written hematopoietic tumor karyotypes have been summarized by Anthony Moorman who has written some programs ([67]; A. Moorman, personal comm.).

The singular virtue of cytogenetics as applied to authentication is that karyotypes, unlike DNA profiles, are increasingly published for new cell lines by their originators and sometimes updated for older cell lines. Recently published guidelines for describing newly established cell lines [23] stress the importance of adequate cytogenetic documentation. By comparison of the detailed and accurate karyotype prepared by its originators for the false DAMI cell line [32] we were able to confirm that samples of DAMI held by the DSMZ and ATCC truly represented the material described by its originators, thus proving that the similarity of DAMI to the classic HEL cell line [59] shown by DNA profiling was due to cross-contamination by its originators [56]. In this way, comparative cytogenetics was able to fill the gap occasioned by the loss of stored early passage ampoules of DAMI which its originators claimed was caused by the untimely breakdown of a storage freezer.

## 2.6.  False Cell Lines

In our experience, the most likely contaminants to invade and take over primary cell cultures are well known, long-established "classic" cell lines [57]. It is scarcely coincidental that the "classic" T-cell line, CCRF-CEM, has been instrumental in spawning so many impostors (Table 1). In addition to the above-mentioned karyotypically and genetically distinct MKB-1, a further authenticated subclone is available which grows adherently, almost unprecedented among hematopoietic cell lines. In addition to seniority and ubiquity, CCRF-CEM displays the unusual, though dubious, virtue of thriving at clonal cell densities, heightening the risk of transmission by, for example, aerosols. Other prolific cross-contaminants affecting cell lines of interest in hematology (Table 1) include U-937 and the SK-HEP-1 liver cancer cell line, which we have shown to be the source of a large series of supposed thymic epithelial cell lines [57].

In the absence of previously published data, information relevant to authentication is also afforded by characterization (see below). Table 3 lists hematopoietic cell lines with recurrent primary chromosome changes and shows how, within the limits of incompleteness, some recurrent translocations are sufficiently rarely represented to be of value in identification.

## 2.7.  Interspecies Cross-Contamination

Among the earliest well-known instances of cross-contamination affecting hematopoietic cell lines is the curious case of several purported HD cell lines, FQ, Rb, and SpR, which were subsequently instead shown to be of simian origin [34]. According to our survey, although the vast majority of cross-contamination affecting human tumor cell lines occurs intraspecies, there is some evidence that interspecies contamination may come to predominate downstream when a greater variety of cell lines than those present within the parent laboratory may be encountered [39]. Equally, interspecies contamination should be easier to detect than that occurring intraspecies. A simple and relatively inexpensive method for its detection by isoenzyme analysis is discussed by Steube et al. [85]. As with most electrophoretic methods, the cost effectiveness of isoenzyme analysis disappears unless batch testing is performed. For those wishing to detect individual instances therefore, cytogenetics is the obvious alternative, though some familiarity is required with the chromosome banding patterns of human and the more common animal cell lines, including rodent, ungulate and primate. Murine cell lines are statistically the most likely animal contaminant: mouse chromosomes distinguish themselves from human chromosomes at a glance by virtue of their morphology: telocentric (or Robertsonian fusions thereof) with unmistakeably dense G-band positive paracentromeric heterochromatin, apparent after G-banding [81].

## 3.  CYTOGENETIC CHARACTERIZATION

The notion that tumor cells tend to carry chromosome changes long predates the advent of molecular cytogenetics. Over a century ago, von Hansemann proposed that aberrant mitoses might be typical of cancer cells [92]. The

*Table 3.* Hematopoietic malignant cell lines with recurrent chromosome and gene changes

| Recurrent alteration | Genes fused | Cell lines |
| --- | --- | --- |
| **(a) BCP (ALL-derived)** | | |
| t(1;19)(q23;p13) | *PBX1/E2A* | 697*, ALL-2, KMO-90, KOPN63, LC1;19, LILA-1, LK63, MHH-CALL-3*, PER-278, SUP-B27 |
| t(4;11)(q21;q23) | *AF4/MLL* | AN4;11, B1, KOCL-45, KOCL-58, KOCL-69, RS4;11, SEM |
| t(8;14)(q24;q32) | *MYC/IGH* | 380* (3-way translocation with 18q21), KHM-2B, MHH-PREB-1* |
| t(9;22)(q34;q11) | *ABL/M-BCR* | ALL-1, NALM-27/28 |
| t(9;22)(q34;q11) | *ABLm-BCR* | ALL/MIK, KOPN-30bi/57bi, MHH-TALL1, NALM-20, PALL-1/-2, SUP-B13/15, TOM-1 |
| t(11;19)(q23;p13) | *MLL/ENL* | BS, KOCL-33, KOCL-44, KOCL-50, KOCL-51, KOPN-1, KOPN-8 |
| t(12;21)(p13;q21) | *TEL/AML1* | REH *(3-way translocation with 4q32 and 16q24), SUP-B26, UOC-B4 |
| t(17;19)(q22;p13) | *HLF/E2A* | HAL-01, UOC-B1 |
| **(b) BCP (CML-derived)** | | |
| t(9;22)(q34;q11) | *ABL/M-BCR* | BV-173*, NALM-1* |
| **(c) B-cell (ALL- or lymphoma-derived)** | | |
| t(3;4)(q27;p11–13) | *BCL6/LAZ3* | VAL |
| t(8;14)(q24;q32) | *MYC/IGH* | BALL2, BALM-6/7/8,MC-116*, MN-60*, OCI-LY18, TANOUE* |
| t(14;18)(q32;q21) | *IGH/BCL2* | FL-218 (mcr), FL-218 (mbr), FL-318 (mbr), KARPAS-422* (mbr), SU-DHL6 (mbr), WSU-NHL* (mbr) |
| t(8;14;18)(q24;q32;q21) | *MYC/IGH/BCL2* | DOHH2* (mbr), ROS-50 (mcr), SU-DUL5 (mcr), VAL (mcr) |
| t(8;22)(q24;q11) | *MYC/IGL* | BALM-16 |
| t(11;14)(q13;q32) | *BCL1/IGH* | GRANTA-519*, JVM-2*, NCEB-1 |

*Continued on next page*

*Table 3.* (continued)

| Recurrent alteration | Genes fused | Cell lines |
|---|---|---|
| **(d) B-cell (Burkitt's lymphoma-derived)** | | |
| t(8;14)(q24:q32) | *MYC/IGH* | BL-41*, BL-70*, CA-46*, DAUDI*, DG-75*, EB-1*, RAJI*, WIEN-133 |
| t(8;22)(q24:q11) | *MYC/IGL* | BL-2 |
| **(e) T-cell (ALL- or lymphoma-derived)** | | |
| t(1;7)(p34:q34) | *LCK/TCRB* | CCRF-HSB-2, CTV-1*, SUP-T12 |
| t(1;14)(p32:q11) | *TAL1/TCRA/D* | DU.528 |
| t(7;9)(q34:q34) | *TCRB/TAL2* | SUP-T1* |
| t(9;22)(q34:q11) | *ABL/M-BCR* | CML-T1* |
| t(11;14)(p13:q11) | *TTG2/TCRA/D* | KOPT-K1, LALW-2, TALL-104 |
| t(11;14)(p15:q11) | *TAL1/TCRA/D* | RPMI-8402* |
| inv(14)(q11q32) or | *TCRA/IGH* | HT-1, SUP-T1*, TALL-106 |
| t(14;14)(q11:q32) | | |
| **(f) myelomonocytic (AML- or CML-derived)** | | |
| t(4;11)(q21:q23) | *AF4/MLL* | KOCL-48, MV4;11* |
| inv(3)(q21q26) or | *EVI1/?* | MUTZ-3*, UCSD/AML1 |
| t(3;3)(q21:q26) | | |
| t(6;11)(q27:q23) | *AF-6/MLL* | CTS, ML-1/2* |
| t(9;11)(p21:q23) | *AF9/MLL* | IMS-M1, MOLM-13/14 (cryptic insertion), MONO-MAC-1/6*, THP-1* |
| t(9;22)(q34:q11) | *ABL/M-BCR* | EM-2*/3* (myelocytic); JK-1*, K-562* (erythrocytic); LAMA-84*/87* (erythro-megakaryocytic); MEG-01* (megakaryocytic) |

*Continued on next page*

*Table 3.* (continued)

| Recurrent alteration | Genes fused | Cell lines |
|---|---|---|
| (f) myelomonocytic (AML- or CML-derived) (continued) | | |
| t(9;22)(q34;q11) | ABL/m-BCR | AR230 (CML-derived), KOPM30, MR-87 – (both AML-derived) |
| t(10;11)(p13;q14) | AF10/CALM | U-937* |
| t(15;17)(q22;q21) | PML/RARA | NB-4* |
| inv(16)(p13q22) | MYH11/CBFB | ME-1 |
| (g) ALCL-derived | | |
| t(2;5)(p22;q35) | ALK/NPM | DEL* (masked translocation), KARPAS-299*, SU-DHL-1* |

The table lists recurrent chromosome changes and gene fusions in hematopoietic cell lines classified according to cell type. (*) available from DSMZ. Abbreviations: m/M-BCR, minor/major breakpoint cluster region – t(9;22); mbr, major breakpoint region; mcr, minor (breakpoint) cluster region – t(14;18). Table modified from Drexler et al. [18,20].

introduction of chromosome banding in the 1970s soon revealed the identities of consistent chromosome changes identified in the previous decade in CML and the precise localization of the chromosome breakpoints involved. With hindsight, it is clear that those who looked for consistent chromosome changes among hematopoietic malignancies were fortunate in their choice of disease as, along with tumors of the nervous system, these display specific changes with the greatest frequency and consistency [36]. It seems that one class of consistent chromosome rearrangement, the balanced translocation, whereby material is reciprocally exchanged between chromosomes from different homologous pairs with highly conserved breakpoints on each of the two chromosomes involved, is typical of hematopoietic neoplasms [64]. Investigation of these translocations (including the functionally equivalent inversions) has illuminated the central role played by genic alteration in cancer. Recurrent translocations effect the fusion of specific pairs of oncogenes, one of which is usually a transcriptional activator. These translocations exhibit remarkable specificity for hematopoietic cells blocked at different stages of differentiation [53], explaining the special power of cytogenetics in characterizing this class of cells. A list of cell lines known to carry recurrent translocations leading to known gene fusions is presented in Table 3.

Two types of fusion may be distinguished:

(a) juxtapositional, whereby proteins (typically transcription factors) are activated by being brought under the influence of the promoter regions of constitutively active genes. Examples are the immunoglobulin heavy chain genes – at 2p12 (*IGK*), 14q32 (*IGH*), or 22q11 (*IGL*) – in B-cell leukemia-lymphoma, and the T-cell receptor genes – at 7q35 (*TCRB*) or 14q11 (*TCRA/D*) – in T-cell leukemia-lymphoma.

(b) chimeric, whereby some of the exons from both participant genes are transcribed into a single mRNA and translated into a single novel protein. The latter are found across a wide variety of hematopoietic, and a few solid, tumors and derived cell lines. Examples are REH with t(12;21)(p12;q21) causing *TEL-AML1* fusion in BCP-ALL [89]; MONO-MAC-6 with t(9;11)(p21;q23) causing *MLL-AF9* fusion in AML FAB-M5 [55]; and KARPAS-299 with t(2;5)(p23;q35) causing *NPM-ALK* fusion in ALCL [14].

The simultaneous demonstration of identical recurrent chromosome rearrangements in both the biopsy and in a derived cell line is overwhelming evidence of a shared clonal relationship. The availability of hematopoietic malignant cell lines bearing recurrent translocations has played an important role in advancing these studies. In some cases, cell lines known to carry the chromosomal rearrangement in question were used for cloning the translocation breakpoints. Examples are K-562 for *BCR-ABL* in t(9;22)(q34;q11)

[37], NB-4 for *PML* in t(15;17)(q22;q12-21) [24], and KASUMI-1 for *ETO* in t(8;21)(q22;q22) [65].

It is not possible to discuss how cytogenetic characterization has contributed to the characterization of all types of cell line listed in Table 3. However, consideration of one of the largest cytogenetic groups, cell lines with t(9;22), a typical cytogenetic feature of two distinct neoplasms, CML and ALL, is instructive and illustrates how the characterization of hematopoietic cell lines is related to chromosome and related gene changes [21]. In addition, the way in which cytogenetics may be used to characterize cell lines derived from tumors in which no recurrent change is known is discussed with reference to Hodgkin's disease. Finally, some cell lines which have been incorrectly or incompletely characterized cytogenetically are referred to briefly.

## 3.1. t(9;22)(q34;q11) Cell Lines as Models

An example of a cytogenetic change occurring in more than one distinct category of hematopoietic neoplasm is the t(9;22) which, together with t(8;14)(q24;q32) in B-cell leukemia- lymphoma, is the translocation studied in most detail at the molecular level. The der(22) partner of this reciprocal translocation was the first somatic chromosome change specifically associated with any tumor [71] and designated the "Philadelphia chromosome" (Ph). However, it was not until the advent of chromosome banding that the rearrangement was seen to be a reciprocal translocation [79], marking a milestone in cancer cytogenetics.

t(9;22) is primarily associated with CML and occurs in more than 95% of all cases, including a minority where the translocation may be cryptic (due to a genomic insertion) or masked by the involvement of one or more additional chromosomes [43]. However, the t(9;22) rearrangement is also the single most frequent acquired cytogenetic alteration in ALL [4] and has been also recorded in rare cases of AML (Table 3). In addition, two B-lymphoblastoid cell lines with t(9;22) have been described: SD-1 from blood taken from a patient at diagnosis of ALL [13], and PhB1 from a patient with CML in blast crisis [47]. No other leukemic translocation targets such a wide variety of cells, implying the occurrence of t(9;22) at primitive stages during hematopoietic differentiation, a conjecture supported by the observation that transgenic mice with *BCR-ABL* may develop either myeloid, or B/T-cell lymphoid leukemias [75]. This lineage diversity is also an indication that *BCR-ABL* fusion is in itself insufficient to induce full neoplastic transformation with differentiation arrest.

Following the results of molecular investigation, we now know that it is possible to distinguish three different types of t(9;22) based on the breakpoint cluster region (BCR) location of the breakpoints in chromosome 22 band q11: M(ajor)-*BCR*, m(inor)-*BCR* and $\mu$-*BCR* are translated into proteins of 210, 190 and 230 kDa, respectively [62]. By means of RT-PCR, these various possibilities may be readily distinguished, while cytogenetic probes are commercially available to identify the M/m-*BCR* variants. Following such studies, it has become clear that *BCR-ABL* variants are unequally partitioned between CML and ALL and derived cell lines. Almost all CML, about half of all ALL and some AML patients carry M-*BCR*. A few CML, about half of all ALL and remaining AML patients carry m-*BCR*, while $\mu$-*BCR* has only been recorded so far in a subtype of CML, termed "chronic neutrophilic leukemia". As far as cell lines are concerned, more than 40 have been established from CML and more than 20 from ALL, while only three have been established from AML [21]. CML-derived cell lines have been shown to reflect the distribution of the different *BCR* breakpoints *in vivo*, almost all carrying the M-*BCR* breakpoint, the exceptions including AR230 with $\mu$-BCR [93]. In contrast only 2/17 molecularly characterized ALL-derived cell lines carry M-*BCR*, less than a quarter of the number expected. These are ALL-1 [48] and the autologous pair NALM-27/28 [1]. Of the Ph positive B-LCL, SD1 carries m-*BCR* and PhB1 M-*BCR*. It should be noted that in the best known *BCR- ABL* positive cell line of all, K-562, the original Ph is no longer recognizable, having undergone secondary rearrangement, resulting in an approximate thirty-fold coamplification of the fusion gene via tandem duplication. This rearrangement is unprecedented in both patients and cell lines. SPI-801/802, which also carries this change, is now known to be merely a subclone of K-562 (Table 1). The rearrangement has been recorded in several independent samples of K-562 [18,78,96], implying its origin *in vivo* or at early passage.

A less obvious feature of t(9;22) cell lines concerns the numbers of Ph-fusion genes present and their relationship to unrearranged alleles of *ABL* and to ploidy levels. The question is more than of theoretical interest. For example, where cell lines are being employed as positive controls for cytogenetic or PCR detection of *BCR-ABL*, it is important to know whether, or how many, rearranged and unrearranged copies of each allele are present, particularly for cytogenetic detection among interphase cells. In addition, it has been sugggested that *ABL* and *BCR-ABL* may act antagonistically [33], reminiscent of *TEL-AML1* fusion in BCP-ALL accomplished via the recurrent t(12;21)(p12;q12). The latter rearrangement is normally accompanied by deletion of the remaining allele of TEL [76], and this antagonistic pairing has been also observed in a t(12;21) BCP-ALL cell line, REH [89]. Similarly, loss of the unrearranged allele of ABL has been noted in some *BCR-ABL* cell lines (R.A.F. MacLeod, unpublished). Among *BCR-ABL* cell lines (excluding the B-LCL which tend to undergo tetraploidization at late passage irrespective of their origins), all 20 examples established from patients with ALL remain diploid, while 14/40 cell lines established from CML patients are trip-loid or tetraploid [21] refuting the assertion that *BCR-ABL* promotes genetic instability per se. In primary CML, the t(9,22) is present at disease onset. Nevertheless, none of the forty-odd available CML- derived cell lines was established prior to blast crisis, revealing a second type of representational disparity when compared to the primary disease.

## 3.2. Hodgkin's Disease

Hodgkin's disease is the most enigmatic of all hematopoietic neoplasms on account of unresolved questions regarding the origin of the Hodgkin/Reed-Sternberg (H-RS) cell and its relationship to other hematopoietic lineages. It might be expected that continuous HD cell lines could play a valuable role, and the relatively small number of HD cell lines are, indeed, the subjects of much inquiry [17]. Yet their relevance is controversial. In addition to the cross-contaminations discussed above, several so-called HD cell lines have been misclassified. Establishment of continuous cell lines from tumor mater-ial taken from HD patients is difficult, and many of the cell lines resulting are lymphoblasts immortalized by EBV. Unfortunately, several of the cell lines described as having originated from HD are merely B-LCL, including HS-445 and RPMI-6666. Nevertheless, these continue to be cited as HD in some publications.

The major problem in characterizing HD cell lines is that, unlike all other major common hematopoietic neoplasms, no consistent chromosome trans-location has been identified. This is attributed to the difficulty of isolating scarce H-RS cells from reactive normal tissue. In the absence of any recurrent primary translocation to identify the malignant clone and allow cloning of a putative oncogene target, cytogenetics has added little to the positive iden-tification of the putative HD precursor. While immunological studies have implicated a lymphocyte precursor, molecular biological studies at the level of single cells have suggested that a substantial fraction of cases of classical HD may represent clonal expansion of B-cells in which immunoglobulin expression is disabled by mutation during the germinal center reaction. A minority of cases are derived from T-cells, in contrast to the HD cell lines where both T- and B-cell phenotypes are equally represented. A second problem is the sheer karyotypic complexity of H-RS cells, in which subtle rearrangements might be unnoticed and for which the technical advances necessary for their analysis, such as M-FISH and SKY (discussed above), are only now becoming available. Nevertheless, with every patient series failing to reveal the existence of a common translocation, it is becoming increasingly likely that some other kind of primary tumorigenic change may underly HD, such as the tumor suppressor gene deletions commonly found in epithelial solid tumors.

A number of cell lines have been established from HD and made available to other investigators, such as HDLM-1/2/3, HD-MyZ, KM-H2, L-480 [20]. Given the possibility that host cells become immortalized by EBV often present, candidate cell lines should be characterized thoroughly prior to being accepted as suitable *in vitro* models for HD. As cytogenetics is uninformative in HD, evidence linking the cell lines to putative H-RS precursors has mainly rested on immunophenotyping. In particular, to be classed as HD, cell lines should constitutively express CD30 (Ki-1) at high levels without evidence of monocytic differentiation. On this criterion alone, the affinity with H-RS cells of the Ki-1-negative HD-MyZ cell line, which has a monocytic phenotype, is questionable, whereas those of HDLM- 1/2/3, KM-H2 and L-428 are reinforced. All HD cell lines exhibit extremely high levels of cytogenetic rearrangement. According to our initial findings, this appears to follow a non-random pattern resembling that reported among HD patients [2,63], inviting the conjecture, supported by twin studies, that HD may be a type of chromosome instability disease, as recently proposed [25]. We have observed "jumping translocations" involving certain types of DNA repeat in HD cell lines [57a]. HDLM-1/2/3, KM-H2, and L-428 all display several landmark bands recurrent in HD [57a]. On the other hand, HD-frequent breakpoints were no more likely than HD-infrequent breakpoints to be rearranged in other types of hematopoietic cell lines, whose breakage patterns HD-MyZ resembled. Consistency between the results of immunophenotyping and cytogenetics serves to confirm the validity of HDLM-1/2/3, KM-H2 and L-428, and suggests that when full and accurate karyotyping of HD becomes feasible by the routine application of SKY and M-FISH, sufficiently strong non-random patterns of breakpoints may emerge to justify their molecular cloning, perhaps with the help of these very cell lines.

### 3.3.  Cell Lines with Revised Karyotypes

Several cell lines, more particularly those described prior to the routine use of image analysis and molecular cytogenetic methods, have been subsequently shown to carry recurrent primary chromosome translocations [20,55]. Some of these are listed in Table 4. In most cases the rearrangements are quite subtle, for example t(9;11) as in MONO-MAC-1, MONO-MAC-6, and THP-1, which is notoriously difficult to spot in suboptimal preparations. t(12;21), as present in REH, involving the reciprocal exchange of visually identical G-banding regions, is undetectable without FISH. An additional factor which hampers detection of subtler changes in myeloid leukemias is their ka-

Table 4. Cell lines with revised karyotypes

| Cell line | Type | Ref. | Original karyotype | Revised karyotype (structural changes only) | Key change |
|---|---|---|---|---|---|
| 697 | BCP-ALL | 27 | −46, XY, t(7;19)(q11;q13) | −46, XY, t(1;19)(q23;p13), del(6)(q21) | t(1;19) |
| CTV-1 | T-ALL | 9 | −47, X, −X, −11, t(12;11), +3mar | −47, X, t(1;7)(p34.2;q34), i(6)(q10), del(10)(p13), t(12;16)(q24;q11) | t(1;7) |
| DOHH-2 | B-NHL | 46 | −48, XY, +7, del(12)(q24), t(14;18)(q23;q21) | −47, XY, der(8)t(8;18)(q24;q21), der(14)t(8;14)(q24;q32), der(18)t(14;18)(q32;q21) | t(8;14;18) |
| MHH-TALL-1 | BCP-ALL | Tomeczkowski, unpublished | −46, XY, BCR–ABL negative | −48, XY, t(1;9;22)(q32;q34;q11)inv(9)(p11q13), del(7)(p15), del(19)(p13) | t(9;22) |
| MONO-MAC-1 | AML-M5 | 97 | −41 −43, XY, +3, −12, −17, +mar | −49(43–52)XY, dup(3)(q21q27), t(9;11)(p13;q23), t(10;12, 17)(q24;q13;q11), del(13)(q13q21), t(16;21)(q13;q22.2) | t(9;11) |
| MONO-MAC-6 | AML-M5 | 97 | −41 −43, XY, +3, −12, −17, +mar | −84 −90, XX/XXX, t(9;11)(p22;q23)x2, add(10)(p11) x2, add(12)(q?21), del(13)(q13q14der(13)t(13;14)(p11;q;12)x2 | t(9;11) |
| ML-1/2 | AML-M4 | 73 | −91, XX, del(1)(p23), add(1)(p22), del(6)(q23), der(11)t(11;13)(q23;q32), der(13)t(11;13)(q23;q32), add(14)(q13), del(17)(q23), der(16)t(1;16)(p13;p35) | −92(84–94)XX, der(1)t(1;?)(p21;?)x2, del(6)(q23)x2, der(6)t(6;11)(q27;q23)x2, der(11)t(6;11)(q27;q23), add(11)(p11;q23)x2, add(11)(q11–13), dup(13)(q32→qter)x2, der(18)t(15;?;18)(q21;?;q11)x2 | t(6;11) |
| THP-1 | AML-M5 | 88 | −46, XY. Normal karyotype | −94, XY/XXYadd(1)(p11), del(1)(q42.2), i(2q), del(6)(p21), i(7)(p10), der(9)t(9;11)(p22;q23)ii(9)(p10)x2, der(11)t(9;11)(p22;q23)x2, add(12)(q24), der(13)t(8;13)(p11;p12)x2, add(?18)(q21) | t(9;11) |

*Continued on next page*

*Table 4.* (continued)

| Cell line | Type | Ref. | Original karyotype | Revised karyotype (structural changes only) | Key change |
|---|---|---|---|---|---|
| U-937 | histiocyt. NHL | 86 | −58, XXY, t(1q;14q), t(1p;13q), add(2)(q?), der(3)t(1;3)(q21;q27), der(5)t(1;5)(p13;q31), t(6p;12q) | −63, XX/XXY, der(5)t(1;5)(p32;q3?), t(1;12)(q21;p13), der(2;3)(q10;p10), add(9)(p22), t(10;11)(p12;q14), i(11)(q10), i(12)(p10), add(16)(q24), del(17)(p13), add(19)(p13), add(21)(p1) | t(10;11) |
| REH | BCP-ALL | 91 | −45, XX, −2B, +1C | −46, X, del(3)(p22), t(4;12;21;16)(q32;p13;q22;q24.3), inv(12)(p13q22), t(5;12)(q31–32;p12), der(16)t(16;21)(q24.3;q22) | t(12;21) |

The table lists cell lines carrying primary key changes overlooked or misinterpreted in original descriptions. Key changes are primary recurrent translocations effecting gene fusions (see Table 3). For brevity, revised karyotypes show structural rearrangements only.

ryotypic complexity, combined with the variety and number of different translocations effecting gene fusions within this group.

# 4. CONCLUSION

Amongst the various ways of detecting and identifying cross-contamination, cytogenetics displays unique versatility, complementing both DNA profiling (intraspecies) and isoenzyme analysis (interspecies). It combines with immunophenotyping to enable the characterization of hematopoietic cell lines to an unrivalled degree.

## Abbreviations

ALCL – anaplastic large cell lymphoma;
ALL – acute lymphoblastic leukemia;
AMegL – acute megakaryocytic leukemia;
ATCC – American Type Culture Collection;
AML – acute myeloid leukemia;
B-LCL – B-lymphoblastoid cell line;
BCP – B-cell precursor;
BCR – breakpoint cluster region;
CML – chronic myeloid leukemia;
DSMZ – Deutsche Sammlung von Mikroorganismen und Zellkulturen;
EBV – Epstein-Barr virus;
FISH – fluorescence *in situ* hybridization;
HD – Hodgkin's disease;
H-RS – Hodgkin's/Reed-Sternberg;
ISCN – International System for Human Chromosome Nomenclature;
LOH – loss of heterozygosity;
NHL – non-Hodgkin's lymphoma;
PCR – polymerase chain reaction;
RFLP – restriction fragment length polymorphism;
RT – reverse transcriptase;
SKY – spectral karyotyping;
VNTR – variable number tandem repeats.

## References

1.  Ariyasu T et al. *Human Cell* 11: 43, 1998.
2.  Atkin NB. *Cytogenet Cell Genet* 80: 23, 1998.
3.  Bennett JM et al. *Br J Haematol* 33: 451, 1976.
4.  Berger R et al. *Cancer Genet Cytogenet* 44: 143, 1990.
5.  Berger R et al. *Genes Chrom Cancer* 3: 332, 1991.
6.  Boerwinkle E et al. *Proc Natl Acad Sci USA* 86: 212, 1989.
7.  Budowle B et al. *Am J Hum Genet* 48: 137, 1991.
8.  Cahill DP et al. *Nature* 392: 300, 1998.

9.   Chen PM et al. *Gann* 75: 660, 1984.
10.  Collins SJ et al. *Nature* 270: 347, 1977.
11.  Dalton WT et al. *Blood* 71: 242, 1988.
12.  Delabie J et al. *Br J Haematol* 94: 198, 1996.
13.  Dhut S et al. *Leukemia* 5: 49, 1991.
14.  Dirks WG et al. *Leukemia* 10, 142, 1996.
15.  Dirks WG et al. *In Vitro Cell Dev Biol Animal* 35: 558, 1999.
16.  Dirks WG et al. *Cell Mol Biol* 45: 841, 1999.
17.  Drexler HG. *Leukemia Lymphoma* 9: 1, 1993.
18.  Drexler HG et al. *Leukemia* 9: 480, 1995.
19.  Drexler HG et al. *Human Cell* 11: 51, 1998.
20.  Drexler HG et al. *DSMZ Catalogue of Human and Animal Cell Lines*, 7th edn., Braunschweig, 1999.
21.  Drexler HG et al. *Leukemia Res* 23: 207, 1999.
22.  Drexler HG et al. *Leukemia* 13: 1601, 1999.
23.  Drexler HG, Matsuo Y. *Leukemia* 14: 777, 2000.
24.  Du Thé H et al. *Nature* 347: 558, 1990.
25.  Falzetti D et al. *Haematologica* 84: 298, 1999.
26.  Fernandez E et al. *Blood* 83: 3245, 1994.
27.  Findley HW et al. *Blood* 60: 1305, 1982.
28.  Foley GE et al. *Cancer* 18: 522, 1965.
29.  Gallagher R et al. *Blood* 54: 713, 1979.
30.  Gazitt Y et al. *Leukemia* 7: 2034, 1993.
31.  Gignac SM et al. *Leukemia Lymphoma* 10: 359, 1993.
32.  Greenberg SM et al. *Blood* 72: 1968, 1988.
33.  Guo XY et al. *Leukemia Lymphoma* 30: 225, 1998.
34.  Harris NL et al. *Nature* 289: 228, 1981.
35.  Hay RJ. et al. *J Cell Biochem* 24: 107, 1996.
36.  Heim S, Mitelman F. *Cancer Cytogenetics*, Wiley, 1995.
37.  Heisterkamp N et al. *Nature* 306: 293, 1983.
38.  Hirata J et al. *Leukemia* 4: 365, 1990.
39.  Hukku B et al. *Adv Exp Med Biol* 172: 13, 1984.
40.  Hummel M et al. *N Engl J Med* 333: 901, 1986.
41.  ISCN. In: Mitelman F (ed.), *An International System for Human Cytogenetic Nomenclature*, S Karger, Basel, 1995.
42.  Jeffreys AJ et al. *Nature* 314: 67, 1985.
43.  Kantarjian HM et al. *Blood* 82: 691, 1993.
44.  Kanzler H et al. *J Exp Med* 184: 1495, 1986.
45.  King BL et al. *Am J Pathol* 144: 486, 1994.
46.  Kluin-Nelemann JM et al. *Leukemia* 5: 221, 1991.
47.  Knuutila S et al. *Acta Hematol* 90: 190, 1994.
48.  Lange B et al. *Blood* 70: 192, 1987.
49.  Larson RA et al. *Am J Med Genet* 76: 827, 1984.
50.  Lengauer C et al. *Nature* 386: 623, 1997.
51.  Li R et al. *Proc Natl Acad Sci USA* 94: 14506, 1997.
52.  Lichter P. *Trends in Genetics* 13: 475, 1997.
53.  Look AT. *Science* 278, 1997.
54.  MacLeod RAF et al. *In vitro Cell Dev Biol* 28A: 591, 1992.
55.  MacLeod RAF et al. *Blood* 82: 3221, 1993.

56. MacLeod RAF et al. *Leukemia* 11: 2032, 1997.
57. MacLeod RAF et al. *Int J Cancer* 83: 555, 1999.
57a. MacLeod RAF et al. *Leukemia* 14: 1803, 2000.
58. Markovic O, Markovic N. *In vitro Cell Dev Biol* 34A: 1, 1998.
59. Martin P, Papayannopoulou T. *Science* 216: 1233, 1982.
60. Masters JRW et al. *Br. J. Cancer* 57: 284, 1988.
61. Matsuo Y et al. *Human Cell* 2: 423, 1989.
62. Melo JV. *Blood* 88: 2375, 1996.
63. Mitelman F. *Catalog of Chromosome Aberrations in Cancer*, 5th edn., Wiley-Liss, 1994.
64. Mitelman F et al. *Nat Genet* 15: 417, 1997.
65. Miyoshi H et al. *EMBO J* 12: 2715, 1993.
66. Molenaar JJ et al. *Blood* 92: 230, 1998.
67. Moorman A et al. *Cytogenet Cell Genet* 77: 35, 1997.
68. Muller S et al. *Human Genet* 101: 149, 1997.
69. Muller S et al. *Cytogenet Cell Genet* 78: 260, 1998.
70. Nakamura Y et al. *Science* 235: 1616, 1987.
71. Nowell PC, Hungerford DA. *Science* 132: 1497, 1960.
72. Ohta M et al. *Cancer Res* 46: 3067, 1986.
73. Ohyashiki K et al. *Cancer Genet Cytogenet* 37: 103, 1989.
74. Ono Y et al. *Mol Cell Biol* 18: 6939, 1998.
75. Pasternak G et al. *J. Cancer Res Clin Oncol* 24: 643, 1998.
75a. Pepinski W et al. *Rocz Akad Med Bialynst* 41: 316, 1996.
76. Raynaud S et al. *Blood* 87: 2891, 1996.
77. Reid YA et al. *J Leuk Biol* 57: 804 (letter), 1985.
78. Rodley P et al. *Genes Chrom Cancer* 19: 36, 1995.
79. Rowley JD. *Nature* 243: 290, 1973.
79a. Schäfer R et al. *Nucleic Acids Res* 16: 5196, 1988.
80. Schröck E et al. *Science* 273: 494, 1996.
81. Seabright M. *Lancet* ii: 971, 1971.
82. Shimada Y et al. *Cancer* 69: 277, 1992.
83. Speicher MR et al. *Nat Genet* 12: 368, 1996.
84. Stacey GN et al. *Cytotechnology* 8: 13, 1992.
85. Steube KG et al. *In vitro Cell Dev Biol* 31A: 115, 1995.
86. Sundström K et al. *Int J Cancer* 13: 808, 1974.
87. Thacker J et al. *Somat Cell Mol Genet* 14: 519, 1988.
88. Tsuchiya S et al. *Int J Cancer* 26: 171, 1980.
89. Uphoff CC et al. *Leukemia* 11: 441, 1997.
90. Uphoff CC, Drexler HG. In: Spiers RE (ed.), *Encyclopedia of Cell Technology*, Wiley, in press, 2000.
91. Venuat AM et al. *Cancer Genet Cytogenet* 3: 327, 1981.
92. Von Hanseman D. *Virchows Arch A Pathol Anat* 119: 299, 1890.
93. Wada H et al. *Cancer Res* 55: 3192, 1995.
94. Wlodarska I et al. *Blood* 89: 1716, 1997.
95. Wolff RK et al. *Genomics* 3: 347, 1988.
96. Wu SQ et al. *Leukemia* 9: 858, 1995.
97. Ziegler-Heitbrock HW et al. *Int J Cancer* 41: 456, 1988.
98. Zischler H et al. *Human Genet* 82: 227, 1989.

# Human Cell Culture

1.  J.R.W. Masters and B. Palsson (eds.): *Human Cell Culture, Vol. I.* 1998
    ISBN 0-7923-5143-6

2.  J.R.W. Masters and B. Palsson (eds.): *Human Cell Culture, Part 2. Vol. II.* Cancer Cell Lines. 1999
    ISBN 0-7923-5878-3

3.  B.O. Palsson and J.R. Masters (eds.): *Human Cell Culture, Part 3.* Cancer Continuous Cell Lines: Leukaemias and Lymphomas. 2001
    ISBN 0-7923-6225-X

4.  M.R. Koller, B.O. Palsson and J. R. W. Masters (eds.): *Human Cell Culture, Volume IV.* Primary Hematopoietic Cells. 1999
    ISBN 0-7923-5821-X

5.  M.R. Koller, B.O. Palsson and J.R.W. Masters (eds.): *Human Cell Culture, Part 5.* Primary Mesenchymal Cells. 2001
    ISBN 0-7923-6761-8

KLUWER ACADEMIC PUBLISHERS – DORDRECHT / BOSTON / LONDON